Lecture Notes in Computer Science 776

Edited by G. Goos and J. Hartmanis

Hans Jürgen Schneider Hartmut Ehrig (Eds.)

# Graph Transformations in Computer Science

International Workshop
Dagstuhl Castle, Germany, January 4-8, 1993
Proceedings

Springer-Verlag
Berlin Heidelberg New York
London Paris Tokyo
Hong Kong Barcelona
Budapest

Series Editors

Gerhard Goos
Universität Karlsruhe
Postfach 69 80
Vincenz-Priessnitz-Straße 1
D-76131 Karlsruhe, Germany

Juris Hartmanis
Cornell University
Department of Computer Science
4130 Upson Hall
Ithaca, NY 14853, USA

Volume Editors

Hans Jürgen Schneider
Lehrstuhl für Programmiersprachen, Universität Erlangen-Nürnberg
Martensstraße 3, D-91058 Erlangen, Germany

Hartmut Ehrig
Institut für Software und Theoretische Informatik, Technische Universität Berlin
Franklinstraße 28/29, D-10587 Berlin, Germany

CR Subject Classification (1991): G.2.2, F.4.2, D.2.2, E.1, I.3.5

ISBN 3-540-57787-4 Springer-Verlag Berlin Heidelberg New York
ISBN 0-387-57787-4 Springer-Verlag New York Berlin Heidelberg

CIP data applied for


Printed in Germany

Typesetting: Camera-ready by author
SPIN: 10131942 45/3140-543210 - Printed on acid-free paper

## Preface

The research area of graph grammars and graph transformations is a relatively young discipline of computer science. Its origins date back to the early 1970s. Nevertheless methods, techniques, and results from the area of graph transformation have already been studied and applied in many fields of computer science such as formal language theory, pattern recognition and generation, compiler construction, software engineering, concurrent and distributed systems modelling, database design and theory, and so on.

This wide applicability is due to the fact that graphs are a very natural way to explain complex situations on an intuitive level. Hence they are used in computer science almost everywhere, e.g., as data- and control flow diagrams, entity relationship diagrams, Petri nets, visualization of soft- and hardware architectures, evolution diagrams of non-deterministic processes, SADT diagrams, and many more. Like the "token game" for Petri nets, graph transformation brings dynamics to all these descriptions, since it can describe the evolution of graphical structures. Therefore graph transformation becomes attractive as a "programming paradigm" for complex structured software and graphical interfaces. In particular graph rewriting is promising as a comprehensive framework in which the transformation of all these very different structures can be modelled and studied in a uniform way.

This Dagstuhl Seminar was prepared by a program committee consisting of

Bruno Courcelle (Bordeaux)
Hartmut Ehrig (Berlin)
Grzegorz Rozenberg (Leiden)
Hans Jürgen Schneider (Erlangen)

During the seminar 33 lectures and 3 system demonstrations where presented by the participants from 8 European countries, U.S.A. and Japan in the following areas:

- Foundations of graph grammars and transformations
- Applications of graph transformations to
  - Concurrent computing
  - Specification and programming
  - Pattern generation and recognition

The system demonstrations in the evening showed efficient implementations of the algebraic approach to graph transformations (AGG-System), of a software specification language based on graph rewriting (PROGRESS) and of a functional programming language (Concurrent Clean) based on term graph rewriting. In each case the theoretical techniques of the underlying approach and typical applications were demonstrated in corresponding lectures during the day. In addition, interesting new applications of graph transformations were presented in several lectures in the following areas: concurrent constraint programming; actor systems; specification of languages for distributed systems, of hybrid database languages, and of an efficient narrowing machine; and – last but not least – pretty pattern generation and recognition ranging from graphical modelling for CAD to abstractions of modern art including Escher and Picasso.

In the lectures concerning foundations, on one hand new results concerning graph languages and graph automata and their connections to decision problems were presented. On the other hand new concepts and results for the algebraic approach to graph transformations based on double and single pushouts were shown. Notions for abstraction and semantical constructions were given leading to canonical derivation sequences, true concurrency and event structures, and also extensions of results from graph grammars to HLR (High Level Replacement) systems. The HLR approach is a categorical unification of different approaches with several interesting new applications including those to rule based modular system design and transformation and refinement of Petri nets.

Altogether it was a very fruitful interaction between theory, applications, and practical demonstrations.

At the end of the seminar, we invited the participants to submit final versions of their papers to a regular refereeing process. We wish to thank all the people who served as referees - and have done a good job and spent a lot of time - for conscientiously reading preliminary and final versions. Especially, we would like to express our warmest thanks to the members of the Program Committee and in particular to Ugo Montanari, who additionally joined us and took over some part of handling the refereeing process. After all, we could accept only 24 papers for publication in these proceedings.

Last but not least, we owe thanks to Springer-Verlag for producing this volume in the usual outstanding quality.

January 1994

Hartmut Ehrig
Hans Jürgen Schneider

## Contents


Path-Controlled Graph Grammars for Multiresolution Image Processing and Analysis
*K. Aizawa/A. Nakamura* ........ 1

Syntax and Semantics of Hybrid Database Languages
*M. Andries/G. Engels* ........ 19

Decomposability Helps for Deciding Logics of Knowledge and Belief
*S. Arnborg* ........ 37

Extending Graph Rewriting with Copying
*E. Barendsen/S. Smetsers* ........ 51

Graph-Grammar Semantics of a Higher-Order Programming Language for Distributed Systems
*K. Barthelmann/G. Schied* ........ 71

Abstract Graph Derivations in the Double Pushout Approach
*A. Corradini/H. Ehrig/M. Löwe/U. Montanari/F. Rossi* ........ 86

Note on Standard Representation of Graphs and Graph Derivations
*A. Corradini/H. Ehrig/M. Löwe/U. Montanari/F. Rossi* ........ 104

Jungle Rewriting: an Abstract Description of a Lazy Narrowing Machine
*A. Corradini/D. Wolz* ........ 119

Recognizable Sets of Graphs of Bounded Tree-Width
*B. Courcelle/J. Lagergren* ........ 138

Canonical Derivations for High-Level Replacement Systems
*H. Ehrig/H.-J. Kreowski/G. Taentzer* ........ 153

A Computational Model for Generic Graph Functions
*M. Gemis/J. Paredaens/P. Peelman/J. Van den Bussche* ........ 170

Graphs and Designing
*E. Grabska* ........ 188

ESM Systems and the Composition of Their Computations
*D. Janssens* ........ 203

Relational Structures and Their Partial Morphisms in View of Single Pushout Rewriting
*Y. Kawahara/Y. Mizoguchi* ........ 218

Single Pushout Transformations of Equationally Defined Graph Structures with Applications to Actor Systems
*M. Korff* ........ 234

Parallelism in Single-Pushout Graph Rewriting
*M. Löwe/J. Dingel* ........ 248

Semantics of Full Statecharts Based on Graph Rewriting
*A. Maggiolo-Schettini/A. Peron* ........ 265

Contextual Occurrence Nets and Concurrent Constraint Programming
*U. Montanari/F. Rossi* ........ 280

Uniform-Modelling in Graph Grammar Specifications
*M. Nagl* ........ 296

Set-Theoretic Graph Rewriting
*J.-C. Raoult/F. Voisin* ........ 312

On Relating Rewriting Systems and Graph Grammars to Event Structures
*G. Schied* ........ 326

Logic Based Structure Rewriting Systems
*A. Schürr* ........ 341

Guaranteeing Safe Destructive Updates Through a Type System with Uniqueness Information for Graphs
*S. Smetsers/E. Barendsen/M. v. Eekelen/R. Plasmeijer* ........ 358

Amalgamated Graph Transformations and Their Use for Specifying AGG - an Algebraic Graph Grammar System
*G. Taentzer/M. Beyer* ........ 380

List of authors ........ 395

# Path-Controlled Graph Grammars for Multiresolution Image Processing and Analysis

Kunio Aizawa[1] and Akira Nakamura[2]

[1]Department of Applied Mathematics
Hiroshima University
Higashi-Hiroshima, 724 Japan

[2]Department of Computer Science
Meiji University
Kawasaki, Kanagawa, 214 Japan

**Abstract:** In this paper, we define graph compression rules for the graphs representing two-dimensional rectangular grids with black and white pixels by making use of the PCE way of embedding. The compression rules rewrite four nodes having same label and forming a square into a node with the label. It also inserts and deletes nodes with special labels to preserve the neighborhood relations in the original image. Then we introduce an image compression algorithm using the concept of our graph compression rules. We show that the time complexity of our algorithm is $O(N\log_2 N)$, where N is the number of the black nodes of input graph, which is same as the case of the best quadtree representation.

**Keywords:** graph grammars, path-controlled embedding, quadtrees, normalized quadtrees, region representation

CONTENTS

1. Introduction
2. Basic definitions
3. Image compression rules on graphs
4. Optimal compression algorithm
5. Conclusions

## 1. Introduction

The graph structure is a strong formalism for representing pictures in syntactic pattern recognition. Many models for graph grammars have been proposed as a kind of hyper-dimensional generating systems (see e.g., [1], [2], and [3]), whereas the use of such grammars for pattern recognition is relatively infrequent. As one of such graph grammars, we introduced node-replacement path-controlled embedding graph grammars (nPCE graph grammars) in [4] for describing uniform structures. Originally, such embedding mechanism using paths of edges was introduced in [5] and [6]. These works are collected in [7]. Our grammars utilize partial path groups to define their embedding function.

On the other hand, region representation on digital spaces is an important issue in image processing and computer graphics. The quadtree representation of a digital image provides a variable resolution encoding of a region according to the sizes and number of maximal nonoverlapping blocks. It also provides easy computation for topological relations such as adjacency, connectedness, and borders. Samet [8], [9] provide good tutorial and bibliography of the researches on quadtrees as well as their applications. Recently, in [10], quadtree representation are used as a data structures for parallel image recognition.

In this paper, we define graph compression rules for the graphs representing two-dimensional rectangular grids with black and white pixels by making use of the PCE way of embedding. We refer to the subset of black pixels as the "region" and to the subset of white pixels as the "region's background." The compression rules rewrite four nodes having same label and forming a square into a node with the label. It also inserts and deletes nodes with special labels to preserve the neighborhood relations in the original image. Then we introduce an image compression algorithm using the concept of our graph compression rules. We show that the time complexity of our algorithm is $O(N\log_2 N)$, where N is the number of the black nodes of input graph, which is same as the case of the best quadtree representation.

## 2. Basic definitions

In this section, we review the definitions of the string descriptions of graphs [11] and nPCE graph grammars [12].

**Definition 2.1.** A *directed node- and edge-labelled graph* (*EDG-graph*) over $\Sigma$ and $\Gamma$ is a quintuple $H = \langle V, E, \Sigma, \Gamma, \varphi\rangle$, where V is the finite, nonempty set of nodes, $\Sigma$ is the finite, nonempty set of node labels, $\Gamma$ is the finite nonempty set of edge labels, E is the set of edges of the form $\langle v, \lambda, w\rangle$, where $u, w \in V$, $\lambda \in \Gamma$, $\varphi: V \rightarrow \Sigma$ is the node labelling function.

Let us take a set of edge labels as shown in Fig. 1 representing "EAST", "WEST", "NORTH", and "SOUTH", and ordered $-h \leq -v \leq h \leq v$.

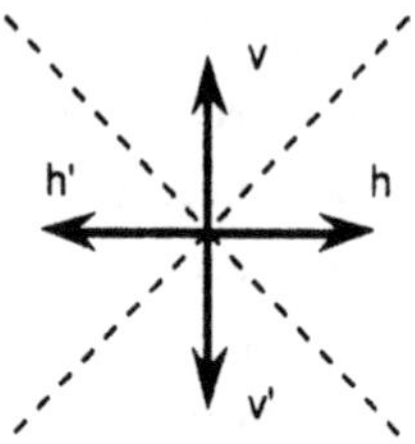


Fig. 1. The set of edge labels.

**Definition 2.2.** An EDG-graph H is called an *OS-graph* if

(1) for each $\lambda \in \Gamma$ there exists an inverse edge label $\lambda^{-1} \in \Gamma$,

(2) $\Gamma$ is simply ordered by a relation $\leq$,

(3) for each $v \in V$, if there exists $<v, \lambda, w> \in E$ then there does not exist $<v, \gamma, z>$ such that $\lambda = \gamma$ or $<z, \beta, v> \in E$ such that $\lambda^{-1} = \beta$.

Now we review the definitions of nPCE graph grammars. These definitions of nPCE graph grammars are slightly different from the former version [4] since the definitions of the path group are different. Then we examine their powers for generating OS-graphs. At first we review the definitions of the path groups describing the square grid [13].

**Definition 2.3.** A *discrete space* is a finitely presented abelian path group $\Gamma = (X/D)$, where X has 2n generators $s_1, s_2, ..., s_n, s_1^{-1}, s_2^{-1}, ..., s_n^{-1}$, and D contains all relations other than the commutativity $(s_i s_j s_i^{-1} s_j^{-1} = 1)$ and the inverse iterations $(s_i s_i^{-1} = 1)$. The *square grid* is a discrete space described by a four generators $s_1 =$ (north), $s_2 =$ (east), $s_1^{-1} =$ (south), $s_2^{-1} =$ (west), and $D = \emptyset$.

Note that the path groups defined above can also be defined on a graph generated by a graph grammar by regarding the edge labels of the generated graph as the generators.

**Definition 2.4.** For any given OS-graph H, its node P, and a string $\pi = \{c_1 c_2 ... c_i\}$ of its edge labels, $P\pi$ is *realizable on H* if and only if there exists a set of nodes $\{P_0, P_1, ..., P_i\}$ such that $P_0 = P$ and $<P_j, c_j, P_{j-1}>$ or $<P_{j-1}, c_j^{-1}, P_j>$ is an edge of H $(1 \leq j \leq i)$.

We review briefly the definition of nPCE grammars using abelian path groups to control embedding mechanism.

**Definition 2.5.** *A node-replacement graph grammar using path controlled embedding with 4 generators abelian path groups, (nPCE$_4$ grammar)*, is a construction $G=<\Sigma_N, P, \psi, Z, \Delta_N, \Delta_E>$, where $\Sigma_N$ is a finite nonempty set of node labels, $\Delta_N$ is a finite nonempty subset of $\Sigma_N$, called terminal node labels, $\Delta_E=\{h', v', h, v\}$, called terminal edge labels, P is a finite set of productions of form $(v_a, \beta)$, where $v_a$ is a graph consisting of only one node labelled with $a\in \Sigma_N$, $\beta$ is a connected OS-graph, $\psi$ is a mapping from $\Delta_E^+$ into $\Delta_E$ provided that for any $\pi\in \Delta_E^+$, $\psi$ maps $\pi$ into c, the first label of $\pi$, i.e., there exists a $\sigma\in \Delta_E^*$ such that $\pi=c\sigma$, Z is a connected OS-graph over $(\Sigma_N, \Delta_E)$ called *the axiom*.

A direct derivation step of a nPCE$_4$ grammar G, $\Rightarrow_G$, is performed as follows:

Let H be an OS-graph. Let $p = (v_a, \beta)$ be a production in P. Let $\beta'$ be isomorphic to $\beta$ (with h, an isomorphism from $\beta'$ into $\beta$), where $\beta'$ and $H-v_a$ have no common nodes. Then the result of the application of p to H (by using h) is obtained by replacing $v_a$ with $\beta'$ and adding edges $<u, \lambda, w>$ between every nodes u in $\beta'$ and every w in $H-v_a$ such that if the path from node 1 to node u on $\beta'$ is $\sigma$ then w is the node of H defined by $v_ac\sigma$ or its equivalent path under abelian path group with four generators and if $\psi(c\sigma)$=h or v then the added edges are $<u, \psi(c\sigma), w>$, otherwise $<w, \psi(c\sigma)^{-1}, u>$, where c is an element of $\Delta_E$.

We will denote the reflexive and the transitive closure of $\Rightarrow_G$ by $\Rightarrow_G^*$ and the transitive closure of $\Rightarrow_G^+$. The language of G, denoted as L(G), is defined by $L(G) = \{H \mid H$ is an OS-graph over $(\Delta_V, \Delta_E)$ and $Z \Rightarrow_G^* H\}$.

## 3. Image compression rules on graphs

Now we define a graph compression rule which works on the OS-graphs representing two-dimensional rectangular arrays. The compression rules rewrite four nodes having same label and forming a unit square into a node with the same label to get hierarchical graph representation. From now on we call the four nodes "unit square grid". Such rules are almost equal to the compression law in the quadtrees. These graph rewriting rules, however, need some intermediate nodes to preserve the neighborhood relations of the original graph and to restrict both of the indegree and outdegree within two. These intermediate nodes are labelled with ↑ or ↓, which mean the ascending and descending compression levels according to the direction of edges attached to the intermediate nodes and labelled with "h" or "v". If we traverse such edges against their direction, the meaning of the labels will be inverted. They also have at most two edges labelled with $h_U$, $h_D$, $v_R$, or $v_L$. These labels mean the upper horizontal, lower horizontal, right-hand vertical, left-hand vertical neighbors, respectively, according to the descending direction of the compression levels.

**Example 3.1.** The quadtree representation of the 8×8 binary picture of Fig. 2a is given in Fig. 2b. The corresponding graph representation is given in Fig. 2c.

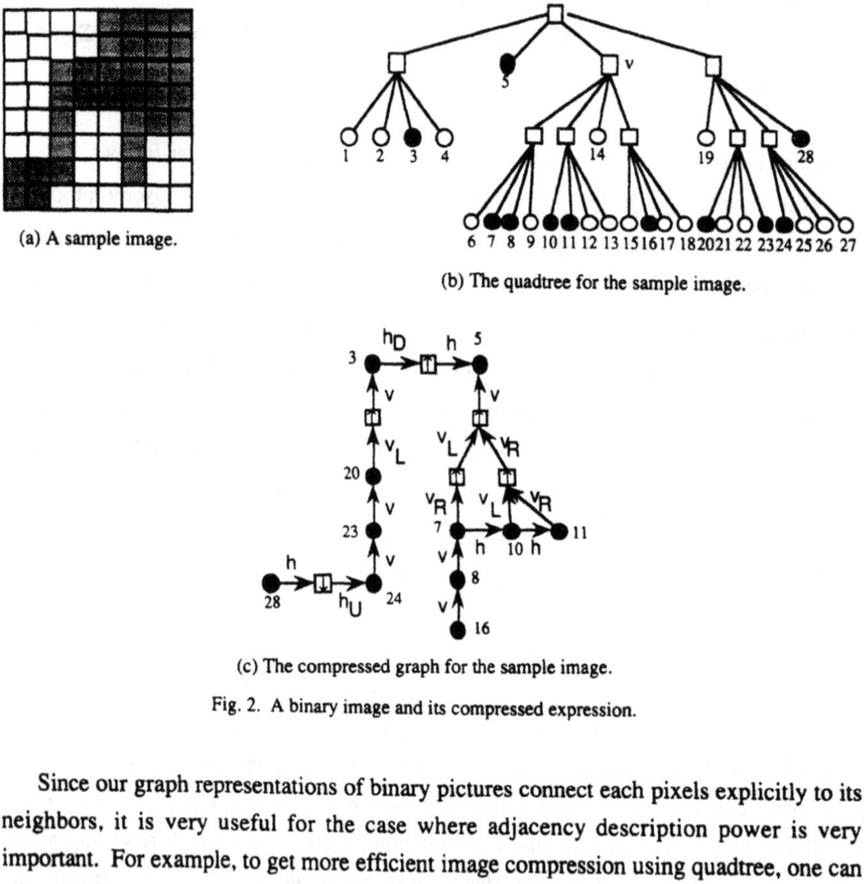


(a) A sample image.

(b) The quadtree for the sample image.

(c) The compressed graph for the sample image.

Fig. 2. A binary image and its compressed expression.

Since our graph representations of binary pictures connect each pixels explicitly to its neighbors, it is very useful for the case where adjacency description power is very important. For example, to get more efficient image compression using quadtree, one can move partition lines from boundaries of quadrants to more critical parts of the input image. In such cases, quadtrees no longer assure the neighborhood relations (see Fig. 3). Our graph representation still works correctly for such cases. Our graph representation uses path-controlled rules to embed the newly replaced graph, it can treat other structures than the square array without any significant changes. It also can ignore the region's background (i.e., white pixels) as in the case of linear quadtrees (see [14].) Furthermore, the neighborhood relations between nodes of different size are represented by trees whose root is the node representing bigger region and leaves are its neighbors of the specific direction. Such representations may present a good data structure for various image processing.

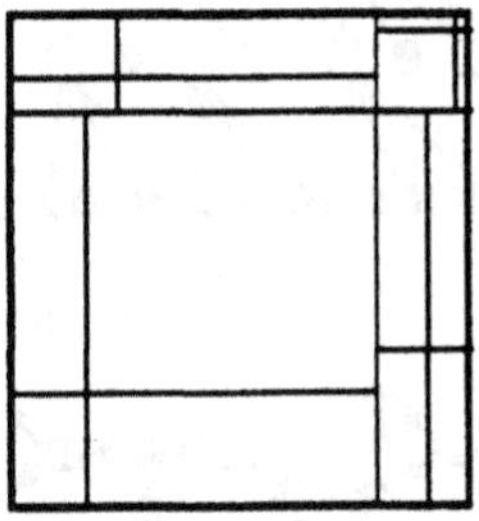

Fig. 3. Moved partition lines destroys adjacency of quadtree.

Of course our representations need more space to store adjacency relations as in the case of "nets and ropes" (see [15]), and need additional intermediate nodes to represent compression levels. So they do not always give better compression result than usual quadtree expression even if we count the number of nodes only. For an example, please see the case of Fig. 4a-c. In the case, the quadtree representation is better than the graph in contrast of the case in Fig.2.

In this section, we define graph rewriting rules for getting a compressed graph representation for a region and show that such transformation preserve the adjacency of the given image. We also show that the relative sizes multiplication of two regions represented by two different nodes are enumerated by the number of the intermediate nodes and their labels on the arbitrary path between these nodes.

At first, we extend the path-controlled embedding rules to context-sensitive rewriting rules. They need node identifiers to specify the nodes from which $\pi$ starts and to which new edges are connected.

**Definition 3.1.** A graph production rules is a construction $(\alpha, \beta, \psi)$, where $\alpha$ and $\beta$ are connected OS-graphs, $\psi$ is a mapping from $(v \times \Delta_E^+)$ into $(v \times \Delta_E)$ provided that for any $\pi \in \Delta_E^+$, $\psi$ maps $\pi$ into c, the first label of $\pi$, i.e., there exists a $\sigma \in \Delta_E^*$ such that $\pi = c\sigma$

Note here that the extended embedding rules still do not depend on node labels. By making use of the extended embedding rules, the definition of graph rewriting rules for image compression are as follows:

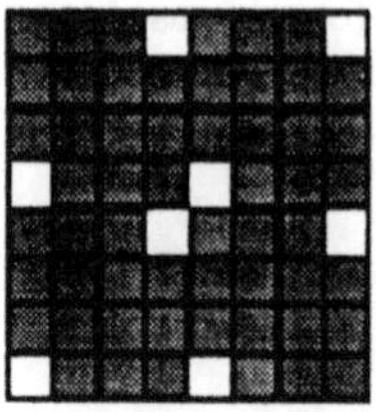

(a) A worse case image.

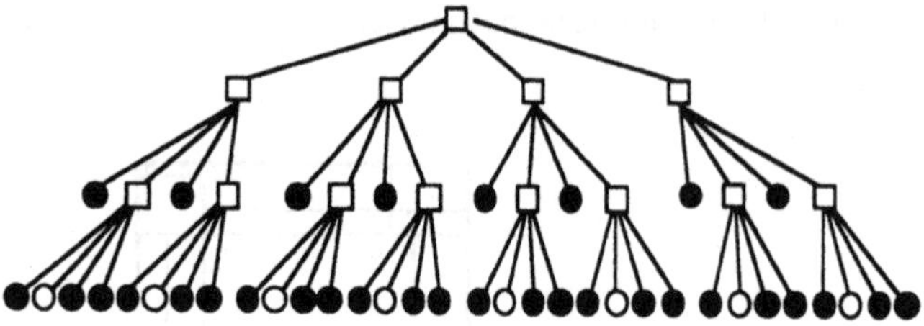

(b) A quadtree expression.

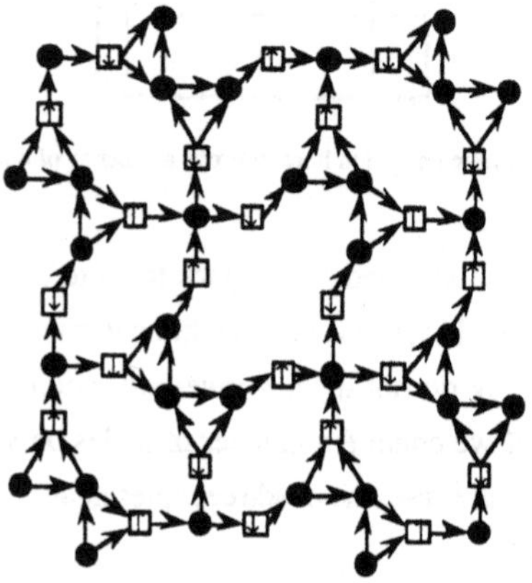

(c) The compressed graph expression.

Fig. 4. A worse case for graph expression.

**Definition 3.2.** Graph rewriting rules for image compression have a form scheme (α, β, ψ), where α→β is a following rewriting rule:

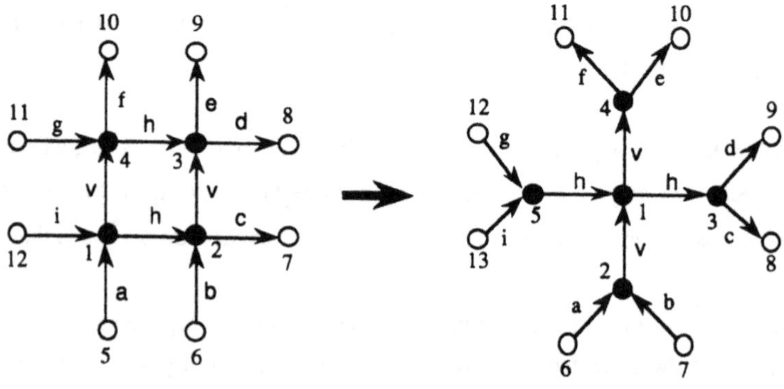


such that nodes 1, 2, 3, 4 must exist, their labels are all "black", there must exist four edges labelled h or v between them as shown above, and the label for node 1' is also "black". All other nodes and edges, if exist, must satisfy one of the following:

If labels for nodes 5 and 6 are "black", then $a=b=v$, $a'=v_L$, $b'=v_R$,
the label for node 2' is ↑,
labels for nodes 6', 7' are "black", and
$\psi(5, v^{-1})=(6', v^{-1})$, $\psi(5, h^{-1})=(6', h^{-1})$
$\psi(5, h)=(6', h)$, $\psi(6, v^{-1})=(7', v^{-1})$
$\psi(6, h^{-1})=(7', h^{-1})$, $\psi(6, h)=(7', h)$.

If labels for nodes 5 and 6 are ↑, then $a=b=v$, $a'=v_L$, $b'=v_R$,
the label for node 2' is ↑,
labels for nodes 6', 7' are ↑, and
$\psi(5, v_L^{-1})=(6', v_L^{-1})$,
$\psi(5, v_R^{-1})=(6', v_R^{-1})$
$\psi(6, v_L^{-1})=(7', v_L^{-1})$,
$\psi(6, v_R^{-1})=(7', v_R^{-1})$.

If labels for nodes 5 and 6 are ↓, then $a=v_L$, $b=v_R$, node 5 is identical to node 6,
node 2', 6', 7' do not exist,
$\psi(5, v^{-1})=(1', v^{-1})$.

If labels for nodes 7 and 8 are "black", then $c=d=h$, $c'=h_D$, $d'=h_U$,
the label for node 3' is ↓,
labels for nodes 8', 9' are "black", and
$\psi(7, h)=(8', h)$, $\psi(7, v^{-1})=(8', v^{-1})$
$\psi(7, v)=(8', v)$, $\psi(8, h)=(9', h)$
$\psi(8, v^{-1})=(9', v^{-1})$, $\psi(8, v)=(9', v)$.

If labels for nodes 7 and 8 are ↓, then $c=d=h$, $c'=h_D$, $d'=h_U$,
the label for node 3' is ↓,
labels for nodes 6', 7' are ↓, and
$\psi(7, h_D)=(8', h_D)$, $\psi(7, h_U)=(8', h_U)$
$\psi(7, h_D)=(8', h_D)$, $\psi(7, h_U)=(8', h_U)$.

If labels for nodes 7 and 8 are ↑, then $c=h_D$, $b=h_U$, node 7 is identical to node 8,
node 3', 8', 9' do not exist,
$\psi(7, h)=(1', h)$.

If labels for nodes 9 and 10 are "black", then $e=f=v$, $e'=v_R$, $f'=v_L$,
the label for node 4' is ↓,
labels for nodes 10', 11' are "black", and
$\psi(9, v)=(10', v)$, $\psi(9, h^{-1})=(10', h^{-1})$
$\psi(9, h)=(10', h)$, $\psi(10, v)=(11', v)$
$\psi(10, h^{-1})=(11', h^{-1})$,

$\psi(10, h)=(11', h)$.

If labels for nodes 9 and 10 are ↓, then $e=f=v$, $e'=v_R$, $f'=v_L$,
the label for node 4' is ↓,
labels for nodes 10', 11' are ↓, and
$\psi(9, v_L)=(10', v_L)$, $\psi(9, v_R)=(10', v_R)$
$\psi(10, v_L)=(11', v_L)$,
$\psi(10, v_R)=(11', v_R)$.

If labels for nodes 9 and 10 are ↑, then $e=v_L$, $f=v_R$, node 9 is identical to node 10,
node 4', 10', 11' do not exist,
$\psi(9, v)=(1', v)$.

If labels for nodes 11 and 12 are "black", then $g=i=h$, $g'=h_U$, $i'=h_D$,
the label for node 5' is ↑,
labels for nodes 12', 13' are "black", and
$\psi(11, h^{-1})=(12', h^{-1})$,
$\psi(11, v^{-1})=(12', v^{-1})$
$\psi(11, v)=(12', v)$, $\psi(12, h^{-1})=(13', h^{-1})$
$\psi(12, v^{-1})=(13', v^{-1})$,
$\psi(12, v)=(13', v)$.

If labels for nodes 11 and 12 are ↑, then $g=i=h$, $g'=h_U$, $i'=h_D$,
the label for node 5' is ↑,
labels for nodes 12', 13' are ↑, and
$\psi(11, h_D)=(12', h_D)$,
$\psi(11, h_U)=(12', h_U)$
$\psi(11, h_D)=(12', h_D)$,
$\psi(11, h_U)=(12', h_U)$.

If labels for nodes 11 and 12 are ↓, then $g=h_U$, $i=h_D$, node 11 is identical to node 12,
node 5', 12', 13' do not exist,
$\psi(11, h^{-1})=(1', h^{-1})$.

In the above definition, all neighbors of the compressed node in a direction (i.e., east, west, north, or south) must represent the regions in the same size. One can remove such restriction to define more general graph compression rules. In general, they yield more efficient compressed results. But they introduce nodes with more than two incoming and/or outgoing edges. Thus we do not treat them in this section but give a few comments for the situation they bring in the last section.

Now we prove that the applications of rewriting rule defined above to graphs representing two-dimensional digital images do not change the neighborhood relation of each pair of nodes.

**Lemma 3.1.** For any given graph H representing a digital image on two-dimensional square grid and its compressed graph H', any two nodes $n_1, n_2$ of H are adjacent in H if and only if there exists a path between $n_1$ and $n_2$ in H' such that all nodes on the path other than $n_1, n_2$ are labelled with $\uparrow$ or $\downarrow$.

*Proof*: Obviously, any nodes and edges rewriting described in rules explicitly do not change any adjacency of rewritten nodes since they add and/or delete some intermediate nodes labelled with $\uparrow$ or $\downarrow$ only. If given two nodes $n_1, n_2$ of H are adjacent, the edge addition by embedding rules do not change their adjacency either. Assume here that two nodes $n_1, n_2$ of H are not adjacent and the embedding rules add an edge labelled with c between them. Since the embedding rules add an edge between the two nodes $n_1, n_2$, there exists a path $\pi$ between them which is equivalent to c under abelian path group with four generators. However, since graph H represents a digital image on square grid, there must be an edge labelled with c between $n_1, n_2$ in H. This is a contradiction.

**Lemma 3.2.** Assume that $n_1, n_2$ are two nodes of compressed graph H, and $r_1$ and $r_2$ are the square regions represented by $n_1, n_2$, respectively. Then the ratio of the side length of $r_1$ to $r_2$ is $2^{\sigma(\pi)}$, where $\pi$ is an arbitrary acyclic path from $n_1$ to $n_2$ in H and $\sigma$ is a function defined as follows:

$\sigma(\pi)$ = (the number of nodes labelled with $\uparrow$ on $\pi$) – (the number of nodes labelled with $\downarrow$ on $\pi$),

where $\pi$ is traced from $n_1$ to $n_2$ and the label c of a node n on $\pi$ is c if $\pi$ enters n according with the edge direction, $c^{-1}$ if $\pi$ enters n against the edge direction.

*Proof*: If H is a non-compressed graph, there exists no node labelled with $\uparrow$ or $\downarrow$. So lemma holds for such graphs. At the first application of our compressing rule to H, since there is no node labelled with $\uparrow$ or $\downarrow$, any acyclic path between the newly replaced node and other nodes includes only one node labelled with $\uparrow$ or $\downarrow$. The label of such node is $\uparrow$ if the path go into the newly replaced node (according with edge direction) and $\downarrow$ if the path go from it (also according with edge direction). Thus lemma holds for the first application of the rule. Assume here that lemma holds until the (k-1)th application. Let $n_1$ and $n_2$ be the nodes such that the ratio of side length of regions $r_1$ to $r_2$ (regions represented by $n_1, n_2$, respectively) is determined by a path $\pi$ passing through the newly replaced graph by k-th application of the rule. Assume that, before the k-th application, the path defining the ratio enters the square grid of four black nodes (which will be rewritten by k-th application) through a node labelled with ↑ (according with edge direction) and leaves through a node labelled with ↓. Then, after k-th application, from the definition of rewriting rule, path $\pi$ must pass through one more node labelled with ↑ (according with edge direction) and lose one node with ↓. So the ratio is not affected by k-th application. For any other combinations of entering and leaving nodes, situations are as follows:

| **Entering** | **Leaving** | **After k-th application** |
|---|---|---|
| ↑ | none or ↓ | Increase one ↑ when entering, increase one ↓ when leaving. |
| none | ↑ | Increase one ↑ when entering, decrease one ↑ when leaving. |
| none | none or ↓ | Increase one ↑ when entering, increase one ↓ when leaving. |
| ↓ | ↑ | Decrease one ↓ when entering, decrease one ↑ when leaving. |
| ↓ | none or ↓ | Decrease one ↓ when entering, increase one ↓ when leaving. |

Therefore the ratio between nodes not rewritten by k-th application does not affected by k-th application. Proof for the cases of the ratio between rewritten node and not rewritten one is almost same as the above cases.

From Lemmas 3.1 and 3.2, it is obvious that the adjacency tree of a node n in a quadtree corresponds to the tree of intermediate nodes with black nodes as leaves which is directly connected to node n by an edge with a label of adjacency direction in a compressed graph if all rewriting rules are applied to the same places as in the case of a quadtree and all white nodes are ignored. For example, for node 5 in Fig. 2b, southern neighbor is gray node v with adjacency tree as shown in Fig. 5a, which corresponds to the tree of Fig. 5b, a part of compressed graph of Fig. 2c.

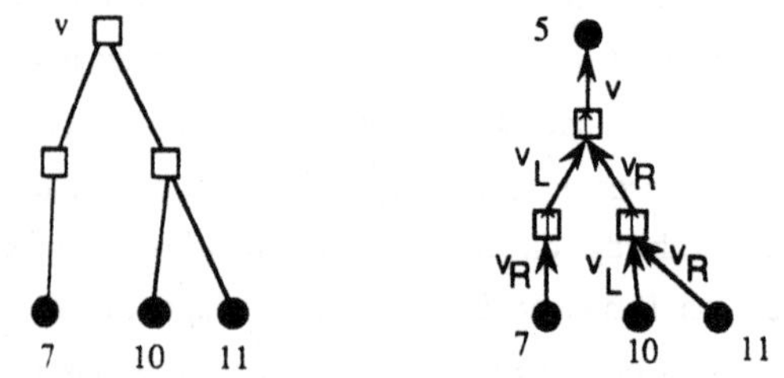


(a) Adjacency tree for node 5. (b) A part of graph expression.

Fig. 5. Trees of adjacency.

From Lemmas 3.1, 3.2 and the above fact, it is not so difficult to see the following theorem:

**Theorem 3.1.** For any given region R and its quadtree expression Q, there exists a graph H corresponding to Q. Graph H is obtained from a graph expression corresponding to R by making use of graph rewriting rules in Definition 3.2.

A part of relationships between quadtree expressions and compressed graph expressions is illustrated in this theorem. From the result, it can be shown that there exists an algorithm to reconstruct the original image from given compressed graph expression which corresponds to the quadtree expression. In fact, the result can be extended to arbitrary compressed expression. We will show it in a next paper under preparing.

**4. Optimal compression algorithm**

From the results of previous section, it has been shown that there exists a compressed graph expression which corresponds to a quadtree for a region. However, we cannot always get such graph expression since the applications of graph rewriting rules are executed nondeterministically. In this section, we introduce a region compressing algorithm using graph rewriting rules defined in previous section. It generates compressed graph expression which corresponds to the normalized quadtree expression for a given graph expression of a region. The normalized quadtree was introduced in [16] and it provides a minimal cost quadtree expression for a given $2^K \times 2^K$ binary image. Its main idea is moving image of size $2^K \times 2^K$ on the grid of size $2^{K+1} \times 2^{K+1}$. An algorithm for finding the normalized quadtree was given in the same paper and it was shown that its time complexity is $O(S^2 \log_2 S)$, where S is the length of the grid. We show that the time complexity of our algorithm for compressed graph expression is $O(N \log_2 N)$, where N is the number of black nodes of input graph. For the worst case, it has same complexity as the normalized quadtree.

In [16], it was shown that the translations by at most $2^K$-1 pixels to the east and to the south is enough to get the normalized quadtree if the size of the image is $2^K \times 2^K$. The result can be improved to $2^i$-1, where i is of the size $2^i \times 2^i$ of the largest square component represented by a black leaf. Since every other black leaves represent square components of size $2^j \times 2^j$ for some $j \leq i$, the quadtree is invariant to any translations by multiples of $2^i$ pixels. Thus our algorithm proceeds by translating the graph four times, i.e., by 0 node, by 1 node to the east, by 1 node to the north, and 1 node to the north east. After each translation, it traverse the graph by making use of usual breadth-first search algorithm and lists all black nodes each of which is the southwest node of a unit square graph. For the traverse, it uses edges labelled with h, v and their inverse only since other edge labels represent gaps of compression levels. After all southwest nodes are listed, it applies rewriting rules for each unit square graph represented by these southwest nodes. It subtracts 3 from *counter*, which is initialized to the number of black nodes, when a rewriting rule is applied to a unit square graph to simulate leaf decreasing of a quadtree representation. Then it calls itself recursively on new compressed graph. As stated in [16], translating the graph by 1 node at level d recursion corresponds to a translation of the original graph by $2^d$ nodes. Recursion halts when there is no candidate for applying rules.

Our algorithm is as follows:

```
procedure normalize (G, CG: GRAPH; leaves: integer);
  const  maxlength=100 {sufficient constant to store G, CG, G1};
  type   GRAPH=array [1..maxlength] of record
             n_label: NODELABEL;
             west, north, east, south: array [11, 12, 1] of integer;
             visited: boolean
         end;
         LIST=array [1..maxlength] of integer;
         NODE=record
             node_id, dif_h, dif_v: integer
         end;
  var    G, CG: GRAPH;
         L1, L2: LIST;
         last1, last2, i: integer;

function southwest (v: NODE): boolean;
  var    m: integer;
  begin {southwest}
      if v.dif_h=0 and v.dif_v=0 then
         if G[v.node_id].east[12]≠0 then
            if G[v.node_id].north[12]≠0 then begin
               m:=G[v.node_id].north[12];
               if G[m].east[12]≠0 then return (true)
            end
         end
      end;
      return (false)
  end; {southwest}

procedure findnext (v: node);
{QUEUE of NODE is a usual queue of data type NODE. ENQUEUE(v, Q) adds element v
at the last of Q. DEQUEUE(Q) deletes the first element of Q.}
```

```
var    Q: QUEUE of NODE;
       x, y: node;
begin {findnext}
    G[v.node_id].visited:=true;
    ENQUEUE(v, Q);
    while not EMPTY(Q) do begin
        x:=FRONT(Q);
        if SOUTHWEST(x)=true then begin
            last2:=last2+1;
            L2[last2]:=x.node_id
        end;
        DEQUEUE(Q);
        if G[x.node_id].west≠0 then begin
            y.node_id:=G[x.node_id].west;
            y.dif_h:=mod(x.dif_h+1, 2);
            y.dif_v:=x.dif_v;
            if G[y.node_id].visited=false then begin
                G[y.node_id].visited:=true;
                ENQUEUE(y, Q);
            end
        end
        if G[x.node_id].south≠0 then begin
            y.node_id:=G[x.node_id].south;
            y.dif_h:=x.dif_h;
            y.dif_v:=mod(x.dif_v+1, 2);
            if G[y.node_id].visited=false then begin
                G[y.node_id].visited:=true;
                ENQUEUE(y, Q);
            end
        end
        if G[x.node_id].east≠0 then begin
            y.node_id:=G[x.node_id].east;
            y.dif_h:=mod(x.dif_h+1, 2);
            y.dif_v:=x.dif_v;
            if G[y.node_id].visited=false then begin
                G[y.node_id].visited:=true;
                ENQUEUE(y, Q);
            end
        end
        if G[x.node_id].north≠0 then begin
            y.node_id:=G[x.node_id].north;
            y.dif_h:=x.dif_h;
```

```
            y.dif_v:=mod(x.dif_v+1, 2);
            if G[y.node_id].visited=false then begin
                G[y.node_id].visited:=true;
                ENQUEUE(y, Q);
            end
        end
    end
  end; {findnext}

procedure apply_rule (m: integer);
{Apply rewriting rule of Definition 3.1 to a unit square graph whose southwest node is
node G1[m]}

procedure compress (G: GRAPH; L1, L2: LIST; last1, last2, counter: integer);
  var   G1: GRAPH;
        x: NODE;
        counter1, i, m, dh, dv: integer;
  begin {compress}
    for dv:=0 to 1 do
        for dh:=0 to 1 do begin
            last2:=0;
            G1:=G;
            counter1:= counter;
            for i:=1 to last1 do begin
                x.node_id:=L1[i];
                x.dif_h:=dh;
                x.dif_v:=dv;
                if G[x.node_id].visited=false the FINDNEXT(x)
            end;
            if last2≠0 then begin
                for i:=1 to last2 do begin
                    m:=L2[i];
                    apply_rule(m);
                    counter1:=counter1-3;
                    G[m]. visited:=false
                end;
                compress (G1, L2, L2, last2, 0, counter1)
            else
                if counter1<leaves then begin
                    leaves:=counter1;
                    CG:=G1
                end
        end
```

```
end; {compress}

begin {normalize}
    for i:= 1 to leaves do L1[i]:=i;
    CG:=G;
    last1:= leaves;
    last2:=0;
    compress (G, L1, L2, last1, last2, counter)
end; {normalize}
```

Now we analyze the time complexity of our algorithm.

**Theorem 4.1.** The time complexity of our algorithm in the worst case is $O(N\log_2 N)$, where N is the number of nodes (all black nodes) of original graph.

*Proof*: For an arbitrary graph with N black nodes (all of them are stored in list L1) representing square grid, bandwidth first search needs time proportional to N to find out all southwest nodes and to store them in list L2. We also need time proportional to N for applying rewriting rules to all unit square grid represented by southwest nodes in L2. Then all nodes in L2 are stored to L1 to represent all black nodes which is ready for further compression. The number of these nodes do not exceed 1/4 of the number of previous members of L1. So the maximum depth of recursion is $(\log_2 N)/2$. These steps repeat for each of four different translating directions at each recursion level. Thus the total time to execute our algorithm is proportional to

$$\underbrace{4\left(N+4\left(\left(\frac{1}{4}\right)N+4\left(\left(\frac{1}{4}\right)^2 N+\cdots\right)\right)\right)}_{\frac{\log_2 N}{2}\ \text{times}}.$$

The sum is equal to $2N\log_2 N$. Therefore our algorithm is of time complexity $O(N\log_2 N)$.

## 5. Conclusions

We have introduced a compressed graph expression for two-dimensional binary image. It is obtained by applying graph rewriting rules to an original (non-compressed) graph representing square grid. We also have presented an algorithm for creating graph expression corresponding to the normalized quadtree of an input image. The worst-case time complexity of this algorithm is $O(N\log_2 N)$, where N is the number of black nodes of the original graph.

In this paper, we restrict ourself to treat unit square grids stand in a row. But, in some cases, unit square grids stand in diagonal positions bring more efficient compression (see Fig. 6a-b). This method can be used to generalize the notion of "forest of quadtrees" in [17] to our graph expression.

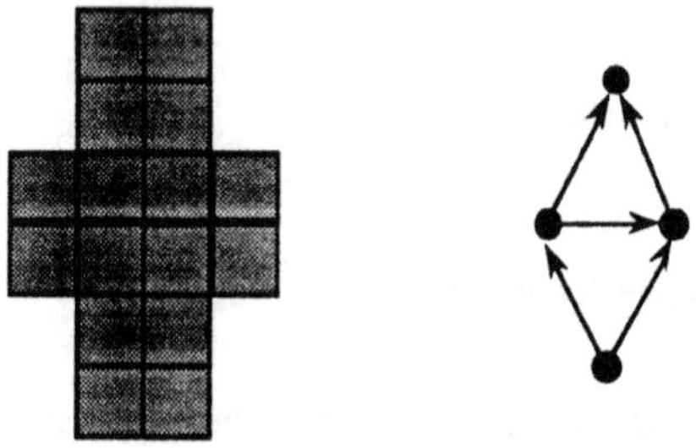

(a) Sample image. (b) Compressed graph expression using diagonal position.

Fig. 6. More compact expression using diagonal position.

From the standpoint of graph grammars, the membership problem for compressed graph languages seems to be very interesting. It will be investigated in our future work.

## References


[1] Ehrig, H., M. Nagl and G. Rozenberg (ed.): *Graph-Grammars and Their Application to Computer Science*, Lecture Notes in Computer Science, 153, Springer-Verlag, Berlin, 1983.

[2] Ehrig, H., M. Nagl, G. Rozenberg and A. Rosenfeld (ed.): *Graph-Grammars and Their Application to Computer Science*, Lecture Notes in Computer Science, 291, Springer-Verlag, Berlin, 1987.

[3] Ehrig, H., H.-J. Kreowski and G. Rozenberg (ed.): *Graph Grammars and Their Application to Computer Science*, Lecture Notes in Computer Science, 532, Springer-Verlag, 1991.

[4] Aizawa, K. and A. Nakamura: Graph grammars with path controlled embedding, *Theoretical Computer Science*, 88, pp. 151-170, 1991.

[5] Nagl, M.: Eine praezisierung des Pfaltz/Rosenfeldschen produktionsbegriffs bei mehrdimensionalen grammatiken, Arbeitsber. d. Inst. f. Math. Masch. u. Datenver, 6, 3, pp. 56-71, Universität Erlangen-Nürnberg, 1973.

[6] Nagl, M.: Formal sprachen von marikierten graphen, Arbeitsber. d. Inst. f. Math. Masch. u. Datenver., 7, 4, Universität Erlangen-Nürnberg, 1974.

[7] Nagl, M.: *Graph-Grammatiken: Theorie Implementierung Anwendeungen*, Vieweg, 1979.

[8] Samet, H.: A tutorial on quadtree research, in A. Rosenfeld (ed.), *Multiresolution Image Processing and Analysis*, Springer-Verlag, Berlin, pp. 1984.

[9] Samet, H.: *The Design and Analysis of Spatial Data Structures*, Addison Wesley, New York, 1990.

[10] Saudi, A.: Parallel recognition of multidimensional images using regular tree grammars, *Lecture Notes in Computer Science*, 654, pp. 231-239, 1992.

[11] Shi, Q. Y. and K. S. Fu: Parsing and translation of attributed expansive graph languages for scene analysis, *IEEE Transaction on Pattern Analysis and Machine Intelligence*, PAMI-5, pp. 472-485, 1983.

[12] Aizawa, K. and A. Nakamura: Pathe-controlled graph grammars for syntactic pattern recognition, *Lecture Notes in Conputer Science*, 654, pp. 37-53, 1992.

[13] Mylopoulos, J. P. and T. Pavlidis: On the topological properties of quantized spaces, *Journal of the Association for Computing Machinery*, 18, pp. 239-254, 1971.

[14] Gargantini, I.: An effective way to represent quadtrees, *Cmmunications of the ACM*, 25, pp. 905-910, 1982.

[15] Hunter, G. M. and K. Steiglitz: Operations on images using quad trees, PAMI-1, pp. 145-153, 1979.

[16] Li, M., W. I. Grosky and R. Jain: Normalized quadtrees with respect to translations, *Computer Graphics and Image Processing*, 20, pp. 72-81, 1982.

[17] Jones, L. and S. S. Iyengar: Representation of regions as a forest of quadtrees, *Proc. of Pattern Recognition and Image Processing Conference*, Dallas, pp. 57-59, 1981

# Syntax and Semantics of Hybrid Database Languages*

Marc Andries and Gregor Engels

Leiden University, Dept. of Comp. Science
Niels Bohrweg 1, 2333 CA Leiden, The Netherlands
E-mail: {andries,engels}@wi.leidenuniv.nl

**Abstract.** We present the hybrid query language HQL/EER for an Extended Entity-Relationship model. As its main characteristic, this language allows a user to freely mix graphical and textual formulation of a query. We show how syntax and semantics of this hybrid language are formally defined by means of a slightly extended version of PROGRES, a specification formalism based on programmed and attributed graph rewriting systems.

## 1 Introduction

Ever since the introduction of the first graph grammar-like models in the late sixties, efforts are made to validate these models by using them for *specification* purposes in various areas of computer science. One of these areas is the specification of databases and the allowed operations on them [7, 9, 26]. Another area in which graph grammars, and especially graph rewriting systems, have been successfully applied, is that of *visual languages* [4, 13, 16, 19].

The work presented in this paper is related to both of the above areas, and was triggered by a particular line of research in the database area. In the late eighties, the observation that both structure and contents of an object-oriented database allow for a natural graphical representation, inspired a number of researchers to develop *graph oriented* database models, in which notions from graph theory are used to uniformly define both the data representation part of the model and its manipulation and retrieval language [3, 6, 9, 17, 24, 27, 32]. In several of these models, it is investigated to what extent arbitrary data manipulations may be expressed in a *purely* graphical way. One can conclude from this research that there is no limit to the expressive power that may be obtained with pure graph based manipulation languages [2].

However, one also gets the impression that some of this research overshoots its mark in the sense that the pure graphical formulation of a query sometimes even looks more complex than its textual equivalent. The obvious solution to this problem is to try and combine the "best of both worlds", i.e., to develop languages that allow those parts of an operation that are most clearly specified graphically resp. textually, to be indeed specified graphically resp. textually.

* Work supported by COMPUGRAPH II, ESPRIT BRWG 7183.

In view of this, we developed a ***hybrid query language***, in which almost any component of a given query may be expressed either textually or graphically, according to the users taste.[2] Our Hybrid Query Language for the Extended Entity Relationship model (HQL/EER, the main concepts of which are illustrated in Section 4) is an extension of SQL/EER, a textual query language for an extended version of the Entity Relationship model [10] (discussed briefly in Sections 3 and 2). It is our aim in this paper to show how both syntax (cf. Section 5) and semantics (cf. Section 6) of the graphical part of an HQL/EER query may be formalized using the formalism for attributed graph rewriting systems called PROGRES [31]. Syntax is defined by means of suitable PROGRES productions, while the semantics of the language is formally defined by translating a hybrid query into a fully textual one, by means of attribute evaluations within these productions.

An important contribution of this paper is that, in the course of our efforts to model this database language using PROGRES, we came across a number of features which are desirable to properly model our language, but are unfortunately missing from PROGRES. In most cases, these features concern the unequal treatment of edges with respect to nodes. We digress on this matter in the introduction to Section 5, where we discuss several extensions to PROGRES, in order to obtain a language which is better suited for our particular application, i.e., the definition of syntax and semantics of a hybrid query language.

## 2 The Extended Entity-Relationship Model

In this Section, we sketch briefly the main concepts of our Extended Entity-Relationship (EER) model [10, 22, 23]. It is based upon the classical Entity-Relationship model [5] and extended with the following concepts known from semantic data models [25]:

- components, i.e., object-valued attributes to model complex structured entity types:
- multivalued attributes and components to model association types;
- type construction in order to support specification and generalization;
- several structural restrictions like the specification of keys, cardinality constraints,..., which are, however, of no interest to this paper.

Let us illustrate the EER model and its features with a small example. It models the world of surfing people who surf on different kinds of waters (cf. Figure 1).

First of all, one easily recognizes the basic concepts of the ER model. These are entity types like **PERSON**, relationship types like **surfs_on_river**, and attributes like **Name** (of **PERSON**) or **Times/Year** (of **surfs_on_river**).

The concept of *type construction* provides means to construct new entity types (called *output types*) from already existing entity types (called *input types*).

[2] The term "hybrid" is in fact inspired by hybrid syntax directed editors, where the user can freely choose between a syntax-directed and a free style of editing [12].

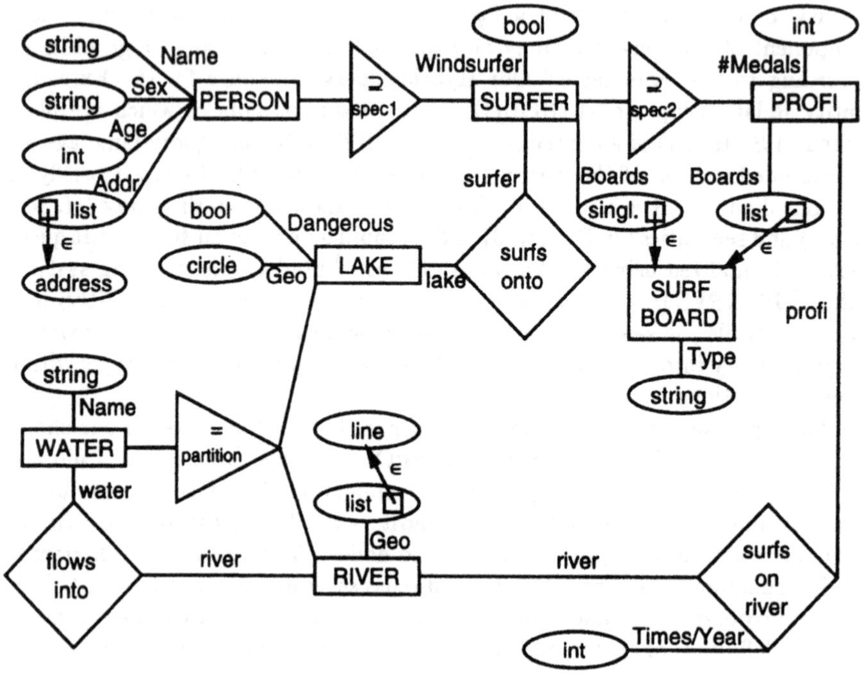


**Fig. 1.** An EER diagram, modelling the world of surfers

This means that each object in the set of instances of a constructed entity type also belongs to the set of instances of the input types. Type constructions are represented by triangles, where all input types are connected by edges with the baseline, and the output types with the opposite point. For instance, the type construction **spec1** represents the special case of a specialization. It has one input type **PERSON** and one output type **SURFER**, i.e. **PERSON** is specialized to **SURFER**. Then, **SURFER** in turn is specialized to **PROFI**, this time by **spec2**. This means that each **PROFI** is a surfer and therefore also a person.

A type construction with only one input type is called specialization. Another example is **WATER**, which is specialized into **LAKE** and **RIVER**. We require the output types to have disjoint sets of instances. This means that in our example modelling, a water can be either a lake, or a river. In the case of specialization, we assume that all attributes are implicitly inherited from the input types to the output types. For instance, each instance of type **LAKE** also has the attribute **Name** defined for the entity type **WATER**.

We do not allow entity types to be constructed by several type constructions. Every entity type can only be constructed once. Furthermore, every constructed type must not, directly or indirectly, be input type of its own type construction.

Generally, an instance of the input type(s) of a type construction is not necessarily a member of the instance set of (one of) the output type(s). For instance, there could be surfers who are not profis. If a total partition of the input instances is desired, the type construction triangle in the diagram is labeled with

'=' instead of '$\supseteq$'. For example, each instance of type **WATER** must be an instance of **LAKE** or **RIVER**, but nothing else.

Complex structured entity types can be modeled by components. Roughly speaking, components can be seen as object-valued attributes. For instance, **Boards** is a component of **PROFI**, which consists of a list of references to instances of type **SURFBOARD**. Each profi possesses one or more surfboards. Components are always represented by an oval, even if they are single valued, in which case we label the oval with "singl.". Both attributes and components can be multivalued, i.e., set-, bag-, or list-valued. In this case, we write a square into the oval that is connected to the corresponding entity type or atomic value type via an arrow, which is always labeled $\in$. E.g., note that both the entity type **PROFI** and its ancestor in the construction hierarchy (i.e., the entity type **SURFER**) have a **Boards**-component, but with different types. Since an ordinary surfer can only possess one surf board, the **Boards**-attribute for the entity type **SURFER** is single-valued. Since professional surfers can own several surf boards, the **Boards**-attribute for the entity type **PROFI** is a *list* of surf boards. This way, the known concept of *overriding* (both of attributes and components) is incorporated in the EER model.

# 3 SQL for the Extended Entity-Relationship Model

Based on the data model introduced in Section 2, we now informally repeat the main concepts of the textual query language SQL/EER. A complete description of this language can be found in [21]. Its formal semantics is based on a formally defined calculus for the EER model [15]. Both syntax and semantics of SQL/EER are defined by means of an attributed string grammar [20].

SQL/EER directly supports all the concepts of the EER model, plus a number of well known features that are an integral part of nowadays query languages:

1. relationships, attributes of relationships, components and type constructions;
2. arithmetic;
3. aggregate functions;
4. nesting of the output;
5. subqueries as variable domains.

Analogous to relational SQL, SQL/EER uses the **select-from-where** clause.

As a first example, consider the SQL/EER query of Figure 2 (over the scheme of Figure 1). It retrieves the name and age of all persons older than 18.

**select** p.Name, p.Age
**from** p **in** PERSON
**where** p.Age $\geq$ 18

**Fig. 2.** Name and age of adults (SQL/EER version)

In this query, the variable p is declared. It ranges over the set of currently stored persons. This variable can now be used to build terms like p.Name and p.Age, to compute the name and age of the person p, respectively. The formula "p.Age $\geq$ 18" uses the predicate "$\geq$", defined for the integer data type.

Besides entity types and relationship types, any multi-valued term can also be used as range in a declaration. For instance, in the SQL/EER query of Figure 3, the variable a is bound to the finite list of addresses of person p1.

```
select  p1.Name
from    a in p1.Addr, p1 in PERSON, b in p2.Addr, p2 in PERSON
where   a = b and p2.Name = 'John'
```

**Fig. 3.** Name of persons sharing an address with John (SQL/EER version)

This query retrieves the name of all persons who share an address with a person called "John". Note that the result of an SQL/EER is a multiset. This means that the same name may appear several times in the answer of this query. By placing the reserved word **distinct** in front of the term list in the **select**-clause, a set of distinct names is computed.

The last example shows the use of inheritance and the use of relationship types as predicates in SQL/EER. Suppose we want to know the names of those professional surfers who surf on rivers that flow into lakes on which they also surf. Figure 4 shows the corresponding SQL/EER query.

```
select  p.Name
from    r in RIVER, l in LAKE, p in PROFI
where   p surfs_onto l and p surfs_on_river r and r flows_into l
```

**Fig. 4.** Name of profis, surfing on rivers, flowing into lakes they surf on (SQL/EER version)

Here, the variable p is declared of type `PROFI`. As profis are "specialized" persons, the attribute `Name` is also defined for them. Thus, p.Name is a correct term. Furthermore, relationship types can be used as predicate names in formulas. In the case of relationships with more than two participating entity types, prefix notation is used instead of infix. Participation in a relationship is inherited too. Therefore, a variable of type `LAKE` (like l) is allowed as participant in relationship `flows_into` within the (sub-)formula "`r flows_into` l".

## 4 Specification of Hybrid Queries

In the previous section, we discussed a fully textual query language for the EER model. In this section, we show informally how this language is extended with *graphical alternatives* for some of its language constructs. The resulting language of this extension is called the *hybrid* query language HQL/EER.

Briefly, a query in HQL/EER consists of a piece of text (obeying the syntax of SQL/EER) and/or an attributed labeled graph. As it is the case with the textual part, the graph generally consists of declarations (as in the **from**-clause of the text), conditions (as in the **where**-clause of the text), as well as selections (as in the **select**-clause of the text).

For expressing these declarations and conditions graphically, we restrict ourselves to the same graphical symbols used for representing EER-schemes. For instance, variables for a river and a water may be declared by drawing two (rectangular) nodes labeled resp. `RIVER` and `WATER`.

The structural constraints applying to the construction of EER schemes, apply to the graphical part of hybrid queries as well. For instance, the condition that we are only interested in pairs of a river and a water such that the river flows into the water, is indicated by drawing a (diamond shaped) node labeled `flows_into` with edges to both other nodes (cf. Figure 5).

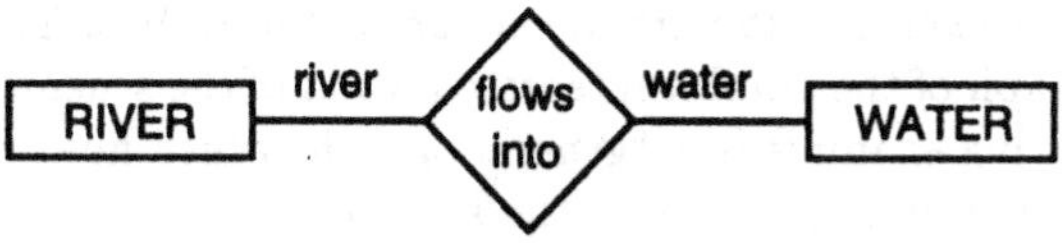


**Fig. 5.** A sample graphical part of an HQL/EER query

In other words, the graphical part of a hybrid query always consists of subgraphs of a graphical representation of the EER-scheme, on two exceptions. First, nodes in the graphical part may be "identified" to express sharing. Second, in order to deal with inheritance, whenever a node labeled with a certain entity type is expected, a node labeled with a subtype of that type may also be used. In the example of Figure 5, this means that we could replace the `WATER`-node by a `LAKE`-node, since there is a type construction in our example scheme with `WATER` as input and `LAKE` as output.

As a consequence of the above restriction, arithmetic and aggregate functions, and hence many formulas, cannot be expressed in HQL/EER. However, it is easy to come up with graphical notations for these concepts. Additionally, since the graphical part of a hybrid query is only related to the "outermost" query, graphs cannot be associated to subqueries. Independent of these restrictions, the expressive power of HQL/EER is obviously equal to that of SQL/EER.

Since (some) nodes in the graphical part of an HQL/EER query correspond to declared variables, "references" to such nodes may be used as variables in the textual part. This is the way in which the textual and graphical part of a hybrid query are linked together.

We now illustrate these concepts on some simple examples over the scheme of Figure 1. Figure 6 shows a possible expression of the query of Figure 2 in HQL/EER. Intuitively, the `PERSON`-node corresponds to the declaration of the variable p in the textual expression, while e.g., the `int`-node corresponds to the term p.age in the textual expression, since it is linked to the node corresponding to the variable p by means of an `Age`-edge.

**select** x, y
**where** y ≥ 18

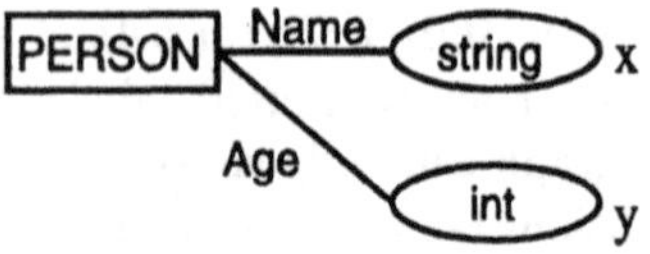


**Fig. 6.** Name and age of adults (HQL/EER version)

Note that in the textual part of the hybrid query, two variables are used but not declared. Instead, they refer to nodes in the graph. Hence the variable x ranges over the names of all persons, while y ranges over their ages. The textual part specifies that the values assigned to x and y are retrieved, if and only if the value for y, i.e., the person's age, is larger than 18.

In the following sections, we formalize this correspondence between declarations, terms and formulas in a textual expression on one hand, and subgraphs of a graphical expression on the other hand.

In the previous example, the graphical part of the HQL/EER query consists simply of a subgraph of the graphical representation of the scheme. The graphical part of the following example is a "general" graph, which however still satisfies the structural constraints imposed by the scheme.

Figure 7 shows a way of expressing the query of Figure 3 in HQL/EER. The fact that a *single* `address`-node is linked to the `Addr`-attributes of *both* `PERSON`-nodes indicates that we are interested in people *sharing* an address.

The graphical part of this query contains two illustrations of constructs whose graphical expression may be considered more natural than their textual counterpart. The aforementioned sharing of the `address`-node as opposed to the join predicate "a=b" in the textual query, is one example. Second, the graphical arrangement of the various nodes shows the interconnection of the persons and their respective address lists in a more straightforward manner than the declarations in the textual version of this query.

**select** p1.Name
**where** p2.Name = `'John'`

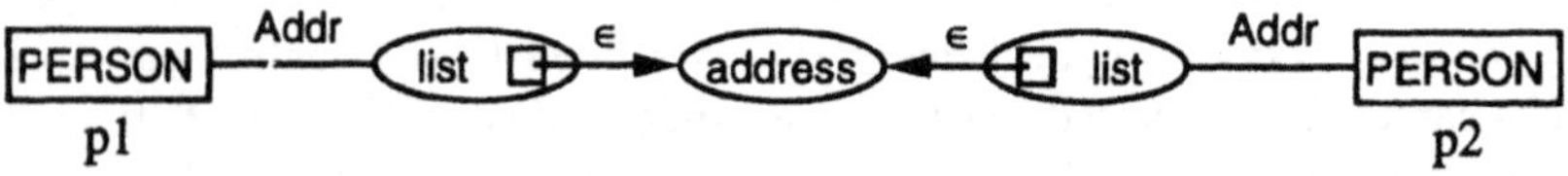


**Fig. 7.** Name of persons sharing an address with John (HQL/EER version I)

With another version of the SQL/EER query of Figure 3 (cf. Figure 8), we illustrate hybrid queries consisting *merely* of a graph. The fact that the `Name`-attribute of one of the persons should have the string value 'John' is indicated by adding this value under the corresponding node. The shading of the other `string`-node indicates the information to be retrieved, i.e., the names of the persons who share an address with John.

Since HQL/EER queries may be totally graphical, our formalism subsumes the purely graphical query languages mentioned in the Introduction. More pre-

**Fig. 8.** Name of persons sharing an address with John (HQL/EER version II)

cisely, the outlook of the graphical part of HQL/EER queries was inspired by the view-tool of the prototype visual database interface described in [14], based on the Graph-Oriented Object Database model [18].

Figure 9 shows another example of an entirely graphical specification of a query, namely that of Figure 4. Note that an entity of type `LAKE` plays the role of a water in the relationship `flows_into`, illustrating how inheritance is used, in a manner similar to SQL/EER. Since `PROFI` (resp. `LAKE`) inherits the participation in the `surfs_onto` (resp. `flows_into`) relationship from `SURFER` (resp. `WATER`), the graphical part of this query is still considered to satisfy the structural constraints imposed by the scheme.

Note also how each of the graph increments consisting of a diamond node and the two rectangles it is connected to, corresponds to a conjunct in the **where**-clause of the textual expression.

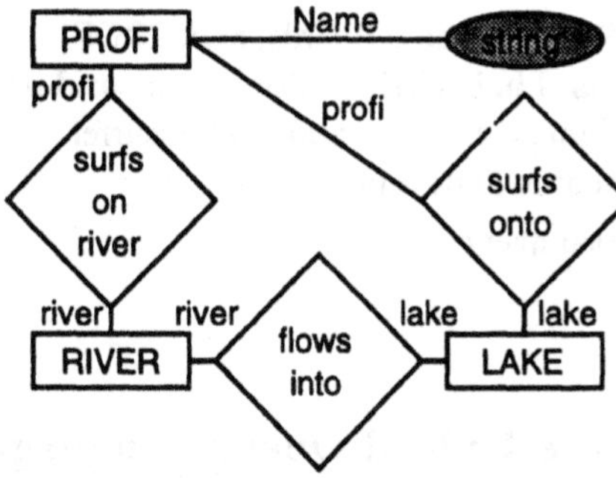


**Fig. 9.** Names of profis, surfing on rivers, flowing into lakes they surf on (HQL/EER version)

We conclude this collection of examples with a slightly more involved one, illustrating our claim that HQL/EER allows those parts of an operation that are most clearly specified graphically resp. textually, to be indeed specified graphically resp. textually. The hybrid query of Figure 10 retrieves the address lists of surfers, who have a relative (i.e., a person with the same name)

- being a professional windsurfer;
- owning a board of the same type as the (single) board owned by the surfer;
- not surfing on dangerous lakes.

As it is hard to express negation graphically, we leave the part of the query involving this negation in the textual part, and express everything else graphically.

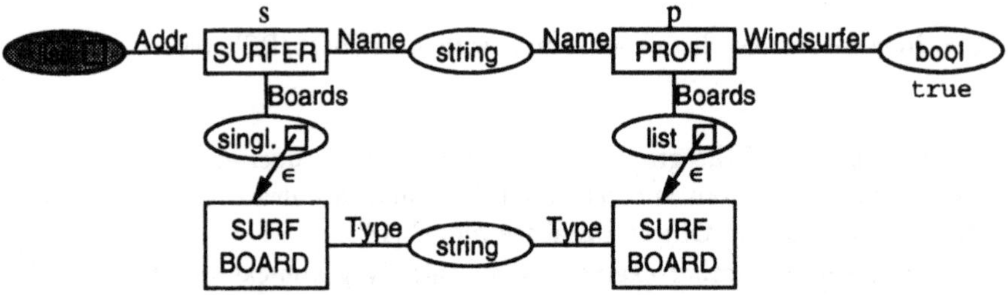


**where** **not exists** l **in** LAKE : ( l.Dangerous **and** p surfs_onto l )

**Fig. 10.** "Involved" hybrid query

In contrast, Figure 11 shows an SQL/EER version of the same query. Note how in this textual query, related information is again dispersed over e.g., the declarations in the **from**-clause and the join-predicates in the **where**-clause.

**select** s.Addr
**from** s **in** SURFER, pb **in** p.Boards, p **in** PROFI
**where** **not exists** l **in** LAKE : ( l.Dangerous **and** p surfs_onto l )
**and** p.Name = s.Name **and** p.Windsurfer
**and** s.Boards.Type = pb.Type

**Fig. 11.** "Involved" textual query

## 5 Definition of the Hybrid Query Language

In Section 4, we discussed HQL/EER by means of examples. In the following sections, we explain how syntax and semantics of HQL/EER are formally defined using graph rewriting.

In this section, we concentrate on the syntax definition. On one hand, the formalization of HQL/EER involves the *representation* of the graphical part of hybrid queries as labeled, attributed graphs. Node and edge labels correspond to scheme elements, like entity types and values. Node attributes are used for storing non-structural information which is part of the query, such as atomic values. On the other hand, we formalize the *construction* of an HQL/EER query as a sequence of applications of graph rewriting rules.

As an example, consider Figure 12, which shows the attributed graph corresponding to the graphical part of the hybrid query of Figure 8. In the upper part of the rectangles, resp. near the arrows, we indicate the label of nodes resp. edges. In the bottom part of the rectangles, we specify the name and value of node attributes. Formula and Decl will be referred to as attributes of the graph itself (formally, these might be looked upon as attributes of some uniquely labeled node, present in any graph). The precise meaning of types and attributes is explained in the remainder of this Section.

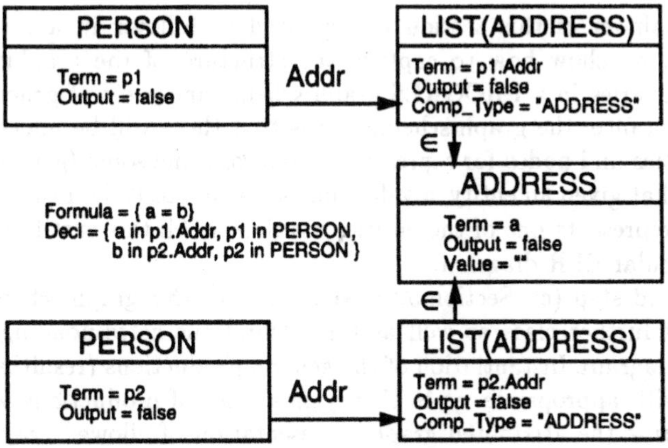


**Fig. 12.** Graph representation of the hybrid query of Figure 8

The need for both an expressive graph model and a powerful graph rewriting formalism, motivates our choice of the graph rewriting formalism called PROGRES [30, 31, 33]. PROGRES is a very high level language based on the concepts of PROgrammed Graph REwriting Systems, and was originally developed in the context of modelling software development environments [11].

Graph rewriting rules (or *productions*) in PROGRES are specified over a *graph scheme*. Such a graph scheme is a set of graph properties (i.e., structural integrity constraints and attribute dependencies) common to a certain collection of graphs. The following components of a graph scheme are distinguished:

- type declarations: these are used to introduce labels for nodes and edges in the considered category of graphs, to declare and initialize node attributes, and to specify the node type of sources and targets of edges;
- node class declarations: these denote coercions of node types with common properties by means of multiple inheritance, hence they play the role of second order types. Class declarations also include attribute declarations.

*Productions* specify when and how graphs are constructed by substituting an isomorphic occurrence of one graph (called the *left-hand side* of the production) by an isomorphic copy of another graph (called the *right-hand side* of the production). Productions are *parametrized* by node and edge types. Furthermore, we call productions *generic*, if they are specified using (among others) classes.

Generic productions represent the *set* of parametrized productions, obtained by instantiating these classes with any of their types.

In addition, productions may specify application conditions (in their **condition**-clause) in terms of structural and attribute properties of the isomorphic occurrence of the left-hand side. The embedding transformation is expressed in the **embedding**-clause, while attribute-computations are performed in the **transfer**-clause.

The remainder of this section is organized as follows: in a first step (cf. Section 5.1), we show how to capture the structure of the graphical part of HQL/EER queries in a PROGRES graph scheme and a set of generic productions. For instance, the graph scheme shows that there will be nodes for representing entities and nodes for representing values, while some (generic) production shows that given an entity, a value may be linked to it by means of an edge, which then represents one of the entities attributes. All this is still independent of any particular EER diagram.

In a second step (cf. Section 5.2), we show how this graph scheme is to be extended by further class as well as type definitions corresponding to a concrete EER diagram. Instantiation of the generic productions (resulting from the first step) with appropriate types then yields a set of productions which completely describe the attributed graph representation of allowed graphical parts of HQL/EER queries.

Given the above two-step method for the formalization of HQL/EER, we now digress on some characteristics we found were missing in PROGRES [29] to enable us to properly formalize HQL/EER, and whose incorporation into other graph rewriting formalisms may add considerably to their fitness as a specification language.

First, since we want to be able to specify all productions in the scheme independent part, we have to be able to parametrize productions in terms of edge types, even though there is no such concept as edge classes. Treating edges and nodes on totally equal grounds by introducing the concept of edge classes into PROGRES, is one solution we considered, but as this appeared to raise some additional problems, outside the scope of our research on hybrid database languages, we chose to restrict ourselves to enhancing the concept of production in two ways:

1. edge type parameters are allowed in the header of a production;
2. redirection of edges in the **embedding clause** of productions is allowed without stating the edge labels: this will e.g., allow us to "merge" two nodes of a graph merely on the basis of their label, without having to specify the labels of eventual incoming or outgoing edges.

A second extension concerns the fact that if one wants to translate a graphical representation of a specification (in our case, a query) into a textual format, one often needs the label of some graph element (in our case, e.g., the type of an entity) as a string. In the current version of the PROGRES model, this might be expressed using extra attributes, but in view of the frequency with which we need this kind of information, we chose to incorporate a kind of "built-in" function **to_string** which delivers the type of a node as a string.

## 5.1 HQL/EER in terms of PROGRES classes

The starting point for the definition of HQL/EER is a formalization of the graphical notation introduced in Section 4 in terms of a PROGRES graph scheme. Figure 13 shows part of this scheme.

```
section NODE_CLASSES;
    node class NODE
        external Term: String := "";
                 RefVar: String := "";
                 Output: Boolean:=false;
    end;
    node class ENT_REL is a NODE end;
    node class ENTITYTYPE is a ENT_REL end;
    node class VALUE is a NODE end;
    node class ATOMIC_VALUE is a VALUE
        external Value: String := "";
    end;
    node class COMPLEX_VALUE is a VALUE
        derived  Comp_Type: String := "";
    end;...
end;
```

**Fig. 13.** Definition of a PROGRES graph scheme, EER diagram independent part

The node class NODE is the root of our node class hierarchy. In the **external**-section, attributes of nodes of types of this class (or of its subclasses) are declared, whose value depends on a user-supplied parameter of the production which creates the node. If the value of an attribute (like Comp_Type in the class COMPLEX_VALUE) is derived from that of other attributes, we use the keyword **derived**. The class definition also specifies a default value for all attributes.

The classes ENT_REL and VALUE are direct descendants of the NODE-class. ENT_REL is an extra common ancestor for classes ENTITYTYPE and RELSHIPTYPE, which is useful for the definition of edges ending in attribute nodes. The class VALUE has two direct descendants, namely ATOMIC_VALUE and COMPLEX_VALUE. Nodes of types of the former class have an attribute in which the actual value is stored. For simplicity, we assume all values are stored as strings.

We now come to the productions specifying the syntax of the graphical part of HQL/EER. Since building this graph always starts with the creation of one or more isolated nodes, similar to the declaration of variables in the textual part, we must first specify a graphical alternative for some of the EBNF-rules for variable declarations.

Consider e.g., the graphical counterpart for the rule for declaring variables of an entity type.

**production** CreateEntity ( EType : **type in** ENTITYTYPE ; VarName : **string** ) =

$$\varepsilon ::= \boxed{\begin{array}{l} 1': \\ \text{EType} \end{array}}$$

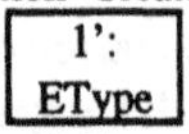

**transfer** 1'.Term := VarName;

    Decl := Decl $\cup$ { 1'.Term & 'in' & **to_string** (EType) };

**end;**

The left hand side of this PROGRES production is the empty graph (represented as $\varepsilon$), while the right hand side consists of a single new node of a type of class ENTITYTYPE. The quoted number is used as a node identifier in e.g., the **transfer**-clause. When creating a new node of some entity type, the value of the Term-attribute (i.e., the name of a variable) must be provided by the user. The corresponding declaration (built using the new nodes Term-attribute as well as its type, obtained by applying the built-in function **to_string**) is inserted in the graph-attribute **Decl**.

An analogous production may be given for declaring variables ranging over relationship types. Note that there is no need to specify a graphical alternative for the rule which states that a variable is a term. The act of creating an isolated node in a graph corresponds to the declaration of a variable, while the node itself may readily be used as a term.

In some cases, a graphical alternative can also be used for declaring a variable ranging over something else than an entity or relationship type. Concretely, if a variable ranges over a list-, bag- or set-valued attribute, the production shown below may be used.

    **production** VarInTerm ( CVType : **type in** COMPLEX_VALUE ;

                        AVType : **type in** ATOMIC_VALUE ;

                          $\in$ : CVType -> AVType ; VarName : **string** ) =

$$\boxed{\begin{array}{l} 1': \\ \text{CVType} \end{array}} ::= \boxed{\begin{array}{l} 1': \\ \text{CVType} \end{array}} \xrightarrow{\in} \boxed{\begin{array}{l} 2': \\ \text{AVType} \end{array}}$$

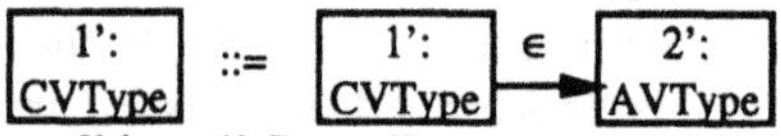

**condition**   1'.Comp_Type = **to_string** (AVType);

**transfer**     2'.Term := VarName;

            Decl := Decl $\cup$ { 2'.Term in 1'.Term };

**end;**

In an application of this production, the formal parameter named **CVType** will be instantiated with the type of the attribute over which the variable will range, while the formal parameter named **AVType** will be instantiated with the type of the variable itself.

This production also illustrates a convention in the usage of quoted numbers. The same number is used for corresponding nodes in the left and right hand side of a production. A number preceded by a quote then refers to the node in the left hand side, while a number followed by a quote refers to the node in the right hand side.

The PROGRES part of the rule which states that a constant string is a term, is quite straightforward.

**production** AddValue ( VType : **type in** ATOMIC_VALUE ; Actual_Value : **string** ) =

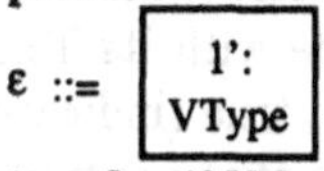


**transfer** 1'.Value := Actual_Value;
1'.Term := 1'.Value;
**end;**

This rule allows the addition of an atomic value to the graphical part of an HQL/EER query. The actual value must of course be provided as a parameter. This value is then assigned to both the Value- and the Term-attribute of the newly created node.

The graphical alternative for the rule which states that a term followed by an attribute is also a term, also has an obvious semantics: to a node of a type of class ENT_REL, a newly created node of a type of class VALUE is linked by means of an edge of a type with the proper source and target type. The term of this new node is assigned the string concatenation (expressed using the operator "&") of the term of the existing node, a dot and the string representation of the type of the edge.

**production** AddAttribute ( ERType : **type in** ENT_REL ; ValType : **type in** VALUE ;
AttName : ERType -> ValType ) =

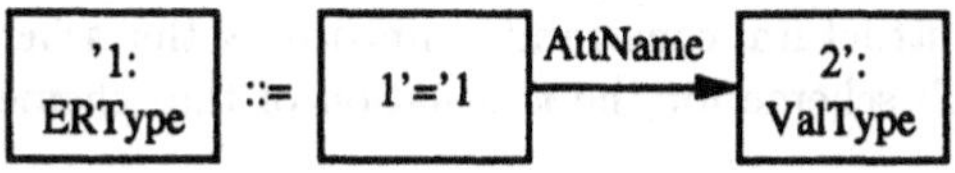


**transfer** 2'.Term := 1'.Term & "." & **to_string** (AttName);
**end;**

An analogous production may be given for roles of relationships instead of attributes.

We now illustrate graphical alternatives for formulas, using the rule for equality predicates. A major advantage of graphical over textual expression of conditions, is that object sharing is expressed much more clearly in a graph. For instance, in order to express textually that two persons should have the same name, a join-predicate like "$p_1$.name = $p_2$.name" must be specified. In a graph representation, we simply draw a node for each person, and one single node for their name, which is then linked by means of edges to each of the Person-nodes. One way to build such a pattern, is first to create two disjoint graphs each representing a person and his name, and then to *identify* the two nodes representing the names, using the PROGRES production outlined below.

**production** IdentifyNodes () =

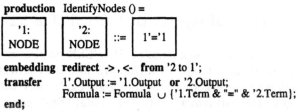

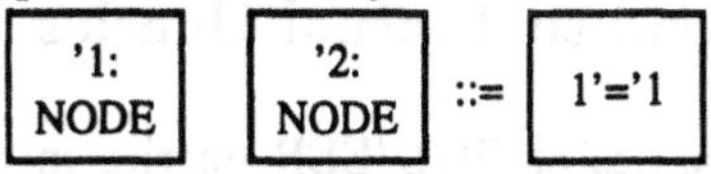


**embedding** **redirect** ->, <- **from** '2 to 1';
**transfer** 1'.Output := '1.Output **or** '2.Output;
Formula := Formula ∪ {'1.Term & "=" & '2.Term};
**end;**

In this production, by means of the **embedding**-clause, we redirect all incoming and outgoing edges from the two nodes that are to be identified, to the newly created node. The formula which specifies that the terms associated to the two input nodes must be equal, is inserted into the variable Formula. Consequently, we loose no information by deleting the second node with its Term-attribute. The Output-attribute of the remaining node is set to true (indicating that the corresponding entity, relationship or value will be part of the output of the query) if the Output-attribute of at least one of the two identified nodes was also true.

### 5.2 HQL/EER in terms of PROGRES types

In the foregoing subsection, we have shown how to define the syntax of the graphical part of HQL/EER queries in terms of a PROGRES graph scheme and associated PROGRES productions.

As classes play the role of second order types, productions specified using classes cannot be applied to concrete graphs without the declaration of a set of related first order types. Given such a set of type definitions, a rule specified in terms of classes actually represents the set of rules, obtained by substituting any class by a type of this class. If a given EER scheme is mapped to an extension of the graph scheme outlined in the foregoing Section, it is guaranteed that any application of a PROGRES rule from the syntax definition of HQL/EER results in a graph that obeys the structural constraints imposed by this scheme. Figure 14 shows part of the graph scheme for the translation of the scheme of Figure 1.

The node classes whose name is suffixed with a "C" are introduced to cope with inheritance. On one hand it is not possible to specify inheritance relationships between node types, so we have to use a class for each entity type. On the other hand, actual nodes have to belong to a type, so for each class we have to declare a type of each of these classes. Note also how attributes and roles are uniformly modeled using edge types.

Given these extra definitions, the production for declaring variables of an entity type may for instance be used to create a node of type PERSON, since this type is declared (indirectly) of class ENTITY. Analogously, the production for the addition of attributes may be used to link a new node of type NUMBER to an existing node of type PERSON by means of an edge of type Age or Length. The "instantiated" production for specifying the age of a person is shown in Figure 15.

## 6 Transformation of Hybrid to Textual Queries

In this section, we define the semantics of HQL/EER queries in terms of the formally defined semantics of SQL/EER [20]. We do this by providing a translation algorithm which transforms a hybrid HQL/EER query consisting of a textual and a graphical part into a purely textual SQL/EER query. Consider an HQL/EER query with a textual part "**select** T **from** D **where** F" (with T a term list, D a list of declarations, and F a formula), and graphical part G. Then the following algorithm shows how to augment the textual part with information

```
section NODE_CLASSES;
    node class PERSON_C is a ENTITYTYPE end;
    node class SURFER_C is a PERSON_C end;
    node class WATER_C is a ENTITYTYPE end;
    node class RIVER_C is a WATER_C end;...
end;
section NODE_TYPES;
    node type SURFER : SURFER_C end;
    node type PERSON : PERSON_C end;
    node type FLOWS_INTO : RELSHIPTYPE end;
    node type STRING : ATOMIC_VALUE end;
    node type ADDRESS : ATOMIC_VALUE end;
    node type LIST_OF_ADDRESS : ORDERED_VALUE
        derived Comp_Type:String := "ADDRESS";
    end;...
end;
section EDGE_TYPES;
    edge type Name : PERSON→STRING;
    edge type Addr : PERSON→LIST_OF_ADDRESS;
    edge type Water : FLOWS_INTO→WATER;
    edge type River : FLOWS_INTO→RIVER;...
end;
```

**Fig. 14.** Definition of a PROGRES graph scheme, EER diagram dependent part

**production** AddAttribute =

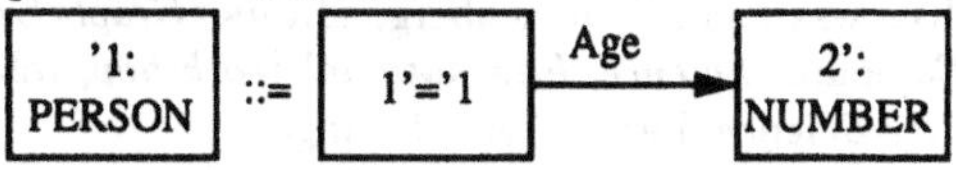


**transfer** 2'.Term := 1'.Term & '.' & 'Age';

**Fig. 15.** An instantiation of a PROGRES rule

from the attributes of G's nodes and G itself, resulting in an SQL/EER query. The semantics of the HQL/EER query is then defined as the semantics of this SQL/EER query.

1. In both T, D and F, substitute any variable referring to a node in the graphical part, by the Term-attribute of this node;
2. Add to T (using commas) the Term-attribute of any node in G whose Output-attribute is true;
3. Add to D (using commas) all declarations in Decl;
4. Add to F (using conjunctions) all formulas in Formula.

E.g., an application of this algorithm to the hybrid query of Figure 8 (whose graph-representation is given in Figure 12) results in the textual query of Figure 3.

## Acknowledgments

Thanks go to Andy Schürr for our discussions concerning the extensions to be made to his formalism PROGRES, in order to accommodate our needs in modeling HQL/EER.

## References

1. *Proceedings of the Ninth ACM Symposium on Principles of Database Systems.* ACM Press, 1990.
2. M. Andries and J. Paredaens. A Language for Generic Graph-Transformations. In Schmidt and Berghammer [28], pages 63–74.
3. M. Angelaccio, T. Catarci, and G. Santucci. *QBD*: A Graphical Query Language with Recursion. *IEEE Trans. Softw. Eng.*, 16(10):1150–1163, 1990.
4. Brandenburg, F.J. Layout Graph Grammars: The Placement Approach. In Ehrig et al. [8], pages 144–156.
5. P. Chen. The entity-relationship model—toward a unified view of data. *ACM Trans. Database Syst.*, 1(1):9–36, 1976.
6. M. Consens and A. Mendelzon. GraphLog: a visual formalism for real life recursion. In ACM [1], pages 404–416.
7. H. Ehrig and H.-J. Kreowski. Applications of Graph Grammar Theory to Consistency, Synchronization, and Scheduling in Data Base Systems. *Information Systems*, 5:225–238, 1980.
8. H. Ehrig, H.-J. Kreowski, and G. Rozenberg, editors. *Graph-Grammars and Their Application to Computer Science, International Workshop*, volume 532 of *Lecture Notes in Computer Science*, Berlin, 1990. Springer.
9. G. Engels. Elementary actions on an extended entity-relationship database. In Ehrig et al. [8], pages 344–362.
10. G. Engels, M. Gogolla, U. Hohenstein, K. Hülsmann, P. Löhr-Richter, G. Saake, and H.-D. Ehrich. Conceptual modelling of database applications using an extended ER model. *Data & Knowledge Engineering*, 9(2):157–204, Dec. 1992.
11. G. Engels, C. Lewerentz, and W. Schäfer. Graph Grammar Engineering – A Software Specification Method. In H. Ehrig, M. Nagl, and G. Rozenberg, editors, *Graph-Grammars and Their Application to Computer Science, International Workshop*, volume 291 of *Lecture Notes in Computer Science*, pages 186–201, Berlin, 1987. Springer.
12. G. Engels and W. Schäfer. *Programmentwicklungsumgebungen, Konzepte und Realisierung.* Leitfäden der Angewandten Informatik. B.G.Teubner, Stuttgart, 1989.
13. F. Fracchia and P. Prusinkiewicz. Physically-Based Graphical Interpretation of Marker Cellwork L-Systems. In Ehrig et al. [8], pages 363–377.
14. M. Gemis, J. Paredaens, and I. Thyssens. A visual database managment interface based on GOOD. In *Proceedings of the International Workshop on Interfaces to Database Systems*, 1992. To appear.
15. M. Gogolla and U. Hohenstein. Towards a semantic view of an extended entity-relationship model. *ACM Trans. Database Syst.*, 16(3):369–416, 1991.
16. H. Göttler, J. Günther, and G. Nieskens. Use Graph Grammars to Design CAD-Systems! In Ehrig et al. [8], pages 396–410.

17. M. Gyssens, J. Paredaens, and D. Van Gucht. A graph-oriented object database model. In ACM [1], pages 417–424.
18. M. Gyssens, J. Paredaens, and D. Van Gucht. A graph-oriented object model for end-user interfaces. In H. Garcia-Molina and H. Jagadish, editors, *Proceedings of the 1990 ACM SIGMOD International Conference on Management of Data*, volume 19:2 of *SIGMOD Record*, pages 24–33. ACM Press, 1990.
19. A. Habel and H.-J. Kreowski. Collage Grammars. In Ehrig et al. [8], pages 411–429.
20. U. Hohenstein and G. Engels. Formal Semantics of an Extended Entity-Relationship Query Language. In S. Spaccapietra, editor, *Proceedings of the 9th International Conference on Entity-Relationship Approach*, 1990.
21. U. Hohenstein and G. Engels. SQL/EER – Syntax and Semantics of an Entity-Relationship-Based Query Language. *Information Systems*, 17(3):209–242, 1992.
22. U. Hohenstein and M. Gogolla. A Calculus for an Extended Entity-Relationship Model Incorporating Arbitrary Data Operations and Aggregate Functions. In C. Batini, editor, *Proceedings of the 7th International Conference on Entity-Relationship Approach*, pages 129–148, 1988.
23. U. Hohenstein, L. Neugebauer, G. Saake, and H.-D. Ehrich. Three-level specification using an extended entity-relationship model. In R. R. Wagner, R. Traunmüller, and H. C. Mayr, editors, *Informationsbedarfsermittlung und -analyse für den Entwurf von Informationssystemen*, volume 143 of *Informatik-Fachberichte*, pages 58–88. Springer, 1987.
24. T. Houchin. Duo: Graph-based database graphical query expression. In Q. Chen, Y. Kambayashi, and R. Sacks-Davis, editors, *Proceedings of The Second Far-East Workshop on Future Database Systems*, volume 3 of *Advanced Database Research and Development Series*, pages 286–295, Singapore, Apr. 1992. World Scientific.
25. R. Hull and R. King. Semantic database modeling: Survey, applications, and research issues. *ACM Comput. Surv.*, 19(3):201–260, 1987.
26. M. Nagl. *Graph-Grammatiken: Theorie, Anwendungen, Implementierung.* Vieweg, 1979.
27. P. Peelman, J. Paredaens, and L. Tanca. G-Log: A declarative graphical query language. In C. Delobel, M. Kifer, and Y. Masunaga, editors, *Proceedings 2nd International Conference on Deductive and Object-Oriented Databases*, number 566 in Lecture Notes in Computer Science, pages 108–128, Berlin, Dec. 1991. Springer.
28. G. Schmidt and R. Berghammer, editors. *Proceedings of the 17th International Workshop on Graph-Theoretic Concepts in Computer Science*, volume 570 of *Lecture Notes in Computer Science*, Berlin, 1992. Springer.
29. A. Schürr. Private communication.
30. A. Schürr. Introduction to PROGRESS, an Attribute Grammar Based Specification Language. In M. Nagl, editor, *Proceedings of the 15th International Workshop on Graph-Theoretic Concepts in Computer Science*, volume 411 of *Lecture Notes in Computer Science*, pages 151–165, Berlin, 1989. Springer.
31. A. Schürr. *Operationales Spezifizieren mit programmierten Graphersetzungssystemen.* PhD thesis, RWTH Aachen, 1991. Deutsche Universitäts Verlag, Wiesbaden. (in German).
32. K.-Y. Whang, A. Malhotra, G. Sockut, L. Burns, and K.-S. Choi. Two-dimensional specification of universal quantification in a graphical database query language. *IEEE Trans. Softw. Eng.*, 18(3):216–224, Mar. 1992.
33. A. Zündorf and A. Schürr. Nondeterministic Control Structures for Graph Rewriting Systems. In Schmidt and Berghammer [28], pages 48–62.

# Decomposability helps for deciding logics of knowledge and belief

Stefan Arnborg *
The Royal Institute of Technology

**Abstract**

We show that decision problems in modal logics (logics of knowledge and belief) are easy for decomposable formulas. Satisfiability of a formula of size $n$ and treewidth $k$ can be decided in time $O(nf(k))$, where $f$ is a double exponential function. This result holds not only for the logics S5 and KD45 with NP-complete decision problems, but also for extensions to multiple agents as in the standard logics $K_n$, $T_n$, $S4_n$, $S5_n$ and $KD45_n$, whose decision problems are PSPACE complete for arbitrary formulas. Moreover, the method works for these logics extended with operators for distributed and common knowledge, which otherwise cause a complexity increase to exponential time for the satisfiability problem.

## 1 Introduction

Applications involving reasoning of one or several agents with beliefs and knowledge about the external world can be formalized in various modal logics, like the standard modal logics known as K, T, S4, S5, KD45 and their extensions to multiple agents, $K_n$, $T_n$, $S4_n$, $S5_n$ and $KD45_n$. For a selfcontained description of these logics, see Halpern and Moses[11]. Reasoning in a logic can be reduced to validity or satisfiability checking of a formula. But decision problems of modal propositional logics are of high complexity, and consequently the vision of a perfect mechanized reasoner can probably not be realized – something must be given up. One can either give up accuracy – sometimes getting an incorrect decision, or completeness – solving only a restricted class of formulas. Sometimes, one 'gives up' by realizing that too much time has been spent without finding the answer. But 'giving up' can be realized in many ways. With a decision method full of heuristics one can usually solve many cases not solvable with standard direct methods – but it is not in general possible to say when the method will work. Restricting the formula to a conjunction of Horn clauses works well in some cases, but not all since, *e.g.*, the model counting problem is $\#P$-complete, and the restriction is not acceptable in many applications.

Decomposition methods solve restricted classes of problems, namely those with small treewidth. Treewidth is an integer parameter of a graph or a formula and the

*email:stefan@nada.kth.se; nada, kth, s-100 44 Stockholm, Sweden

cost of solving a formula of size $n$ and treewidth $k$ is typically $nf(k)$, where $f$ is exponential or double (even triple in hypothetical applications) exponential. Thus, decomposition methods have a soft constraint in that the treewidth cannot be large (the practical limit is normally 4 to 14 depending on which problem is solved, see *e.g.,* [6]). Decomposition methods have a long history in Operations Research[7,5], Reliability Engineering[17] and Artificial Intelligence[13,18]. They have recently been extensively studied in Theoretical Computer Science and Mathematics. A common criticism of these methods is that no non-empirical evidence is available to show that the methods are useful in real world applications. It has turned out that many application areas exist where the technique has proven useful, but a special problem that has withstood serious efforts by mathematicians or computer scientists, like factoring large numbers or breaking cryptographic systems, does usually not seem to be decomposable.

Without going into the usefulness argument, we want to show that decision problems in modal logics are indeed easy for decomposable formulas. This has been known for some time in the case of classical propositional logic where satisfiability is simply the existence of a global solution to a number of local constraints, each constraint demanding that the truth value of a formula is consistent with the truth-values of its immediate subformulas. It is also known that decomposability helps for some logics based on classical propositional logic, like the logics of Bayesian inference, probabilistic logic, circumscriptive propositional logic, and fuzzy propositional logic[13,1,10].

For modal logics, the Kripke style semantics implies that satisfiability is not simply the existence of a global solution satisfying local constraints, but also a global relation between several, sometimes exponentially many, global solutions. Some modal logics have PSPACE complete decision problems, and then standard methods for decomposable problems based on monadic second order logic[5] do not apply. Only a few examples are known of PSPACE complete problems shown easy for decomposable graphs[8]. For this reason the current problem has some intrinsic interest for understanding the limits of decomposability, besides the possibility of a practical application.

We will show that decomposability helps for logics of knowledge and belief, in the sense that linear time decision methods exist for any formula family of uniformly bounded treewidth. The methods we use are fairly direct and do not rely on the monadic second order logic approach, but are similar in spirit to Shenoy and Shafers approach to similar problems[18]. This gives a good grasp of these logics and their relationship with decomposability. It must also be said that the technique used is not optimized in other ways and cannot be used in practical applications without further tuning.

## 2 The propositional logic of $n$-agent knowledge

Formulas in our logic are built from propositional variables $p$, $q$, $r$, ..., logical constants $\wedge$ and $\neg$, and the knowledge operator $K_i$ for $i = 1, 2, \ldots, n$. $K_i f$ can be interpreted as 'agent $i$ knows $f$'. Thus, a propositional variable is a formula and if

$f_1$ and $f_2$ are formulas, then $\neg f_1$, $f_1 \wedge f_2$ and $K_i f_1$, for $1 \leq i \leq n$ are formulas, and there are no other formulas.

The logics $K_n$, $T_n$, $S4_n$, $S5_n$ and $KD45_n$ have complete axiomatizations distinguished by different axiom schemes, usually called A1, A2, A3, A4, A5, A6. Table 1, upper part, shows the axioms and which axiom sets characterize the standard modal logics. For $n = 1$ we omit the subscript and get the standard modal logics K, T, S4, S5 and KD45. The semantics of the logics can be defined with Kripke structures. Such a structure $M$ consist of a set of possible worlds (states) and a binary possibility relation $\mathcal{K}_i$ for each agent (sometimes the term accessibility is used for possibility in the literature). In each possible world, every formula is either true or false. We use notation $(M, s) \models f$ to say that $f$ is true in world $s$ of structure $M$. If $(s,t) \in \mathcal{K}_i$, then agent $i$ considers world $t$ possible when in world $s$, and whatever he knows when in world $s$ must be true in world $t$. In other words,

- $(M, s) \models p$ for propositional variable $p$ if and only if $p$ is true in world $s$
- $(M, s) \models f_1 \wedge f_2$ if and only if $(M, s) \models f_1$ and $(M, s) \models f_2$
- $(M, s) \models \neg f$ if and only if not $(M, s) \models f$
- $(M, s) \models K_i f$ if and only if for all $t$ such that $(s,t) \in \mathcal{K}_i$, $(M,t) \models f$

Modal logics can be extended with operators for distributed and common knowledge. Intuitively, $Df$ means that $f$ is known by an agent who knows everything that is known by someone, $Ef$ that $f$ is known by everyone, and $Cf$ that not only $f$ is known by everyone, but this fact is also known by everyone and everyone knows this latter fact, *etc.*. Let $\mathcal{D} = \cap_i \mathcal{K}_i$ and $\mathcal{E} = \cup_i \mathcal{K}_i$ and $\mathcal{C} = \mathcal{E}^* = \mathcal{E}^0 \cup \mathcal{E} \cup \mathcal{E}^2 \cup \ldots$ Then

- $(M, s) \models Df$ if and only if, $(M, t) \models f$ for every world $t$ such that $(s,t) \in \mathcal{D}$.
- $(M, s) \models Ef$ if and only if, $(M, t) \models f$ for every world $t$ such that $(s,t) \in \mathcal{E}$.
- $(M, s) \models Cf$ if and only if, $(M, t) \models f$ for every world $t$ such that $(s,t) \in \mathcal{C}$.

Depending on application, one will normally require some properties on the relations $\mathcal{K}_i$. In the standard modal logics these are combinations of properties *reflexive, symmetric, transitive, Euclidean* and *serial*. The standard logics can be defined with Kripke semantics so that a formula is satisfiable in the logic if and only if it is satisfiable in a structure with the properties indicated by the lower part of Table 1. The set of properties will be denoted $\mathcal{P}$ and the standard modal logics are thus characterized by a set of properties $P \subset \mathcal{P}$.

It is known[11] that, when analyzing a set of formulas, it is only necessary to consider worlds giving truth-values to these formulas and all their immediate and indirect subformulas. For a subformula closed set $\mathcal{F}$, such worlds will be represented by those subsets of $\overline{\mathcal{F}}$, where $\overline{\mathcal{F}} = \mathcal{F} \cup \{\overline{f} \mid f \in \mathcal{F}\}$, which contain exactly one of $f$ and $\overline{f}$ for every $f \in \mathcal{F}$. The interpretation is that world $w$ contains $f$ if $(M, w) \models f$, otherwise it contains $\overline{f}$. By the above definitions, a world in a structure must be

| Axiom | $K_n$ | $T_n$ | $S_4$ | $S_5$ | $KD45$ | Definition |
|---|---|---|---|---|---|---|
| A1 | * | * | * | * | * | all propositional tautologies |
| A2 | * | * | * | * | * | $(K_i\phi \wedge K_i\phi \rightarrow \psi) \rightarrow \psi$ |
| A3 | | * | * | * | | $K_i\phi \rightarrow \phi$ |
| A4 | | | * | * | * | $K_i\phi \rightarrow K_iK_i\phi$ |
| A5 | | | | * | * | $\neg K_i\phi \rightarrow K_i\neg K_i\phi$ |
| A6 | | | | | * | $\neg K_i\bot$ |
| Property | | | | | | |
| *reflexive* | | * | * | * | | $(s,s) \in \mathcal{K}_i$, all $s$ and $i$ |
| *transitive* | | | * | * | * | if $(s,t),(t,u) \in \mathcal{K}_i$, then $(s,u) \in \mathcal{K}_i$ |
| *symmetric* | | | | * | | if $(s,t) \in \mathcal{K}_i$, then $(t,s) \in \mathcal{K}_i$ |
| *Euclidean* | | | | | * | if $(s,t),(s,u) \in \mathcal{K}_i$, then $(t,u) \in \mathcal{K}_i$ |
| *serial* | | | | | * | $(s,t) \in \mathcal{K}_i$, all $s$ and $i$, and some $t$ |

Table 1: Standard modal logics defined with axiom schemes and Kripke semantics

consistent *wrt* the propositional operators $\wedge$ and $\neg$. The ***projection*** of a world $w$ on $X \subset \mathcal{F}$ is $w \downarrow X = w \cap \overline{X}$. The ***transpose*** $\tau S$ of a set of sets $S$ is the set of sets obtainable by selecting one element from each member of $S$. As an example, if $S = \{\{a,b\},\{a,c\}\}$, then $\tau S = \{\{a\},\{a,c\},\{a,b\},\{b,c\}\}$. Likewise, if $\{\} \in s$, then $\tau s = \{\}$, and if all members of $S$ are singletons, then $\tau S$ is also a singleton.

# 3 Tree-decomposition of formula set

Consider a given set of formulas $\mathcal{F}$, which is closed under taking subformulas (*i.e.*, if $f \in \mathcal{F}$, then all immediate and indirect subformulas of $f$ are also in $\mathcal{F}$). A ***tree-decomposition*** of $\mathcal{F}$ is a tree $T$ and a map $X_t$ from nodes of $T$ to subsets of $\mathcal{F}$ such that

- every member of $\mathcal{F}$ is in some $X_t$,
- for every non-atomic formula $f \in \mathcal{F}$, there is some $X_t$ that contains $f$ and its immediate subformulas,
- if $f \in X_t \cap X_s$, then $f \in X_w$ for every node $w$ on the path in $T$ from $s$ to $t$.

The ***width*** of a tree-decomposition is $\max_{t \in T} |X_t| - 1$. The minimum width over all tree-decompositions is the ***treewidth*** of $\mathcal{F}$. Tree decompositions were introduced for graphs by Robertson and Seymour[16], and treewidth is equivalent to many parameters introduced for graphs, such as being a partial graph of a $k$-tree[2]. A (graph or) formula is decomposable if it has a small treewidth, and the cost of applying decomposition methods is at least exponential in the treewidth. If a formula set has small treewidth $k$, an optimal (minimum width) treedecomposition is easy to find in time $O(n^{k+2})$, $O(f(k)n \log n)$ or $O(f(k)n)$[3,14,12,15,4], although for arbitrary treewidth the problem of finding an optimal tree-decomposition is NP-hard[3].

In our application, we only have to consider a formula set $\mathcal{F}$ consisting of a given formula $\varphi$ and all its subformulas. A *binary tree-decomposition* is a tree-decomposition with the following additional properties:

- The tree $T$ is a binary tree
- The generating formula $\varphi$ for $\mathcal{F}$ is in $X_t$ where $t$ is the root of $T$.
- Every $f \in \mathcal{F}$ has a designated leaf $l_f$ in $T$ such that $X_{l_f} = \{f\}$. For every leaf $l$ which is not a designated leaf, $X_l = \{\}$. For no non-leaf $n$ is $X_n = \{\}$.
- For every $f \in \mathcal{F}$ there is a designated node $n_f$ of $T$. If $f$ is non-atomic, $X_{n_f}$ contains $f$ and its immediate subformulas. If $f$ is atomic, $n_f = l_f$.
- For no $f$ in $\mathcal{F}$ is $f$ in $X_n$ unless $l_f$ or $n_f$ is in the subtree of $T$ rooted at $n$.

For a given tree-decomposition it is easy to get a binary tree-decomposition of the same width by the operations: replicating a node of degree larger than 3, adding a leaf $l$ adjacent to a node $n$ with $X_l \subset X_n$, and orienting the tree.

A non-deterministic tree automaton is defined on a binary decomposition tree $T$ as follows: The automaton will accept consistent subsets of $\mathcal{F}$ which correspond to propositional interpretations of $\mathcal{F}$ in such a way that the true formulas are in the subset and the false ones are not. In such an interpretation, formula-subformula consistency must hold for operators $\wedge$ and $\neg$, but not for operator $K_i$.

The automaton has states that are subsets of $\overline{\mathcal{F}}$. Indeed, node $n$ will have state set $S_n$ whose members are consistent subsets of $\overline{X_n}$, where a consistent set contains either $f$ or $\overline{f}$, but not both, for every $f \in X_n$. The automaton is characterized by two function families, the selection functions $\lambda_l : \{0,1\} \rightarrow S_l$ for $l$ a leaf in $T$ and the transition functions $\delta_n : S_l \times S_r \rightarrow 2^{S_n}$, for $l$ and $r$ subnodes of $n$ in $T$.

The selection functions are defined by:

$$\begin{aligned} \lambda_{l_f}(0) &= \{\{\overline{f}\}\} \quad , \\ \lambda_{l_f}(1) &= \{\{f\}\} \quad , \\ \lambda_l(i) &= \{\{\}\} \quad , \text{if } X_l = \{\}. \end{aligned}$$

The procedure for computing $\delta_n(s_l, s_r)$ is as follows:

1. let $s := (s_l \cup s_r) \cap \overline{X_n}$. If $\{f, \overline{f}\} \subset s$ for some $f$, then set $\delta_n(s_l, s_r) = \{\}$ and exit.
2. Set $S := \{s\}$. For every $f$ such that $n_f = n$, do:
   - If $f = \neg f_1$, then replace every $s$ in $S$ by $s \cup \{f, \overline{f_1}\}$ and $s \cup \{\overline{f}, f_1\}$.
   - If $f = f_1 \wedge f_2$, then replace $s$ in $S$ by $s \cup \{f_1, f_2, f\}$, $s \cup \{f_1, \overline{f_2}, \overline{f}\}$, $s \cup \{\overline{f_1}, f_2, \overline{f}\}$ and $s \cup \{\overline{f_1}, \overline{f_2}, \overline{f}\}$.
   - If $f = K_i f_1$, then replace $s$ in $S$ by $s \cup \{f_1, f\}$, $s \cup \{f_1, \overline{f}\}$, $s \cup \{\overline{f_1}, f\}$ and $s \cup \{\overline{f_1}, \overline{f}\}$.
3. Remove from $S$ every $s$ such that $s$ contains both $f$ and $\overline{f}$ for some $f$.
4. Let $\delta_n(s_l, s_r) = S$.

Intuitively, step 1 combines two projections of worlds on overlapping formula sets and projects the combination on $X_n$. Then it checks that the result is consistent (and removes it otherwise). In step 2 we enforce propositional constraints for $\neg$ and $\wedge$ in the projection of a world on $X_n$. The third substep in step 2 is just adding formulas ($f_1$ and $f$) to the state descriptors so that we later can enforce modal constraints between different states in a projected Kripke structure. Note, in particular, that the combination $\{K_i f, \overline{f}\}$ is not ruled out here, since it is forbidden in a world only when $\mathcal{K}_i$ is required to be reflexive, and in those cases the constraint will be taken care of later. We can note here that it is not strictly necessary to demand that every subformula has a designated leaf in the binary tree decomposition. One could also let the selection function for subformula $f$ act on any node $n$ in the decomposition tree with $f \in X_n$. The resulting computation scheme would be messier but possibly just a little faster.

The tree automaton can check an interpretation $F \subset \overline{\mathcal{F}}$, where $F$ contains exactly one of $f$ and $\overline{f}$ for every $f \in \mathcal{F}$, for consistency as follows: Assign state set $R_l = \lambda_l(|X_l \cap F|)$ to each leaf $l$. Propagate the state sets to the root, assigning $R_n = \delta'_n(R_l, R_r)$ to node $n$ with subnodes $l$ and $r$, where $\delta'_n = \bigcup_{r_l \in R_l \wedge r_r \in R_r} \delta_n(r_l, r_r)$. The root $t$ will be assigned $\{\}$ if the interpretation is inconsistent, otherwise it will be assigned the projection of the interpretation on $X_t$. The tree automaton can also compute the projection on $X_t$, $t$ the root of $T$, of all consistent worlds. This is done with the same computation modified to assign $R_l = \lambda_l(\{0,1\})$ for every leaf $l$. The result will be found in $R_t$. Moreover, by assigning state sets top-down in $T$ we can find the projection on each $X_n$ of all consistent worlds.

**Example**: The formula used as an example in [11], $(p \wedge \neg(p \wedge q)) \wedge (K_1 \neg p \wedge K_1 K_2 q)$, has the subformula set shown in Figure 1. One of its decompositions is shown in Figure 2 and a binary tree-decomposition derived from the tree-decomposition is shown in Figure 3. The treewidth of the formula set is 3, *i.e.*, the decompositions are optimal. The transition function for the node marked $\{d, e, h, p\}$, containing the constraint $d = \neg e$, is shown in Figure 4. In this case all values for $\delta_n$ are singletons or empty. This is advantageous in the decision algorithms we will present, but does not affect our worst-case timing analysis.

# 4 Deciding satisfiability in S5 and KD45

The following is proven as Theorem 2.7 in [11]:

**Proposition 4.1** *a)If a formula $\varphi$ in S5 is satisfiable, then it is satisfiable in a structure $S$ with a universal possibility relation, $\mathcal{K}_1 = \{(s,t) \mid s,t \in S\}$*

*b)If a formula $\varphi$ in KD45 is satisfiable, then it is satisfiable in a structure $S \cup \{s_0\}$ with the possibility relation $\mathcal{K}_1 = \{(s,t) \mid s,t \in S\} \cup \{(s_0, s) \mid s \in S\}$.*

The tree automaton for a tree-decomposition of a formula $\varphi$ can be used to compute, inductively over the decomposition tree and for each of its nodes $n$, the

$$
\begin{array}{rrcl}
a: & b \wedge c & = & (p \wedge \neg(p \wedge q)) \wedge (K_1 \neg p \wedge K_1 K_2 q) \\
b: & p \wedge d & = & (p \wedge \neg(p \wedge q)) \\
c: & f \wedge h & = & (K_1 \neg p \wedge K_1 K_2 q) \\
d: & \neg e & = & \neg(p \wedge q) \\
e: & & = & p \wedge q \\
f: & K_1 g & = & K_1 \neg p \\
g: & & = & \neg p \\
h: & K_1 i & = & K_1 K_2 q \\
i: & & = & K_2 q \\
p: & & = & p \\
q: & & = & q
\end{array}
$$

Figure 1: Subformula set of $(p \wedge \neg(p \wedge q)) \wedge (K_1 \neg p \wedge K_1$

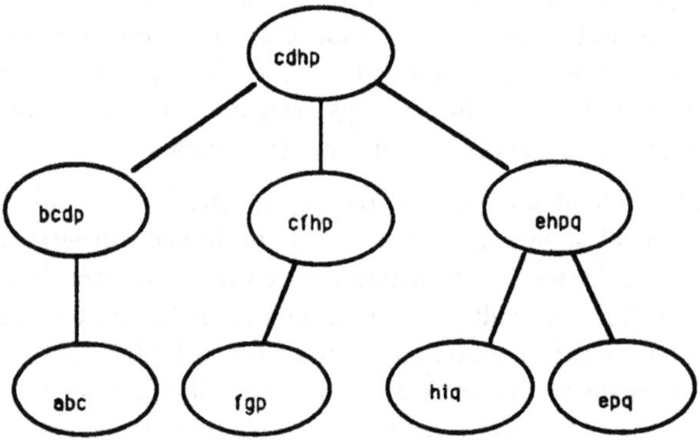


Figure 2: Tree-decomposition of formula set

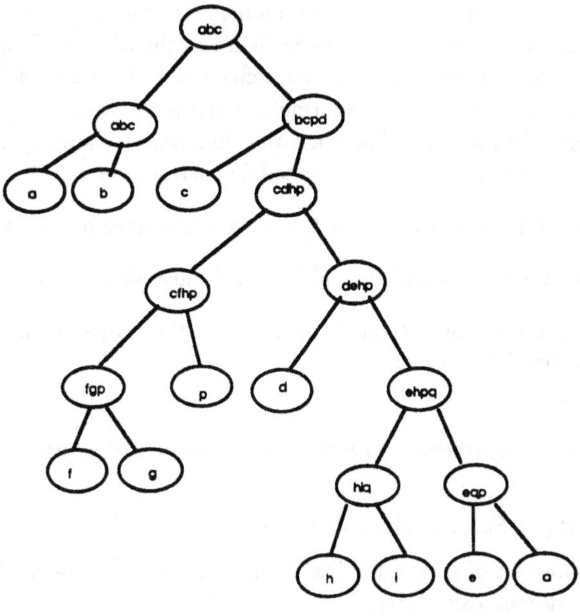


Figure 3: Binary tree-decomposition of formula set

| $\delta$ | $\{d\}$ | $\{\overline{d}\}$ |
|---|---|---|
| $\{e,h,p,q\}$ | $\{\}$ | $\{\{d,e,h,p\}\}$ |
| $\{e,\overline{h},p,q\}$ | $\{\}$ | $\{\{d,e,\overline{h},p\}\}$ |
| $\{\overline{e},h,p,\overline{q}\}$ | $\{\{d,\overline{e},h,p\}\}$ | $\{\}$ |
| $\{\overline{e},h,\overline{p},q\}$ | $\{\{d,\overline{e},h,\overline{p}\}\}$ | $\{\}$ |
| $\{\overline{e},h,\overline{p},\overline{q}\}$ | $\{\{d,\overline{e},h,\overline{p}\}\}$ | $\{\}$ |
| $\{\overline{e},\overline{h},p,\overline{q}\}$ | $\{\{d,\overline{e},\overline{h},p\}\}$ | $\{\}$ |
| $\{\overline{e},\overline{h},\overline{p},q\}$ | $\{\{d,\overline{e},\overline{h},\overline{p}\}\}$ | $\{\}$ |
| $\{\overline{e},\overline{h},\overline{p},\overline{q}\}$ | $\{\{d,\overline{e},\overline{h},\overline{p}\}\}$ | $\{\}$ |

Figure 4: Transition function for node marked $\{d,e,h,p\}$

projections on $X_n$ of all maximal sets, and minimal sets containing a world projecting to a given state in $S_n$, of worlds that can figure as the set $S$ of Proposition 4.1 *wrt* the propositional and modal constraints defined in the subtree of $T$ rooted at $n$. A constraint for an operator of the formula which is the outermost operator of formula $f$ is defined in node $n_f$. The transition functions $\delta_n$ already take care of the propositional operators $\wedge$ and $\neg$. The modal constraints are:

**Max constraint** If $S$ has a world with $Kf$, then it can have no world with $\overline{f}$.

**Min constraint** If $S$ has a world with $\overline{Kf}$, then it must also have a world with $\overline{f}$.

The modal constraints for $Kf$ can be enforced on the projection of $S$ on $X_{n_{Kf}}$, since this set contains both $Kf$ and $f$.

**Algorithm S5SAT**

INPUT: A formula $\varphi$ represented as a tree-decomposition of width $k$ of its set of subformulas

OUTPUT: YES if $\varphi$ is S5 satisfiable, otherwise NO

METHOD: Compute set of state sets $\text{Max}(n)$ and $\text{Min}(n,s)$, $n$ a node of $T$, $s \in S_n$, bottom up in the decomposition tree:

For leafs of $T$ we have:

$$\begin{array}{rcll} \text{Max}(l) & = & \{\{\{\overline{f}\},\{f\}\}\} & ,l=l_f \\ \text{Min}(l,\{f\}) & = & \{\{f\}\} & ,l=l_f \\ \text{Min}(l,\{\overline{f}\}) & = & \{\{\overline{f}\}\} & ,l=l_f \\ \text{Max}(l) & = & \{\{\}\}, & ,X_l=\{\} \\ \text{Min}(l,\{\}) & = & \{\{\}\}, & ,X_l=\{\} \end{array}$$

For a non-leaf $n$ of $T$ with sons $l$ and $r$ we have:

1. Compute $\text{Ma} := \tau\{\delta_n(m_l,m_r) \mid m_l \in \text{Max}(l), m_r \in \text{Max}(r)\}$ and $\text{Mi}(s) := \tau\{\delta_n(m_l,m_r) \mid s \in \delta_n(s_l,s_r), m_l \in \text{Min}(l,s_l), m_r \in \text{Min}(r,s_r)\}$ (a maximal model must be contained in maximal models for the subnodes and a minimal model must contain minimal models for the subnodes).

2. For every member max of Ma and every formula $Kf$ such that $n_{Kf}=n$:

   - If max contains a member with $\{Kf,\overline{f}\}$, then delete it from max,
   - if max contains both a member with $Kf$ and a member with $\overline{f}$, then replace max by two sets in Ma: max with all members containing $Kf$ deleted and max with all members containing $\overline{f}$ deleted.

3. For every member min of $\text{Mi}(s)$, $s \in S_n$, and every formula $Kf$ such that $n_{Kf}=n$:

   - If min contains a member with $\overline{Kf}$ but no member with $\overline{f}$, then replace min in $\text{Mi}(s)$ by the sets $\{\text{min} \cup \{u\} \mid u \in S_n, \overline{f} \in u\}$.

4. Remove from Ma every set that is not a superset of a set in $\text{Mi}(s)$, $s \in S_n$.

5. Remove from Mi($s$) every set that is not a subset of a member of Ma.

6. Remove from Mi($s$) every set that does not contain $s$.

7. Remove non-minimal sets of each Mi($s$) and non-maximal sets of Ma.

8. Let Max($n$) := Ma and Min($n, s$) := Mi($s$), $s \in S_n$.

Output YES if some Min($t, s$), $t$ the root of $T$, contains a set with a state containing $\varphi$, otherwise output NO.
[End of algorithm]

**Correctness argument for S5SAT**

A set min in Min($t, s$) can be extended to the set $S$ of an S5 structure of the type described in Proposition 4.1 as follows. Suppose $l$ and $r$ are the sons of $t$ in $T$. By the construction process, min contains some set $\delta_t(\min_l, \min_r)$, where $\min_l$ and $\min_r$ are members of the Min sets of the subnodes of $t$. The sets in $S' = \{m \cup m_l \cup m_r \mid m \in \min, m_l \in \min_l, m_r \in \min_r, m \downarrow X_l = m_l \downarrow X_t, m \downarrow X_r = m_r \downarrow X_t\}$ are propositionally consistent over $X_t \cup X_l \cup X_r$ and $S'$ itself satisfies all modal constraints defined in nodes $t$, $l$ and $r$. This construction can be continued downwards to the leaves of $T$. The projection of the resulting set on any $X_n$, $n$ in $T$, will be a set of states between a minimal and a maximal set of states for node $n$, *i.e.*, all propositional and modal constraints will be satisfied.

**Timing analysis for S5SAT**

The state sets $S_n$ are of size at most $|2^{X_n}| = 2^{|X_n|} \leq 2^{k+1}$, *i.e.*, single exponential in $k$. The dominating quantities computed for each node of the decomposition tree are the Min and Max sets. These have size at worst double exponential in the treewidth $k$, and can also be computed in time proportional to their size. Therefore, the time required for each node is bounded by a function double exponential in $k$ but independent of the formula size. The size of the decomposition tree is $O(N)$ for a formula of size $N$, since at least half its leafs can be mapped bijectively to the subformulas of $\varphi$. Thus, for fixed treewidth $k$, the computation cost is $O(N)$.

## 4.1 Satisfiability in KD45

The related logic KD45 also has an NP-complete decision problem. An algorithm for decomposable formulas can be obtained by a small modification of the above algorithm. The modification consists in relating $s$ in Min($n, s$) to the projection on $X_n$ of world $s_0$ of in Proposition 4.1, case b), and similarly replacing Max($n$) by Max($n, s$). The rules for computing minimal and maximal state sets must account for possibility of worlds in $S$ when in $s_0$.

# 5 Deciding satisfiability in general logics of knowledge and belief

The general algorithm for $n$ agents and possibility relations constrained by a property $P$ in $\mathcal{P}$ must construct projections of Kripke structures on nodes of the decomposition tree. These structures must have the combination of properties for the possibility relations as shown in Table 1. How can we know that the projections we consider can (cannot) be combined into structures with the required properties? Fortunately, global structures exist precisely when the projections have the required properties:

**Definition 5.2** *A projection of a Kripke structure* $(M, \mathcal{K}_1, \ldots, \mathcal{K}_n)$ *on a set* $X$ *is the structure* $(M \downarrow X, \mathcal{K}_1 \downarrow X, \ldots, \mathcal{K}_n \downarrow X)$, *where for a relation* $R$, $R \downarrow X = \{(s \downarrow X, t \downarrow X) \mid (s,t) \in R\}$ *and for a set* $t$, $t \downarrow X = t \cap X$.

**Projection Lemma**: For a Kripke structure with property $P \in \mathcal{P}$ for the possibility relations, all its projections also have property $P$.
**Proof**: Since every property in $\mathcal{P}$ is a conjunction of properties from Table 1, it suffices to prove the lemma for the properties making up the conjunctions: reflexive, transitive, Euclidean, symmetric, serial. But each of these is easily seen to be invariant. As an example, seriality of a relation $R$ implies that for all $s$ there is a $t$ such that $(s,t) \in R$. But every element $s'$ of the projection $R'$ is the image of an element $s$, for which there is a $t$ such that $(s,t) \in R$ by seriality of $R$. But for the projection $t'$ of $t$ we have $(s', t') \in R'$. Since $s'$ was arbitrary, $R'$ must also be serial ∎

**Extension Lemma**: If relations $R_1$ and $R_2$ on consistent worlds of $X_1$ and $X_2$ exist, both with some property $P \in \mathcal{P}$, and $R_1 \downarrow X_2 = R_2 \downarrow X_1$, then there is a relation $R$ with property $P$ on consistent worlds of $X = X_1 \cup X_2$ such that $R \downarrow X_1 = R_1$ and $R \downarrow X_2 = R_2$
**Proof:** Assume the conditions of the Lemma satisfied and let $R' = R_1 \downarrow X_2$. A world of $X$ can be obtained from worlds $w_1$ of $X_1$ and $w_2$ of $X_2$ as $w_1 \cup w_2$ if and only if $w_1 \downarrow X_2 = w_2 \downarrow X_1$. Consider all pairs $(s,t) \in R'$ and all $(s_{ij}, t_{ij}) \in R_i$, $i = 1, 2$, $j = 1, \ldots, n_{sti}$, such that $s_{ij} \downarrow X_{3-i} = s$ and $t_{ij} \downarrow X_{3-i} = t$, $i = 1, 2$. The relation $R$ sought is the union over all $s, t$ of $(s_{1j} \cup s_{2j'}, t_{1j} \cup t_{2j'})$, $j = 1, \ldots, n_{st1}$, $j' = 1, \ldots, n_{st2}$. We must show that this relation has property $P$ if $R_1$ and $R_2$ have property $P$. This is done by considering each of the constituent properties of $P$. Take transitive as an example. Thus, we can assume $R_1$ and $R_2$ transitive. Let $s, t, u$ be consistent states over $X_1 \cup X_2$, with $(s,t), (t,u) \in R$. We must show that $(s,u) \in R$. Let $s_i = s \downarrow X_i$, $t_i = t \downarrow X_i$, $u_i = u \downarrow X_i$, $i = 1, 2$. Then, by the construction of $R$, $(s_i, t_i) \in R_i$ and $(t_i, u_i) \in R_i$ for $i = 1, 2$. Since $R_1$ and $R_2$ are transitive, we also have $(s_i, u_i) \in R_i$ for $i = 1, 2$, and, by the construction of $R$, $(s, u) = (s_1 \cup s_2, t_1 \cup t_2) \in R$, so $R$ is also transitive. The other properties are similarly verified ∎

The general algorithm computes, over the nodes of a decomposition tree $T$, the maximal and minimal state sets possible by agent $i$ from a given state. Then we find every Kripke structure of worlds of $X_t$ where $t$ is the root of $T$, compatible with the maximality and minimality constraints and with the required properties $P \in \mathcal{P}$ for the logic at hand. If one of these structures has a state with the target formula $\varphi$, then, and only then, $\varphi$ is satisfiable in the particular logic.

**Algorithm KnSAT**

INPUT: A formula $\varphi$ represented as a tree-decomposition of width $k$ of its set of subformulas

OUTPUT: YES if $\varphi$ is $K_n$ satisfiable, otherwise NO

METHOD: Compute set of state sets $\text{Max}_i(m,s)$ and $\text{Min}_i(m,s)$, for all nodes $m$ in $T$, every state $s$ of $m$ and all $i$, $1 \le i \le n$, bottom up in the decomposition tree:
For leafs of $T$ we have:

$$\begin{aligned}
\text{Max}_i(l,\{f\}) &= \{\{\{\overline{f}\},\{f\}\}\} &, l = l_f \\
\text{Min}_i(l,\{f\}) &= \{\} &, l = l_f \\
\text{Max}_i(l,\{\overline{f}\}) &= \{\{\{\overline{f}\},\{f\}\}\} &, l = l_f \\
\text{Min}_i(l,\{\overline{f}\}) &= \{\} &, l = l_f \\
\text{Max}_i(l,\{\}) &= \{\{\}\}, &, X_l = \{\} \\
\text{Min}_i(l,\{\}) &= \{\{\}\}, &, X_l = \{\}
\end{aligned}$$

For a non-leaf $m$ of $T$ with sons $l$ and $r$ we have:

1. Compute
   $\text{Ma}_i(s) := \bigcup_{s\in\delta_m(s_l,s_r)} \tau\{\delta'_m(m_l,m_r) \mid m_l \in \text{Max}_i(l,s_r), m_r \in \text{Max}_i(r,s_r)\}$
   and
   $\text{Mi}_i := \bigcup_{s\in\delta_m(s_l,s_r)} \tau\{\delta'_n(m_l,m_r) \mid m_l \in \text{Min}_i(l,s_l), m_r \in \text{Min}_i(r,s_r)\}$.

2. For every state $s$ in $S_m$, every $i$, every member max of $\text{Ma}_i(s)$ and every formula $K_i f$ such that $n_{K_i f} = m$:

   - If $s$ contains $K_i f$, then delete from max in $\text{Ma}_i(s)$ every state containing $\overline{f}$.
   - If $s$ contains $\overline{Kf}$ but min has no member with $\overline{f}$, then replace min in $\text{Mi}_i(s)$ by the sets $\{\text{min} \cup \{u\} \mid u \in S_m, \overline{f} \in u\}$.
   - Remove from $\text{Ma}_i(s)$ every set that is not maximal in $\text{Ma}_i(s)$ or not a superset of a set in $\text{Mi}_i(s)$.
   - Remove from $\text{Mi}_i(s)$ every set that is not minimal in $\text{Mi}_i(s)$ or not a subset of a member of $\text{Ma}_i(s)$.

3. While there is a set min in some $\text{Mi}_i(s)$ with $t \in \text{min}$, $\text{Mi}_i(t) = \{\}$, remove min from $\text{Mi}_i(s)$

4. Let $\text{Max}_i(s) := \text{Ma}_i(s)$ and $\text{Min}_i(s) := \text{Mi}_i(s)$.

Output YES if, for $t$ the root of $T$ and some $s \in S_t$, $\varphi \in s$, $\text{Min}_i(t,s)$ is nonempty, otherwise output NO.
[End of algorithm]

For general modal logics $T_n$, $S4_n$, $S5_n$ and $KD45_n$, characterized by a property $P \in \mathcal{P}$ for its Kripke structures, we can get a decision procedure by modifying the computation of maximal and minimal state sets in algorithm **KnSAT**. We add the following operations to step 2 and repeat step 2 until no more changes occur (since we always replace sets by larger sets in $\text{Mi}_i(s)$ and by smaller in $\text{Ma}_i(s)$, the repetition will terminate):

*reflexive:* If $s$ contains $\{K_i f, \overline{f}\}$, then set $\text{Ma}_i(s) := \{\}$. Otherwise, add $s$ to every member of $\text{Mi}_i(s)$.

*transitive:* If $s$ contains $K_i f$, then remove from each member of $\text{Ma}_i(s)$ all $s'$ containing $\overline{K_i f}$.

*transitive and symmetric:*. Remove from every member max of $\text{Ma}_i(s)$ all states $s'$ with $K_i f \in s \neq K_i f \in s'$.

*transitive and Euclidean:* Remove from every member max of $\text{Max}_i(s)$ every $t$ such that $\{K_i f, \overline{f}\} \subset t$. For every pair of members $t, t'$ of max in $\text{Ma}_i(s)$ such that $K_i f \in t$ and $\overline{f} \in t'$, replace max in $\text{Ma}_i(s)$ by $\{u \mid K_i f \notin u, u \in \max\}$ and $\{u \mid \overline{f} \notin u, u \in \max\}$.

*serial:* If $\text{Mi}_i(s) = \{\{\}\}$, then set $\text{Mi}_i(s) = \{\{u\} \mid u \in S_n\}$.

**Correctness and timing analysis for algorithm KnSAT**

The argument that a Kripke structure with property $P$ can be produced for every state $s$ with $\text{Min}_i(t, s) \neq \{\}$ closely follows the arguments for tableaux constructions in Ch 6 of [11]. Indeed, select any member of $\text{Min}_i(t, s)$ as the $i$-successors of state $s$, and continue to select analogously the $i$-successors of every state so selected. If $P$ implies seriality or reflexivity, then the possibility relations produced will also have the same property. For properties symmetry, transitivity and Euclidianicity, the added rules for the Max set computation ensure that the $P$-closure of the created possibility relations will satisfy all modal constraints. The timing analysis gives the same performance, linear in $N$ and double exponential in $k$, as algorithm **S5SAT**.

## 5.1 Distributed and common knowledge

Logics with distributed knowledge ($D$ operator) can be easily treated as suggested in [11], whereas common knowledge is trickier. The constraint imposed by the $C$ operator is local in the decomposition tree, but we seem to loose the possibility to choose the $i$-successors of states independently of each other. Apparently, we must compute all Kripke structures satisfying the constraints for the $C$ operator in every node, and keep only those that are compatible with some kept structures of its subnodes. The set obtained for the root of $T$ will then be the set of projections of Kripke structures satisfying all constraints. The cost of this method is $f(k)N$, where $f(k)$ is the worst case cost of finding and sorting all Kripke structures for a world of $k+1$ elements. With a straightforward implementation, this cost is double exponential as in previous algorithms, but with significantly larger exponents. For decomposable formulas it is however a theoretical improvement to the alternatives, which seem to be at least double exponential in $N$.

# References


[1] S. ARNBORG. Graph decompositions and tree automata in reasoning with uncertainty. *submitted*

[2] S. ARNBORG. Efficient Algorithms for Combinatorial Problems on Graphs with Bounded Decomposability — A Survey, *BIT 25 (1985), 2-33.*

[3] S. ARNBORG, D.G. CORNEIL AND A. PROSKUROWSKI, Complexity of Finding Embeddings in a $k$-tree, *SIAM J. Alg. and Discr. Methods 8(1987), 277-284.*

[4] S. ARNBORG, B. COURCELLE A. PROSKUROWSKI AND D. SEESE, An Algebraic Theory of Graph Reduction, *to appear*, JACM.

[5] S. ARNBORG, J. LAGERGREN AND D. SEESE, Easy Problems for Tree-decomposable graphs *J. of Algorithms* **12**(*1991) 308-340.*

[6] S. ARNBORG AND A. PROSKUROWSKI, Linear Time Algorithms for NP-hard Problems on Graphs Embedded in $k$-trees, *Discr. Appl. Math. 23(1989) 11-24;*

[7] U. BERTELE AND F. BRIOSCHI, *Nonserial Dynamic Programming*. Academic Press, New York, 1972.

[8] H.L. BODLAENDER, The complexity of path-forming games. *RUU-CS-89-29, University of Utrecht 1989.*

[9] B. COURCELLE, The monadic second order logic of graphs I: Recognizable sets of finite graphs, *Information and Computation* **85**(1) March 1990, 12-75;

[10] D. DUBOIS AND H. PRADE Inference in possibilistic hypergraphs. *Proc. 3rd Int. Conf. Information Processing and Management of Uncertainty in Knowledge based Systems (IPMU), Paris, July 1990. (B. B. Bouchon-Meunier, R. Yager, L.A. Zadeh, eds) Springer Verlag, Lecture Notes in Computer Science*

[11] J.Y. HALPERN AND Y. MOSES, A guide to completeness and complexity for modal logics of knowledge and belief. *Artificial Intelligence* **54**(1992) 319-379,

[12] J. LAGERGREN, Efficient parallel algorithms for tree-decomposition and related problems *Proc. IEEE FoCS, 1990* 173-182.

[13] S.L. LAURITZEN AND D.J. SPIEGELHALTER Local computations with probabilities on graphical structures and their applications to expert systems *J. Royal Statist. Soc. Ser. B* **50** *157-224.*

[14] J. MATOUŠEK AND R. THOMAS, Algorithms finding tree-decompositions of graphs, *J. of Algorithms* **12**(1991) 1-22;

[15] B. REED, Finding approximate separators and computing treewidth quickly. *Proc. ACM SToC 1992* 221-228.

[16] N.ROBERTSON AND P.D.SEYMOUR, Graph Minors II. Algorithmic Aspects of Tree Width. Journal of Algorithms 7 (1986), 309-322.

[17] A. ROSENTHAL, Computing the Reliability of a Complex Network, *SIAM J. Appl. Math. 32 (1977) 384-393.*

[18] P. SHENOY, G. SHAFER, Axioms for probability and belief function propagation, in *uncertainty in Artificial Intelligence* 4 (R.D. Shachter, T.S. Levitt, L.N. Kanal, J.F.Lemmer, Eds.) Elsevier Science Publishers B.V. 1990.

# Extending Graph Rewriting with Copying

Erik Barendsen Sjaak Smetsers

University of Nijmegen, Computing Science Institute
Toernooiveld 1, 6525 ED Nijmegen, The Netherlands
e-mail erikb@cs.kun.nl, sjakie@cs.kun.nl, fax +31.80.652525.

**Abstract.** The notion of term graph rewrite system (TGRS) is extended with a *lazy copying* mechanism. By analyzing this mechanism, a confluence result is obtained for these so-called *copy term graph rewrite systems* (C-TGRS). Some ideas on the use of lazy copying in practice are presented.

## 1 Introduction

There are several models of computation that can be viewed as a basis for functional programming. Traditional examples are the *lambda calculus* and *term rewrite systems*. Graph rewriting is a relatively new concept, providing a model that is sufficiently elegant and abstract, but at the same time incorporates mechanisms that are more realistic with respect to actual implementation techniques.

Graph rewrite systems were introduced in Barendregt et al. [1987b]. The present paper deals with a restricted form of GRS's: the so-called *term graph rewrite systems* (TGRS's, see Barendregt et al. [1987a]). An alternative approach to graph rewriting, based on category theory, can be found in Ehrig et al. [1987]. TGRS's are very well suited as a basis for implementation of functional languages. This is demonstrated by the 'intermediate language' CLEAN that is based on TGRS's (see Brus et al. [1987]). Sequential implementations based on graph rewriting turned out to be successful (see Smetsers et al. [1991]). Although pure TGRS's are convenient to model *sequential* computations, they do not capture some aspects that are important in *parallel* implementations.

In a functional program the evaluation result of any expression is independent of the chosen reduction order by the absence of side effects. This makes functional languages attractive for implementation on parallel hardware. An important class of parallel machines is formed by systems of loosely coupled processors, each equipped with its own local memory. An implementation of functional languages on such a machine has to deal with the exchange of data during program execution. This is a nontrivial matter. In the literature, however, the communication mechanism is often considered as a pure implementation issue and therefore handled *ad hoc*.

In the 'overall view' of a computation, the communication between two processors involves duplication of data. In order to investigate theoretical properties of such a computation, graph rewriting systems will be extended with a general copying mechanism. The information transfer is determined by the moment a processor needs data, and by the amount of information that is available (i.e. not subject to a present computation) at that moment. In terms of copying this means that it should be possible to defer copying at some points in the subject graph until these parts of the computation have been finished. This way of copying is called *lazy copying*. Lazy copying can be used to model communication *channels*, as proposed in van Eekelen et al. [1991]. Together with a mechanism for creating parallel reduction processes, the lazy copying formalism provides a specification method for arbitrary process topologies with various kinds of communication links.

In section 5 some approaches to copying are discussed, and the concept of lazy copying is formalized, resulting in the notion of *copy term graph rewrite system* (C-TGRS). Then some properties of C-TGRS are investigated, leading to the proof of a confluence result. Finally the use of lazy copying in practice is considered. Although we consider a restricted class of TGRS's, we expect that our results can be generalized further.

It turns out that the extension of TGRS's with copying can be described in an elegant way. It is also attractive from a theoretical point of view: such extended TGRS's provide an intermediate formalism between TRS's (where inputs of functions are duplicated during a rewrite step $\mathbf{F}(x) \to \mathbf{G}(x,x)$) and TGRS's (where this duplication does not take place).

The notion of TGRS is still very fresh. Due to the lack of standard terminology, notation, and basic theory it was necessary to include a rather extensive section on the foundations of graph rewriting. We aim at an operational description in the style of existing syntactical work in lambda calculus and term rewriting. For more details and examples the reader is referred to Barendsen and Smetsers [1992]. The theory developed there provides an useful framework for the description and the analysis of complex graph operations.

## 2 Graphs

This section summarizes some basic notions that were presented in Barendsen and Smetsers [1992].

The objects of our interest are directed graphs in which each node is labeled with a specific symbol. The number of outgoing edges of a node is determined by its symbol. In the sequel we assume that $\mathcal{N}$ is some basic set of *nodes* (infinite; one often takes $\mathcal{N} = \mathbb{N}$), and $\Sigma$ is a (possibly infinite) set of *symbols* with *arity* in $\mathbb{N}$.

**Definition 1.** (i) A *labeled graph* (over $\langle \mathcal{N}, \Sigma \rangle$) is a triple

$$g = \langle N, symb, args \rangle$$

such that

(a) $N \subseteq \mathcal{N}$; $N$ is the set of *nodes* of $g$;
(b) $symb : N \to \Sigma$; $symb(n)$ is the *symbol* at node $n$;
(c) $args : N \to N^*$ such that $\text{length}(args(n)) = \text{arity}(symb(n))$.

Thus $args(n)$ specifies the outgoing edges of $n$. The $i$-th component of $args(n)$ is denoted by $args(n)_i$.

(ii) A *rooted graph* is a quadruple

$$g = \langle N, symb, args, r \rangle$$

such that $\langle N, symb, args \rangle$ is a labeled graph, and $r \in N$. The node $r$ is called the *root* of the graph $g$.

(iii) The collection of all finite rooted labeled graphs over $\langle \mathcal{N}, \Sigma \rangle$ is indicated by $\mathbb{G}$.

*Convention.* (i) $m, n, n', \ldots$ range over nodes; $g, g', h, \ldots$ range over (rooted) graphs.

(ii) If $g$ is a (rooted) graph, then its components are referred to as $N_g$, $symb_g$, $args_g$ (and $r_g$) respectively.

(iii) To simplify notation we usually write $n \in g$ instead of $n \in N_g$.

**Definition 2.** (i) A *path* in a graph $g$ is a sequence $p = (n_0, i_0, n_1, i_1, \ldots, n_\ell)$ where $n_0, n_1, \ldots, n_\ell \in g$, and $i_0, i_1, \ldots, i_{\ell-1} \in \mathbb{N}$ are 'edge specifications' such that $n_{k+1} = args(n_k)_{i_k}$ for all $k < \ell$. In this case $p$ is said to be a *path from* $n_0$ *to* $n_\ell$ (notation $p : n_0 \rightsquigarrow n_\ell$).

(ii) Let $p$ be as above. The *length* of $p$ (notation $|p|$) is $\ell$. $p$ *contains* $n$ (notation $n \in p$) if $n_k = n$ for some $k \leq \ell$. $p$ is a *rooted path* if it starts with the root of the graph (i.e. $p : r_g \rightsquigarrow n$ for some $n \in g$). The collection of rooted paths in $g$ is denoted by RP$(g)$. $p$ is *cyclic* if $n_j = n_k$ for some $j \neq k$.

(iii) Let $m, n \in g$. $m$ is *reachable from* $n$ (notation $n \rightsquigarrow m$) if $p : n \rightsquigarrow m$ for some path $p$ in $g$.

**Definition 3.** Let $g$ be a rooted graph.

(i) $g$ is *coherent* if $\forall n \in g \; \exists p \;\; p : r_g \rightsquigarrow n$.

(ii) $g$ is a *tree* if $\forall n \in g \; \exists! p \;\; p : r_g \rightsquigarrow n$.

**Definition 4.** Let $g$ be a graph and $n \in g$. The *subgraph of* $g$ *at* $n$ (notation $g \mid n$) is the rooted graph $\langle N, symb, args, n \rangle$ where $N = \{m \in g \mid n \rightsquigarrow m\}$, and $symb$ and $args$ are the restrictions (to $N$) of $symb_g$ and $args_g$ respectively.

Below we describe the relation $\preceq$, induced by structure preserving mappings called graph homomorphisms.

**Definition 5.** Let $g, h$ be graphs, and $\varphi : N_g \to N_h$.

(i) $\varphi$ is a *(graph) homomorphism* from $g$ to $h$ (notation $\varphi : g \to h$) if for all $n \in g$

$$symb_h(\varphi(n)) = symb_g(n),$$
$$args_h(\varphi(n))_i = \varphi(args_g(n)_i).$$

(ii) $\varphi$ is a *rooted homomorphism* (notation $\varphi : g \underset{r}{\to} h$) if $\varphi : g \to h$ and moreover $\varphi(r_g) = r_h$.

**Definition 6.** The ordering $\preceq$ on $\mathbb{G}$ is defined as follows.

$$g \preceq h \quad \text{if} \quad \varphi : h \underset{r}{\rightarrow} g \text{ for some } \varphi.$$

Here $h$ can be seen as the result of partially *unraveling* $g$. Sometimes we write $g \preceq_\varphi h$ to indicate the homomorphism explicitly.

*Example 1.* Our suggestive drawings of graphs will be self-explanatory.

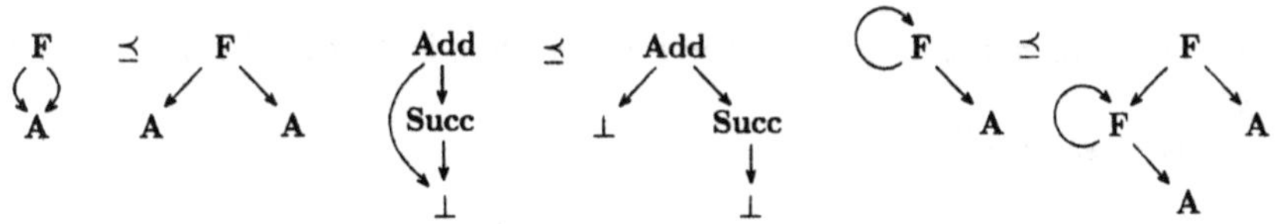


To each graph $g$ one can associate a (possibly infinite) tree $\mathrm{U}(g)$ by an operation called *complete unraveling*. By $g \sim_{\mathrm{T}} g'$ we denote that $g$ and $g'$ are *tree equivalent*, i.e. have the same unraveling.

**Definition 7.** $g$ is *equivalent* to $h$ (notation $g \equiv h$) if there exists a bijective rooted homomorphism from $g$ to $h$. We will usually identify equivalent graphs and replace $\equiv$ by $=$.

**Definition 8.** Let $g, g' \in \mathbb{G}$.

(i) $g$ and $g'$ are *upward compatible* (notation $g \uparrow g'$) if for some $h \in \mathbb{G}$ one has $g \preceq h$ and $g' \preceq h$.

(ii) $g$ and $g'$ are *downward compatible* (notation $g \downarrow g'$) if for some $h \in \mathbb{G}$ one has $h \preceq g$ and $h \preceq g'$.

In Barendsen and Smetsers [1992] it is proved that the set of *regular graphs*

$$\mathbb{G}^* = \{g \in \text{GRAPHS} \mid g' \preceq g \;\; \text{for some } g' \in \mathbb{G}\},$$

equipped with the ordering $\preceq$, forms an unpointed cpo. Using the same proof technique one can show the following.

**Theorem 9.** (i) $g \sim_{\mathrm{T}} g' \;\Rightarrow\; g \uparrow g'$.

(ii) $g \downarrow g' \;\Rightarrow\; g \uparrow g'$.

*Proof.* See Barendsen and Smetsers [1992]. In fact something stronger is proved: one can construct a $\preceq$-smallest common unraveling. □

This result allows us to reason about compatible graphs entirely within the collection of finite graphs: a translation to the (possibly infinite) complete unravelings is not needed.

# 3 Graph rewriting

This section briefly describes graph rewriting by means of certain elementary operations.

**Definition 10.** Let $\perp$ be a special symbol in $\Sigma$ with arity 0. Let $g$ be a graph.

(i) The set of *empty nodes* of $g$ (notation $g^\circ$) is the collection

$$g^\circ = \{n \in g \mid symb_g(n) = \perp\}.$$

(ii) The set of *non-empty nodes* (or *interior*) of g is denoted by $g^\bullet$. So $N_g = g^\circ \cup g^\bullet$.

(iii) $g$ is *closed* if $g^\circ = \emptyset$.

The objects on which computations are performed are closed graphs; the others are used as auxiliary objects, e.g. for defining graph rewrite rules.

**Definition 11.** (i) A *term graph rewrite rule* (or *rule* for short) is a triple $R = \langle g, l, r\rangle$ where $g$ is a (possibly open) graph, and $l, r \in g$ (called the *left root* and *right root* of $R$), such that

(a) $(g \mid l)^\bullet \neq \emptyset$;

(b) $(g \mid r)^\circ \subseteq (g \mid l)^\circ$.

(ii) If $symb_g(l) = \mathbf{F}$ then $R$ is said to be a *rule for* $\mathbf{F}$.

(iii) $R$ is *left-linear* if $g \mid l$ is a tree.

Here condition (1) expresses that the left-hand side of the rewrite rule should not be just a variable. Moreover condition (2) states that all variables occurring on the right-hand side of the rule should also occur on the left-hand side.

*Notation.* We will write $R \mid l$, $R \mid r$ for $g_R \mid l_R$, $g_R \mid r_R$ respectively.

**Definition 12.** Let $p, g$ be graphs.

(i) A *matching homomorphism* (or *match* for short) from $p$ to $g$ is a function $\mu : N_p \to N_g$ such that for all $n \in p^\bullet$

$$symb_g(\mu(n)) = symb_p(n),$$
$$args_g(\mu(n))_i = \mu(args_p(n)_i).$$

In this case we write $\mu : p \xrightarrow[\text{m}]{} g$.

(ii) $\mu$ is a *rooted match* from $p$ to $g$ (notation $\mu : p \xrightarrow[\text{rm}]{} g$) if $\mu : p \xrightarrow[\text{m}]{} g$ and $\mu(r_p) = \mu(r_g)$.

**Definition 13.** Let $s, g \in \mathbb{G}$. Then $s$ is a *segment* of $g$ (notation $s \subseteq g$) if $s$ is coherent, and $s$ results (up to isomorphism) from $g$ by replacing some subgraphs by $\perp$. More formally, $s \subseteq g$ if there exists an injective match $\iota : s \xrightarrow[\text{rm}]{} g$. Observe that such an injection is unique; it will be referred to as $\iota_{s,g}$, or just $\iota$ if there is no danger of confusion.

Intuitively, $s \subseteq g$ if $s$ is an initial ('rooted') part of $g$. The selection of a segment of a given graph is indicated as follows. The construction itself is left implicit.

**Definition 14.** Let $g$ be a graph and $S \subseteq N_g$. The *segment* of $g$ *up to* $S$ (notation $g \rtimes S$) is obtained by replacing the $S$-nodes in $g$ by $\perp$.

A segment is obtained by cutting off subgraphs: the symbol $\rtimes$ is meant to resemble a pair of scissors.

*Example 2.*

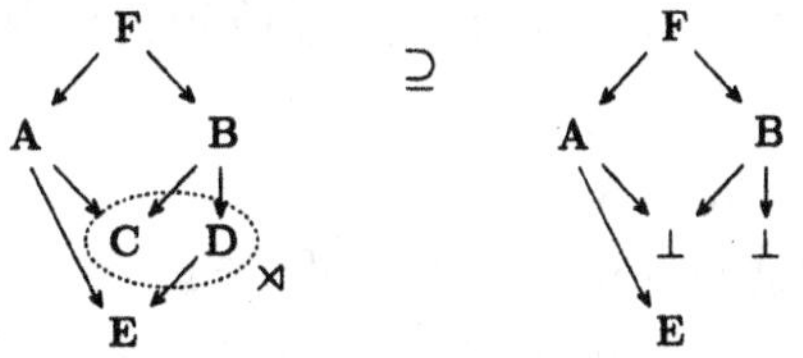


Traditionally, a redirection in a graph $g$ is an operation which replaces all references to a given node by references to another one. This notion is generalized in the following definition.

**Definition 15.** (i) A *redirection map* (or simply *redirection*) on $g$ is a function $\rho : N_g \to N_g$.

(ii) $\rho$ is *projective* if $\rho \circ \rho = \rho$.

(iii) Let $N \subseteq N_g$. $N$ is *$\rho$-closed* if $\rho(N) \subseteq N$.

(iv) Let $h$ be a (sub)graph. Then $\rho$ is *applicable* to $h$ if $N_h$ is $\rho$-closed.

(v) The *field* of $\rho$ (notation $\text{Fld}(\rho)$) is the set

$$\text{Fld}(\rho) = \{n \in N_g \mid \rho(n) \neq n\}.$$

(vi) $\rho$ is a *single redirection* if $\text{Fld}(\rho)$ is a singleton. In this case $\rho$ is written as $[m \mapsto n]$ if $\text{Fld}(\rho) = \{m\}$ and $\rho(m) = n$.

A redirection map on $g$ induces a transformation of $g$, as indicated before.

**Definition 16.** (i) Let $\rho$ be a redirection on $g$. The result of *applying* $\rho$ to $g$ (notation $\rho(g)$) is the graph

$$\rho(g) = \langle N_g, symb_g, \overline{\rho} \circ args_g, \rho(r_g) \rangle$$

where $\overline{\rho}$ is the extension of $\rho$ to sequences, i.e. $\overline{\rho}(n_1, \ldots, n_k) = (\rho(n_1), \ldots, \rho(n_k))$.

(ii) The application $[m \mapsto n](g)$ is traditionally denoted as $g[m := n]$. (This notation will also occasionally be used in the degenerated case $m = n$.)

Below a method for extending graphs will be described. In order to allow broader applications the description is more general than is needed for pure graph rewriting.

**Definition 17.** Let $N, A$ be sets of nodes. The *disjoint set extension* of $N$ with $A$ (notation $N \cup^{+} A$) is the set $N \cup A^{*}$, where $A^{*} = \{a^{*} \mid a \in A\}$ is a set of fresh nodes associated with $A$.

*Convention.* There is a canonical way of extending an operation $f$ on $N$ or $A$ to an operation $f^{*}$ on $N \cup^{+} A$. If no confusion can arise, we keep writing $f$, $a$ for $f^{*}$, $a^{*}$ respectively. The same will be done for subsets of $N, A$ and the corresponding subsets of $N \cup^{+} A$.

**Definition 18.** Let $g$ be a graph.

(i) Let $h$ be a graph, $H \subseteq N_h$, and $\mu : (N_h - H) \to N_g$. The *extension* of $g$ with $\langle h, H\rangle$ according to $\mu$ (notation $g +_{\mu} \langle h, H\rangle$) is defined as follows. Set $g +_{\mu} \langle h, H\rangle = k$, where

$$\begin{aligned}
N_k &= N_g \cup^{+} H; \\
\mathit{symb}_k(n) &= \mathit{symb}_g(n) && \text{if } n \in g, \\
&= \mathit{symb}_h(n) && \text{otherwise;} \\
\mathit{args}_k(n)_i &= \mathit{args}_g(n)_i && \text{if } n \in g, \\
&= \mathit{args}_h(n)_i && \text{if } n, \mathit{args}_h(n) \in H, \\
&= \mu(\mathit{args}_h(n)_i) && \text{otherwise;} \\
r_k &= r_g.
\end{aligned}$$

(ii) An *extension specification* (or *extension* for short) for $g$ is a triple $E = \langle h, H, \mu\rangle$ with $h$, $H$ and $\mu$ as above. Then the extension of $g$ with $E$ is

$$g + E = g +_{\mu} \langle h, H\rangle.$$

(iii) The *extension root* of $E$ (notation $\mathrm{r}(E)$) is defined as follows.

$$\begin{aligned}
\mathrm{r}(E) &= r_h && \text{if } r_h \in H, \\
&= \mu(r_h) && \text{otherwise.}
\end{aligned}$$

*Notation.* We use $\langle h, \mu\rangle$ as an abbreviation for $\langle h, h^{\bullet}, \mu\rangle$. Analogously we write $g +_{\mu} h$ for $g +_{\mu} \langle h, h^{\bullet}\rangle$.

**Definition 19.** Let $g$ be a rooted graph. The result of performing *garbage collection* on $g$ (notation $\mathrm{GC}(g)$) is the graph obtained from $g$ by deleting all nodes that are not reachable from its root, i.e. $\mathrm{GC}(g) = g \mid r_g$.

The three basic operations extension, redirection, and garbage collection are combined into a single operation called *replacement*. This mechanism corresponds to *substitution* in lambda calculus and term rewriting.

**Definition 20.** Let $E$ be an extension of $g$, and let $n \in g$. The result of *replacing* $n$ by $E$ in $g$ (notation $g[n := E]$) is defined by

$$g[n := E] = \mathrm{GC}((g + E)[n := \mathrm{r}(E)]).$$

**Definition 21.** Let $g$ be a graph, and $\mathcal{R}$ a set of rewrite rules.
(i) An $\mathcal{R}$-*redex* in $g$ (or just *redex*) is a tuple $\Delta = \langle R, \mu \rangle$ where $R \in \mathcal{R}$, and $\mu : (R \mid l) \underset{\mathrm{m}}{\rightarrow} g$. The node $\mu(l)$ is called the *redex root* of $\Delta$.
(ii) The extension associated with $\Delta$ is

$$E_\Delta = \langle R \mid r,\ (R \mid r)^\bullet - (R \mid l)^\bullet,\ \mu \rangle.$$

We write $g + \Delta$ instead of $g + E_\Delta$, and $g[n := \Delta]$ instead of $g[n := E_\Delta]$. Moreover $\mathrm{r}(\Delta)$ means $\mathrm{r}(E_\Delta)$.
(iii) Let $\Delta = \langle R, \mu \rangle$ be a redex. The result $g'$ of *contracting* $\Delta$ in $g$ is given by the replacement

$$g' = g[\mu(l) := \Delta].$$

In this case we say that $g$ reduces to $g'$ via $\Delta$ (notation $g \overset{\Delta}{\underset{\mathcal{R}}{\rightarrow}} g'$, or just $g \underset{\mathcal{R}}{\rightarrow} g'$).

The following results concern the interaction between the basic graph operations.

**Lemma 22.** *Let $g$ be a graph, $E = \langle h, I, \mu \rangle$ an extension of $g$, and let $\rho$ be a redirection.*
(i) $\mathrm{GC}(\rho(\mathrm{GC}(g))) = \mathrm{GC}(\rho(g))$.
(ii) $\mathrm{GC}(\rho(\mathrm{GC}(g) + E)) = \mathrm{GC}(\rho(g) + E)$.

**Proposition 23.** *Let $\langle h, H, \mu \rangle$ be an extension of $g$, and let $\rho_1, \rho_2$ be redirections of $g, h$ respectively such that $N_h - H$ is $\rho_2$-closed. Then*

$$\rho_1 \circ \mu \circ \rho_2(g +_{\mu \circ \rho_2} \langle h, H \rangle) = \rho_1(g) +_{\rho_1 \circ \mu} \langle \rho_2(h), H \rangle.$$

(Here $\mu$ is tacitly extended to all nodes of the extended graph, by defining $\mu(n) = n$ if $n \notin N_h - H$. The same is done for $\rho_1$ and $\rho_2$.)

## 4 Graph rewrite systems

A collection of graphs and a set of rewrite rules can be combined into a graph rewrite system. A special class of so-called orthogonal graph rewrite systems is the subject of further investigations.

**Definition 24.** A *term graph rewrite system* (TGRS) is a tuple $\mathcal{T} = \langle \mathcal{G}, \mathcal{R} \rangle$ where $\mathcal{R}$ is a set of rewrite rules, and $\mathcal{G} \subseteq \mathbb{G}$ is a set of closed graphs which is closed under $\mathcal{R}$-reduction.

It appears that under certain circumstances the order of reduction can be changed without influencing the final result. This is made more explicit below.

**Definition 25.** Let $\Delta_1 = \langle R_1, \mu_1 \rangle$ and $\Delta_2 = \langle R_2, \mu_2 \rangle$ be redexes in $g$.
(i) $\Delta_1$ and $\Delta_2$ are *disjoint* if $\mu_1(l_1) \notin \mu_2(R_2 \mid l_2)^\bullet$ and $\mu_2(l_2) \notin \mu_1(R_1 \mid l_1)^\bullet$.
(ii) $\Delta_1$ and $\Delta_2$ *interfere* if $\mu_1(l_1) = \mathrm{r}(\Delta_2)$ and $\mu_2(l_2) = \mathrm{r}(\Delta_1)$.

**Definition 26.** Let $\mathcal{T} = \langle \mathcal{G}, \mathcal{R} \rangle$ be a TGRS.

(i) $\mathcal{T}$ is *left-linear* if each $R \in \mathcal{R}$ is left-linear.

(ii) $\mathcal{T}$ is *regular* if for each $g \in \mathcal{G}$ the $\mathcal{R}$-redexes in $g$ are pairwise disjoint.

(iii) $\mathcal{T}$ is *orthogonal* if $\mathcal{T}$ is both left-linear and regular.

(iv) $\mathcal{T}$ is *interference-free* if $\mathcal{G}$ does not contain any graph with interfering redexes.

**Theorem 27.** *Let $\mathcal{T} = \langle \mathcal{G}, \mathcal{R} \rangle$ be orthogonal and interference-free. Then $\mathcal{R}$-reduction satifies the Church-Rosser property. In a picture:*

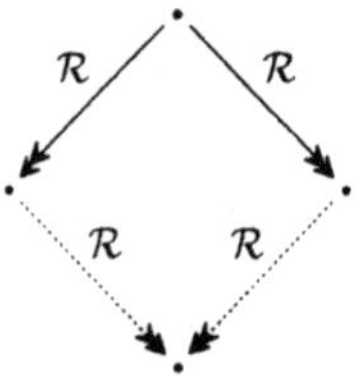


**Remark 28.** If an orthogonal TGRS $\mathcal{T}$ is cycle-free, i.e. the graphs of $\mathcal{T}$ do not contain cycles, then $\mathcal{T}$ is interference-free.

Unraveling a graph might influence the redexes of that graph. The following result implies that unraveling and ordinary reduction can be combined preserving the Church-Rosser property (see Barendsen and Smetsers [1992]).

**Proposition 29.** *Let $g$ be cycle-free. Suppose $g \preceq g'$ and $g \to_{\mathcal{R}} h$. Then $g' \twoheadrightarrow_{\mathcal{R}} h'$ for some $h' \succeq h$.*

In order to decribe the combination of unraveling and ordinary reduction we denote unraveling as a rewrite step, i.e. $g \to_{\preceq} g'$ if $g \preceq g'$.

**Theorem 30.** *Let $\mathcal{T} = \langle \mathcal{G}, \mathcal{R} \rangle$ be orthogonal and cycle-free. Then $\mathcal{R}\preceq$-reduction is Church-Rosser.*

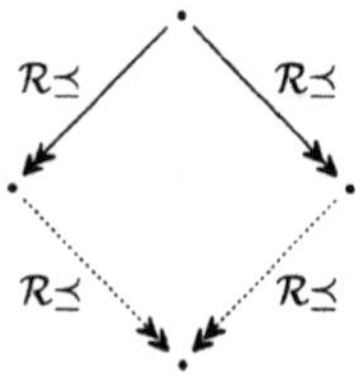


## 5 Copying

Extending graph rewriting with a copying mechanism can be done in several ways. Below some of these are discussed.

Intuitively, copying a graph causes a new graph to be created that has the same structure as, but no nodes in common with the original graph. Let **C** be a unary symbol in $\Sigma$. A copy node in a graph $g$ is a node with symbol **C**. The intended meaning is that evaluation of a copy node leads to duplication of the graph it refers to. We wish to describe how such a **C**-node can be handled.

The first possibility is to extend an existing TGRS $\mathcal{T} = \langle \mathcal{G}, \mathcal{R} \rangle$ to a system $\mathcal{T}^{C}$ by adding new rewrite rules for copy reduction to $\mathcal{R}$, thus obtaining $\mathcal{R}^{C}$. This can be done in various ways, the simplest of which is adding a rule

$$\mathbf{C}(\mathbf{F}(x_1, \ldots, x_n)) \rightarrow \mathbf{F}(\mathbf{C}(x_1), \ldots, \mathbf{C}(x_n))$$

for each symbol **F** with arity $n$ in $\mathcal{G}$. Although this extension can be described easily, there are several disadvantages. Firstly, the number of extra rewrite rules may be infinite if in $\mathcal{G}$ infinitely many symbols are used. Secondly, mixing copy rules with the reduction rules in $\mathcal{R}$ may destroy properties of $\mathcal{R}$-reduction such as confluence (see section 7), or at least make it very difficult to check whether known properties of $\mathcal{T}$ extend to $\mathcal{T}^{C}$. This is reinforced by the fact that copying is done stepwise and therefore cannot be considered as a basic action such as redirection or extension. Another unwanted effect is that sharing cannot be maintained, as is illustrated below. The loose arrows indicate edges coming in from the surrounding graph.

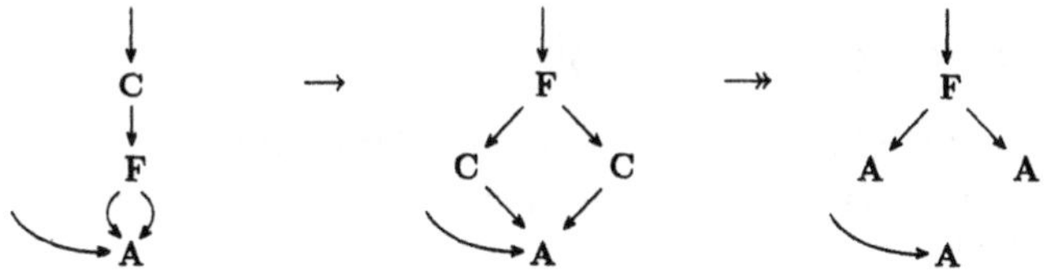


Thus shared redexes below **F** are duplicated. This last drawback could in principle be overcome by using more sophisticated rewrite rules.

The second possibility is the addition of a new copy mechanism to graph rewriting. A rude solution involves replacing the **C**-node by a copy of the whole subgraph it refers to, resulting in so-called *eager copying*. It is visualized in the following figure.

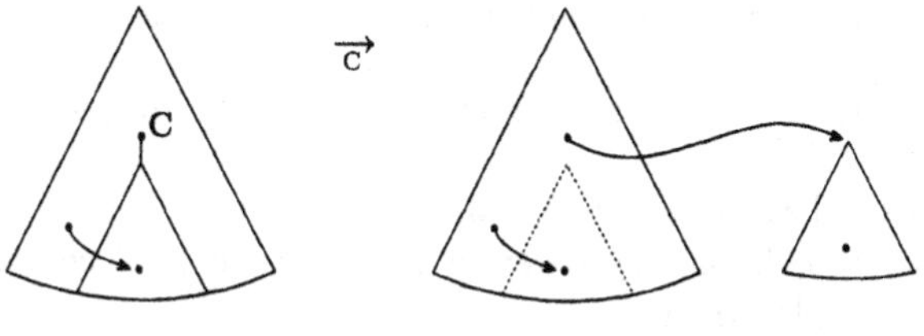


Thus sharing below the copy node remains intact. However, all redexes below that node are duplicated, involving 'duplication of work'. It may therefore be convenient to duplicate only some redexes of the subgraph in question and postpone the rest of the copy action. This is called *lazy copying*. New copy nodes are added where the copy procedure has been interrupted. In a picture this looks as follows.

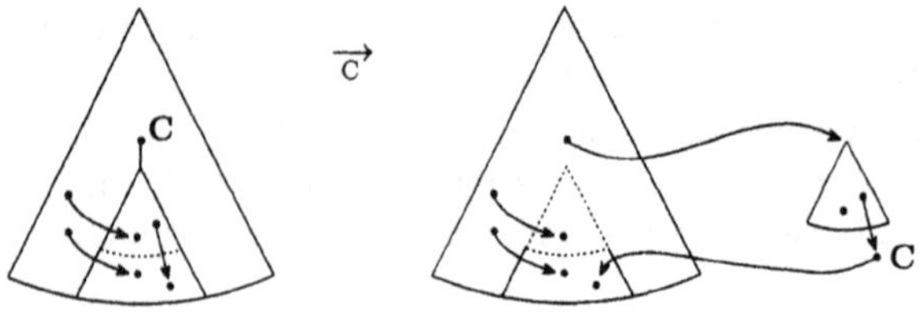


Of course, this mechanism is more complicated to describe than the previous one. Moreover, if one uses a reduction strategy to select a specific reduction order matters become complicated because one has to decide at which point to interrupt the copying. We will go into this in section 8.

The rest of this section is devoted to a formal description of lazy copying.

**Definition 31.** Let $g$ be a graph.

(i) Let $n \in g$ and $s \subseteq g \mid n$. Note that $\langle s, \iota_{s,g} \rangle$ is an extension of $g$. The *s-expansion of g at n* (notation $g[n := s]$) is defined by

$$g[n := s] = g[n := \langle s, \iota_{s,g} \rangle].$$

(ii) $g'$ is an *expansion* of $g$ if $g' = g[n := s]$ for some $n$ and $s$.

**Lemma 32.** *Let* $s \subseteq g \mid n$. *Then* $g \preceq g[n := s]$.

**Definition 33.** Let $\mathbf{C}$ be a special symbol in $\Sigma$ with arity 1.

(i) A node $n \in g$ is a *copy node* if $symb_g(n) = \mathbf{C}$. The set of copy nodes of $g$ is denoted by $\mathrm{C}(g)$.

(ii) The definition of *term graph rewrite rule* (11) is extended with

(3) $\mathrm{C}(g \mid l) = \emptyset$, if $R = \langle g, l, r \rangle$ is a rule.

This requirement states that only the right-hand side of a rewrite rule may contain copy nodes. Thus copy nodes may not be used as part of a pattern.

**Definition 34.** Let $g$ be a graph, and $M \subseteq N_g$. The *copy indirection* of $M$ in $g$ (notation $\mathrm{CI}(g, M)$) is the graph that is obtained from $g$ by replacing each node $n \in M$ to a new copy node with $n$ as argument.

*Notation.* If $n$ is a unary node in $g$ then $args_g(n)_1$ is written as $succ(n)$.

**Definition 35.** (i) A *copy redex* in a graph $g$ is a pair $\langle n, s \rangle$ where

(a) $n \in \mathrm{C}(g)$

(b) $s \subseteq g \mid succ(n)$ and $s^\bullet \neq \emptyset$.

(ii) Let $\Delta = \langle n, s \rangle$ be a copy redex in $g$. The result $g'$ of *contracting* $\Delta$ in $g$ is defined by the replacement

$$g' = g[n := \mathrm{CI}(s, s^\circ)],$$

which is shorthand for $g[n := \langle \mathrm{CI}(s, s^\circ), \iota \rangle]$.

(iii) We write $g \xrightarrow[\mathrm{C}]{\Delta} g'$ to indicate that $g$ *copy-reduces* to $g'$ via redex $\Delta$.

**Definition 36.** A *copy term graph rewrite system* (C-TGRS) is a structure $\mathcal{T} = \langle \mathcal{G}, \mathcal{R} \rangle$ where $\mathcal{G} \subseteq \mathbb{G}$ is a set of graphs, closed under $\mathcal{R}$-reduction and C-reduction.

**Remark 37.** Notions introduced for TGRS's (such as left-linearity and orthogonality) are used for C-TGRS's as well by considering them as ordinary TGRS's.

*Example 3.* For the indicated graph $g$ and segment $s \subseteq g$, the copy indirection of $s$ and the result of contracting copy redex $\langle n, s \rangle$ in $g$ is displayed below.

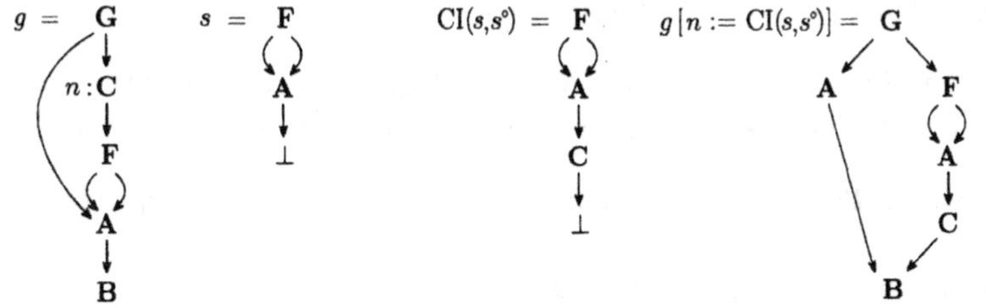


## 6 Analyzing copy reduction

Because lazy copying is a complicated concept, we will decompose copy reduction into three operations, namely
(1) partial unraveling (expansion),
(2) addition of **C**-nodes,
(3) elimination of **C**-nodes.
The main part of this section addresses the **C**-elimination mechanism and its interaction with the three basic graph operations mentioned in section 3, and with unraveling. Then it is shown that copy reduction can indeed be splitted in the above-mentioned way.

The elimination of **C**-nodes can be viewed as a redirection. This is made precise in the following.

**Definition 38.** Let $g$ be a rooted graph, and $D$ be a collection of of nodes of $g$. A path $p$ in $g$ is *within* $D$ (notation $p \subseteq D$) if $n \in D$ for all $n \in p$. $D$ is *cycle-free* (*in* $g$) if each path of $g$ within $D$ are a-cyclic.

**Remark 39.** Let $g$ be graph. Let $D \subseteq g$ be a finite set of unary nodes. Then every a-cyclic path $p$ in $D$ is uniquely determined by its begin and end points, for each node of $p$ has exactly one successor. Consequently, if $D$ is cycle-free then for any $n \in g$ there exists exactly one $p : n \rightsquigarrow m$ such that

$$m \notin D,$$
$$\forall m' \in p \ \ [m' \neq m \Rightarrow m' \in D].$$

The latter expression is abbreviated to $p - m \subseteq D$.

In the sequel we will only consider finite sets $D$ consisting of nodes with arity 1. Remark 39 enables us to define the following.

**Definition 40.** Let $g$ be a graph, and $D \subseteq \mathrm{C}(g)$ cycle-free. Define the redirection $\mathrm{e}_{D,g}$ as follows. Let $n \in g$. Then $\mathrm{e}_{D,g}(n)$ is the uniquely determined $m \notin D$ such that $p : n \rightsquigarrow m$ for some $p$ with $p - m \subseteq D$. Then $\mathrm{e}_{D,g}$ is obviously projective, and $\mathrm{Fld}(\mathrm{e}_{D,g}) = D$.

Now the actual elimination of **C**-nodes in $g$ is obtained by applying the corresponding redirection operation to $g$.

**Definition 41.** Let $g$ be a graph and $D \subseteq \mathrm{C}(g)$. $\mathrm{E}_D(g)$ denotes the graph that is obtained by deleting all copy nodes of $D$ from $g$, i.e. $\mathrm{E}_D(g) = \mathrm{e}_{D,g}(g)$.

**Proposition 42.** *Let $g$ be a graph, $D_1, D_2 \subseteq \mathrm{C}(g)$ such that $D_1 \cup D_2$ is cycle-free.*
(i) $\mathrm{e}_{D_1 \cup D_2, g} = \mathrm{e}_{D_2, \mathrm{E}_{D_1}(g)} \circ \mathrm{e}_{D_1, g}$.
(ii) $\mathrm{E}_{D_1 \cup D_2}(g) = \mathrm{E}_{D_2}(\mathrm{E}_{D_1}(g))$.

The interaction between **C**-elimination and reduction and unraveling respectively can now be described. We spell out a proof in detail to illustrate the usage of the properties of the basic operations.

**Proposition 43.** *Let $g$ be a graph, $m, n \in g$, and let $D \subseteq g$ be cycle-free in both $g$ and $g[n := m]$ such that $n \notin D$.*
(i) $[n \mapsto \mathrm{e}_{D,g}(m)] \circ \mathrm{e}_{D,g} = \mathrm{e}_{D,g[n:=m]} \circ [n \mapsto m]$.
(ii) $\mathrm{E}_D(g[n := m]) = \mathrm{E}_D(g)[n := \mathrm{e}_{D,g}(m)]$.

**Proposition 44.** *Let $g \underset{\mathcal{R}}{\rightarrow} h$. Let $D \subseteq \mathrm{C}(h)$ be cycle-free. Then*

$$\mathrm{GC}(\mathrm{E}_D(g)) \underset{\mathcal{R}}{\rightarrow} \mathrm{GC}(\mathrm{E}_D(h)).$$

*Proof.* It will be shown that

$$\mathrm{GC}(\mathrm{e}_{D,g}(g)) \underset{\mathcal{R}}{\rightarrow} \mathrm{GC}(\mathrm{e}_{D,h}(h)).$$

Say $g \overset{\Delta}{\underset{\mathcal{R}}{\rightarrow}} h$ with $\Delta = \langle R, \mu \rangle$. Then

$$h = \mathrm{GC}((g + \Delta)[\mu(l) := \mathrm{r}(\Delta)]).$$

Observe that $\mu(R \mid l)^\bullet \cap D = \emptyset$ since $(R|l)^\bullet$ is **C**-free. So $\mathrm{e}_{D,g}(n) = n$ for any $n \in \mu(R \mid l)^\bullet$. It follows that $\mathrm{e}_{D,g} \circ \mu : (R|l)^\bullet \underset{\mathrm{m}}{\rightarrow} \mathrm{e}_{D,g}(g)$. Write $\mathrm{e}_{D,g}(\Delta) = \langle R, \mathrm{e}_{D,g} \circ \mu \rangle$. Then $\mathrm{e}_{D,g}(\Delta)$ is a redex in $\mathrm{e}_{D,g}(g)$. Note that $\mathrm{r}(\mathrm{e}_{D,g}(\Delta)) = \mathrm{e}_{D,g}(\mathrm{r}(\Delta))$, and $\mathrm{e}_{D,g}(\mu(l)) = \mu(l)$. Contracting $\mathrm{e}_{D,g}(\Delta)$ in $\mathrm{GC}(\mathrm{e}_{D,G}(g))$ results in

$\mathrm{GC}(\ (\ \mathrm{GC}(\mathrm{e}_{D,g}(g)) + \mathrm{e}_{D,g}(\Delta)\ )[\mu(l) := \mathrm{r}(\mathrm{e}_{D,g}(\Delta))]\ )$
$= \mathrm{GC}(\ (\ \mathrm{e}_{D,g}(g) + \mathrm{e}_{D,g}(\Delta)\ )[\mu(l) := \mathrm{r}(\mathrm{e}_{D,g}(\Delta))]\ )$, by lemma 22 (ii)
$= \mathrm{GC}(\ (\ \mathrm{e}_{D,g}(g + \Delta)\ )[\mu(l) := \mathrm{r}(\mathrm{e}_{D,g}(\Delta))]\ )$, by proposition 23

$= \mathrm{GC}(\ (\ \mathrm{e}_{D,g}(g+\Delta)\ )[\mu(l) := \mathrm{e}_{D,g}(\mathrm{r}(\Delta))]\ )$
$= \mathrm{GC}(\ (\ \mathrm{e}_{D,\mathrm{GC}(g+\Delta)}(g+\Delta)\ )[\mu(l) := \mathrm{e}_{D,g}(\mathrm{r}(\Delta))]\ ),$
$= \mathrm{GC}(\ \mathrm{e}_{D,\mathrm{GC}(g+\Delta)[\mu(l):=\mathrm{r}(\Delta)]}(\ (g+\Delta)[\mu(l) := \mathrm{r}(\Delta)]\ )\ ),$ by proposition 43 (i)
$= \mathrm{GC}(\ \mathrm{e}_{D,\mathrm{GC}(g+\Delta)[\mu(l):=\mathrm{r}(\Delta)]}(\ \mathrm{GC}(\ (g+\Delta)[\mu(l) := \mathrm{r}(\Delta)]\ )\ )\ ),$ by lemma 22 (i)
$= \mathrm{GC}(\ \mathrm{e}_{D,h}(h)\ ).$ □

**Proposition 45.** *Let $g \preceq_\varphi h$. Let $D \subseteq \mathrm{C}(h)$ be cycle-free.*
(i) $\mathrm{e}_{\varphi^{-1}(D),g}(g) \preceq_\varphi \mathrm{e}_{D,h}(h)$.
(ii) $\mathrm{GC}(\mathrm{E}_{\varphi^{-1}(D)}(g)) \preceq_\varphi \mathrm{GC}(\mathrm{E}_D(h))$.

Some technical work has to be done in order to split copy reduction into more basic actions.

**Lemma 46.** *Let $g$ be a graph, $n \in g$, and $s \subseteq g \mid n$*
(i) *If $n$ is unary then* $\mathrm{GC}(\mathrm{E}_{\{r_s\}}(g[n := s])) = \mathrm{GC}(g[n := s \mid succ(n)])$.
(ii) *If $s^\bullet \neq \emptyset$ then* $\mathrm{GC}(\mathrm{E}_{(s^\circ)^\bullet}(g[n := \mathrm{CI}(s, s^\circ)])) = \mathrm{GC}(g[n := s])$.

For convenience some new reduction relations are introduced. Remember $g \underset{\preceq}{\rightarrow} h$ iff $g \preceq h$.

**Definition 47.** The reduction relations $\underset{-}{\rightarrow}$ and $\underset{+}{\rightarrow}$ are defined by

$$g \underset{-}{\rightarrow} h \Leftrightarrow h = \mathrm{GC}(\mathrm{E}_D(g)) \quad \text{for some nonempty } D \subseteq \mathrm{C}(g),$$
$$g \underset{+}{\rightarrow} h \Leftrightarrow \mathrm{C}(g) \neq \emptyset \text{ and } g = \mathrm{GC}(\mathrm{E}_D(h)) \quad \text{for some nonempty } D \subseteq \mathrm{C}(h).$$

**Proposition 48.** $g \underset{C}{\rightarrow} g' \;\Rightarrow\; g \underset{\preceq}{\rightarrow} \cdot \underset{+}{\rightarrow} \cdot \underset{-}{\rightarrow} g'$.

*Proof.* Let $\langle n, s\rangle$ be the copy redex of $g$ that is reduced in order to obtain $g'$. Hence $s \subseteq g \mid succ(n)$ and $g' = \mathrm{GC}(g +_\iota \mathrm{CI}(s, s^\circ)[n := r_s])$. Let $s_c \subseteq g \mid n$ such that $s = s_c \mid succ(n)$. Then

$$\begin{aligned} g &\underset{\preceq}{\rightarrow} \mathrm{GC}(g[n := s_c]), && \text{by lemma 32} \\ &\underset{+}{\rightarrow} \mathrm{GC}(g[n := \mathrm{CI}(s_c, s_c^\circ)]), && \text{by lemma 46 (ii)} \\ &\underset{-}{\rightarrow} \mathrm{GC}(g[n := \mathrm{CI}(s, s^\circ)]), && \text{by lemma 46 (i)} \\ &= g'. \ \square \end{aligned}$$

# 7 Confluence

This section starts with the introduction of some standard terminology.

**Definition 49.** (i) A *reduction relation* on $\mathcal{G}$ is just a binary relation $\boldsymbol{R}$ on $\mathcal{G}$.
(ii) If $\boldsymbol{R}_1, \boldsymbol{R}_2$ are reduction relations, then $\boldsymbol{R}_1\boldsymbol{R}_2$ is $\boldsymbol{R}_1 \cup \boldsymbol{R}_2$.
(iii) A graph $g$ *$\boldsymbol{R}$-reduces in one step* to $g'$ (notation $g \underset{R}{\rightarrow} g'$) if $(g, g') \in \boldsymbol{R}$.
(iv) The relations $\underset{R}{\twoheadrightarrow}$, $\underset{R}{\rightarrow}^+$ and $\underset{R}{\rightarrow}^=$ denote the reflexive transitive closure, the transitive closure and the reflexive closure of $\underset{R}{\rightarrow}$ respectively.

We assume the reader is familiar with the notions *normal form*, *reduction path*, etcetera.

Let $\mathcal{T} = \langle \mathcal{G}, \mathcal{R} \rangle$ a orthogonal cycle-free C-TGRS. We want to obtain a Church-Rosser-like result for $\mathcal{T}$. A pure Church-Rosser result for $\mathcal{R}$C-reduction is impossible, as is shown in the following example.

*Example 4.* Let $\mathbf{I}(x) \rightarrow x$ be the usual rule for the identity function. Then one has two 'divergent' reduction paths.

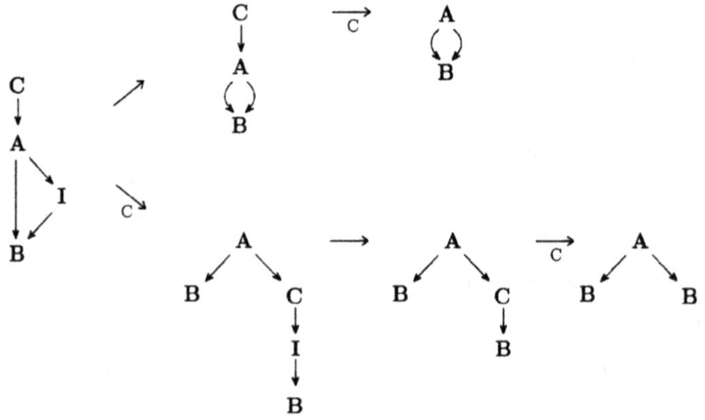

For this reason we study confluence of the system in which the copy action is decomposed. Our aim is to show that $\mathcal{R}{\preceq}{+}{-}$-reduction is confluent.

A useful tool to prove the confluence of a combined reduction relation is the so-called Hindley-Rosen lemma, due to Hindley [1964] and Rosen [1973].

**Lemma 50.** (i) (ii) (iii)

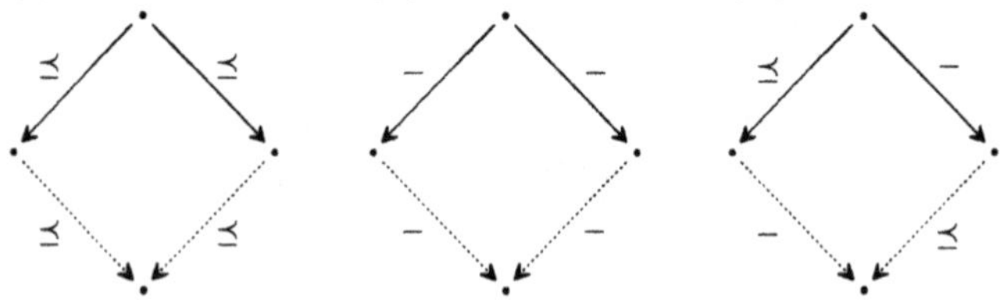

*Proof.* (i) By theorem 9 (ii).
(ii) By proposition 42 (ii).
(iii) By proposition 45 (ii). □

**Proposition 51.** $\preceq-$ *is Church-Rosser.*

*Proof.* By the Hindley-Rosen lemma, using the facts that $\xrightarrow[\preceq]{}$ and $\xrightarrow[-]{}$ are Church-Rosser (lemma 50 (i, ii)), and that $\xrightarrow[\preceq]{}$ and $\xrightarrow[-]{}$ commute (lemma 50 (iii)). □

**Lemma 52.**

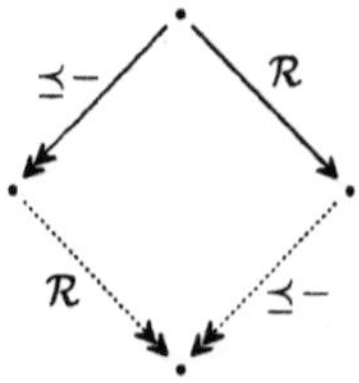


*Proof.* By propositions 29 and 44 and a diagram chase. □

**Theorem 53.** *$\mathcal{R}\preceq-$ is Church-Rosser.*

*Proof.* Again by the Hindley-Rosen lemma, using proposition 51, lemma 52 and theorem 27. □

The confluence proof is carried out in two main steps. First it is shown how +-reduction can be dealth with, and next the Church-Rosser result for $\mathcal{R}\preceq-$-reduction is used. Below, $\rightarrow$ stands for $\mathcal{R}\preceq+-$-reduction.

**Neutralization Lemma 54.**

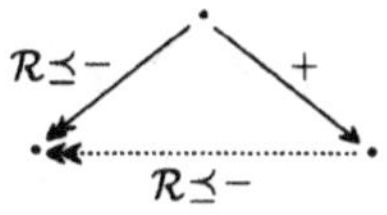


**Proposition 55.**

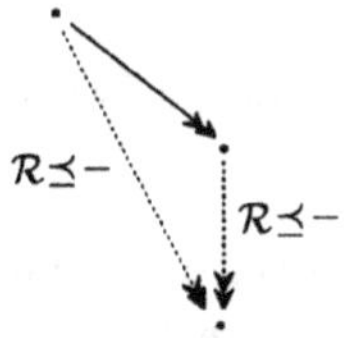


*Proof.* Using the neutralization lemma (triangles) and theorem 53 (quadrangles) one can erect the following diagram.

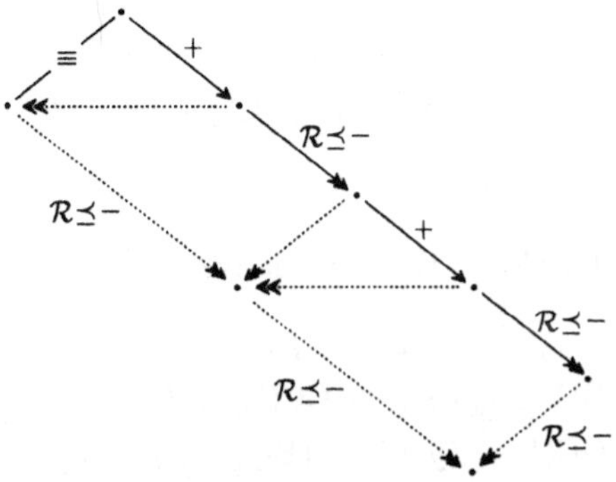


□

**Theorem 56.**

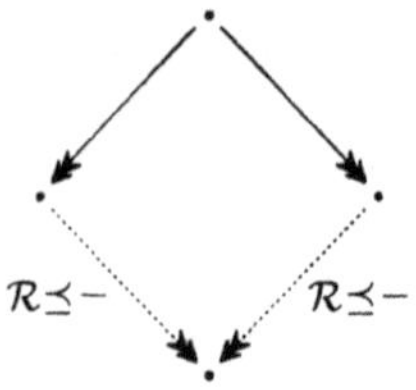


*Proof.* By proposition 55 and theorem 53.

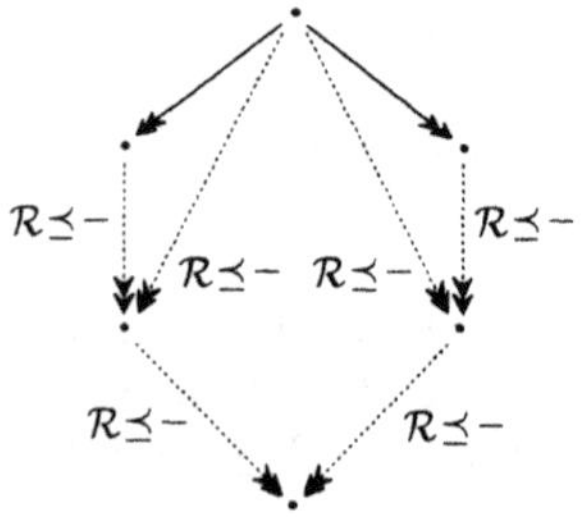


□

**Corollary 57.**

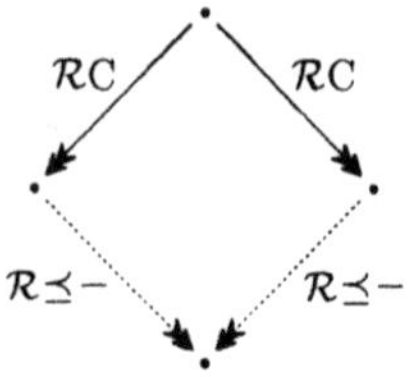


The relations between various normal forms can now be expressed.

**Definition 58.** Let $g, g'$ be graphs.

(i) $g$ and $g'$ are *copy tree equivalent* (notation $g \sim_{\mathrm{CT}} g'$) if they are tree equivalent after elimination of all present copy nodes, i.e.

$$\mathrm{E}_{\mathrm{C}(g)}(g) \sim_{\mathrm{T}} \mathrm{E}_{\mathrm{C}(g')}(g').$$

(ii) $\sim_{\preceq-}$ is the reflexive symmetric transitive closure of the reduction relation $\preceq-$.

**Theorem 59.** *Let $g, g' \in \mathbb{G}$. Suppose $\mathrm{C}(g), \mathrm{C}(g')$ are cycle-free. Then*

$$g \sim_{\mathrm{CT}} g' \Leftrightarrow g \sim_{\preceq-} g'.$$

*Proof.* ($\Rightarrow$) Using $\mathrm{U}(\mathrm{GC}(h)) = \mathrm{U}(h)$ one derives

$$\begin{aligned} g \sim_{\mathrm{CT}} g' &\Rightarrow \mathrm{E}_{\mathrm{C}(g)}(g) \sim_{\mathrm{T}} \mathrm{E}_{\mathrm{C}(g')}(g') \\ &\Rightarrow \mathrm{GC}(\mathrm{E}_{\mathrm{C}(g)}(g)) \sim_{\mathrm{T}} \mathrm{GC}(\mathrm{E}_{\mathrm{C}(g')}(g')) \\ &\Rightarrow \mathrm{GC}(\mathrm{E}_{\mathrm{C}(g)}(g)) \uparrow \mathrm{GC}(\mathrm{E}_{\mathrm{C}(g')}(g')), \quad \text{by theorem 9 (i).} \end{aligned}$$

($\Leftarrow$) By induction on the generation of $\sim_{\preceq -}$. The crucial case is $g \underset{\preceq}{\rightarrow} g'$. Then by lemma 50 (iii) one has $\mathrm{GC}(\mathrm{E}_{\mathrm{C}(g)}(g)) \preceq \mathrm{GC}(\mathrm{E}_{\mathrm{C}(g')}(g'))$ and therefore $\mathrm{E}_{\mathrm{C}(g)}(g) \sim_{\mathrm{T}} \mathrm{E}_{\mathrm{C}(g')}(g')$. □

As a corollary of the Church-Rosser theorem, normal forms turn out to be unique up to copy tree equivalence.

**Theorem 60.** (i) *If $g$ has $g_1, g_2$ as $\mathcal{R}$C-normal forms, then $g_1 \uparrow g_2$.*
(ii) *If $g$ has $g_1, g_2$ as $\mathcal{R}$-normal forms, then $g_1 \sim_{\mathrm{CT}} g_2$.*

## 8 Towards an implementation of lazy copying

When using term rewriting systems or term graph rewriting systems as a model of computation, these systems are usually equipped with a reduction strategy.

In order to determine *redexes* to be contracted one often uses an auxiliary function selecting *nodes* that are intended to be rewritten in the subject graph. For ordinary orthogonal TGRS's such a node uniquely determines a redex. In the case of C-TGRS's this node can be considered as a partial specification of a redex, since copy redexes consist of a copy node together with an appropriate segment. So a copy node actually specifies a *set* of copy redexes. In this section we will describe a natural method for choosing an appropriate segment, thus obtaining a proper redex selecting procedure.

To have some control over the choice of segments, we extend the language of graph rewriting with the possibility of attaching a so called *defer* attribute to nodes. Intuitively, this attribute temporarily prevents the node in question of being copied. Defer attributes are also allowed to appear in the right-hand side of a rewrite rule. In van Eekelen et al. [1991] these attributes are used to define primitives for specifying communication for various kinds of parallelism.

During graph rewriting, defer attributes are *added* and *propagated* while performing the graph operations extension, redirection and garbage collection, according to the following rules.

- If $n$ is deferred in $g$, then $n$ is deferred in $g + E$.
- If $n$ is deferred in $E$, then $n^*$ is deferred in $g + E$.
  (This applies e.g. to $g + \Delta$ for redexes $\Delta = \langle R, \mu \rangle$ in which $R \,|\, r$ contains deferred nodes.)
- If $n$ is deferred in $g$, then $\rho(n)$ is deferred in $\rho(g)$.
- If $n$ is deferred in $g$, and $r_g \rightsquigarrow n$, then $n$ is deferred in $\mathrm{GC}(g)$.

The above-mentioned node selection is made precise in the following. Let $\mathcal{T} = \langle \mathcal{G}, \mathcal{R} \rangle$ be a C-TGRS with defer attributes.

**Definition 61.** (i) $\mathrm{D}(g)$ denotes the collection of deferred nodes of $g$.

(ii) A *pre-strategy* (for $\mathcal{T}$) is a function $\sigma$ which takes each $g \in \mathcal{G}$ to a set of $\mathcal{R}$-redex roots, and/or copy nodes in $g$, such that for all $n \in \sigma(g)$

$$n \in \mathrm{C}(g) \;\Rightarrow\; succ(n) \notin \mathrm{D}(g),$$

i.e. $\sigma$ will not attempt to copy a 'deferred graph'.

(iii) A graph is in *σ-normal form* if $\sigma(g) = \emptyset$.

Now let $\sigma$ be a pre-strategy for $\mathcal{T}$. Actual rewriting according to $\sigma$ proceeds as follows. If $\sigma$ selects an $\mathcal{R}$-redex root, the corresponding redex is contracted. If $\sigma$ points out a copy node then a copy reduction step is done. The subgraph $g \mid succ(n)$ is copied up to the deferred nodes of $g$. Thus

$$\begin{aligned} g \underset{\sigma}{\rightarrow} g' \text{ if for some } n \in \sigma(g) & \\ & n \in \mathrm{C}(g), \text{ and } g \overset{\langle n,s\rangle}{\underset{\mathrm{C}}{\rightarrow}} g', \\ & \text{where } s = (g \mid succ(n)) \rtimes \mathrm{D}(g), \qquad \text{or} \\ & n \notin \mathrm{C}(g), \text{ and } g \overset{\Delta_n}{\underset{\mathcal{R}}{\rightarrow}} g', \\ & \text{where } \Delta_n \text{ is the } \mathcal{R}\text{-redex associated with } n. \end{aligned}$$

Defer attributes are *removed* during reduction according to the following rule.

- Once a rewrite step is performed, a node $n \in g$ loses its defer attribute if $g \mid n$ is in $\sigma$-nf.

As a consequence, each graph in $\mathcal{R}$C-normal form does not have any deferred nodes. Note that the above $s$ is nonempty since $succ(n)$ is not deferred.

The concept of lazy copying, combined with the above segment selection, has been incorporated in the programming language CONCURRENT CLEAN (see Nöcker et al. [1991]). In the implementation of this language on a distributed multi-processor system this technique is used in combination with a parallel version of the so-called *functional* reduction strategy. Communication channels are expressed by using copy nodes. The definition of $\underset{\sigma}{\rightarrow}$ then guarantees that no 'work' is communicated (i.e. no nodes that are marked as 'deferred' are copied from one processor to another). Moreover, the specific definition of the parallel functional reduction strategy guarantees that eventually the necessary information is available by reducing any deferred part of the graph.

The following result formulates the adequacy of reduction in CONCURRENT CLEAN. The details will not be worked out here.

**Theorem 62.** *In* CONCURRENT CLEAN, *reduction is sound and complete, i.e. normal forms are unique up to tree equivalence, and normal forms can be reached via the above reduction method.*

## References


BARENDREGT, H.P., M.C.J.D. VAN EEKELEN, J.R.W. GLAUERT, J.R. KENNAWAY, M.J. PLASMEIJER and M.R. SLEEP

[1987a] Term graph reduction, in: *Proceedings of Parallel Architectures and Languages Europe (PARLE'87), Eindhoven, The Netherlands*, LNCS 259, volume II, Springer-Verlag, Berlin, pp. 141–158.

[1987b] Towards an intermediate language based on graph rewriting, in: *Proceedings of Parallel Architectures and Languages Europe (PARLE'87), Eindhoven, The Netherlands*, LNCS 259, volume II, Springer-Verlag, Berlin, pp. 159–175.

BARENDSEN, E. and J.E.W. SMETSERS

[1992] Graph rewriting and copying, Technical report no. 92-20, Department of Computer Science, University of Nijmegen.

BRUS, T., M.C.J.D. VAN EEKELEN, M.O. VAN LEER and M.J. PLASMEIJER

[1987] Clean: a language for functional graph rewriting, in: *Proceedings of the 3rd International Conference on Functional Programming Languages and Computer Architecture, Portland, Oregon*, LNCS 274, Springer-Verlag, Berlin, pp. 364–384.

VAN EEKELEN, M.C.J.D., M.J. PLASMEIJER and J.E.W. SMETSERS

[1991] Parallel graph rewriting on loosely coupled machine architectures, in: *Proceedings of Conditional and Typed Rewriting Systems (CTRS'90), Montreal, Canada*, LNCS 516, Springer-Verlag, Berlin, pp. 354–369.

EHRIG, H., M. NAGL, G. ROZENBERG and A. ROSENFELD (eds.)

[1987] *Proceedings of the 3rd International Workshop on Graph-Grammars and the Application to Computer Science, Warrenton, Virginia, USA, December 1986*, LNCS 291, Springer-Verlag, Berlin.

HINDLEY, J.R.

[1964] The Church-Rosser property and a result in Combinatory Logic, dissertation, University of Newcastle-upon-Tyne.

NÖCKER, E.G.J.M.H., J.E.W. SMETSERS, M.C.J.D. VAN EEKELEN and M.J. PLASMEIJER

[1991] Concurrent Clean, in: *Proceedings of Parallel Architectures and Languages Europe (PARLE'91), Eindhoven, The Netherlands*, LNCS 505, Springer-Verlag, Berlin, pp. 202–219.

ROSEN, B.K.

[1973] Tree manipulation systems and Church-Rosser theorems, *J. Assoc. Comput. Mach.* 20, pp. 160–187.

SMETSERS, J.E.W., E.G.J.M.H. NÖCKER, J.H.G. VAN GRONINGEN and M.J. PLASMEIJER

[1991] Generating efficient code for lazy functional languages, in: *Proceedings of Conference on Functional Programming Languages and Computer Architecture (FPCA '91), Cambridge, MA*, LNCS 523, Springer-Verlag, Berlin, pp. 592–617.

# Graph-Grammar Semantics of a Higher-Order Programming Language for Distributed Systems

Klaus Barthelmann[1] and Georg Schied[2]

[1] Johannes Gutenberg-Universität Mainz, Institut für Informatik, Staudinger Weg 9, Postfach 3980, 55099 Mainz, Germany, barthel@informatik.mathematik.uni-mainz.de

[2] Institut für Informatik, Abteilung für Programmiersprachen, Breitwiesenstr. 20–22, 70565 Stuttgart, Germany, schied@informatik.uni-stuttgart.de

**Abstract.** We will consider a new tiny, yet powerful, programming language for distributed systems, called DHOP, which has its operational semantics given as algebraic graph rewrite rules in a certain category of labeled graphs. Our approach allows to separate actions which affect several processes from local changes such as variable bindings. We also sketch how to derive an implementation from this specification.

## 1 Introduction

There are already many calculi for programming distributed systems. Roughly, they can be divided into three groups:

1. Fully fledged programming languages (like occam2 [12], POOL [1], and SR [2]) contain a lot of features useful in practical applications. Their semantics is usually given informally.
2. Small languages (like CSP [11], DNP [7], extended typed $\lambda$-calculus [18], and Facile [19]) could be used for programming tasks, although they concentrate on the main ideas. Their semantics is formally defined.
3. Process algebras (like ACP [5], CCS [14], $\pi$-calculus [15], and Plain CHOCS [24]) are designed for theoretical investigations. Only the barest minimum of concepts is accepted. (There are not even variable bindings.) Their semantics is, therefore, quite simple.

We were interested in the specification and verification of distributed algorithms and in experimenting with (prototypes of) distributed systems. Therefore, we sought a simple programming language with a clear semantics that could actually be used. That naturally led us to the second of the above groups. But we did not find a candidate that offered all desired features, among which were the following:

1. communication by passing messages over unbuffered channels with a non-deterministic choice between them, as in CSP. Message passing mechanisms usually lead to a more modular program design than communication primitives for shared memory. Furthermore, asynchronous message passing can be

modeled by adding buffer processes, so synchronous message passing seems more primitive. The nondeterministic choice between several possible communications is necessary, e. g., inside "server" processes. The channels over which certain requests are sent represent the concept of a "service". The actual partners in a communication do not know each other and can therefore be replaced by equivalent ones.

2. parameterized process definitions, as in DNP. An explicit statement for process creation enables us to have several instances of one process, which may differ slightly in their behavior and have different neighbors. Looping statements are redundant in view of possible "recursive" calls.
3. first class processes and channels. Besides giving a nicer calculus, both of them have a practical justification. If a process in a remote place is to be exchanged the new process definition must be sent in a message (see [24]). Channels correspond to port capabilities in modern operating systems. Passing them is necessary to adapt the network structure (see [15]).
4. unidirectional channels that connect exactly two processes. This decision has interesting consequences, which are discussed briefly in the last section. The main advantage lies in a clear network topology.

The original design of a new language called DHOP (Distributed Higher Order Processes) was done by the second author in [21]. Because communication mechanisms are best explained by drawing and manipulating pictures, the semantics is based entirely on graph rewrite systems. Much work has proven already that this approach is very well suited (e. g. [8, 6, 22, 13, 16, 23] and numerous theoretical investigations).

The next section introduces DHOP. Then some technical prerequisites follow. The semantics of DHOP is covered in section four.

## 2 Syntax of DHOP

Let us take a closer look at DHOP. The syntax is given in extended BNF. We start with message passing facilities.

```
ExternalSelection: "[" Guard { "[]" Guard } "]"
            Guard: Communication "=>" Block
    Communication: identifier "!" Parameters
                 | identifier "?" Parameters
```

In a **Communication**, an exclamation mark stands for an attempt to send some **Parameters** over a named channel, whereas with a question mark the data are to be received. (For sake of simplicity, formal and actual parameter lists look the same.) Every **Communication** is embedded in an **ExternalSelection**, where a nondeterministic choice is taken among all the possible communications (i. e. those communications for which there is a partner available). Only the **Block** corresponding to the chosen communication is executed.

Process definitions and invocations come next.

```
          Action: "LET" IdentifierList "=" ExpressionList
                | "CHANNEL" identifier
                | "SPAWN" identifier Parameters
  ExpressionList: Expression { "," Expression }
      Expression: "PROCESS" Parameters Block
```

The LET construct binds each name in the list of identifiers to the corresponding value in the list of expressions. Both lists must have the same length, of course. Process definitions are the only possible expressions. The CHANNEL construct creates a new channel and names it, whereas the SPAWN construct instantiates a named process definition.

The form of the parameter lists indicates that processes and channels (what else?) may be passed in messages.

```
      Parameters: "(" ValuePart InPart OutPart ")"
       ValuePart: [ IdentifierList ]
          InPart: [ "IN" IdentifierList ]
         OutPart: [ "OUT" IdentifierList ]
  IdentifierList: identifier { "," identifier }
```

Because there are no primitive data types like truth values or integers, only the names of process definitions can be passed in the ValuePart. (We will show in a moment that this is not a serious restriction. Confer also [25].) The InPart names input ends of channels, the OutPart names output ends. As soon as the name of a channel appears in an actual parameter list, the corresponding end is given away and is no longer usable.

At last, we give the general program structure.

```
         Program: Block
           Block: { Action ";" } Statement
       Statement: ExternalSelection
                | "STOP"
```

As usual, a Block limits the scope of the identifier bindings it contains. The final Statement determines how to continue the enclosing process. STOP works like an empty ExternalSelection, i. e. it terminates the enclosing process.

Two small examples may illustrate the concepts.

*Example 1.* The process in Fig. 1 behaves like a variable. It accepts a new value via the channel set and shows the current value via the channel ask. The definition of variable contains the definition of a local process bound. In the initial "state", it will not participate in a communication over ask because its value is undefined. After a value was received over set, however, it will enter the "state" bound in which it is fully functional and "loops" forever.

*Example 2.* Processes are sufficient to model all the data types that are usually part of programming languages. An easy exercise are the truth values. One of them is defined in Fig. 2, the other (with identical parameter lists, of course) would be very similar.

```
LET variable = PROCESS (IN set OUT ask)
   LET bound = PROCESS (value IN set OUT ask)
      [
         set?(newval) =>
         SPAWN bound (newval IN set OUT ask);
         STOP
      []
         ask!(value) =>
         SPAWN bound (value IN set OUT ask);
         STOP
      ];
   [
      set?(newval) =>
      SPAWN bound (newval IN set OUT ask);
      STOP
   ]
```

**Fig. 1.** A process that behaves like a variable

A truth value is able to understand a few "commands" which it receives via input channels not, and, and or. An empty message over not tells it to take on the opposite value. Messages over and and or contain another truth value that gets combined with the current "state". The output channels allow other processes to find out the current "state". A `false` process will participate in a communication via `iffalse` at any time, but it will never send anything via `iftrue`.

Figure 3 shows how to find out the value of an identifier `condition` which is known to be bound to a truth value. If it is true, the block `...ThenPart` will be executed, otherwise the block `...ElsePart`. After the process definition is instantiated, the decision is taken in an external selection.

Before we turn to a formal semantics for DHOP, we need a few definitions.

## 3 Technical Prerequisites

We are going to use the well-known double pushout approach to graph rewriting (confer, for example, [9]). Therefore, we have to specify the category in which we are working. We start with plain graphs and label their vertices by elements of certain algebras to arrive at $\Sigma$-labeled graphs.

**Definition 1 (Graphs).** The category of *graphs* contains

- objects of the form $(E, V, s)$, where $E$ and $V$ are finite sets (of *edges* and *vertices*, respectively) and $s: E \to V \times V$ is a mapping (giving the source and target vertices for an edge), and

```
LET false = PROCESS (IN not, and, or OUT iffalse, iftrue)
   [
      not?() =>
      SPAWN true (IN not, and, or OUT iffalse iftrue);
      STOP
   []
      and?(other) =>
      SPAWN false (IN not, and, or OUT iffalse, iftrue);
      STOP
   []
      or?(other) =>
      SPAWN other (IN not, and, or OUT iffalse, iftrue);
      STOP
   []
      iffalse!() =>
      SPAWN false (IN not, and, or OUT iffalse, iftrue);
      STOP
   ]
```

**Fig. 2.** A process that behaves like the truth value false

```
CHANNEL not; CHANNEL and; CHANNEL or;
CHANNEL iffalse; CHANNEL iftrue;
SPAWN condition (IN not, and, or OUT iffalse, iftrue);
[
   iftrue?() => ...ThenPart
[]
   iffalse?() => ...ElsePart
]
```

**Fig. 3.** Executing "IF condition THEN ... ThenPart ELSE ... ElsePart"

- morphisms of the form $(e, v): G \to G'$, where $e: E \to E'$, $v: V \to V'$ are mappings, and the diagram

$$\begin{array}{ccc} E & \xrightarrow{s} & V \times V \\ {\scriptstyle e}\downarrow & & \downarrow{\scriptstyle v \times v} \\ E' & \xrightarrow[s']{} & V' \times V' \end{array}$$

  commutes.

**Definition 2 (Signature).** A *signature* $\Sigma$ consists of

- a finite set $S$ of *sorts*,

- a finite set $\Omega$ of *function symbols*, and
- an *arity* mapping $\tau: \Omega \to S^+$.

**Definition 3 (Algebras).** The category of $\Sigma$-*algebras* for some signature $\Sigma$ contains

- objects of the form $(A, I)$, where the *carrier* $A = \biguplus_{s \in S} A_s$ is the disjoint union of sets $A_s$ (of elements of sort $s \in S$) and $I$ interprets each function symbol $f \in \Omega$, $\tau(f) = s_1 \dots s_n s$, as a mapping $f_A: A_{s_1} \times \dots \times A_{s_n} \to A_s$, and
- morphisms of the form $h: A \to B$, such that for all $s \in S$ and $a \in A_s$, $h(a) \in B_s$, and for all $f \in \Omega$, $\tau(f) = s_1 \dots s_n s$, the diagram

$$\begin{array}{ccc} A^n & \xrightarrow{f_A} & A \\ {\scriptstyle h\times\ldots\times h}\downarrow & & \downarrow{\scriptstyle h} \\ B^n & \xrightarrow[f_B]{} & B \end{array}$$

  commutes.

**Definition 4 ($\Sigma$-Labeled Graphs).** The category of $\Sigma$-*labeled graphs*, which have their vertices labeled with elements of some $\Sigma$-algebra, contains

- objects of the form $(E, V, s, a, A, I)$, where $(E, V, s)$ is a graph, $(A, I)$ is a $\Sigma$-algebra, and $a: V \to A$ is a mapping, and
- morphisms of the form $(e, v, h): G \to G'$, where $(e, v)$ is a graph morphism, $h$ is a $\Sigma$-algebra morphism, and the diagram

$$\begin{array}{ccc} V & \xrightarrow{a} & A \\ {\scriptstyle v}\downarrow & & \downarrow{\scriptstyle h} \\ V' & \xrightarrow[a']{} & A' \end{array}$$

  commutes.

***Proposition 5.*** *The category of $\Sigma$-labeled graphs is finitely cocomplete.*

*Proof.* The category of graphs and the category of $\Sigma$-algebras are finitely cocomplete. The category of $\Sigma$-labeled graphs is the comma-category $(\mathcal{F}_V, \mathcal{A})$ of forgetful functors $\mathcal{F}_V: (E, V, s) \mapsto V, (e, v) \mapsto v$ and $\mathcal{A}: (A, I) \mapsto A, h \mapsto h$, which map graphs to their vertex sets and $\Sigma$-algebras to their carriers, respectively. Since $\mathcal{F}_V$ is cocontinuous we can apply [20, 5.2 Theorem 3]. □

$\Sigma$-labeled graphs are a special case of a general labeling construction. We refer the interested reader to [21].

Now we are ready to define rewrite rules in the category of $\Sigma$-graphs.

**Definition 6 (Rewrite Rule).** A *rewrite rule* has the form

$$(L \overset{l}{\longleftarrow} K \overset{r}{\longrightarrow} R) \ ,$$

where $L$ (*left-hand side*), $K$ (*gluing graph*), $R$ (*right-hand side*) are $\Sigma$-labeled graphs, $l = (e_l, v_l, h_l)$, $r = (e_r, v_r, h_r)$ are morphisms such that the graph components $e_l$, $v_l$, $e_r$, $v_r$ are injective and the algebra components $h_l$, $h_r$ are isomorphisms.

The above restrictions reduce the existence of pushout-complements to the plain graph case (see [21, Theorem 4.2.12]). We already know from Proposition 5 that all pushouts exist. So we are now able to rewrite graphs by double pushouts.

## 4 Semantics of DHOP

We model a process net by a $\Sigma$-labeled graph, where $\Sigma$ is the disjoint union of three signatures given below. The main ingredient is an "abstract syntax" for program fragments. The nonterminals in the context-free grammar of Sect. 2 are turned into sorts, and we introduce an operator for every language construct. In this manner, a program becomes a term. For example, there will be a binary function symbol $\cup$ denoting nondeterministic choice. An `ExternalSelection` of the form

$$[\text{guard}_1 \ [] \ \text{guard}_2 \ [] \ \text{guard}_3\,]$$

looks like

$$(\text{guard}_1 \cup \text{guard}_2) \cup \text{guard}_3 \ .$$

We define an algebra by adding the equations

$$g_1 \cup g_2 = g_2 \cup g_1$$
$$(g_1 \cup g_2) \cup g_3 = g_1 \cup (g_2 \cup g_3) \ .$$

We denote the abstract syntax of a program fragment $pf$ by "$pf$".

Besides that, we need signatures and algebras for identifiers (with equality), nonnegative numbers (with functions zero and successor in the usual notation), and lists (with the empty list $\varepsilon$ and function :: to add a new element in front of the list). All of them are standard, so we omit them.

Having clarified the basic constituents, we now turn to the form of the graphs. There are three kinds of labels corresponding to different types of objects, namely

- process states. The labels are quadruples consisting of
  1. a program fragment $S$ (in abstract syntax),
  2. an environment for process definitions $\rho_p$ (a list of pairs, each consisting of an identifier and a value, see below),
  3. an environment for channels $\rho_{ch}$ (a list of pairs, each consisting of an identifier and a channel label), and
  4. a counter $n$ (a nonnegative number).

The counter value is used only to generate unique labels for process identifications and channels.

- process identifications. The labels are lists of nonnegative numbers.
- channels. The labels are lists of nonnegative numbers.

The initial graph to a program $S$ is depicted in Fig. 4.

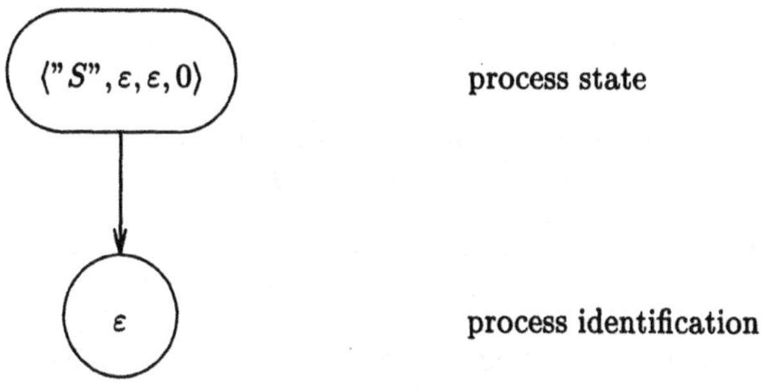


**Fig. 4.** Initial graph for program $S$

The rest of the present section will be devoted to the rewrite rules. As usual, the gluing graph contains no edges and is not shown. The images of the vertices inside the left-hand and right-hand sides are marked with small numbers instead. Because of the restrictions we posed on the morphisms the gluing graph can be uniquely reconstructed.

The rewrite rule for actions starting with LET (see Fig. 5) does not change the graph structure at all. Only the state of the executing process changes, which is expressed in a new label. The program fragment $S$ remains to be executed. The environment $\rho_p$ is extended with the process identifiers $id_1, \ldots, id_n$. (We assume that multiple bindings of the same identifier within one block are forbidden, so the order of succession does not matter here.) The other components $\rho_{ch}$ and $n$ stay as they are. The value of $id_i$, $1 \leq i \leq n$, in the new environment is $\langle "expr_i", \rho'_p \rangle$, that is, the body of $id_i$ is given by the abstract syntax of $expr_i$ and the process environment of $id_i$ is $\rho'_p$ (this recursive definition of the environment ensures that $id_i$ may call itself and the other $id_j$ recursively). We do not need to remember $\rho_{ch}$ because these identifiers will not be visible to instances of $id_i$.

A new channel with both ends attached to the current process is created by an action starting with CHANNEL (see Fig. 6). The identifier $id$ is bound to the channel label, accordingly. The counter is incremented in order to maintain unique identifications.

The following Figs. 7 and 8 stand for an infinite family of rewrite rules according to the length of the parameter lists. But note that only a finite collection of them is needed for any single program.

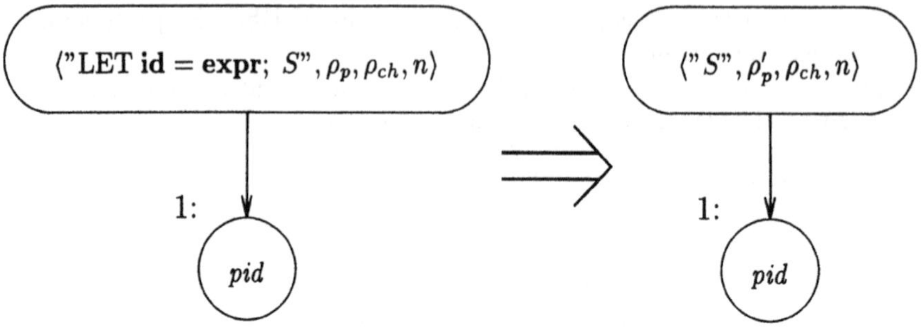


where $\mathbf{id} = id_1, \ldots, id_n$, $\mathbf{expr} = expr_1, \ldots, expr_n$, and
$\rho'_p = \langle id_1, \langle "expr_1", \rho'_p\rangle\rangle :: \ldots :: \langle id_n, \langle "expr_n", \rho'_p\rangle\rangle :: \rho_p$

**Fig. 5.** Rewrite rules corresponding to actions LET

When a new process instance is created via SPAWN, it is necessary that the active process has access to all the channels named in the actual parameter list. So the appearance of all these channels in the left-hand side of the rewrite rules (in Fig. 7 drawn above) acts as an application condition. The process identification of the created process is derived from the counter value. The environments of the created process are filled with the values passed in the parameter list.

To make this precise, we first have to define a function on environments:

$$\begin{aligned} \text{eval}(id, \langle id, value\rangle :: env) &= value \\ \text{eval}(id, \langle id', value\rangle :: env) &= \text{eval}(id, env), \qquad id' \neq id \\ \text{eval}(id, \varepsilon) &= \text{error} \end{aligned}$$

eval$(id, env)$ picks the first value associated with $id$ in the list $env$. With the help of this function, there remains the tedious, but trivial, task to define the passing of parameters. We assume that the actual parameter list $"apl" = "(av\ ai\ ao)"$ is given by $"av" = av_1 :: \ldots :: av_k :: \varepsilon$, $"ai" = ai_1 :: \ldots :: ai_l :: \varepsilon$, $"ao" = ao_1 :: \ldots :: ao_m :: \varepsilon$, and $\text{eval}(ai_i, \rho_{ch}) = in_i$, $1 \leq i \leq l$, $\text{eval}(ao_i, \rho_{ch}) = out_i$, $1 \leq i \leq m$. Furthermore, if $\text{eval}(id, \rho_p) = \langle "\text{PROCESS } fpl\ body", \sigma\rangle$, $"fpl" = "(fv\ fi\ fo)"$, $"fv" = fv_1 :: \ldots :: fv_k :: \varepsilon$, $"fi" = fi_1 :: \ldots :: fi_l :: \varepsilon$, $"fo" = fo_1 :: \ldots :: fo_m :: \varepsilon$, then

$$\begin{aligned} \sigma_p &= \langle fv_1, \text{eval}(av_1, \rho_p)\rangle :: \ldots :: \langle fv_k, \text{eval}(av_k, \rho_p)\rangle :: \sigma \ , \\ \sigma_{ch} &= \langle fi_1, in_1\rangle :: \ldots :: \langle fi_l, in_l\rangle :: \langle fo_1, out_1\rangle :: \ldots :: \langle fo_m, out_m\rangle :: \varepsilon \ . \end{aligned}$$

The most interesting rewrite rules (shown in Fig. 8) concern the exchange of values in communications. To apply it, two processes and a connecting channel are required together with some channels which are passed from the first process to the second. The rewrite rules as they stand do not allow that the vertex $ch$ coincides with one of the $out_i$ because the adjacent edges cannot be forced to merge as well. It is required that $\text{eval}(c_1, \rho_{ch,1}) = ch = \text{eval}(c_2, \rho_{ch,2})$ and $apl$

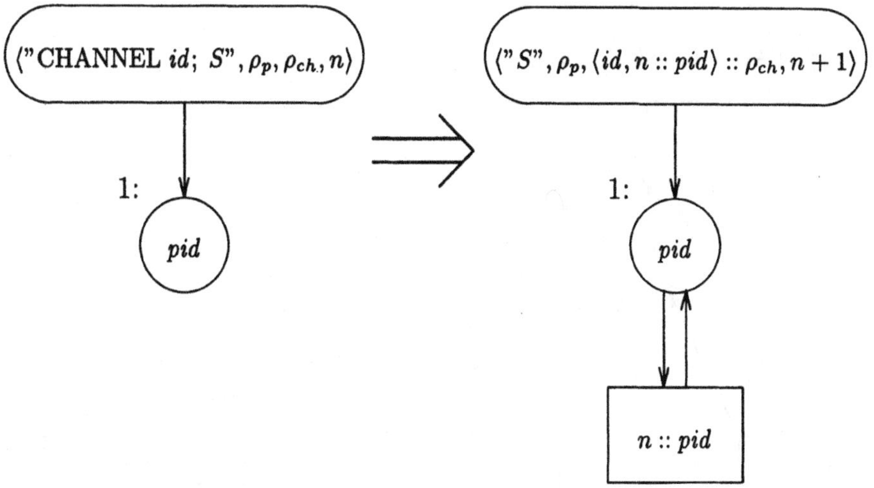


**Fig. 6.** Rewrite rules corresponding to actions **CHANNEL**

is compatible to *fpl*, that is, the corresponding parts have equal lengths. With notation as before, the new environments are constructed as follows:

$$\rho'_{p,2} = \langle fv_1, \text{eval}(av_1, \rho_{p,1})\rangle :: \ldots :: \langle fv_k, \text{eval}(av_k, \rho_{p,1})\rangle :: \rho_{p,2} \ ,$$
$$\rho'_{ch,2} = \langle fi_1, in_1\rangle :: \ldots :: \langle fi_l, in_l\rangle :: \langle fo_1, out_1\rangle :: \ldots :: \langle fo_m, out_m\rangle :: \rho_{ch,2} \ .$$

Interesting with these rewrite rules is that they specify successful communications only. It is possible that several rewrite rules of this family are applicable to one or both of the processes at the same time. (Remember that, according to the commutativity of the operator ∪ stated at the beginning of this section, the sequence of alternatives in an external selection does not matter.) In certain situations, even the same partners could choose among several channels and message formats. The decision to apply a specific rule selects the branch to be taken.

Finally, we could add a few simple clean-up rules that eliminate processes executing the **STOP** statement together with the channels that are no longer accessible. The initial graph for $S$ together with the finite collection of rewrite rules corresponding to the actions and statements appearing in $S$ completely and unambiguously define the semantics of the program $S$.

## 5 Some Remarks on an Implementation

We believe that a programming language must prove the value of its concepts in practical applications. Therefore, we built the operational semantics just given

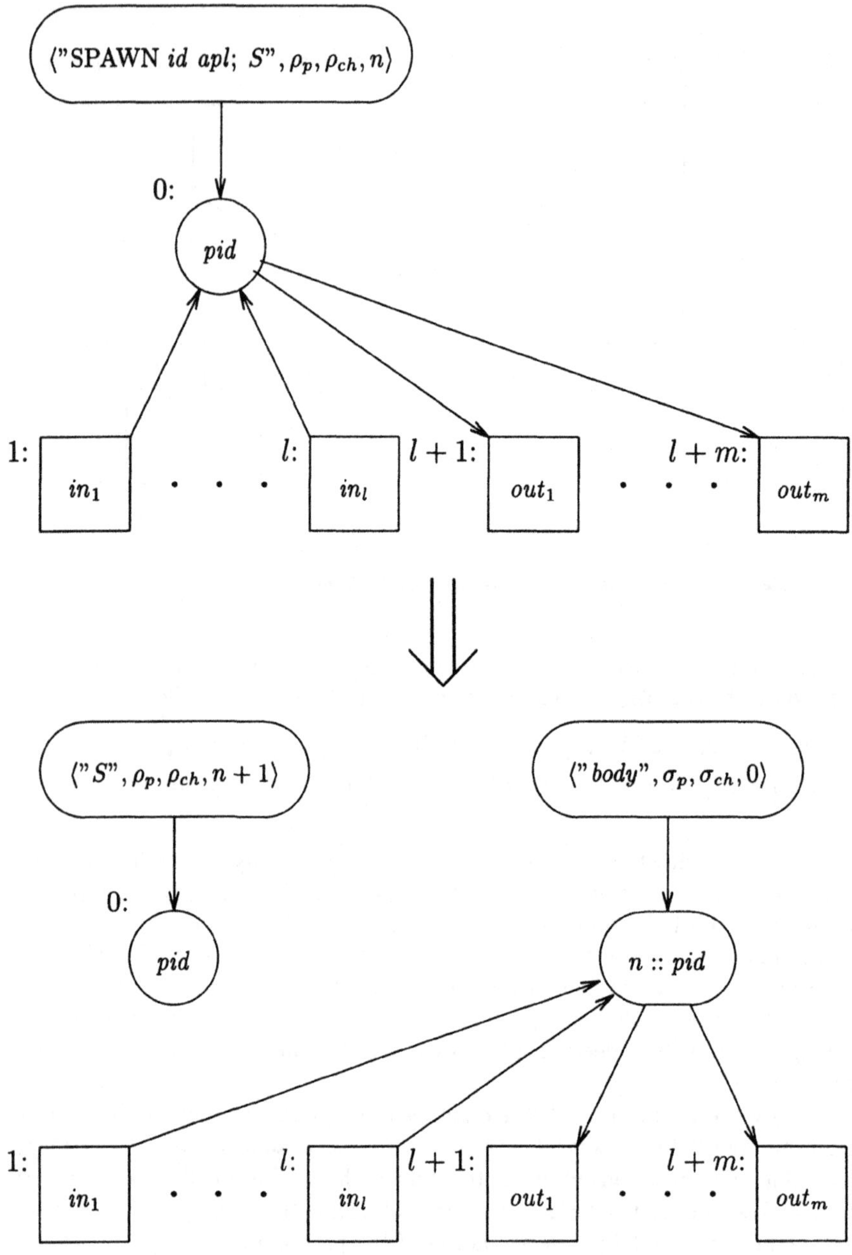


where $in_1, \ldots, in_l, out_1, \ldots, out_m$, *body*, $\sigma_p$, $\sigma_{ch}$ are given in the text.

**Fig. 7.** Rewrite rules corresponding to actions **SPAWN**

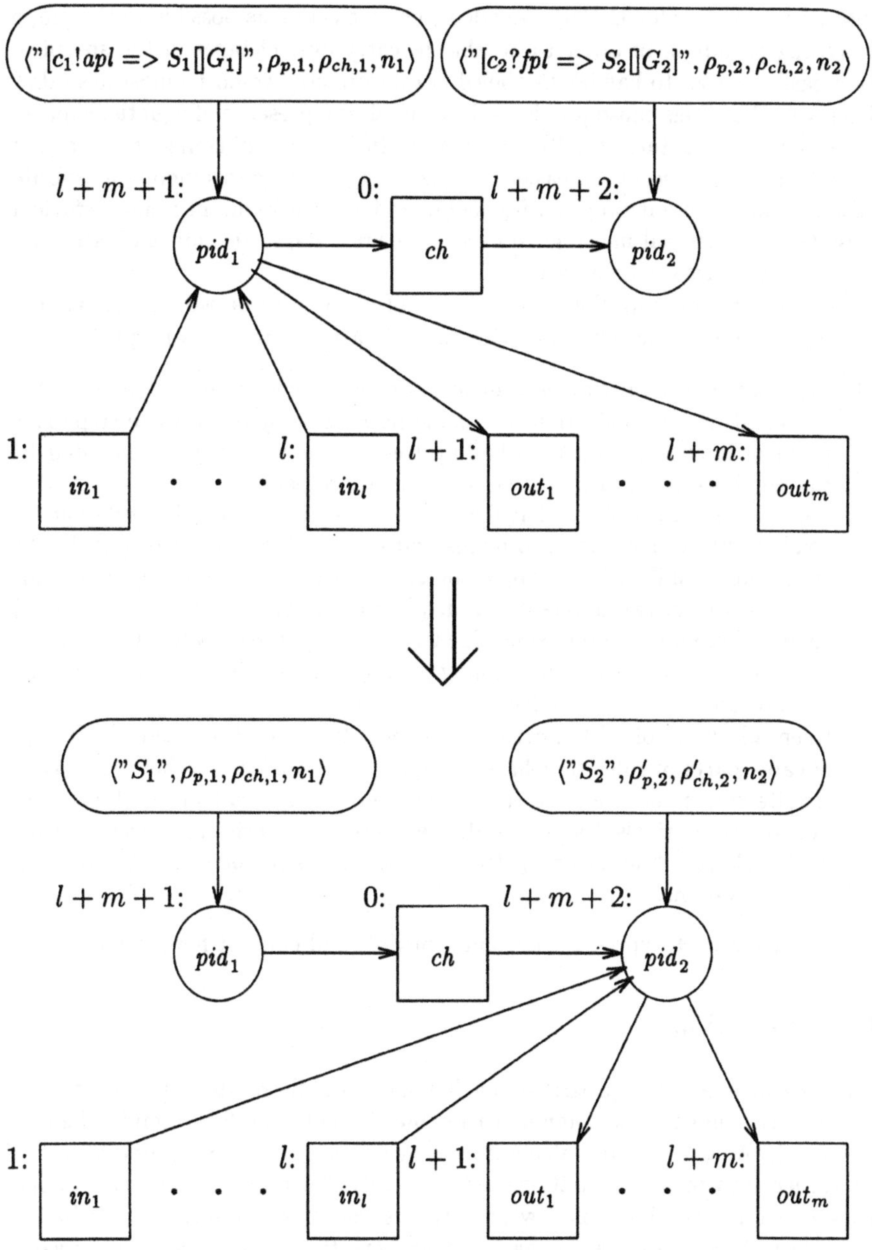


where $ch$, $\rho'_{p,2}$, $\rho'_{ch,2}$ are given in the text and the rest is like in Fig. 7.

**Fig. 8.** Rewrite rules corresponding to external selections

into an interpreter for DHOP. We tried to stay as close as possible to the graph rewriting paradigm in order to be able to carry over changes in the specification easily. A tool to handle $\Sigma$-labeled graphs directly would be most desirable. However, the labels cause problems: In general, the presence of equations makes it undecidable whether two $\Sigma$-terms match. In our case, of course, this was not a problem, but it was even easier to consider the $\Sigma$-algebras merely as specifications and represent program fragments and environments in a more efficient way. Given a general machine to execute rewrite steps, an implementation for DHOP could already be finished.

Up to this, locating the left-hand side of a rewrite rule would still be very costly. To improve efficiency we made use of a few theoretical concepts:

1. Find out which rewrite rules depend on the previous application of other rules and which applications are made impossible by one rewrite step. This problem is closely related to finding the *event structure* [17] generated by the rewrite rules, which is an area of current research.
2. Split those rewrite rules that depend on more than one other rule and/or conflict with other rules into pieces and use *amalgamation* (see [6]). In the specification of DHOP, the applicability of the rewrite rules in Fig. 8 depends on the states of two processes and it possibly conflicts with other choices of communications on either side. The rules were split vertically into two parts with the channels as common gluing points. This step is useful anyway if we want to do rewriting on a distributed system.
3. Keep all the applicable rewrite rules in a data structure, and all the applicable parts of rules that have to be amalgamated in another. After each rewrite step remove all the (parts of) rewrite rules that are no longer applicable (the "rivals") and add all the (parts of) rewrite rules that became applicable (the "successors"). The data structures can be reused for different rewrite systems.

The resulting prototype was sufficiently flexible and efficient for our purposes.

## 6 Discussion

We presented a small programming language for distributed systems with a much more sophisticated communication mechanism than, for example, the rendezvous in Ada [3]. In fact, algebraic specifications of a rather general form can be turned into process definitions automatically. Examples 1 and 2 correspond to very simple specifications (with only one equation $\text{ask}(\text{set}(x)) = x$ in the former case). Nevertheless, giving a formal semantics was quite straightforward, the tedious part being the handling of variable bindings and the like.

A crucial feature in our approach was the labeling mechanism for the graphs used in the semantics description. When applying a rewrite rule, a side effect of the morphism that locates the left-hand side in the host graph is to "evaluate" the variables in the vertex labels by mapping them to elements of an algebra. It is important that identical variables get the same values independently of the vertex (label) inside a rewrite rule to which they belong. So, for our purposes, the mechanism of [23] is not adequate.

The use of a labeling category (that idea apparently did not emerge earlier than [23]) allows to choose what should go into the graph and what should go into the labels. We used graphs to model a net of processes, that is, the level where parallelism is observed. $\Sigma$-algebras as labels are only one possibility (see [10]), one, however, that we found very convenient since the handling of variable bindings, for example, could be done in a common way. It is possible to lay more emphasis on the graph structure. In particular, we could represent the programs as graphs instead of terms. The need for equivalences in the abstract syntax would disappear. On the other hand, we would have to face the problem that graphs (process definitions) must then be stored as part of the labels (process environments). The representation of (hyper-)graphs as terms as in [4] could help us in this case. A $\Sigma$-labeled graph is turned into a term with additional sorts and function symbols for vertices and edges. It is an interesting idea to pull (the term representation of) a graph out of a label and add it to the host graph as soon as a process is created.

Speaking of [4], a second extreme is to turn the double pushouts into term rewrite steps and use equational term rewriting only. This is no practical possibility, however.

A last remark concerns our decision to have channels connect two processes exclusively. This leads to a great amount of channel passing in messages. Let us assume, for example, that input and output is done via dedicated channels. Then it is clear that a process which just finished printing out something has to submit the output channel to the next one requesting it. This is a desired feature since it makes synchronizations explicit. Furthermore, the correct usage of channels can be statically checked. But if we want to store channels in data structures (lists, for example, to make neighborhood more flexible), it seems impossible to maintain this single way paradigm. This is one case where the graph rewrite approach can prove its flexibility! We only need to add a few edges on the right-hand side of Figs. 7 and 8 such that the calling/sending process does not lose connection to its channels. Thus, graph rewriting encourages us to try out various ideas in a programming language for distributed systems.

## Acknowledgements

We would like to thank the anonymous referees for their valuable suggestions to improve this paper.

## References

1. P. America: Designing an Object-Oriented Programming Language with Behavioural Subtyping. In: Foundations of Object-Oriented Languages, LNCS 489 (1991), 60–90.

2. G. Andrews, R. Olsson: The SR Programming Language: Concurrency in Practice. Benjamin/Cummings 1993.
3. ANSI: The Programming Language Ada. Reference Manual. LNCS 155 (1983).
4. M. Bauderon, B. Courcelle: Graph Expressions and Graph Rewritings. Math. Systems Theory 20 (1987), 83–127.
5. J.A. Bergstra, J.W. Klop: Algebra of communicating processes with abstraction. TCS 37 (1985), 77–121.
6. P. Böhm, H.-R. Fonio, A. Habel: Amalgamation of graph transformations: A synchronization mechanism. JCSS 34 (1987), 377–408.
7. A. de Bruin, W. Böhm: The Denotational Semantics of Dynamic Networks of Processes. ACM ToPLaS 7/4 (1985), 656–679.
8. P. Degano, U. Montanari: A model for distributed systems based on graph rewriting. JACM 34/2 (1987), 411–449.
9. H. Ehrig: Tutorial introduction to the algebraic approach of graph grammars. In: GraGra 86, LNCS 291 (1987), 3–14.
10. L. Hess, B.H. Mayoh: Graphics and their grammars. In: GraGra 86, LNCS 291 (1987), 232–249.
11. C. A. R. Hoare: Communicating Sequential Processes. Prentice Hall 1985.
12. INMOS Limited: occam 2 Reference Manual. Prentice Hall 1988.
13. D. Janssens: Equivalence of computations in actor grammars. TCS 109 (1993), 145–180.
14. R. Milner: Communication and Concurrency. Prentice Hall 1989.
15. R. Milner, J. Parrow, D. Walker: A Calculus of Mobile Processes I & II. Information and Control 100 (1992), 1–77.
16. U. Montanari, F. Rossi: Graph rewriting for a partial ordering semantics of concurrent constraint programming. TCS 109 (1993), 225–256.
17. M. Nielsen, G. Plotkin, G. Winskel: Petri nets, event structures and domains. TCS 13 (1981), 85–108.
18. F. Nielson: The Typed $\lambda$-Calculus with First Class Processes. In: PARLE 89, LNCS 366 (1989), 357–373.
19. S. Prasad, A. Giacalone, P. Mishra: Operational and Algebraic Semantics for Facile: A Symmetric Integration of Concurrent and Functional Programming. In: ICALP 90, LNCS 443 (1990), 765–780.
20. D.E. Rydeheard, R.M. Burstall: Computational Category Theory. Prentice Hall 1988.
21. G. Schied: Über Graphgrammatiken, eine Spezifikationsmethode für Programmiersprachen und verteilte Regelsysteme. Dissertation, Arbeitsberichte des IMMD 25, 2, Universität Erlangen-Nürnberg 1992.
22. H.J. Schneider: Describing Distributed Systems by Categorical Graph Grammars. In: WG 89, LNCS 411 (1989), 121–135.
23. H.J. Schneider: On categorical graph grammars integrating structural transformations and operations on labels. TCS 109 (1993), 257–274.
24. B. Thomsen: Plain CHOCS—A second generation calculus for higher order processes. Acta Informatica 30 (1993), 1–59.
25. D. Walker: $\pi$-Calculus Semantics for Object-Oriented Programming Languages. In: TACS 91, LNCS 526 (1991), 532–547.

# Abstract Graph Derivations in the Double Pushout Approach*

A. Corradini[1], H. Ehrig[2], M. Löwe[2], U. Montanari[1], and F. Rossi[1]

[1] Università di Pisa, Dipartimento di Informatica, Corso Italia 40, 56125 Pisa, Italy ({andrea,ugo,rossi}@di.unipi.it)
[2] Technische Universität Berlin, Fachbereich 20 Informatik, Franklinstraße 28/29, 1000 Berlin 10, Germany ({ehrig,loewe}@cs.tu-berlin.de)

**Abstract.** In the algebraic theory of graph grammars, it is common practice to present some notions or results "up to isomorphism". This allows one to reason about graphs and graph derivations without worrying about representation-dependent details.
Motivated by a research activity aimed at providing graph grammars with a truly-concurrent semantics, we front in this paper the problem of formalizing what does it mean precisely to reason about graph derivations "up to isomorphism". This needs the definition of a suitable equivalence on derivations, which should be consistent with the relevant definitions and results, in the sense that they should extend to equivalence classes. After showing that a naive equivalence is not satisfactory, we propose two requirements for equivalences on derivations which allow the sequential composition of derivations and guarantee the uniqueness of canonical derivations, respectively. Three new equivalences are introduced, the third of which is shown to be satisfy both requirements. We also define a new category having the abstract derivations as arrows, which is, in our view, a fundamental step towards the definition of a truly-concurrent semantics for graph grammars.

## 1 Introduction

The algebraic theory of graph grammars [EPS73, Ehr87] comprises many results concerning parallelism and concurrency: for example, two basic constructions are defined, called *analysis* and *synthesis*, which under suitable hypotheses can transform the parallel application of productions to a graph into a sequential derivation using the same productions, and vice versa. Two derivations which are related by a finite number of applications of analysis and synthesis are called *shift-equivalent*. Moreover, in the thesis of Hans-Jörg Kreowski ([Kre77]) it is shown that any graph derivation can be transformed into a shift-equivalent *canonical* derivation, where each production is applied as early as possible. In general, since the analysis and synthesis constructions are not deterministic, from

* Research partially supported by the COMPUGRAPH Basic Research Esprit Working Group n. 7183

a given derivation many distinct canonical derivations can be obtained: nevertheless, all of them are isomorphic. This is formalized by a result in [Kre77] stating that every derivation has, up to isomorphism, a unique shift-equivalent canonical derivation.

The assumption that one can safely reason up to isomorphism about constructions in a category has been widely used (sometimes explicitly, sometimes not) in the literature of the algebraic approach to graph rewriting. Since the "meaning" of such assumptions is perfectly clear from an intuitive point of view, on the one hand the researchers working in the area always assumed that a formalization of what "reasoning up to isomorphism" means was straightforward, on the other hand the need of such a formalization never arose till recently.

In the framework of a research activity recently started by the authors and aimed at providing a truly-concurrent semantics for graph grammars, it turned out that a crucial point in the study of such a semantics is the definition of a satisfactory notion of equivalence on graph derivations. In fact, one has to find the right trade-off between two conflicting requirements. On the one hand, one would like to abstract out from irrelevant, representation-dependent details; for example, one does not want to distinguish between different, but isomorphic results of derivations. On the other hand, the equivalence cannot relate too many derivations, because one should be able to extend relevant properties of (and constructions on) concrete derivations to the equivalence classes, in order to extend to the more abstract setting (some of) the results of the algebraic theory of graph grammars. For example, we require that the sequential composition, as well as the analysis and synthesis constructions, should be defined on equivalence classes of derivations (which will be called "abstract derivations" in the following).

In this paper we consider various equivalences on derivations, studying their properties with respect to these requirements. The first one is the obvious formalization of the idea of reasoning up to isomorphism, by defining abstract graphs as isomorphism classes of graphs and abstract derivations as isomorphism classes of derivations (defined in the obvious way, because derivations are just suitable diagrams in the category of graphs). We pointed out in [CELMR93] that this formalization is not satisfactory, because the sequential composition of abstract derivations cannot be defined properly.

Therefore we introduce a new, finer equivalence based on *standard* isomorphisms i.e., a proper subclass of isomorphisms satisfying certain requirements. We show that defining abstract derivations as standard-isomorphism classes of derivations, they can be composed sequentially. However, the new equivalence is not compatible with the analysis and synthesis constructions, because applying one of these non-deterministic constructions to the same derivation one can obtain derivations which are indeed isomorphic, but not standard isomorphic.

Finally, we present a third equivalence on graph derivations that meets the informal requirements mentioned above. This equivalence will allow us to define a category having abstract graphs as objects and abstract derivations as arrows. Such a category, besides being interesting in itself as a computational model of a grammar, is a fundamental step towards the definition of a truly-concurrent semantics for graph grammars.

The paper is organized as follows. Section 2 is essentially a background section, but some definitions and results of the theory of graph grammars are presented in a way that departs significantly from the traditional presentation, because we strictly avoid to reason up to isomorphism. In Section 3 we discuss the requirements that a notion of equivalence on derivations should satisfy to be satisfactory from the perspective of true concurrency, formulating them as suitable algebraic properties. We require that the equivalence allows one to define the sequential composition of abstract derivations, and that it is compatible with the construction of canonical derivations i.e., that any pair of canonical derivations obtained from two equivalent derivations be equivalent as well. It is worth stressing that these requirements are motivated by the goal of defining a truly-concurrent semantics: if one would extend to the abstract setting other results of the theory (for example those concerning concurrency and amalgamation [Ehr87]), it might be the case that additional requirements should be considered. Section 4 is devoted to the presentation of three different equivalences on derivations: the first two enjoy complementary advantages, while the third one satisfies all the requirements of Section 3. This last equivalence is used in Section 5 to define two categories having abstract graphs as objects and abstract derivations as arrows. These categories are related with two categories of concrete derivations which are defined in Section 2.

The reader is supposed to have some familiarity with the very basic notions of Category Theory (see for example [AHS90]). Most of the proofs are omitted because of space limitations, and will appear in the full version of the paper.

## 2 Background

We introduce here the basic definitions and results of the algebraic theory of graph grammars, following the so-called "double-pushout approach". The contents of this section essentially summarizes [Ehr87]; however, the definitions that are usually introduced up to isomorphism are presented in a significantly different way that avoids any assumption of that kind (see for example Definitions 3 and 11). Moreover, the categories of Definition 14 are original.

In this paper we stick to colored graphs: although most of the definition and results are formulated in pure categorical terms, some specific properties of the category of graphs are sometimes exploited.

**Definition 1 (colored graphs).** Given two fixed alphabets $\Omega_E$ and $\Omega_V$ for edge and vertex labels, respectively, a **(colored) graph (over $(\Omega_E, \Omega_V)$)** is a tuple $G = \langle E, V, s, t, l_E, l_V \rangle$, where $E$ is a finite set of **edges**, $V$ is a finite set of **vertices**, $s, t : E \rightarrow V$ are the **source** and **target** functions, and $l_E : E \rightarrow \Omega_E$ and $l_V : V \rightarrow \Omega_V$ are the **edge** and the **node labeling** functions, respectively.

A **graph morphism** $f : G \rightarrow G'$ is a pair $f = (f_E : E \rightarrow E', f_V : V \rightarrow V')$ of functions which preserve sources, targets, and labels, i.e., such that $f_V \circ t = t' \circ f_E$,

$f_V \circ s = s' \circ f_E$, $l'_V \circ f_V = l_V$, and $l'_E \circ f_E = l_E$. A graph morphism $f$ is an **isomorphism** if both $f_E$ and $f_V$ are bijections. If there exists an isomorphism from graph $G$ to graph $H$, then we write $G \cong H$; moreover, $[G]$ denotes the isomorphism class of $G$, i.e., $[G] = \{H \mid H \cong G\}$. The category having colored graphs as objects and graph morphisms as arrow is called **Graph**. □

**Definition 2 (graph productions, graph grammars, direct derivation).** A **graph production** $p = (L \overset{l}{\leftarrow} K \overset{r}{\rightarrow} R)$ is a pair of injective graph morphisms $l : K \rightarrow L$ and $r : K \rightarrow R$. The graphs $L$, $K$, and $R$ are called the **left-hand side** (lhs), the **interface**, and the **right-hand side** (rhs) of $p$, respectively. Two productions are **isomorphic** if there are three isomorphisms between the corresponding graphs that make the two resulting squares commutative. A **graph grammar** $\mathcal{G} = \langle \{p_i\}_{i \in I}, G_0 \rangle$ is a set of graph productions together with an initial graph $G_0$.

Given a graph $G$, a graph production $p = (L \overset{l}{\leftarrow} K \overset{r}{\rightarrow} R)$, and an **occurrence** (i.e., a graph morphism) $g : L \rightarrow G$, a **direct derivation from** $G$ **to** $H$ **using** $p$ **(based on** $g$**)** exists if and only if the diagram in Figure 1 can be constructed, where both squares are required to be pushouts in **Graph**. In this case, $D$ is called the **context** graph, and we write $G \Rightarrow_p H$.[3] □

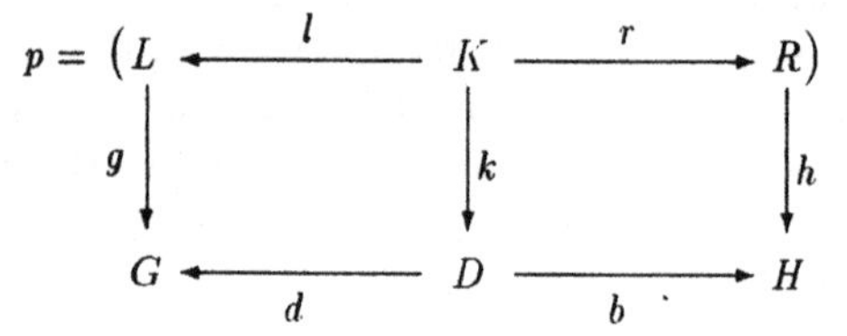

**Fig. 1.** Direct derivation as double-pushout construction.

The notion of direct derivation just introduced is intrinsically sequential: it models the application of a single production of a grammar $\mathcal{G}$ to a given graph. The categorical framework makes easy the definition of the parallel application of more than one production to a graph.

**Definition 3 (parallel productions).** Given a graph grammar $\mathcal{G}$, a **parallel production (over** $\mathcal{G}$**)** has the form $q \equiv \langle (p_1, in^1), \ldots, (p_k, in^k) \rangle = (L \overset{l}{\leftarrow} K \overset{r}{\rightarrow} R)$ (see Figure 2), where $k \geq 0$, $p_i = (L_i \overset{l_i}{\leftarrow} K_i \overset{r_i}{\rightarrow} R_i)$ is a production of $\mathcal{G}$ for each $i \in \{1, \ldots, k\}$, $L$ is a coproduct of $\{L_1, \ldots, L_k\}$, and similarly $R$ is a coproduct of $\{R_1, \ldots, R_k\}$ and $K$ is a coproduct of $\{K_1, \ldots, K_k\}$. Moreover, $l$ and $r$ are the mediating morphisms uniquely determined by the families of

[3] Note that, unlike the related literature, we explicitly label the double pushout diagram with the name of the involved production $p$.

arrows $\{l_i\}_{i\leq k}$ and $\{r_i\}_{i\leq k}$, respectively. Finally, for each $i \in \{1,\ldots,k\}$, $in^i$ denotes the triple of injections $\langle in^i_L : L_i \to L, in^i_K : K_i \to K, in^i_R : R_i \to R\rangle$. A parallel production like the one above is **proper** if $k > 1$; the **empty** production is the (unique) parallel production with $k = 0$.

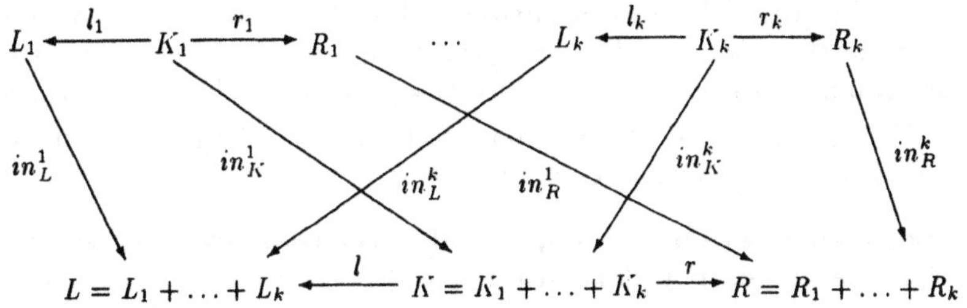


**Fig. 2.** The parallel production $\langle (p_1, in^1), \ldots, (p_k, in^k)\rangle = (L \overset{l}{\leftarrow} K \overset{r}{\rightarrow} R)$.

In the rest of the paper we will assume, without loss of generality, that for any given pair of graphs $A$ and $B$ a "canonical" coproduct diagram $\langle A+B, in_A : A \to A+B, in_B : B \to A+B\rangle$ is fixed once and for all. Then for each pair of (possibly parallel) productions $p = (L \overset{l}{\leftarrow} K \overset{r}{\rightarrow} R)$ and $p' = (L' \overset{l'}{\leftarrow} K' \overset{r'}{\rightarrow} R')$ we denote by $p + p'$ the parallel production $\langle (p, in), (p', in')\rangle = (L + L' \overset{l+l'}{\leftarrow} K + K' \overset{r+r'}{\rightarrow} R + R')$, where $in$ and $in'$ are triples of canonical injections. Moreover, given a list $\langle p_1, \ldots, p_k\rangle$ of productions of $\mathcal{G}$, by $p_1 + p_2 + \ldots + p_k$ we denote the parallel production $((\ldots(p_1 + p_2) + \ldots) + p_k)$. Note that the $+$ operator on productions defined in this way is neither associative nor commutative.

The **graph grammar with parallel productions** $\mathcal{G}^+$ generated by a grammar $\mathcal{G}$ has the same initial graph of $\mathcal{G}$, and as productions all the parallel productions over $\mathcal{G}$. In particular, $\mathcal{G}^+$ includes a standard representative for each production $p$ of $\mathcal{G}$, namely the parallel production $p \equiv \langle (p, \langle id_L, id_K, id_R\rangle)\rangle = (L \overset{l}{\leftarrow} K \overset{r}{\rightarrow} R)$. A **parallel direct derivation over** $\mathcal{G}$ is a direct derivation over $\mathcal{G}^+$; it is **proper** if so is the applied parallel production. □

The last definition slightly differs from the usual one (see for example [Ehr87]), because a parallel production includes here explicitly the injections from the corresponding individual productions. The existence of canonical coproducts is assumed here only to simplify the syntax of parallel productions: it does not imply that coproducts are unique.

As we shall see later, the empty parallel production plays a relevant role in some definitions and results, although this is not so evident in the literature, where often direct derivation steps using the empty production are implicitly removed from a derivation. This is a consequence of reasoning up to isomorphism, because, as the next fact shows, the application of the empty production to a graph produces an isomorphic graph.

**Fact 4 (direct derivations via the empty production).** *The unique empty production is* $\emptyset = (\emptyset \stackrel{id_\emptyset}{\leftarrow} \emptyset \stackrel{id_\emptyset}{\rightarrow} \emptyset)$. *Nevertheless, if* $G \cong H$, *then there exist as many distinct direct derivations* $G \Rightarrow_\emptyset H$ *as there are distinct triples* $\langle D, d : D \to G, b : D \to H \rangle$, *where* $d$ *and* $b$ *are isomorphisms.* □

**Definition 5 (empty direct derivations, identity derivation).** Every direct derivation via the empty production is called an **empty direct derivation.** For each graph $G$, the empty direct derivation having as bottom morphisms $G \stackrel{id_G}{\leftarrow} G \stackrel{id_G}{\rightarrow} G$ (which exists by the above fact) is called the **identity derivation** of $G$, and it is denoted by $G \Rightarrow_\emptyset^{id} G$. □

**Definition 6 (derivations).** A **sequential derivation (over** $\mathcal{G}$**)** is either a graph $G$ (called an **empty derivation,** and denoted by $G : G \Rightarrow^* G$), or a sequence of direct derivations $\rho = \{G_{i-1} \Rightarrow_{p_i} G_i\}_{i \in \{1,\ldots,n\}}$ such that $p_i$ is a production of $\mathcal{G}$ for all $i \in \{1, \ldots, n\}$. In the last case, the derivation is written $\rho : G_0 \Rightarrow_\mathcal{G}^* G_n$ or simply $\rho : G_0 \Rightarrow^* G_n$. If $\rho : G \Rightarrow^* H$ is a derivation (possibly empty), then graphs $G$ and $H$ are called the **starting** and the **ending graph** of $\rho$, and will be denoted by $\sigma(\rho)$ and $\tau(\rho)$, respectively. The **length** of a sequential derivation $\rho$ is the number of direct derivations in $\rho$, if it is not empty, and 0 otherwise. A **(parallel) derivation over** $\mathcal{G}$ is a sequential derivation over $\mathcal{G}^+$. The set of all (parallel) derivations over $\mathcal{G}$ is denoted by $Der(\mathcal{G})$. The **sequential composition** of two derivations $\rho$ and $\rho'$ is defined if and only if $\tau(\rho) = \sigma(\rho')$; in this case it is denoted $\rho ; \rho' : \sigma(\rho) \Rightarrow^* \tau(\rho')$, and it is obtained by identifying $\tau(\rho)$ with $\sigma(\rho')$. □

The empty derivation $G : G \Rightarrow^* G$ is introduced for technical reasons, and should not be confused with any empty *direct* derivation (having indeed length 1 and not 0). By the definition of sequential composition, it follows that for each derivation $\rho : G \Rightarrow^* H$, $G ; \rho = \rho = \rho ; H$. Thus empty derivations will be used as the identities in the categories of derivations introduced below. In the rest of the paper we will not pay much attention to empty derivations, because, unlike empty *direct* derivations, their handling is straightforward.

Derivations are essentially the "computations" of a graph grammar, when regarded as a computational formalism. Two (parallel) derivations should be considered as equivalent when they apply the same productions to the "same" subgraph of a certain graph, although the order in which the productions are applied may be different. This basic idea is formalized in the literature through the *shift equivalence* introduced below. The shift equivalence is based on the possibility of sequentializing a parallel direct derivation and on the inverse construction, which is possible only in the case of *sequential independence*.

**Definition 7 (sequential independence).** Given a derivation $G \Rightarrow_{q'} X \Rightarrow_{q''} H$ (as in Figure 3), it is **sequential independent** iff there exist graph morphisms $f : L_2 \to X_2$ and $t : R_1 \to Y_1$ such that $j \circ f = s$ and $i \circ t = r$. □

The next well-known results state that every parallel direct derivation can be sequentialized in an arbitrary way as the sequential application of the component

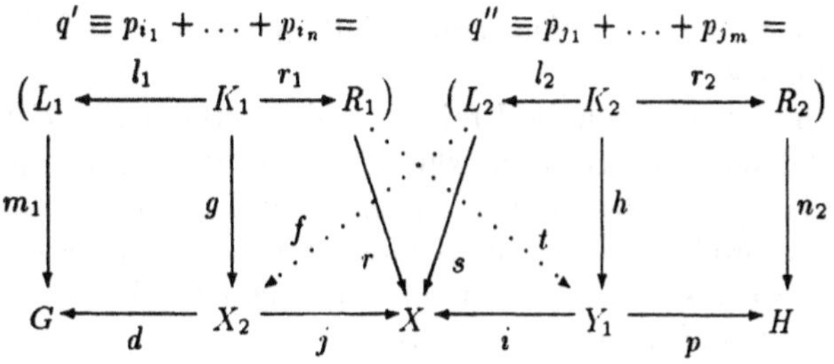


**Fig. 3.** A sequential independent derivation.

productions, and, conversely, that every sequential independent derivation can be transformed into a parallel direct derivation. These constructions are non-deterministic, and are used to define suitable relations among derivations. Once again, the quest for an explicit handling of isomorphisms results in statements that slightly differ from the usual ones.

**Proposition 8 (analysis of parallel direct derivations).** *Let $\rho = (G \Rightarrow_q H)$ be a parallel direct derivation using the parallel production $p_1 + \ldots + p_k = (L \xleftarrow{l} K \xrightarrow{r} R)$. Then for each partition $\langle I = \{i_1, \ldots, i_n\}, J = \{j_1, \ldots, j_m\}\rangle$ of $\{1, \ldots, k\}$ (i.e., $I \cup J = \{1, \ldots, k\}$ and $I \cap J = \emptyset$) there is a constructive way to obtain a sequential independent derivation $\rho' = (G \Rightarrow_{q'} X \Rightarrow_{q''} H)$, called an* **analysis** *of $\rho$, where $q' = p_{i_1} + \ldots + p_{i_n}$, and $q'' = p_{j_1} + \ldots + p_{j_m}$. Such a construction is in general not deterministic (graph $X$ is determined only up to isomorphisms), and thus it induces a relation among derivations. If $\rho$ and $\rho'$ are as above, we shall write $\rho' \in ANAL_I(\rho)$ (or $\langle \rho', \rho \rangle \in ANAL_I$); moreover, if $\rho_1 = \rho_2 ; \rho ; \rho_3$, $\rho_1' = \rho_2 ; \rho' ; \rho_3$, and $\rho_2$ has length $i - 1$ then we will write $\rho_1' \in ANAL_I^i(\rho_1)$, indicating that $\rho'$ is an analysis of $\rho$ at step $i$.* □

**Proposition 9 (synthesis of sequential independent derivations).** *Let $\rho = (G \Rightarrow_{q'} X \Rightarrow_{q''} H)$ be a sequential independent derivation. Then there is a constructive way to obtain a parallel direct derivation $\rho' = (G \Rightarrow_{q'+q''} H)$, called a* **synthesis** *of $\rho$. Also this construction is in general not deterministic and it induces a relation among derivations. If $\rho$ and $\rho'$ are as above, we shall write $\rho' \in SYNT(\rho)$; moreover, if $\rho_1 = \rho_2 ; \rho ; \rho_3$, $\rho_1' = \rho_2 ; \rho' ; \rho_3$ and $\rho_2$ has length $i - 1$ then we will write $\rho_1' \in SYNT^i(\rho_1)$.* □

**Definition 10 (shift equivalence).** Let $ANAL$ be the union of all analysis relations (i.e., $ANAL = \bigcup\{ANAL_I^i \mid i \in \mathbb{N} \wedge I \subseteq \mathbb{N}\}$), and similarly let $SYNT$ be the union of all synthesis relations ($SYNT = \bigcup\{SYNT^i \mid i \in \mathbb{N}\}$). The smallest equivalence relation on derivations containing $ANAL \cup SYNT$ is called **shift equivalence** and is denoted by $\equiv_{SHIFT}$. If $\rho$ is a derivation, by $[\rho]_{SHIFT}$ we denote the equivalence class containing all derivations shift-equivalent to $\rho$. □

From the last definitions it follows immediately that $\rho \equiv_{SHIFT} \rho'$ implies

$\sigma(\rho) = \sigma(\rho')$ and $\tau(\rho) = \tau(\rho')$. The name "shift equivalence" is fully justified in Proposition 12 below.

As said in the Introduction, in [Kre77] it is shown that each derivation has a shift-equivalent canonical derivation, which is unique up to isomorphisms. Such a canonical derivation can be obtained by repeatedly "shifting" the applications of the individual productions as far as possible towards the beginning of the derivation. It is worth stressing that, unlike the original definition, we explicitly consider the deletion of empty direct derivations.

**Definition 11 (shift relation, canonical derivations).** For each pair of natural numbers $\langle i, j \rangle$ greater than 0, the relation $SHIFT_j^i$ is defined as $SHIFT_j^i = SYNT^{i-1} \circ ANAL_{\{j\}}^i$ ($\circ$ is the standard composition of relations). Moreover, for each number $i > 0$, relation $SHIFT_0^i$ is defined as $\langle \rho_1, \rho_2 \rangle \in SHIFT_0^i$ iff $\rho_1$ has length $i$, $\rho_1 = \rho; G \Rightarrow_q H$, $\rho_2 = \rho; G \Rightarrow_q X \Rightarrow_\emptyset H$, and the double pushout $G \Rightarrow_q H$ is obtained from $G \Rightarrow_q X$ by composing the rhs pushout with isomorphism $b \circ d^{-1}$, where $\langle D, d : D \to X, b : D \to H \rangle$ is the triple corresponding to the empty direct derivation $X \Rightarrow_\emptyset H$ by Fact 4.

Relation $<_{SHIFT}$ is defined as the union of relations $SHIFT_j^i$ for each $i$, $j \in \mathbb{N}$. The transitive and reflexive closure of $<_{SHIFT}$, denoted $\leq_{SHIFT}$, is called the **shift relation**. A derivation is **canonical** iff it is minimal with respect to $\leq_{SHIFT}$. □

We have that $\rho_1 <_{SHIFT} \rho_2$ iff $\rho_1$ is obtained from $\rho_2$ either by removing the last direct derivation, if it is empty and $\rho_1$ has length at least one, or by moving the application of a single production one step towards left. It is worth stressing here that for each $i \in \mathbb{N}$, relation $SHIFT_0^i$ is deterministic, in the sense that $\langle \rho_1, \rho_2 \rangle, \langle \rho_1', \rho_2 \rangle \in SHIFT_0^i$ implies $\rho_1 = \rho_1'$.

The next proposition justifies the name of the equivalence introduced in Definition 10.

**Proposition 12 (shift equivalence as the closure of shift relations).** *The smallest equivalence relation containing the $\leq_{SHIFT}$ relation is exactly relation $\equiv_{SHIFT}$ introduced in Definition 10.* □

We present now two important results taken from [Kre77]. The first one states that the shift relation enjoys a Church-Rosser property; the second one guarantees the existence of a shift-equivalent canonical derivation for any given derivation. Since we do not allow to reason up to isomorphism, the theorem of uniqueness of canonical derivations presented in [Kre77] cannot be used. Nevertheless, we will show that similar uniqueness results hold with respect to suitable notions of equivalence among derivations, as discussed in Section 4.

**Proposition 13 (properties of the shift relation).**

1. *Relation $\leq_{SHIFT}$ enjoys the following Church-Rosser property: if $\rho_1 \in SHIFT_j^i(\rho)$ and $\rho_2 = SHIFT_h^k(\rho)$, then there exist a derivation $\rho_3$ and natural numbers $i'$, $j'$, $k'$ and $h'$ such that $\rho_3 \in SHIFT_{j'}^{i'}(\rho_2)$ and $\rho_3 \in SHIFT_{h'}^{k'}(\rho_1)$.*

2. *Relation $\leq_{SHIFT}$ is well-founded, thus for each derivation $\rho$ there exists a canonical derivation $\rho'$ such that $\rho' \leq_{SHIFT} \rho$.*

*Proof.* For the first point, see Lemma 4.7 of [Kre77]. For the second point, see Lemma 4.6 and Theorem 5.3 therein. □

We introduce now two categories having concrete graphs as objects. The first one has all concrete graph derivations as arrows; the second one is obtained from the first one by making a congruence on the arrows with respect to the shift equivalence. These categories will be related in Section 5 with two corresponding categories of abstract derivations.

**Definition 14 (categories of concrete derivations).** Given a graph grammar $\mathcal{G}$, the **category of concrete derivations of** $\mathcal{G}$, denoted $\mathbf{Der}(\mathcal{G})$, is the category having all graphs as objects, and where $\rho : G \to H$ is an arrow iff $\rho$ is a derivation, $\sigma(\rho) = G$ and $\tau(\rho) = H$ (see Definition 6). Arrow composition is defined as the sequential composition of derivations, and the identity of object $G$ is the empty derivation $G : G \Rightarrow^* G$.

The **category of concrete derivations up to shift equivalence**, denoted $\mathbf{Der}(\mathcal{G})/_{SHIFT}$, has graphs as objects, and as arrows equivalence classes of derivations with respect to the shift equivalence. More precisely, $[\rho]_{SHIFT} : G \to H$ is an arrow of $\mathbf{Der}(\mathcal{G})/_{SHIFT}$ iff $\sigma(\rho) = G$ and $\tau(\rho) = H$. The composition of arrows is defined as $[\rho]_{SHIFT} ; [\rho']_{SHIFT} = [\rho ; \rho']_{SHIFT}$, and the identity of $G$ is the equivalence class $[G]_{SHIFT}$, which contains only the empty derivation $G$.

We denote by $[\_]_{SHIFT}$ the obvious functor from $\mathbf{Der}(\mathcal{G})$ to $\mathbf{Der}(\mathcal{G})/_{SHIFT}$ which is the identity on objects, and maps each concrete derivation to its shift equivalence class. □

## 3 Requirements for equivalences on derivations.

Since the pushout object of two arrows is unique up to isomorphism, given a graph $G$ and a production $p$, if $G \Rightarrow_p H$, then $G \Rightarrow_p H'$ for each graph $H'$ isomorphic to $H$. Therefore the application of a production to a graph can produce an unbounded number of different results. This fact is highly counter-intuitive, because in the above situation one would expect a deterministic result, or, at most, some finite set of possible outcomes. Indeed, in the algebraic approach to graph grammars, one usually considers a concrete graph as a specific representation of a "system state", and since any kind of abstract semantics should be representation independent, one handles (more or less explicitly) ***abstract graphs***, i.e., isomorphism classes of concrete graphs: with this choice, a direct derivation becomes clearly deterministic.

This solution is satisfactory if one is interested just in the graphs generated by a grammar. However, if one considers instead a kind of semantics which associates to each grammar all the possible derivations (like the truly-concurrent semantics we are interested in) it becomes mandatory to reason not only in terms of abstract graphs, but also in terms of *abstract derivations*, i.e., equivalence classes of derivations with respect to a suitable equivalence.

The issue of defining a reasonable notion of equivalence on derivations has been addressed in a partial way in the literature. On the one hand, great emphasis has been placed on the shift equivalence presented in Definition 10, which has been recognized to capture the essentials of the semantics of parallel graph derivations. On the other hand, however, the issue of considering derivations in a representation independent way has been addressed only in a naive way. The shift equivalence itself is clearly not representation independent, because it is able to relate only derivations starting from and ending with the same concrete graphs.

In the following we first formalize the notion of "isomorphism of derivations", showing that the obvious definition based on the categorical notion of isomorphism of diagrams is not acceptable from a semantical point of view, because it would identify too many derivations. Next we will introduce some reasonable requirements for equivalences on derivations, formalizing them as suitable algebraic properties that such an equivalence should satisfy. It is worth stressing again that in this and in the next section we consider equivalences on derivations which take care only of abstraction, i.e., which relate only derivations which differ for the concrete representations of the states, and possibly for the representation of the applied productions. Let us introduce the first of these equivalences.

**Definition 15 (isomorphism of derivations).** Let $\rho : G_0 \Rightarrow^* G_n$ and $\rho' : G'_0 \Rightarrow^* G'_{n'}$ be two derivations as depicted in Figure 4. They are **isomorphic** (written $\rho \equiv_0 \rho'$) if $n = n'$ and there exists a family of isomorphisms between corresponding graphs such that the resulting diagram commutes. □

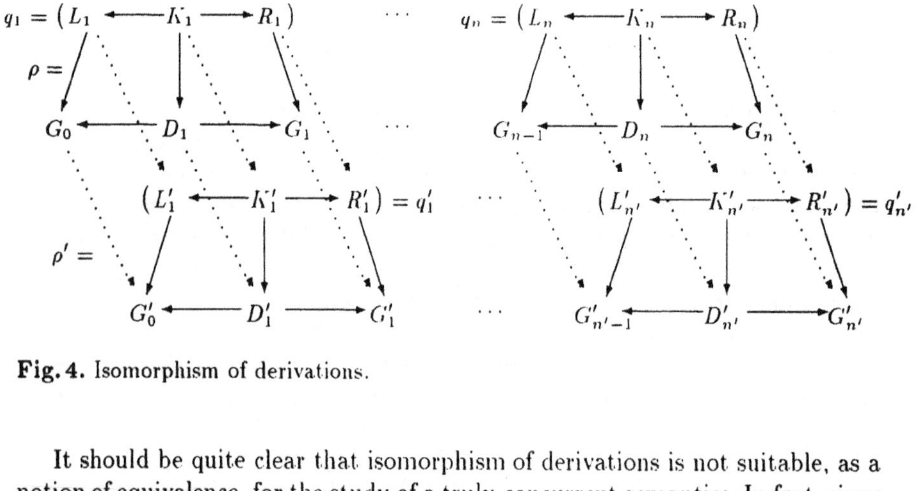


**Fig. 4.** Isomorphism of derivations.

It should be quite clear that isomorphism of derivations is not suitable, as a notion of equivalence, for the study of a truly-concurrent semantics. In fact, since it makes no reference to the productions applied during the derivations, it identifies derivations which can hardly be considered as differing just for representation issues. For example, if a grammar $\mathcal{G}$ contains two isomorphic productions $p$ and $p'$, two direct derivations using $p$ and $p'$ would be considered as equivalent. Even worst, two parallel productions $q = p_1 + \ldots + p_k$ and $q' = p'_1 + \ldots + p'_{k'}$ may happen to be isomorphic also if some of the involved sequential productions are distinct or if $k \neq k'$ i.e., if the number of applied productions is different. Since in any reasonable event-based truly-concurrent semantics of a graph grammar an event would essentially model the application of a production to a graph, this means that an equivalence based on isomorphisms of derivations could be able to identify derivations which have a different number of associated events. As a consequence, in the next section we will propose other equivalences on derivations which take care explicitly of the productions applied at each direct derivation: this is the main reason why in Definition 2, unlike the related literature, we labeled a direct derivation with the (parallel) production applied.

In the rest of this section, we establish some further requirements for a notion of equivalence on derivations which aims at capturing representation independence. Throughout this section $\approx$ denotes an equivalence relation on derivations, and an ***abstract*** derivation, denoted $[\rho]$, is an equivalence class of derivations modulo $\approx$. The first, obvious requirement is that the notion of abstract derivation be consistent with the notion of abstract graph, i.e., any abstract derivation should start from an abstract graph and should end in an abstract graph.

**Definition 16 (well-defined equivalences).** For each abstract derivation $[\rho]$, define $\sigma([\rho]) = \{\sigma(\rho') \mid \rho' \in [\rho]\}$, and similarly $\tau([\rho]) = \{\tau(\rho') \mid \rho' \in [\rho]\}$. Then equivalence $\approx$ is **well-defined** if for each derivation $\rho : G \Rightarrow^* H$ we have that $\sigma([\rho]) = [\sigma(\rho)]$, and $\tau([\rho]) = [\tau(\rho)]$. □

If $\approx$ is well-defined, then for each derivation $\rho : G \Rightarrow^* H$ we can write $[\rho] : [G] \Rightarrow^* [H]$. The second requirement is that we want to be able to compose abstract derivations sequentially, i.e., if $[\rho] : [G] \Rightarrow^* [H]$ and $[\rho'] : [H] \Rightarrow^* [K]$, then we want that also $[\rho] ; [\rho'] : [G] \Rightarrow^* [K]$ be a well-defined abstract derivation. This property will allow us to define in Section 5 a category similar to **Der**($\mathcal{G}$) (see Definition 14), but having abstract graphs as objects and abstract derivations as arrows.

**Definition 17 (equivalence allowing sequential composition).** Given two abstract derivations $[\rho]$ and $[\rho']$ such that $\tau([\rho]) = \sigma([\rho'])$, their **sequential composition** $[\rho] ; [\rho']$ is defined iff for all $\rho_1, \rho_2 \in [\rho]$ and $\rho'_1, \rho'_2 \in [\rho']$ such that $\tau(\rho_1) = \sigma(\rho'_1)$ and $\tau(\rho_2) = \sigma(\rho'_2)$, one has that $\rho_1 ; \rho'_1 \approx \rho_2 ; \rho'_2$. In this case $[\rho] ; [\rho']$ is, by definition, the abstract derivation $[\rho_1 ; \rho'_1]$.

An equivalence $\approx$ **allows for sequential composition** iff $[\rho] ; [\rho']$ is defined for all $[\rho]$ and $[\rho']$ such that $\tau([\rho]) = \sigma([\rho'])$. □

As for the third requirement, we want that uniqueness of canonical derivations holds for abstract derivations, i.e., that each abstract derivation has a unique shift-equivalent abstract canonical derivation. Since for each derivation the existence of a shift-equivalent canonical derivation is ensured by Proposition

13, this property implies that the operation $can_i : Der(\mathcal{G})/_{\approx} \rightarrow Der(\mathcal{G})/_{\approx}$ mapping each abstract derivation to its shift-equivalent abstract canonical derivation is well-defined.

**Definition 18 (uniqueness of canonical derivations).** Equivalence $\approx$ **enjoys uniqueness of canonical derivations** iff for each pair of equivalent derivations $\langle \rho, \rho' \rangle$ and for each pair $\langle \rho_c, \rho'_c \rangle$ of canonical derivations, $\rho \equiv_{SHIFT} \rho_c$ and $\rho' \equiv_{SHIFT} \rho'_c$ implies that $\rho_c \approx \rho'_c$. □

# 4 Towards a satisfactory equivalence on derivations.

In this section we introduce three different equivalences on derivations, and analyze their properties with respect to the requirements established in Section 3. Actually, since by the definitions all the equivalences are clearly well-defined in the sense of Definition 16, we will focus only on the other requirements. The first equivalence we consider is obtained from equivalence $\equiv_0$ of Definition 15 by taking care explicitly of the productions applied at each direct derivation.

**Definition 19 (equivalence $\equiv_1$).** Let $\rho : G_0 \Rightarrow^* G_n$ and $\rho' : G'_0 \Rightarrow^* G'_{n'}$ be two derivations as depicted in Figure 4, and suppose that $q_i = \langle (p_{i1}, in^{i1}), \ldots, (p_{ik_i}, in^{ik_i}) \rangle$ for each $i \in \{1, \ldots, n\}$, and $q'_j = \langle (p'_{j1}, \underline{in}^{j1}), \ldots, (p'_{jk'_j}, \underline{in}^{jk'_j}) \rangle$ for each $j \in \{1, \ldots, n'\}$. Then they are **1-equivalent** (written $\rho \equiv_1 \rho'$) if

1. $n = n'$, i.e., they have the same length;
2. for each $i \in \{1, \ldots, n\}$, $k_i = k'_i$; i.e., the number of productions applied in parallel at each direct derivation is the same;
3. there exists a family of bijective mappings $\{\Pi_1, \ldots, \Pi_n\}$ such that $\Pi_i$ is a permutation of $\{1, \ldots, k_i\}$, and $p_{ij} = p'_{i\Pi_i(j)}$ for each $i \in \{1, \ldots, n\}$ and $j \in \{1, \ldots, k_i\}$; thus the productions applied in corresponding direct derivations must be the same, up to a permutation;
4. there exists a family of isomorphisms $\{\phi_{X_i} : X_i \rightarrow X'_i\}$ with $X \in \{L, K, R\}$ and $i \in \{1, \ldots, n\}$ between corresponding graphs of the parallel productions used in the two derivations such that all triangles formed by the injections from the productions commute: more precisely, $\phi_{X_i} \circ in^{ij}_X = \underline{in}^{i\Pi(j)}_X$ for each $X \in \{L, K, R\}$, $i \in \{1, \ldots, n\}$, and $j \in \{1, \ldots, k_i\}$;
5. there exists a family of isomorphisms $\{\phi_{G_0} : G_0 \rightarrow G'_0, \phi_{X_i} : X_i \rightarrow X'_i\}$ with $X \in \{G, D\}$ and $i \in \{1, \ldots, n\}$ between corresponding graphs of the two derivations, such that all those squares commute, which are obtained by composing two corresponding morphisms $X \rightarrow Y$ and $X' \rightarrow Y'$ of the two derivations and the isomorphisms (fixed by this or by the previous point) $\phi_X$ and $\phi_Y$ relating the source and the target graphs.

We say that $\rho$ **and** $\rho'$ **are 1-equivalent via** $\{\Pi_i\}_{i \leq n}$ if $\{\Pi_i\}_{i \leq n}$ is the family of permutations of point 3 above. Given a derivation $\rho$, its equivalence class with respect to $\equiv_1$ is denoted by $[\rho]_1$, and is called a **1-abstract derivation**. □

As the following proposition shows, the above definition is slightly redundant, because the isomorphisms of point 4 are uniquely determined by the bijective mappings of point 3.

**Proposition 20 (isomorphisms determined by mappings of productions).** *Let $\rho$ and $\rho'$ be two derivations as in Definition 19 which satisfy point 1, 2, and 3. Then there exists a unique family of isomorphisms which satisfies point 4.* □

Thus, in order to prove that two derivations are 1-equivalent, it is sufficient to provide the bijective mappings of point 3 and the family of isomorphisms of point 5 of Definition 19, using for point 4 the unique family of isomorphisms determined by the last result.

Let us consider now the algebraic properties of equivalence $\equiv_1$. We prove below that $\equiv_1$ enjoys uniqueness of canonical derivations. This result is based on an important lemma which states that $\equiv_1$ behaves well with respect to analysis, synthesis, and shift of derivations.

**Lemma 21 (analysis and synthesis preserve equivalence $\equiv_1$).** *Let $\rho \equiv_1 \rho'$ via $\{\Pi_i\}_{i \leq n}$, and let $\phi_{G_0}$ and $\phi_{G_n}$ be the corresponding isomorphisms between their starting and ending graphs (see point 5 of Definition 19).*

1. *If $\rho_1 \in ANAL^i_{\{j\}}(\rho)$, then there exists at least one derivation $\rho_2 \in ANAL^i_{\{\Pi(j)\}}(\rho')$. Moreover, for each $\rho_2 \in ANAL^i_{\{\Pi(j)\}}(\rho')$, it holds that $\rho_2 \equiv_1 \rho_1$ with the same isomorphisms $\phi_{G_0}$ and $\phi_{G_n}$ relating their starting and ending graphs.*
2. *If $\rho_1 \in SYNT^i(\rho)$, then there exists at least one derivation $\rho_2 \in SYNT^i(\rho')$. Moreover, for each $\rho_2 \in SYNT^i(\rho')$, it holds that $\rho_2 \equiv_1 \rho_1$ with the same isomorphisms $\phi_{G_0}$ and $\phi_{G_n}$ relating their starting and ending graphs.*
3. *If $\rho_1 \in SHIFT^i_j(\rho)$, then there exists at least one derivation $\rho_2 \in SHIFT^i_{\Pi_i(j)}(\rho')$. Moreover, for each $\rho_2 \in SHIFT^i_{\Pi_i(j)}(\rho')$, it holds that $\rho_2 \equiv_1 \rho_1$ with the same isomorphisms $\phi_{G_0}$ and $\phi_{G_n}$ relating their starting and ending graphs.*

□

The observation that analysis, synthesis and shift preserve not only equivalence $\equiv_1$, but also the isomorphisms between the starting and ending graphs of equivalent derivations will be used later in Proposition 27.

A direct consequence of the last result is that the following partial function on 1-abstract derivations is well-defined: $\underline{SHIFT^i_j}([\rho]_1) = \mathit{undefined}$ if $SHIFT^i_j(\rho)$ is empty, and $\underline{SHIFT^i_j}([\rho]_1) = [\rho']_1$, if $\rho' \in SHIFT^i_j(\rho)$.[4]

**Theorem 22 ($\equiv_1$ enjoys uniqueness of canonical derivations).** *Let $\rho$, $\rho'$ be two derivations such that $\rho \equiv_1 \rho'$, and let $\rho_c$, $\rho'_c$ be two canonical derivations such that $\rho \equiv_{SHIFT} \rho_c$ and $\rho' \equiv_{SHIFT} \rho'_c$. Then $\rho_c \equiv_1 \rho'_c$.*

[4] Actually, such a definition depends on the chosen representative for the 1-abstract derivation $[\rho]_1$: for a different representative the $j$-th production of step $i$ could be different. Nevertheless, what follows does not depend on this choice.

*Proof.* Let $\sqsubset_{SHIFT}$ be the relation on 1-abstract derivations defined as $[\rho]_1 \sqsubset_{SHIFT} [\rho']_1$ if there exist natural numbers $i$ and $j$ such that $[\rho]_1 = \underline{SHIFT}^i_j([\rho']_1)$ (see the above remark), and let $\sqsubseteq_{SHIFT}$ be the reflexive and transitive closure of $\sqsubset_{SHIFT}$. By the Church-Rosser property and the well-foundedness of relation $\leq_{SHIFT}$ (Proposition 13), and by the fact that the shift relation preserves equivalence $\equiv_1$ of derivations (Lemma 21), it follows immediately that relation $\sqsubseteq_{SHIFT}$ is locally confluent and well-founded, and thus it is confluent.

By Theorem 4.5 of [Kre77], two derivations $\rho_1$ and $\rho_2$ are shift-equivalent iff there exists a derivation $\rho_3$ such that $\rho_3 \leq_{SHIFT} \rho_1$ and $\rho_3 \leq_{SHIFT} \rho_2$. Therefore, in the above hypotheses, since $\rho_c$ and $\rho'_c$ are minimal with respect to $\leq_{SHIFT}$, we have $\rho_c \leq_{SHIFT} \rho$ and $\rho'_c \leq_{SHIFT} \rho'$. Considering now the corresponding 1-abstract derivations, we have $[\rho_c]_1 \sqsubseteq_{SHIFT} [\rho]_1$ and $[\rho'_c]_1 \sqsubseteq_{SHIFT} [\rho']_1$; since $[\rho]_1 = [\rho']_1$ by hypothesis, and since $\sqsubseteq_{SHIFT}$ is confluent, it follows that $[\rho_c]_1 = [\rho'_c]_1$, i.e., $\rho_c \equiv_1 \rho'_c$. □

Despite the last result, equivalence $\equiv_1$ is not completely satisfactory, because it does not allow for sequential composition. In fact, as shown in Counterexample 1.1 of [CELMR93] (in this volume), it is possible to find two 1-equivalent derivations $\rho_1$ and $\rho'_1$ which, when composed with the same derivation $\rho$, yield derivations $\rho\,;\,\rho_1$ and $\rho\,;\,\rho'_1$ which are not 1-equivalent. In that paper we also show that the problem is due to the fact that in the categorical framework many relevant notions cannot be defined "up to isomorphism", even not arrow composition. The solution proposed there is to use a more restrictive notion of equivalence, based on "standard isomorphisms".

**Definition 23 (standard isomorphisms).** A family $s$ of **standard isomorphisms** in category **Graph** is a family of isomorphisms indexed by pairs of isomorphic graphs (i.e., $s = \{s(G, G') \mid G \cong G'\}$), satisfying the following conditions for each $G$, $G'$ and $G'' \in |\mathbf{Graph}|$:

- $s(G, G') : G \to G'$;
- $s(G, G) = id_G$;
- $s(G'', G') \circ s(G, G'') = s(G, G')$. □

**Definition 24 (Equivalence $\equiv_2$).** Let $s$ be an arbitrary but fixed family of standard isomorphisms of category **Graph**, and let $\rho$, $\rho'$ be two derivations as in Definition 19. We say that $\rho$ and $\rho'$ are **2-equivalent** (written $\rho \equiv_2 \rho'$) iff conditions 1 to 5 of Definition 19 are satisfied, and moreover all the isomorphism of point 5 are standard. An equivalence class of derivations with respect to $\equiv_2$ will be denoted by $[\rho]_2$, and will be called a **2-abstract derivation.** □

Now the fact that $\equiv_2$ allows sequential composition follows easily from the results presented in Section 2 of [CELMR93].

**Proposition 25 ($\equiv_2$ allows for sequential composition).** *Let $\rho_1, \rho'_1, \rho_2$ and $\rho'_2$ be four derivations such that $\rho_1 \equiv_2 \rho'_1$, $\rho_2 \equiv_2 \rho'_2$, $\tau(\rho_1) = \sigma(\rho_2)$ and $\tau(\rho'_1) = \sigma(\rho'_2)$. Then $\rho_1\,;\,\rho_2 \equiv_2 \rho'_1\,;\,\rho'_2$.* □

Unfortunately however, uniqueness of canonical derivations is lost for $\equiv_2$, because too many derivations are kept distinct The point is that, as stressed in Proposition 8, in the analysis construction the newly generated intermediate graph is unique only up to isomorphism, and not up to *standard* isomorphism. Therefore it is easy to build a counterexample showing that two different analyses of the same derivation are not 2-equivalent, and that so are the two canonical derivations eventually reached. Thus also equivalence $\equiv_2$ is not satisfactory. However, on the one hand it is easy to see that standard isomorphisms are used, in the sequential composition of abstract derivations, just at the extremes. On the other hand, the shift relation preserves the isomorphisms relating the starting and ending graphs of two equivalent derivations (Lemma 21). These observations motivate the choice of the next equivalence.

**Definition 26 (Equivalence $\equiv_3$).** Let $s$ be an arbitrary but fixed family of standard isomorphisms of category **Graph**, and let $\rho$, $\rho'$ be two derivations as in Definition 19. We say that $\rho$ and $\rho'$ are **3-equivalent** (written $\rho \equiv_3 \rho'$) iff conditions 1 to 5 of Definition 19 are satisfied, and moreover isomorphisms $\phi_{G_0}$ and $\phi_{G_n}$ of point 5 are standard. An equivalence class of derivations with respect to $\equiv_3$ will be denoted by $[\rho]_3$, and will be called a **3-abstract derivation.** □

Thus we require that only the isomorphisms relating the first and the last graphs of two 3-equivalent derivations be standard: all the other isomorphisms can be arbitrary. Equivalence $\equiv_3$ is easily shown to satisfy all the requirements of Section 3.

**Proposition 27 (Properties of equivalence $\equiv_3$).** *Equivalence $\equiv_3$ is well-defined, it allows sequential composition, and it enjoys uniqueness of canonical derivations.* □

## 5 The category of abstract derivations of a grammar.

Summarizing the results of the previous section, we defined an equivalence on derivations (that is $\equiv_3$) which allows us to compose sequentially equivalence classes of derivations, and such that for each derivation there exists a unique equivalence class of canonical derivations which are shift-equivalent to it. Therefore equivalence $\equiv_3$ allows one to abstract from the representation dependent aspects of derivations in a satisfactory way.

Using equivalence $\equiv_3$ we introduce two categories having abstract derivations as arrows, and we relate them to the categories introduced in Definition 14 via suitable functors.

**Definition 28 (the categories of abstract derivations).** Given a graph grammar $\mathcal{G}$, its **category of abstract derivations ADer($\mathcal{G}$)** is defined as follows. The objects of **ADer($\mathcal{G}$)** are abstract graphs, i.e., isomorphism classes of objects of **Graph**. Arrows of **ADer($\mathcal{G}$)** are 3-abstract derivations, i.e., equivalence classes of (parallel) derivations of $\mathcal{G}$ with respect to equivalence $\equiv_3$. The source and target mappings are given by $\sigma$ and $\tau$, respectively; thus $[\rho]_3 : \sigma([\rho]_3) \to \tau([\rho]_3)$. Composition of arrows is given by the sequential composition of the corresponding 3-abstract derivations, and for each object $[G]$ the identity arrow is the 3-abstract derivation $[G : G \Rightarrow^* G]_3$ containing the empty derivation on $G$.

Category $\mathbf{ADer}(\mathcal{G})$ is equipped with an equivalence relation $\equiv_{\underline{SHIFT}}$ on arrows, defined as $[\rho]_3 \equiv_{\underline{SHIFT}} [\rho']_3$ if $\rho \equiv_{SHIFT} \rho'$. The **category of abstract derivations up to shift equivalence** of a grammar $\mathcal{G}$, denoted $\mathbf{ADer}(\mathcal{G})/_{\underline{SHIFT}}$, has the same objects as $\mathbf{ADer}(\mathcal{G})$, and as arrows equivalence classes of arrows of $\mathbf{ADer}(\mathcal{G})$ with respect to equivalence $\equiv_{\underline{SHIFT}}$. □

It is easy to check that category $\mathbf{ADer}(\mathcal{G})$ is well-defined, thanks to the properties of equivalence $\equiv_3$. Also $\mathbf{ADer}(\mathcal{G})/_{\underline{SHIFT}}$, which is the quotient category of $\mathbf{ADer}(\mathcal{G})$ with respect to $\equiv_{\underline{SHIFT}}$, is well-defined, since $\equiv_{\underline{SHIFT}}$ is a congruence with respect to arrow compositions, that is, if $[\rho_1]_3 \equiv_{\underline{SHIFT}} [\rho'_1]_3$ and $[\rho_2]_3 \equiv_{\underline{SHIFT}} [\rho'_2]_3$, then $[\rho_1]_3 ; [\rho_2]_3 \equiv_{\underline{SHIFT}} [\rho'_1]_3 ; [\rho'_2]_3$.

The next definition and result show the relationship among the four categories introduced so far, in terms of functors relating them.

**Definition 29 (functors relating the categories of a grammar).** Let $\mathcal{G}$ be a graph grammar. The categories of derivations of $\mathcal{G}$ introduced in Defintions 14 and 28, are related by the four functors (depicted in Figure 5) defined as follows:

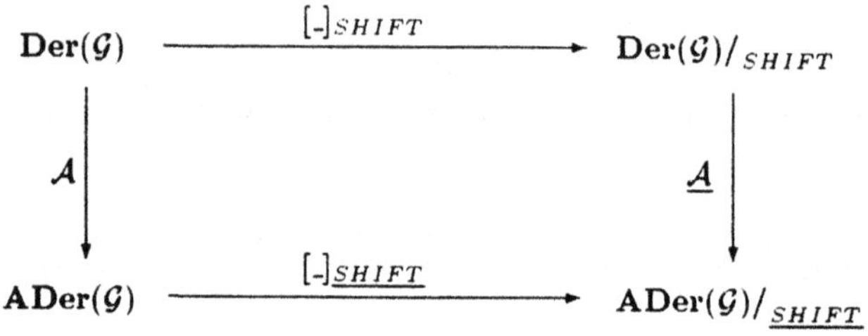


**Fig. 5.** Functors relating categories of derivations of a grammar.

- Functor $[\text{-}]_{SHIFT} : \mathbf{Der}(\mathcal{G}) \to \mathbf{Der}(\mathcal{G})/_{SHIFT}$ is as in Definition 14.
- Functor $[\text{-}]_{\underline{SHIFT}} : \mathbf{ADer}(\mathcal{G}) \to \mathbf{ADer}(\mathcal{G})/_{\underline{SHIFT}}$ is the identity on objects, and maps each 3-abstract derivation to its equivalence class with respect to $\equiv_{\underline{SHIFT}}$.
- Functor $\mathcal{A} : \mathbf{Der}(\mathcal{G}) \to \mathbf{ADer}(\mathcal{G})$ maps each object $G$ to its isomorphism class $[G]$, and each concrete derivation $\rho$ to its 3-abstract equivalence class $[\rho]_3$.
- Functor $\underline{\mathcal{A}} : \mathbf{Der}(\mathcal{G})/_{SHIFT} \to \mathbf{ADer}(\mathcal{G})/_{\underline{SHIFT}}$ maps each object $G$ to its isomorphism class $[G]$, and each arrow $[\rho]_{SHIFT}$ to the $\underline{\text{SHIFT}}$-equivalence class of the corresponding 3-abstract derivation, i.e., to $[[\rho]_3]_{\underline{SHIFT}}$. □

**Theorem 30 (properties of functors).** *All the functors of Definition 29 are well-defined, and the square of Figure 5 commutes.*

*Moreover, functor $\underline{A}$ is an equivalence of categories.* □

Let us shortly comment the relationship among the various categories of derivations. In our view, disregarding the subtleties concerning the explicit handling of the empty (direct) derivations (which were in any case necessary to formalize the present results), the picture summarizes quite well the connections between the theory of graph derivations presented in this paper and the one proposed in the literature till now.

Category $\mathbf{Der}(\mathcal{G})$ is at the most concrete level, and it is introduced just to formalize the fact that the sequential composition of concrete derivations is well-understood and is obtained simply by the juxtaposition of the corresponding diagrams. Since the shift equivalence proposed in the literature relates only derivations between the same graphs, it can be used to make a congruence on the concrete derivations yielding category $\mathbf{Der}(\mathcal{G})/_{SHIFT}$. However, this category is still too concrete, because it has all graphs as objects. Therefore from each graph there are still infinitely many "identical" derivations, differing only for the ending graph. In the other direction, we showed that if one wants to reason on derivations "up to isomorphism" both equivalences $\equiv_2$ and $\equiv_3$ can be used, because they allow for sequential composition. The choice of $\equiv_3$ is motivated by the fact that it is compatible with the shift equivalence.

Finally, thanks to the properties of $\equiv_3$, category $\mathbf{ADer}(\mathcal{G})/_{\underline{SHIFT}}$ can be defined, which seems to represent the derivations at the desired level of abstraction, both independent of representation-dependent details and consistent with the shift equivalence. The fact that functor $\underline{A}$ is an equivalence of categories essentially means that category $\mathbf{Der}(\mathcal{G})/_{SHIFT}$ already has in each homset the "right" structure, although each homset has too many (actually, infinitely many) copies.

## 6 Conclusion and future work.

In the framework of a research activity which is aimed at providing graph-grammars with a satisfactory truly-concurrent semantics, in this paper we addressed the issue of finding a satisfactory notion of equivalence on derivations. Actually, since the "shift equivalence" (defined in the literature a long time ago) is recognized to capture in a completely reasonable way the truly-concurrent aspects of graph derivations, we focused on the definition of a satisfactory equivalence which could take into account only the representation-dependent aspects of derivations. After showing that a naive definition based on isomorphisms of derivations is not satisfactory, we introduced some requirements which were formalized as algebraic properties that an equivalence should satisfy. Through subsequent refinements of the first equivalence, we eventually arrived to the definition of an equivalence satisfying all the requirements.

We considered various categories having graph derivations (at different levels of abstractions) as arrows. Two such categories, both having concrete graphs as objects and concrete derivations as arrows, are proposed as a possible formalization of the state of the art in the theory of graph derivations. Two new categories, both having abstract graphs as objects and abstract derivations as arrows, are defined by exploiting the new equivalence on derivations. Functors relating the four categories establish precise relationship between the theory of graph derivations proposed so far and the one presented in this paper.

Besides being intersting in itself as a computational model for a grammar, the definition of the category $\mathbf{ADer}(\mathcal{G})/_{\underline{SHIFT}}$ is also a fundamental step towards the definition of a truly-concurrent semantics. The definition of such a semantics, which should associate each grammar with an event structure [Win89], is the subject of an ongoing research activity of the authors.

Although in this paper we considered only the double-pushout approach, also the theories of other algebraic approaches to graph rewriting present the same problems addressed here. Therefore analogous investigations should be made for example in the framework of High-Level Replacement Systems ([EHKP91]) and of the single-poshout approach ([Löw90]).

## References


[AHS90] J. Adamek, H. Herrlich, G. Strecker, *Abstract and Concrete Categories*, Wiley Interscience, 1990.

[CELMR93] A. Corradini, H. Ehrig, M. Lowe, U. Montanari, and F. Rossi, *Note on Standard Representation of Graphs and Graph Derivations*, in this volume. Also as Technical Report 92-25, Technische Universität Berlin, 1992.

[Ehr87] H. Ehrig. Tutorial introduction to the algebraic approach of graph-grammars. In H. Ehrig, M. Nagl, G. Rozenberg, and A. Rosenfeld, editors, *Proc. 3rd International Workshop on Graph Grammars and their Application to Computer Science*, volume 291 of *LNCS*, pages 3–14. Springer-Verlag, 1987.

[EHKP91] H. Ehrig, A. Habel, H.-J. Kreowski, F. Parisi-Presicce, *Parallelism and Concurrency in High-Level Replacement Systems*, in Mathematical Structures in Computer Science, 1, 1991, pp. 361–404.

[EPS73] H. Ehrig, M. Pfender, H.J. Schneider, *Graph-grammars: an algebraic approach*, Proc, IEEE Conf. on Automata and Switching Theory, 1973, pp. 167-180.

[Kre77] H.-J. Kreowski, *Manipulation von Graph Transformationen*, Ph.D. Thesis, Technische Universität Berlin, 1977.

[Löw90] M. Löwe, *Extended algebraic graph transformation*, Ph.D. Thesis, Technische Universität Berlin, 1990.

[Win89] G. Winskel, *An Introduction to Event Structures*, in Linear Time, Branching Time and Partial Order in Logics and Models for Concurrency, volume 354 of *LNCS*. Springer-Verlag, 1989. LNCS 354, 1989.

# Note on Standard Representation of Graphs and Graph Derivations*

A. Corradini†, H. Ehrig‡, M. Löwe‡, U. Montanari†, F. Rossi†

†Dipartimento di Informatica
Corso Italia 40
56125 Pisa - ITALY
{andrea,ugo,rossi}@di.unipi.it

‡Technische Universität Berlin
Fachbereich 20, Franklinstr. 28/29
1000 Berlin 10 - GERMANY
{ehrig,loewe}@cs.tu-berlin.de

## Abstract

We show that a naive notion of abstract graph derivations, based on the idea that two derivations are equivalent iff they are isomorphic, does not allow to extend some relevant properties of concrete derivations to abstract ones. This is the main motivation for the introduction of *standard representations* of graphs, which are used to define a (more restricted) notion of equivalence among graph morphisms, direct derivations and graph derivations. The properties of the resulting category of abstract graphs are investigated in depth, and the relationship with skeleton subcategories of **GRAPHS** is worked out.

## Introduction

Considering abstract graphs and abstract graph derivations as isomorphism classes of graphs and graph derivations there are serious problems concerning representation independence in view of an abstract semantics of graph transformations. These problems can be solved by the choice of a standard representation of graphs and of standard isomorphisms, leading to a category **AGRAPHS** of abstract graphs, where abstract graph morphisms are defined as equivalence classes of concrete morphisms up to standard isomorphisms. A standard representation of graphs also determines a skeleton subcategory of **GRAPHS**, called **SGRAPHS**, and a functor F to it, such that **AGRAPHS** is exactly the quotient category of **GRAPHS** with respect to F. As a consequence, categories **SGRAPHS** and **AGRAPHS** are isomorphic, and both are suitable to define in a consistent way a notion of abstract sequential composition of graph transformations, which is an important step towards abstract semantics of graph derivations. Moreover, the equivalence among those categories and **GRAPHS** guarantees the consistency between the result of abstract graph derivations and the corresponding concrete graph derivations in **GRAPHS**. Although in this paper we concentrate on categories of graphs (which provide our motivating examples), the problems and solutions we present here apply to arbitrary categories as well, and in particular to the categories used in the framework of High Level Replacement Systems.
The paper is organized as follows. In Section 1 we analyze the problems that arise if one considers abstract graph derivations as isomorphism classes of concrete graph derivations: some relevant notions like sequential independence cannot be extended to abstract derivations. Section 2 is devoted to the definition of standard representation of graphs and standard isomorphisms, which allow us to define the category of abstract graphs, where abstract graph derivations can be defined is a sound way. In Section 3 we prove the

---

* Research partially supported by the COMPUGRAPH Basic Research Esprit Working Group n. 7183.

existence of standard representations of graphs, and characterize them in terms of functors from **GRAPH** to a skeletal subcategory **SGRAPH.**
This note was initiated during a COMPUGRAPH-meeting of the authors in Pisa concerning abstract semantics of graph transformations and is the first part of a joint work on this subject. The results of this paper are exploited in the paper [CELMR 93] (in this volume), which studies various definitions of abstract graph derivations and their adequacy for the definition of a truly concurrent semantics of graph transformations and grammars using prime algebraic domains and prime event structures.

## Acknowledgements

Once again we are grateful to Helga Barnewitz for excellent typing and drawing of figures.

## 1. Problems with Abstract Graphs and Derivations

The algebraic approach to graph grammars, as introduced in [EPS 73] and [Ehr 79], is based on pushouts in the category **GRAPHS** of graphs. More precisely a graph

$$G: \Omega_E \xleftarrow{l_E} E \underset{t}{\overset{s}{\rightrightarrows}} V \xrightarrow{l_V} \Omega_V \quad \text{in } \mathbf{GRAPHS}$$

consists of sets E (edges), V (vertices) and functions s (source), t (target), $l_E$ (edge labels) and $l_V$ (vertex labels), where $\Omega_E$ and $\Omega_V$ are fixed alphabets for edge and vertex labels respectively. A graph morphism f: $G \to G'$ is a pair $f = (f_E, f_V)$ of functions such that the following diagram commutes componentwise with s, s' and t, t' respectively:

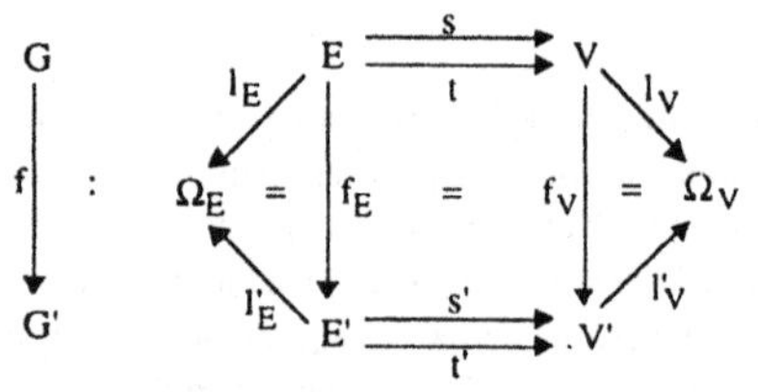

A graph production is given by a pair of graph morphisms $p = (L \xleftarrow{l} K \xrightarrow{r} R)$ where L, R and K are the left-hand side, right-hand side and interface part of the production.

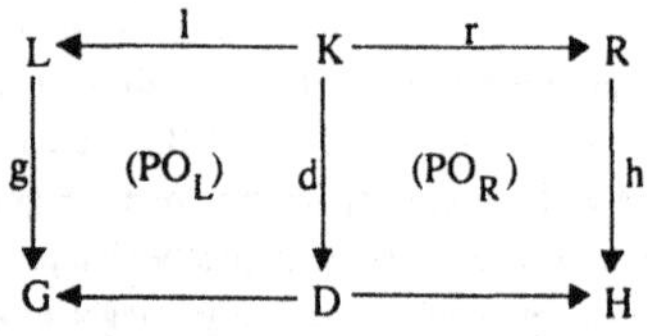

A direct graph derivation from graph G to graph H via production p and morphisms g, d and h, written

$$G \Rightarrow H \text{ via } (PO_L, PO_R),$$

is given by two pushouts $PO_L$ and $PO_R$ in the category **GRAPHS** as in the above picture, where D can be considered as the context graph, G as the gluing of L and D along K (short $G = L +_K D$), and H as the gluing of R and D along K (short $H = R +_K D$). A graph derivation is a sequence of $n \geq 0$ direct graph derivations $G_0 \Rightarrow^* G_n$.

Since pushout constructions are unique up to isomorphism and any kind of abstract semantics should be representation independent, it seems reasonable to consider graphs and direct derivations up to isomorphism and to deal directly with abstract graphs and abstract direct derivations, defined as isomorphism classes of concrete graphs and concrete direct derivations, respectively. The natural consequence of this choice would be to define an abstract graph derivation as a sequence of abstract direct derivations. However, this straightforward approach is not adequate, in the sense that some relevant properties of graph derivations (like sequential independence) cannot be defined for such abstract derivations, as shown in the following counterexample.

**1.1 Counterexample** *(properties of naive abstract derivations are not well-defined)*

The following (concrete) graph derivation sequence

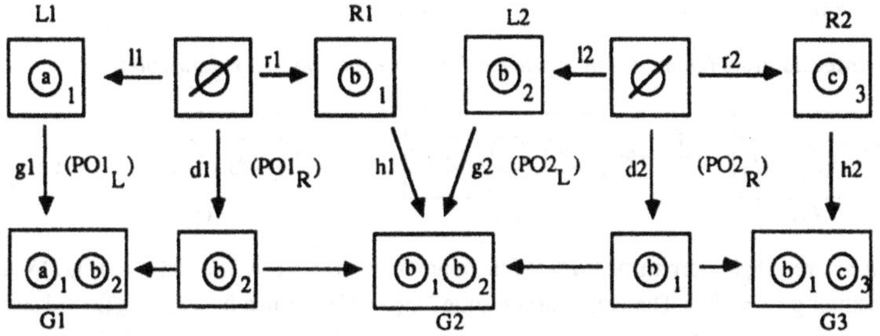


where $h_1(1) = 1$ and $g_2(2) = 2$ is sequentially independent [Ehr 79], because the intersection $h_1(R_1) \cap g_2(L_2)$ in $G_2$ of the occurrence of the right-hand side of the first production ($h_1(R_1)$) with the occurrence of the left-hand side of the second production ($g_2(L_2)$) is empty.

If we replace the second direct derivation $G_2 \Rightarrow G_3$ via ($PO2_L$), ($PO2_R$) by the equivalent (isomorphic) concrete direct derivation $G_2 \Rightarrow G'_3$ via ($PO2'_L$), ($PO2'_R$), where $g_2'(2) = 1$ we obtain the following graph derivation sequence

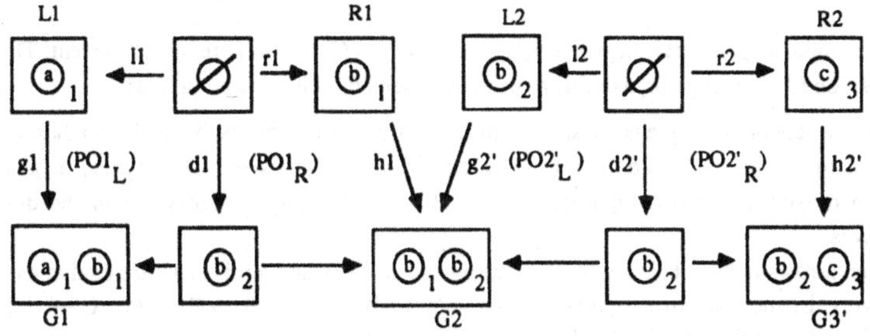


which is sequentially dependent, because the intersection $h_1(R_1) \cap g_2'(L_2)$ consists of the node 1 of $G_2$ which is produced by the first direct derivation and is consumed by the second one.

If one defines abstract direct derivations as isomorphism classes of direct derivations and abstract derivations as sequences of abstract direct derivations, then the two derivation sequences just shown would be distinct representative of the same abstract derivation, because they differ just for the second direct derivation steps, which are indeed isomorphic. As a consequence, there would exist abstract derivation sequences including at the same time representatives which are sequentially independent, and representatives which are not. This shows that the definition of sequential independence cannot be extended to such abstract derivation sequences in the natural way, i.e., considering the corresponding property on concrete representatives. The immediate, negative consequence of this fact is that a relevant part of the theory of graph derivations, which is based on the notion of parallel and sequential independence [Ehr 79], cannot be applied to those abstract derivations. ♦

Counterexample 1.1 shows an undesirable property of abstract graph derivations defined via abstract direct graph derivations. It is worth stressing that this problem is not due to the algebraic approach to graph grammars using pushouts in the category **GRAPHS**, but is instead related to the naive notion of 'abstractness' we have considered. In fact, a related problem arises in a much simpler situation: if we define an abstract graph morphism $[f: G_1 \to G_2]$ as the class of all graph morphisms $f': G'_1 \to G'_2$ such that there are graph isomorphisms $i_1: G_1 \to G'_1$ and $i_2: G_2 \to G'_2$ such that $i_2 \circ f = f' \circ i_1$, then we even fail to define a notion of composition of such abstract morphisms based on representatives, as shown in the next counterexample.

**1.2 Counterexample** *(composition of naive abstract morphisms is not definable)*

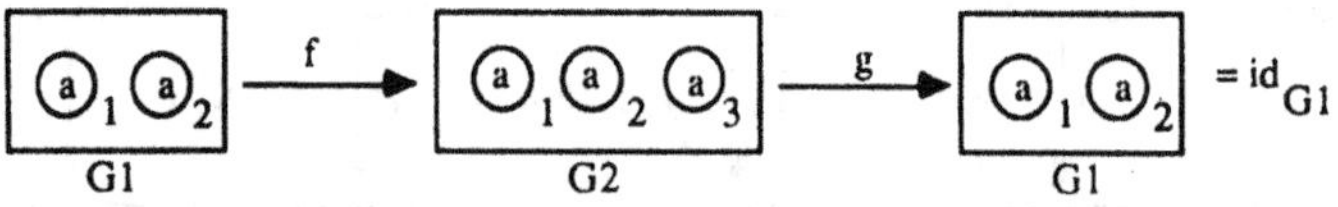


Consider the above graph morphisms (in fact functions) with $f(i) = i$ $(i = 1, 2)$, $g(1) = 1$, $g(2) = g(3) = 2$, with composition $g \circ f = id_{G_1}$. The graph morphism $g': G_2 \to G_1$ defined by $g'(1) = g'(2) = 1$ and $g'(3) = 2$ belongs to the same abstract graph morphism as g (accordingly to the above naive definition), because $i_1 \circ g = g' \circ i_2$ using the isomorphisms $i_1$ on $G_1$ and $i_2$ on $G_2$ defined by $i_1(1) = 2$, $i_1(2) = 1$ and $i_2(1) = 3$, $i_2(2) = 2$, $i_2(3) = 1$. Now, if we replace g by g' in the above diagram, we obtain $g' \circ f = const_1$, with $const_1(1) = const_1(2) = 1$.

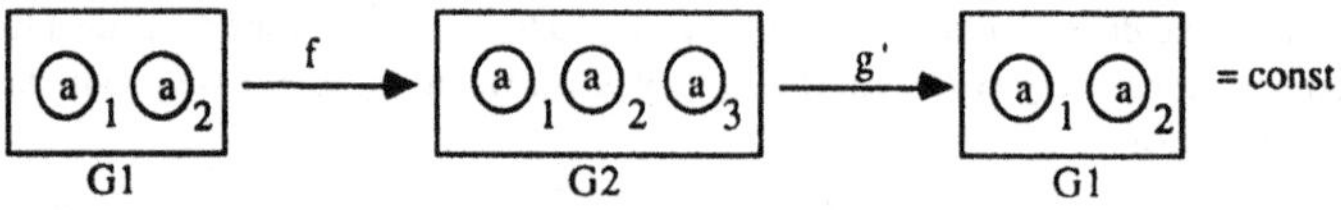


Clearly the composite graph morphisms $id_{G_1}$ and $const_1$ from $G_1$ to itself are not equivalent. This shows that composition of abstract graph morphisms using representatives is not well-defined.

Note that the last counterexample is not specific to abstract graph morphisms, but it also applies to 'abstract functions' in the category **SETS** of sets. Abstracting out from a specific category, in general the composition of abstract arrows (defined as above) in an arbitrary category cannot be defined on representatives.

In fact, given arrows $f: A \to B$ and $g: B \to C$, arrows $f': A' \to B'$ and $g': B' \to C'$ are equivalent to f and g, respectively, iff there are isomorphisms $i_1: A \to A'$, $i_2: B \to B'$, $i_3: B \to B'$ and $i_4: C \to C'$ such that diagrams (1) and (2) below commute. However, in general the outer diagram does not commute because the isomorphisms $i_2$ and $i_3$ may be different.

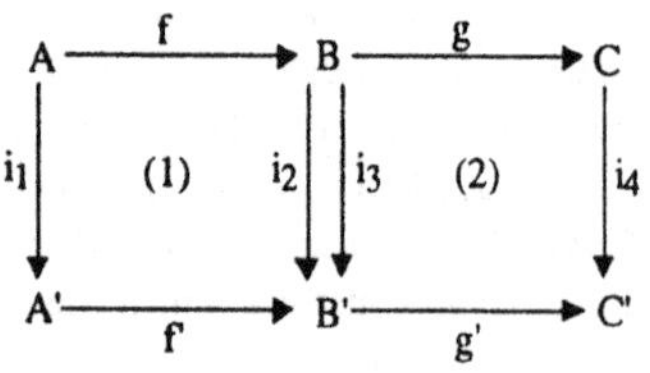

It is worth stressing here that the problems we presented in this section are not new for the Category Theory community [Her 92], although to our knowledge they have not been studied in detail in the literature. Their consequences in the theory of graph grammars are analized here for the first time.

## 2. Abstract Graphs and Derivations based on Standard Representations of Graphs

In order to solve the problems with abstract graphs, graph morphisms and graph derivations discussed in the previous section we introduce here standard representations of graphs and standard isomorphisms between isomorphic graphs, leading to a category **AGRAPHS** of abstract graphs and abstract graph homomorphisms. The problems mentioned above are avoided here by stating that two graph homomorphisms are equivalent (i.e., they are the same arrow in **AGRAPHS**) if and only if they are the same up to a *standard* isomorphism.

**2.1 Definition** *(standard representation of graphs)*

A standard representation (S, s) of graphs consists of a function S defining for each graph G in **GRAPHS** a graph S(G), called standard representation of G, such that

(a) $G \cong S(G)$

(b) $G \cong G'$ implies $S(G) = S(G')$,

and a function s defining for each pair (G, G') of isomorphic graphs in **GRAPHS** an isomorphism $s(G, G'): G \to G'$, called the standard isomorphism of (G, G'), such that

| | | |
|---|---|---|
| (c) | $s(G, G) = id_G$ | for all graphs G |
| (d) | $s(G', G'') \circ s(G, G') = s(G, G'')$ | for all graphs G, G', G'' with $G \cong G' \cong G''$ |
| (e) | $s(G, G') = s(G', G)^{-1}$ | for all graphs G, G' with $G \cong G'$. ♦ |

Note that condition (e) above follows from conditions (c) and (d): it is explicitly mentioned here just to make references easier.

Standard representations of graphs are used to define abstract graphs (abstract graph morphisms) as equivalence classes of graphs (graph morphisms) w.r.t. standard isomorphisms. The category **AGRAPHS**, which has abstract graphs as objects and abstract graph morphisms as arrows, is defined essentially as a quotient category of **GRAPHS** with respect to standard isomorphisms. This definition avoids the problems discussed in the counterexamples in the previous section, as we will show below.

**2.2 Definition** *(abstract graphs and morphisms)*

Given a standard representation (S, s) of graphs we define:

1. An abstract graph [G] is the class

$$[G] = \{G' \,/\, G' \cong G \text{ in } \mathbf{GRAPHS}\}$$

of all graphs G' which are isomorphic to graph G.

2. Two morphisms $f: G_1 \to G_2$ and $f': G_1' \to G_2'$ are called standard isomorphic if we have

$$s(G_2, G_2') \circ f = f' \circ s(G_1, G_1')$$

using the standard isomorphisms $s(G_i, G_i'): G_i \to G_i'$ $(i = 1, 2)$. An abstract graph morphism $[f: G_1 \to G_2]$ is the class

$$[f: G_1 \to G_2] = \{f': G_1' \to G_2' \,/\, f' \text{ is standard isomorphic to } f\}.$$

3. The category **AGRAPHS** of <u>abstract graphs</u> w.r.t. (S, s) consists of all abstract graphs as objects, all abstract graph morphisms $[f: G_1 \to G_2]: [G_1] \to [G_2]$ as morphisms, identities $id_{[G]} = [id_G]$, and composition of morphisms defined by representatives, i.e.

$$[G1] \xrightarrow{[f1]} [G2] \xrightarrow{[f2]} [G3] = [G1] \xrightarrow{[f2 \circ s(G2, G2^*) \circ f1]} [G3]$$

for $f_1: G_1 \to G_2$, $f_2: G_2^* \to G_3$ with $[G_2] = [G_2^*]$ and standard isomorphism $s(G_2, G_2^*)$. ♦

**Remarks**

1. In the definition of abstract graphs [G] we can replace $G' \cong G$ by existence of a standard isomorphism s(G', G), because by definition such an isomorphism exists iff G and G' are isomorphic.

2. In the definition of abstract graph morphisms it is important that we require "f' standard isomorphic to f" instead of "f' isomorphic to f" to avoid the problems discussed in counterexample 1.2. ♦

The next proposition shows that the category **AGRAPHS** of abstract graphs is well-defined.

**2.3 Proposition** *(category of abstract graphs)*

The category **AGRAPHS** of abstract graphs w.r.t. a standard representation (S, s) of graphs as given in 2.2 is well-defined.

**Proof**

For the well-definedness of the category **AGRAPHS** it is sufficient to show that the definition of composition is representation independent, because the other conditions are obviously satisfied. Let $f_1$ and $f_2$ be as in 2.2.3, and let us assume that we have other representatives $f_1': G_1' \to G_2' \in [f_1]$ and $f_2': G_2'^* \to G_3' \in [f_2]$. Then the commutativity of the outer diagram below shows representation independence, where (1) (resp. (2)) commutes by standard equivalence of $f_1$ and $f_1'$ (resp. $f_2$ and $f_2'$) and (3) commutes by the transitive closure of standard isomorphisms (2.1 (d)).

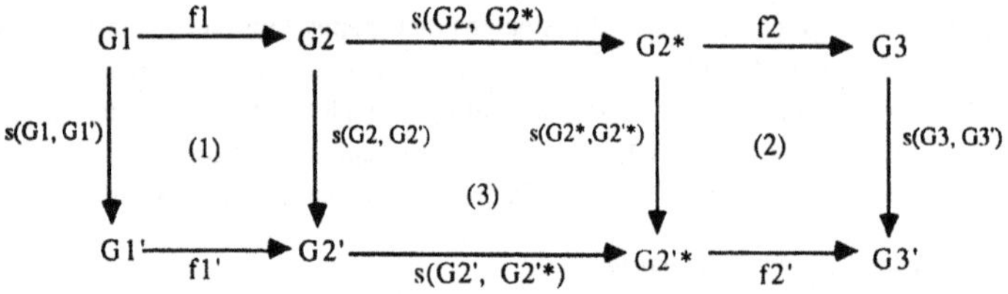


□

**2.4 Definition** *(abstract derivations of graphs)*

Given the category **AGRAPHS** of abstract graphs w.r.t. a standard representation (S, s) of graphs we define:

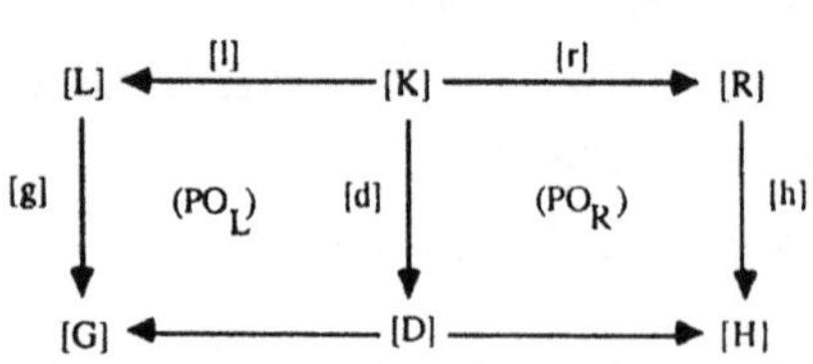

1. A direct abstract derivation $[G] \Rightarrow [H]$ of abstract graphs is a pair of pushouts $PO_L$ and $PO_R$ in the category **AGRAPHS**, as in the above diagram.
2. An abstract derivation $[G_0] \Rightarrow [G_1] \Rightarrow \ldots \Rightarrow [G_n]$, shortly $[G_0] \Rightarrow^* [G_n]$, of abstract graphs is a sequence of $n \geq 0$ direct abstract derivations.
3. The sequential composition of two abstract derivations $[G_1] \Rightarrow^* [G_2]$ and $[G_2] \Rightarrow^* [G_3]$ is the abstract derivation $[G_1] \Rightarrow^* [G_2] \Rightarrow^* [G_3]$. ♦

**Remark**

1. All the constructions above are well-defined because the category **AGRAPHS** is well-defined (see 2.3). In particular, the pushout constructions in **AGRAPHS** can be made on representatives, i.e., on concrete graphs (objects of category **GRAPHS**). The fact that categories **GRAPHS** and **AGRAPHS** are equivalent, as we will show in the next section, guarantees the consistency of pushout objects computed in the two categories. We remind here that two categories **B** and **C** are equivalent if there exist functors $F: \mathbf{B} \to \mathbf{C}$ and $G: \mathbf{C} \to \mathbf{B}$ such that $F \circ G$ is natural isomorphic to the identity functor of category **C**, and similarly $G \circ F$ is natural isomorphic to $Id_{\mathbf{B}}$.
2. Direct abstract derivations and abstract derivations are exactly direct derivations and derivations in the category **AGRAPHS** in the sense of High-Level-Replacement-Systems (see [EHKP 91]). ♦

We revisit now the two counterexample of Section 1, showing that they do not apply to derivations (resp. composition of arrows) performed in category **AGRAPHS**. The essential point is that equivalence of arrows in **AGRAPHS** is not defined w.r.t. arbitrary graph isomorphisms, but only w.r.t. standard isomorphisms.

**2.5 Examples** *(counterexamples of Section 1 revisited)*

1. In counterexample 1.1 we have given two concrete derivation sequences $G_1 \Rightarrow^* G_3$ (via $(PO1_L)$, $(PO1_R)$, $(PO2_L)$, and $(PO2_R)$) and $G_1 \Rightarrow^* G_3$ (via $(PO1_L)$, $(PO1_R)$, $(PO2'_L)$, and $(PO2'_R)$), which were equivalent according to the naive definition of abstract derivation, since $(PO2'_L)$ and $(PO2'_R)$ were isomorphic to $(PO2_L)$ and $(PO2_R)$, respectively. This was undesirable, because the first derivation was sequentially independent while the second was sequentially dependent. The counterexample does not apply anymore if we consider abstract derivations as defined in 2.4.2: in fact, for every choice of a standard representation of graphs (S, s) the two derivations are not equivalent, because $(PO2_L)$ and $(PO2'_L)$ are not standard isomorphic, since the unique standard isomorphism from $G_2$ to $G_2$ is the identity (by 2.1 (c)). Therefore, the two mentioned concrete derivation sequences belong to two distinct abstract derivations. Actually, it is not difficult to prove (but this goes beyond the scope of this paper) that the definition of sequential independence can be extended to abstract derivations as defined in 2.4.2, in the sense that the concrete derivation sequences belonging to an abstract derivation are either all sequential dependent, or they are all sequential independent.
2. In counterexample 1.2 we have shown that using a naive definition of abstract graph morphisms, the composition of such morphism is not well-defined. The counterexample does not apply to abstract graph morphisms as defined in 2.2.2, because morphisms g and g' are isomorphic but not standard isomorphic. In fact, the standard isomorphisms on $G_2$ and $G_1$ are the identities $id_{G_2}$ and $id_{G_1}$. This means that the abstract morphisms [g] and [g'] are different, thus it is not a problem anymore that the abstract compositions $[g] \circ [f] = [id_{G1}]$ and $[g'] \circ [f] = [const_1]$ are leading to different abstract morphisms. ♦

## 3. Existence and Characterization of Standard Representations

In this section we first address the issue of the existence of standard representations of graphs. Actually, since this is not needed for the goals of this paper, the problem of providing an explicit construction of a standard representation of graphs is not addressed here. Anyway this seems to be a non-trivial task.
Next we analyze the relationship between the category **GRAPHS** and the category **AGRAPHS** of abstract graphs with respect to a standard representation of graphs (S, s). We will show that (S, s) defines a functor (actually an equivalence) from category **GRAPHS** to a skeleton subcategory of it, **SGRAPHS**, such that **AGRAPHS** is the quotient category of **GRAPHS** w.r.t. that functor. The fact that **AGRAPHS** is isomorphic to **SGRAPHS** and the equivalence between **GRAPHS** and **AGRAPHS** follow then easily.
We prove that also the converse holds, i.e., that every equivalence between **GRAPHS** and a skeleton subcategory of it determines a standard representation of graphs. This provides an interesting characterization of standard representations of graphs.

**3.1 Theorem** *(existence of standard representations of graphs)*

There exists a standard representation (S, s) of graphs.

**Proof**

By the Axiom of Choice for conglomerates (see [AHS 90]) applied to the equivalence relation "is isomorphic to" on the class of all graphs we obtain for each isomorphism class [G] of graphs a graph S(G) $\in$ [G]. This defines a function S on all graphs such that $G \cong S(G)$, and $G \cong G'$ implies $S(G) = S(G')$, because $[G] = [G']$ (therefore conditions (a) and (b) of Definition 2.1 are satisfied).
Again by the Axiom of Choice we choose for each graph G from the class of all isomorphisms $\{i: S(G) \xrightarrow{\sim} G\}$ one representative, called standard isomorphism s(S(G), G), such that $s(S(G), S(G)) = id_{S(G)}$.
Then we define for each pair of isomorphic graphs $G \cong G'$ the standard isomorphism s(G, G') by

$$s(G, G') = G \xrightarrow{s(S(G),G)^{-1}} S(G') \xrightarrow{s(S(G'), G')} G'$$

which is well defined because S(G) = S(G').

We have to show (c) and (d) of 2.1. In fact,

(c) $s(G, G) = s(S(G), G) \circ s(S(G), G)^{-1} = id_G$
(d) For graphs G, G', G" with $G \cong G' \cong G''$ we have $S(G) = S(G') = S(G'')$ and hence

$$s(G', G'') \circ s(G, G') = G \xrightarrow{s(S(G), G)^{-1}} S(G) = S(G') \xrightarrow{s(S(G'), G')} G'$$

$$\xrightarrow{s(S(G'), G')^{-1}} S(G') = S(G'') \xrightarrow{s(S(G''), G'')} G'' = s(G, G'')$$ □

The next definition associates with every standard representation of graphs a suitable subcategory of **GRAPHS** and a functor to it.

**3.2 Definition** *(category of standard graphs and standard representation functor)*

Given a standard representation of graphs (S, s),

1. the category **SGRAPHS** of standard graphs w.r.t. (S, s) is defined as the full subcategory of the category **GRAPHS**, consisting of all standard representations S(G) of graphs G;
2. the standard representation functor w.r.t. (S, s) is functor F: **GRAPHS** $\rightarrow$ **SGRAPHS** defined as:

$F_O(G) = S(G)$ on objects, and

$F_A(f: G \to G') = s(G', F_O(G')) \circ f \circ s(F_O(G), G): F_O(G) \to F_O(G')$ on arrows.

It is routine to check that F is well-defined. The pair (**SGRAPHS**, F) determined in this way is called the functorial presentation of (S, s) and will be denoted by $\Phi(S, s)$. ♦

Given any category **C**, a skeleton subcategory **S** of **C** is a full, isomorphism-dense subcategory of **C** in which no two distinct objects are isomorphic. **S** is isomorphism-dense in **C** means that every object in **C** is isomorphic to some object in **S** (see [AHS 90]).
The next theorem analyzes the characterizing properties of the functorial presentation of (S, s), providing a nice characterization of standard representations of graphs in terms of skeleton subcategories of **GRAPHS** and of suitable functors to them.

**3.3 Theorem** *(standard representations and skeleton subcategories)*

1. Let (S, s) be a standard representation of graphs, and let $\Phi(S, s)$ = (**SGRAPHS**, F) be its functorial presentation. Then the category of standard graphs **SGRAPHS** is a skeleton subcategory of **GRAPHS**. Moreover, the standard representation functor F: **GRAPHS** → **SGRAPHS** is such that $F \circ I = ID_{\mathbf{SGRAPHS}}$, where I: **SGRAPHS** ↪ **GRAPHS** is the inclusion functor and $ID_{\mathbf{SGRAPHS}}$ is the identity functor. Finally, F maps all and only the standard isomorphism to identity arrows.

2. Let (**SGRAPHS**, F) be such that **SGRAPHS** is a skeleton subcategory of **GRAPHS** and F: **GRAPHS** → **SGRAPHS** is a functor satisfying $F \circ I = ID_{\mathbf{SGRAPHS}}$. Then (**SGRAPHS**, F) determines a standard representation of graphs $\Psi$(**SGRAPHS**, F) = (S, s) defined as

   (a) $S(G) = F_O(G)$ for all $G \in |\mathbf{GRAPHS}|$
   
   (b) $s(G, G') = h$ with $F_A(h) = id_{F_O(G)}$

   that is, the standard isomorphism s(G, G') is defined as the unique arrow from G to G' mapped by F to an identity arrow.

3. Mappings $\Phi$ and $\Psi$ are inverse each other, i.e., $\Psi \circ \Phi(S, s) = (S, s)$ and $\Phi \circ \Psi$(**SGRAPHS**, F) = (**SGRAPHS**, F).

**Proof**

1. The skeletal property of **SGRAPHS** is an immediate consequence of 2.1 and 3.2.1, since there exists exactly one standard representation for every isomorphism class of graphs.
The fact that $F \circ I = ID_{\mathbf{SGRAPHS}}$ follows for objects from the fact that $I(F_O(G)) \cong G$, and thus $F_O(I(F_O(G)) = F_O(G)$; for arrows, if f: G → G' is in **SGRAPHS**, then $F(I(f): I(G) \to I(G')) = s(G', F_O(I(G'))) \circ f \circ s(F_O(I(G)), G) = id_{G'} \circ f \circ id_G = f$, applying 2.1 (c).
Next we show that F maps all the standard isomorphisms to identities. In fact, $F_A(s(G, G')) = s(G', F_O(G')) \circ s(G, G') \circ s(F_O(G), G) = s(F_O(G), F_O(G')) = id_{F_O(G)}$, by 2.1 (c) and (d), since $F_O(G) = F_O(G')$. Finally, suppose that h: G → G' is a graph morphism and that $F_A(h) = id_{F_0(G)}$. Then using 3.2.2 we have $s(G', F_O(G')) \circ h \circ s(F_O(G), G) = id_{F_0(G)}$, and by 2.1 (e) $h = s(F_O(G'), G') \circ s(G, F_O(G))$, which implies, by 2.1 (d), that h is a standard isomorphism as well.

2. We have to show that conditions (a) to (d) of 2.1 hold for the above definition.
   2.1 (a) Suppose that G is not isomorphic to $S(G) = F_O(G)$, and let G' be the unique graph of **SGRAPHS** isomorphic to G (by the skeletal property). Then since two graphs are isomorphic in **SGRAPHS** iff they are the same, we have $G' \cong G \not\cong F_O(G) \Rightarrow G' \neq F_O(G)$; on the other

side, $G' \cong G \Rightarrow I(G') \cong G \Rightarrow F_O \circ I(G') \cong F_O(G) \Rightarrow G' = F_O(G)$ (because $F_O \circ I = ID$ by hypothesis), leading to a contradiction. Thus $G \cong S(G)$.

2.1 (b) Follows immediately from the previous point, since **SGRAPHS** is skeletal.

2.1 (c) - (d) We first need to show that s is well defined, i.e., that for all pair (G, G') of isomorphic graphs of **GRAPHS**, there exists a unique isomorphism $h: G \to G'$ such that $F_A(h) = id_{F_O(G)}$.

We first show the existence of such an h for the pairs of the form (S(G), G). In fact, $S(G) \cong G$ implies that there exists at least one isomorphism, say $f: S(G) \to G$; since the image of an isomorphism through a functor is again iso, $F_A(f): S(G) \to S(G)$ is iso as well, i.e., there exists a morphism $k: S(G) \to S(G)$ such that $F_A(f) \circ k = id_{S(G)} = k \circ F_A(f)$. Then take $h \equiv f \circ I(k)$. Morphism h is clearly iso, and $F_A(h) = F_A(f \circ I(k)) = F_A(f) \circ F_A(I(k)) = F_A(f) \circ k = id_{S(G)}$, exploiting that $F \circ I = ID_{\mathbf{SGRAPHS}}$.

Consider now an arbitrary pair (G, G') of isomorphic graphs. Since there exist isomorphisms $h': S(G) \to G$ and $h'': S(G') = S(G) \to G'$ such that $F_A(h') = id_{S(G)} = F_A(h'')$, we easily get the iso $h = h'' \circ h'^{-1}: G \to G'$ such that $F_A(h) = F_A(h'') \circ F_A(h')^{-1} = id_{S(G)}$.

It remains to show that such morphisms are unique. We prove the stronger fact that functor F is injective on homs, i.e., that for all parallel morphisms $f, g: G \to G'$ in **GRAPHS**, $F_A(f) = F_A(g)$ implies $f = g$. Suppose that $f, g: G \to G'$ are instead distinct, and that $k: S(G) \to G$ and $k': S(G') \to G'$ are any two isomorphisms (which exist by point 2.1 (a)). Then arrows $k'^{-1} \circ f \circ k$, $k'^{-1} \circ g \circ k: S(G) \to S(G')$ are in **SGRAPHS** and must be distinct as well, since they are obtained by composing distinct arrows with the same iso's. On the other hand, since $F \circ I = ID_{\mathbf{SGRAPHS}}$, then $k'^{-1} \circ f \circ k = F_A \circ I(k'^{-1} \circ f \circ k) = F_A(k'^{-1}) \circ F_A(f) \circ F_A(k) = F_A(k'^{-1}) \circ F_A(g) \circ F_A(k) = F_A \circ I(k'^{-1} \circ g \circ k) = k'^{-1} \circ g \circ k$, leading to a contradiction.

Therefore s(G, G') as in point 3.3.2 (b) is well-defined; properties (c) and (d) of 2.1 follow easily by the definition and by the properties of functors, since for all G, $F_A(id_G) = id_{F_O(G)}$ and $F_A(h) = id_{S(G)} = F_A(h')$ implies $F_A(h \circ h') = id_{S(G)}$.

3. Let (S, s) be a standard representation of graphs, let (**SGRAPHS**, F) = $\Phi$(S, s), and let (S', s') = $\Psi \circ \Phi$(S, s). Then using the definitions of $\Phi$ and $\Psi$ we have that $S'(G) = F_O(G) = S(G)$ for all $G \in$ |**GRAPHS**|; moreover, using the last statements of 3.3.1 and 3.3.2 we get s'(G, H) = s(G, H) for all isomorphic graphs G and H, because s(G, H) is mapped by F to the identity and s'(G, H) is chosen as the unique arrow mapped by F to the identity. Thus (S, s) = (S', s').

On the other hand, suppose that (**SGRAPHS**, F) satisfies the hypotheses of 3.3.2, that (S, s) = $\Psi$(**SGRAPHS**, F), and that (**SGRAPHS'**, F') = $\Phi \circ \Psi$(**SGRAPHS**, F). We have immediately that **SGRAPHS** = **SGRAPHS'**, since both are full subcategories of **GRAPHS** and $G \in$ |**SGRAPHS**| $\Leftrightarrow G = F_O(G) \Leftrightarrow G = S(G) \Leftrightarrow G \in$ |**SGRAPHS'**|.

Functors F and F' coincide on objects, because for every graph G, $F_O(G) \cong G \cong F'_O(G)$, and thus $F_O(G) = F'_O(G)$ by the skeletal property of **SGRAPHS**. We show that they coincide on arrows as well, i.e., that if $f: G \to G'$, then $F'_A(f) = F_A(f)$. In fact,

$$
\begin{aligned}
F'_A(f) &= F_A \circ I \circ F'_A(f) && [\text{because } F \circ I = ID_{\mathbf{SGRAPHS}}] \\
&= F_A(s(G', F_O(G')) \circ f \circ s(F_O(G), G)) && [\text{by } 3.2.2] \\
&= F_A(s(G', F_O(G'))) \circ F_A(f) \circ F_A(s(F_O(G), G)) \\
&= id_{F_O(G')} \circ F_A(f) \circ id_{F_O(G)} && [\text{by the definition of } s(G, G') \text{ in } 3.3.2] \\
&= F_A(f)
\end{aligned}
$$

This completes the proof that (**SGRAPHS**, F) = $\Phi \circ \Psi$(**SGRAPHS**, F). □

The bijective correspondence just presented between standard representations of graphs and their functorial

presentations allows us to characterize the category **AGRAPHS** of abstract graphs w.r.t. a given standard representation of graphs (see 2.2) as a quotient category.
We recall that given any functor F: **C** $\rightarrow$ **D**, the <u>quotient category</u> of **C** w.r.t. F, denoted **C**/$_F$, has as objects the equivalence classes of objects of **C** w.r.t. the equivalence relation $c \sim_{F_O} c' \Leftrightarrow F_O(c) = F_O(c')$, and as arrows the equivalence classes of arrows of **C** w.r.t. the equivalence relation $f \sim_{F_A} f' \Leftrightarrow F_A(f) = F_A(f')$. Moreover, it can be easily checked that if functor F is surjective then **C**/$_F$ and **D** are isomorphic.

**3.4 Proposition** *(AGRAPHS as quotient category w.r.t. the standard representation functor)*

Let (S, s) be a standard representation of graphs, let **AGRAPHS** be the corresponding category of abstract graphs, and let $\Phi$(S, s) = (**SGRAPHS**, F) be its functorial presentation. Then **AGRAPHS** is the quotient category of **GRAPHS** w.r.t. the standard representation functor F: **GRAPHS** $\rightarrow$ **SGRAPHS**, i.e., **AGRAPHS** = **GRAPHS**/$_F$. Moreover, F is surjective and thus categories **AGRAPHS** and **SGRAPHS** are isomorphic.

**Proof**

Both categories **AGRAPHS** and **GRAPHS**/$_F$ have equivalence classes of graphs as objects and equivalence classes of graph morphisms as arrows. Thus it is sufficient to show that the corresponding equivalence relations are the same.
For objects we have that $F_O(G) = F_O(G') \Leftrightarrow S(G) = S(G') \Leftrightarrow G \cong G'$, by the definition of F (3.2.2) and exploiting 2.1 (a) and (b). For arrows,

$F_A(f: G \rightarrow H) = F_A(f': G' \rightarrow H') \Leftrightarrow$
$s(H, F_O(H)) \circ f \circ s(F_O(G), G) = s(H', F_O(H')) \circ f' \circ s(F_O(G'), G') \Leftrightarrow$ [by 3.2.2]
$s(F_O(H'), H') \circ s(H, F_O(H)) \circ f \circ s(F_O(G), G) \circ s(G, F_O(G)) =$
$s(F_O(H'), H') \circ s(H', F_O(H')) \circ f' \circ s(F_O(G'), G') \circ s(G, F_O(G)) \Leftrightarrow$
$s(H, H') \circ f = f' \circ s(G, G')$ [by 2.1 (e), since $F_O(H) = F_O(H')$, $F_O(G) = F_O(G')$] $\Leftrightarrow$
f and f' are standard equivalent.

The fact that functor F is surjective directly follows from $F \circ I = ID_{\mathbf{SGRAPHS}}$. This implies that **AGRAPHS** and **SGRAPHS** are isomorphic. □

It is well known that every category is equivalent to any skeleton subcategory of it [AHS 90, ML 71]. The next proposition shows that every functor F: **GRAPHS** $\rightarrow$ **SGRAPHS** such that $F \circ I = ID_{\mathbf{SGRAPHS}}$ provides an equivalence (F, I) of categories. Moreover, if F is the functor associated with a standard representation of graphs (S, s) as in 3.2.1, then all the components of the unit (and of the counit) associated with the equivalence (F, I) are standard isomorphism. This proposition fully formalizes a short observation by Mac Lane about the equivalence between a category and any skeleton subcategory of it (see [ML 71], p. 91).

**3.5 Proposition** *(functorial presentations are equivalences of categories)*

1. If I: **SGRAPHS** $\hookrightarrow$ **GRAPHS** is the inclusion of a skeleton subcategory and F: **GRAPHS** $\rightarrow$ **SGRAPHS** is such that $F \circ I = ID_{\mathbf{SGRAPHS}}$, then (F, I) is an equivalence of categories, i.e., there exist two natural isomorphisms $\varphi: ID_{\mathbf{SGRAPHS}} \Rightarrow F \circ I$ and $\psi: ID_{\mathbf{GRAPHS}} \Rightarrow I \circ F$.

2. If (S, s) is a standard representation of graphs and $\Phi$(S, s) = (F, **SGRAPHS**), then the natural isomorphism $\psi: ID_{\mathbf{GRAPHS}} \Rightarrow I \circ F$ is defined as $\psi_G = s(G, S(G))$ for all $G \in |\mathbf{GRAPHS}|$.

**Proof**

1. Since $F \circ I = ID_{\mathbf{SGRAPHS}}$ by hypothesis, $\varphi$ is simply the identity natural transformation of $ID_{\mathbf{SGRAPHS}}$, i.e., $\varphi_H = id_H$. For $\psi$, we define the component of $\psi$ on a graph $G \in |\mathbf{GRAPHS}|$, $\psi_G$, as the unique isomorphism h: G to $F_O(G)$ such that $F_A(h) = id_{F_O(G)}$. The existence and uniqueness of such an isomorphism has been proved in 3.3.2. For the naturality, we have to show the commutativity of the diagram below (where ID stands for $ID_{\mathbf{GRAPHS}}$). Thanks to the injectivity of F on homs shown in 3.3.2, it is sufficient to show that $F_A(I(F_A(f)) \circ \psi_G) = F_A(\psi_{G'} \circ f)$. Indeed:

$$
\begin{aligned}
F_A(I(F_A(f)) \circ \psi_G) &= F_A(f) \circ F_A(\psi_G) && \text{[because } F \circ I = ID\text{]}\\
&= F(f) \circ id_{F_O(G)} && \text{[by the definition of } \psi_G\text{]}\\
&= F(f) = id_{F(G')} \circ F(f) &&\\
&= F(\psi_{G'}) \circ F(f) && \text{[by the definition of } \psi_{G'}\text{]}\\
&= F(\psi_{G'} \circ f). &&
\end{aligned}
$$

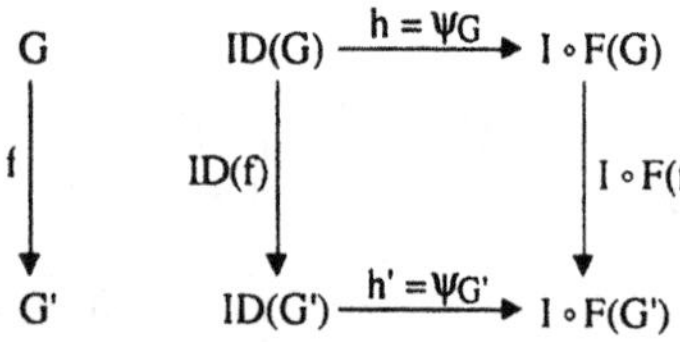


2. This is an immediate consequence of the fact that F maps standard isomorphisms to identities (see 3.3.1) and of the definition of $\psi$. □

The following corollary summarizes the previous results stressing the relationship among the various categories of graphs considered in this paper.

**3.6 Corollary** *(relationship among categories of concrete and abstract graphs)*

Let (S, s) be a standard representation of graphs, let **AGRAPHS** be the corresponding category of abstract graphs, and let $\Phi(S, s)$ = (**SGRAPHS**, F) be the corresponding functorial presentation. Then the relationship among categories **GRAPHS**, **AGRAPHS**, and **SGRAPHS** are summarized in the following diagram.

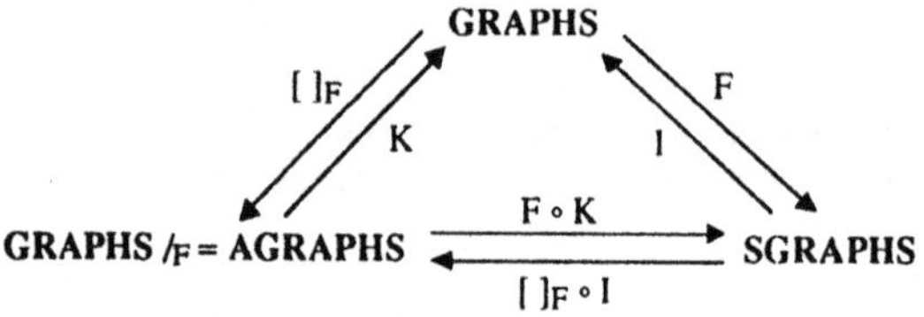


Making reference to the picture, we have that:

- **AGRAPHS** is exactly the quotient category of **GRAPHS** with respect to F, i.e., **AGRAPHS** = **GRAPHS**/$_F$ (see 3.4).
- (F, I) is an equivalence of categories (see 3.5.1)
- Functor $[\_]_F$: **GRAPHS** → **AGRAPHS** maps every object and arrow to its equivalence class.
- Functor K: **AGRAPHS** → **GRAPHS** is defined as $K_O([G]) = S(G)$, and $K_A([f\colon G \to G']) = s(G', S(G')) \circ f \circ s(S(G), G)$.

- $F \circ K$: **AGRAPHS** $\rightarrow$ **SGRAPHS** is the isomorphism whose existence is stated in 3.4, and its inverse is $[\_]_F \circ I$: **AGRAPHS** $\rightarrow$ **SGRAPHS**.
- $([\_]_F, K)$ is an equivalence of categories. ♦

A natural question about standard representations of graphs is whether they are 'unique' in some meaningful sense. Clearly, the categories of standard graphs **SGRAPHS** and **SGRAPHS'** corresponding to two distinct standard representations (S, s) and (S', s') are isomorphic, since they are both skeleton categories of **GRAPHS**. The following result shows in addition that if $\Phi(S, s) =$ (**SGRAPHS**, F) and $\Phi(S', s') =$ (**SGRAPHS'**, F'), given any isomorphism J: **SGRAPHS** $\rightarrow$ **SGRAPHS'** there exists a unique automorphism of **GRAPHS** which is the identity of objects making the resulting square commutative.

**3.7 Proposition** (relating standard representations of graphs)

Let (S, s) and (S', s') be two standard representations of graphs, and let $\Phi(S, s) =$ (**SGRAPHS**, F) and $\Phi(S', s') =$ (**SGRAPHS'**, F') be the corresponding functorial presentations. Moroever, let J: **SGRAPHS** $\rightarrow$ **SGRAPHS'** be an isomorphism of categories. Then there exists a unique automorphism J*: **GRAPHS** $\rightarrow$ **GRAPHS** which is the identity on objects and such that the diagram below commutes

**Proof**

Clearly, on objects $J^*(G) = G$ by the above requirements. On arrows J* is uniquely determined by the commutativity requirement. In fact, if $f: G \rightarrow G'$, we have:

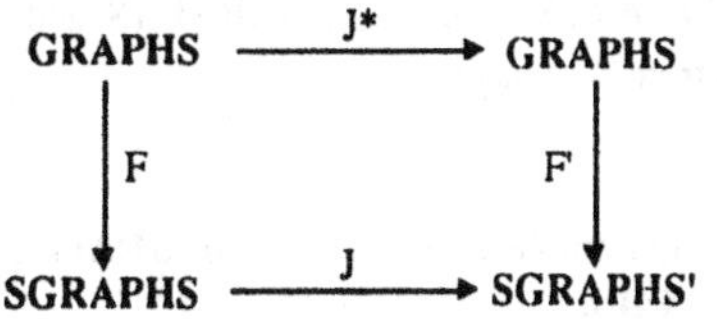


$F'(J^*(f)) = J(F(f)) \Rightarrow$ [by 3.2.2]

$s'(G', S'(G')) \circ J^*(f) \circ s'(S'(G), G) = J(s(G', S(G')) \circ f \circ s(S(G), G)) \Rightarrow$

$J^*(f) = s'(S'(G'), G') \circ J(s(G', S(G')) \circ f \circ s(S(G), G)) \circ s'(G, S'(G))$.

It remains to check that J* defined in this way is a functor J*: **GRAPHS** $\rightarrow$ **GRAPHS**: then the fact that it is indeed an isomorphism follows easily by a symmetric argument, observing that J has an inverse by hypothesis. To show that J* is a functor we check that it preserves identities and composition. In fact,

$J^*(id_G) = s'(S'(G), G) \circ J(s(G, S(G)) \circ id_G \circ s(S(G), G)) \circ s'(G, S'(G)) =$

$s'(S'(G), G) \circ J(id_{S(G)}) \circ s'(G, S'(G)) = s'(S'(G), G) \circ id_{J(S(G))} \circ s'(G, S'(G)) =$

$s'(S'(G), G) \circ id_{S'(G)} \circ s'(G, S'(G)) = id_G$,

and if $f: G \rightarrow G'$ and $g: G' \rightarrow G''$, then

$J^*(g) \circ J^*(f) =$

$s'(S'(G''), G'') \circ J(s(G'', S(G'')) \circ g \circ s(S(G'), G')) \circ s'(G', S'(G')) \circ$

$\circ\, s'(S'(G'), G') \circ J(s(G', S(G')) \circ f \circ s(S(G), G)) \circ s'(G, S'(G)) =$

$s'(S'(G''), G'') \circ J(s(G'', S(G'')) \circ g \circ s(S(G'), G')) \circ id_{S'(G')} \circ$

$\circ\, J(s(G', S(G')) \circ f \circ s(S(G), G)) \circ s'(G, S'(G)) =$

$s'(S'(G''), G'') \circ J(s(G'', S(G'')) \circ g \circ s(S(G'), G') \circ id_{S(G')} \circ$

$\circ\, s(G', S(G')) \circ f \circ s(S(G), G)) \circ s'(G, S'(G)) =$

$s'(S'(G''), G'') \circ J(s(G'', S(G'')) \circ g \circ f \circ s(S(G), G)) \circ s'(G, S'(G)) =$

$J^*(g \circ f)$ □

The equivalence among the categories **GRAPHS**, **AGRAPHS**, and **SGRAPHS** for a given standard representation of graphs (see 3.6) implies that all categorical constructions (like limits and colimits) can be performed in any of such categories yielding compatible results. However, the simpler structure of categories **AGRAPHS** and **SGRAPHS** allows us to give a precise characterization of the number of distinct limit or colimit cones that exist for a given diagram. This fact is very relevant for the algebraic approach to graph grammars, because as suggested in Counterexample 1.1, distinct pushout diagrams may represent distinct computations. Although the next result holds for arbitrary limits and colimits, we show it only for the case of pushouts, since pushouts are the building blocks of graph derivations.

**3.8 Theorem** *(characterization of distinct abstract pushout diagrams)*

Let $C \xleftarrow{c} A \xrightarrow{b} B$ be a diagram in the category of abstract graphs **AGRAPH** (w.r.t. (S, s)). Then:

1. There exists an abstract graph D and abstract graph morphisms k: C → D and h: B → D such that the following is a pushout diagram in **AGRAPHS**;

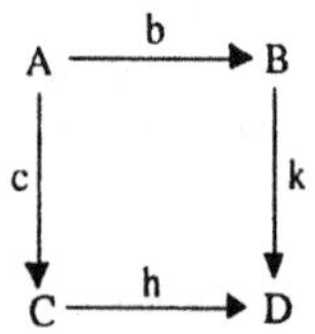


2. If D is as in point 1, then there exists a bijective correspondence between the set of distinct pushout diagrams based on $C \xleftarrow{c} A \xrightarrow{b} B$ in category **AGRAPHS**, and the set of distinct automorphism of D (i.e., isomorphisms from D to itself).

**Proof**

1. This follows immediately from the cocompleteness of **GRAPHS** and from the equivalence between **GRAPHS** and **AGRAPHS**.

2. We prove point 2 for the isomorphic category **SGRAPHS**.
Let PO be the pushout depicted above, and let PO(b, c) be the set PO(b, c) = {(h', D', k') / $D' \xleftarrow{h'} C \xleftarrow{c} A \xrightarrow{b} B \xrightarrow{k'} D'$ is a pushout diagram in **SGRAPHS**}. Moreover, let Auto(D) = {$\phi$: D → D / $\phi$ is an isomorphism}. Then we show that the function $\Theta$: Auto(D) → PO(b, c) defined as $\Theta(\phi) = (\phi \circ h, D, \phi \circ k)$ is a bijection. First, note that since any two pushout objects for the same diagram are isomorphic, and in **SGRAPHS** any two isomorphic objects are the same, we have that D' = D for all triples (h', D', k') in PO(b, c). Moreover, again by general properties of pushouts, if (h', D, k') $\in$ PO(b, c) (since also (h, D, k) $\in$ PO(b, c)) there exists a unique isomorphism $\phi$: D → D such that $\phi \circ h = h'$ and $\phi \circ k = k'$. Existence of such a $\phi$ proves surjectivity of $\Theta$, while uniqueness proves injectivity. □

## 4 Conclusion

We started this note by observing that a naive notion of abstract graph derivations, based on the idea that two derivations are equivalent iff they are isomorphic, does not allow to extend some relevant properties of concrete derivations to abstract ones. This was the main motivation for the introduction of standard representation of graphs, which were used to define a (more restricted) notion of equivalence among graph morphisms, direct derivations and graph derivations. The properties of the resulting category of abstract

graphs have been investigated in depth, and the relationship with skeleton subcategories of **GRAPHS** was worked out.
The problems with abstract graphs and abstract graph derivations discussed in Section 1 are surely problems in the theory of graph grammars, but they are not specific problems for the category **GRAPHS**. In fact, exactly the same problems arise in the theory of High Level Replacement Systems [EHKP 91] as well, where category **GRAPHS** is replaced by an arbitrary category. Moreover, the impossibility of defining abstract objects and abstract morphisms as isomorphism classes of object and morphisms is clearly relevant to any category.
Although all the definitions and results presented in this note concern categories of graphs, it should be clear that most of them hold for arbitrary categories as well. More precisely, given a category **C** one can define a 'standard representation' of objects of **C**, which induces a category **AC** of abstract objects of **C**, isomorphic to a skeleton subcategory **SC** of **C**. Then all the results summarized in Corollary 3.6 still hold if **GRAPHS**, **AGRAPHS** and **SGRAPHS** are replaced by **C**, **AC** and **SC**, respectively. In this note the emphasis was placed on categories of graphs because graph derivations were the motivating example, and because the definitions and results presented here will be used in a research activity aimed at defining a new, truly-concurrent semantics of graph derivations based on event structures (see also [CELMR 93]).
Finally it is worth stressing that some of the observations and results presented here may be not new to category theorists. Skeletal categories and their properties are well known in literature [AHS 90], and the use of standard isomorphisms could be already known in Category Theory (maybe under a different name), as it can be inferred from the observation in [ML 71], page 91. In any case, we are not aware of any paper where the problems described in Section 1 are addressed in an explicit way, and it might be the case that some of the results of Section 3 concerning existence and characterization of standard representations are of interest for other application categories.

## 5 References


[AHS 90] J. Adamek, H. Herrlich, G. Strecker: Abstract and Concrete Categories, Wiley Interscience (1990)

[CELMR 93] Corradini, A., Ehrig, H., Löwe, M., Montanari, U., Rossi, F., *Abstract Graph Derivations in the Double Pushout Approach*, in this volume.

[Ehr 79] H. Ehrig: Introduction to the algebraic theory of graph grammars (A Survey) in: Graph Grammars and Their Application to Computer Science and Biology, Springer LNCS 73, (1979), 1-69

[EHKP 91] H. Ehrig, A. Habel, H.-J. Kreowski, F. Parisi-Presicce: Parallelism and Concurrency in High-Level Replacement Systems, Math. Struct. in Comp. Sci. (1991), vol. 1, pp. 361-404

[EPS 73] H. Ehrig, M. Pfender, H.J. Schneider: Graph Grammars: An Algebraic Approach, Proc. IEEE Conf. SWAT'73, Iowa City 1973, p. 167 - 180

[Her 92] H. Herrlich, Private communication, December 1992.

[ML 71] S. Mac Lane, *Categories for the Working Mathematician*, Springer Verlag, New York, 1971.

# Jungle Rewriting: an Abstract Description of a Lazy Narrowing Machine*

Andrea Corradini[1] and Dietmar Wolz[2]

[1] Università di Pisa, Dipartimento di Informatica, Corso Italia 40, 56125 Pisa, Italy (andrea@di.unipi.it)

[2] Technische Universität Berlin, Fachbereich 20 Informatik, Franklinstraße 28/29, 1000 Berlin 10, Germany (dietmar@cs.tu-berlin.de)

**Abstract.** We propose to use Jungle Rewriting, a graph rewriting formalism, as a formal specification tool for the LANAM, an abstract machine implementing lazy narrowing. Three jungle rewriting systems which model a narrowing program at different levels of abstraction are presented. The first system is proved to be essentially equivalent (from an operational point of view) to the given program. The second one contains only rules which correspond to elementary actions of a narrowing interpreter, while the third one also models the control issues of the abstract machine. We formally prove the equivalence of the first and second systems, while the relationship between the second and the third one is sketched. The last system provides an abstract description of the compiled code of the LANAM, which is outlined.

## 1 Introduction.

*Jungle rewriting* is a graph rewriting formalisms based on the "double-pushout" approach ([EPS 73]), which has been proposed by the first author in joint work with Francesca Rossi. In [CR 93], they showed that jungle rewriting faithfully models both Logic Programming ([Ll 87]) and Term Rewriting Systems ([DJ 90]), also in the case of non-left-linear rewrite rules.

The *Narrowing Calculus* generalizes both Logic Programming and Term Rewriting Systems. Using unification, an instance of the left-hand side of a rule is computed, such that this instance matches the term to be rewritten. In this way instances of the logical variables of the term are generated. Terms not only represent data values as in Logic Programming but also nested function applications. Term Rewriting is a special case of narrowing where no logical variable is bound. The second author designed and implemented an abstract machine (called LANAM) for *lazy narrowing*, i.e., an interpreter of the Narrowing Calculus which uses a lazy evaluation strategy ([Wo 91]).

The overall goal of an ongoing research activity of the authors is to show that jungle rewriting can be used as a flexible and formal specification language, which can be exploited for proving formally the correctness of the LANAM implementation of the Narrowing Calculus. At the same time, jungle rewriting should provide an abstract description of the LANAM code, useful for a better understanding of the abstract machine.

The idea is to introduce various jungle rewriting systems, which describe narrowing and the LANAM at different levels of abstractions. The various jungle rewriting systems are then related by formal results, which prove that each system is implemented in a faithful way by a system corresponding to a more concrete level. When the whole picture

* Research partially supported by the COMPUGRAPH Basic Research Esprit Working Group n. 7183

will be completed, one should obtain a formal proof of the correctness of the LANAM implementation of narrowing.

After a background section, in Section 3 we introduce the first jungle rewriting system, denoted $\mathcal{HR}(\mathcal{P})$, which is shown to model narrowing (with respect to a given program $\mathcal{P}$) essentially at the same level of abstraction. In Section 4, we first show that the rules of $\mathcal{HR}(\mathcal{P})$ do not correspond to the elementary actions of a narrowing interpreter. Then we introduce a restriction on the form of rules and an applicability conditions which, together, guarantee that a jungle rewriting step corresponds to an elementary action of the interpreter. Thus we introduce a new jungle rewriting system, denoted $\mathcal{E}(\mathcal{P})$ and obtained via suitable transformations of the rules in $\mathcal{HR}(\mathcal{P})$, which is shown to implement in a faithful way $\mathcal{HR}(\mathcal{P})$.

Both the systems introduced till now do not consider explicitly the control issues of narrowing. These aspects are considered in Section 5, where jungle rewriting is extended with a notion of "marking", allowing to specify which nodes can be reduced. These markings can be used for the specification of various narrowing strategies: we consider a lazy strategy, similar to the one of the LANAM. We shortly discuss how a new jungle rewriting system that implements the lazy narrowing strategy can be obtained by transforming the rules of $\mathcal{E}(\mathcal{P})$.

Finally, in Section 6 we give a short overview of the LANAM, trying to stress the main relationships with the various jungle rewriting systems introduced so far. The formal relationship between the LANAM code and the jungle rewriting systems is the argument of a forthcoming paper by the same authors. Because of space limitations, most of the proofs are not reported. They will be included in the full version of the paper to appear as a technical report.

## 2 Background: Narrowing Calculus and Jungle Rewriting.

### 2.1 The Narrowing Calculus

Narrowing subsumes term rewriting and unification, and can be efficiently implemented by combining compilation techniques from logic and functional programming [Pad 88, War 83, Jos 87, Lo 92]. Its purpose is to find solutions for sets of equations (**goals**), i.e., substitutions that unify the left- and right-hand sides of all the equations in the set, or to detect that there is no solution. Narrowing can be seen as a combination of reduction and unification: non-variable prefixes of reduction redices are completed to full redices by substituting into variables. We introduce here the basic definitions of the Narrowing Calculus.

Since we will deal with (conditional) equations, we will consider terms built over a many-sorted signature containing the (overloaded) equality operator '=' of type bool. Operator = will be used only for the representation of equations, and its operational behaviour will be defined in an *ad hoc* way.

**Definition 1 (signatures).** A (many-sorted) **signature** $\underline{\Sigma}$ **(with equality)** is a triple $\underline{\Sigma} = (S, \Sigma, \Gamma)$, where $S$ is a set of **sorts** such that $\mathtt{bool} \in S$, $\Sigma = \{\Sigma_{\mathtt{w},\mathtt{s}}\}$ is a family of sets of **functions** indexed by $S^* \times S$, such that $= \in \Sigma_{\mathtt{s}\cdot\mathtt{s},\mathtt{bool}}$ for each $\mathtt{s} \in S$, and $\Gamma$ is a family of sets of **constructors** indexed by $S^* \times S$ such that $\mathtt{true} \in S_{\epsilon,\mathtt{bool}}$. Functions and constructors are also called **operators**. If $\mathtt{f} \in \Sigma_{\mathtt{w},\mathtt{s}} \cup \Gamma_{\mathtt{w},\mathtt{s}}$ then s is called the **type** of f, w is its **arity**, and $|\mathtt{w}|$ (the length of w) is the **rank** of f. □

**Definition 2 (terms, substitutions, unifiers).** Let $X = \{X_{\mathtt{s}}\}_{\mathtt{s} \in S}$ be a family of pairwise disjoint sets of variables indexed by $S$. Then a **term of sort s** is an element of

$T_{\underline{\Sigma}_s}(X)$ (the carrier of sort $s$ of the free $\underline{\Sigma}$-algebra generated by $X$), that is a variable of $X_s$, or a constant in $\Sigma_{\epsilon,s} \cup \Gamma_{\epsilon,s}$, or $f(t_1,\ldots,t_n)$, if $f \in \Sigma_{w,s} \cup \Gamma_{w,s}$, $w = s_1 \cdot \ldots \cdot s_n$, and $t_i$ is a term of sort $s_i$ for each $1 \le i \le n$. A term is a **data term** if all the operators appearing in it are constructors. A term is **linear** if no variable occurs more than once in it. The sort of a term $t$ will be denoted by $sort(t)$. The set of variables occurring in a term $t$ is defined in the usual way and will be denoted by $var(t)$. We require that the special equality operators introduced in Definition 1 always appear as top operators of a term. Given two families of $S$-indexed sets of variables $X = \{X_s\}$ and $Y = \{Y_s\}$, a **substitution $\sigma$ from $X$ to $Y$** is a function $\sigma : X \to T_{\underline{\Sigma}}(Y)$ (used in postfix notation) which is sort preserving. When $X$ is finite, $\sigma$ can be represented as $\sigma = [x_1 \leftarrow t_1,\ldots,x_n \leftarrow t_n]$, where $t_i = x_i\sigma$. If $X_1 \subseteq X$ and $\sigma$ is a substitution from $X$ to $Y$, then $\sigma \uparrow X_1$ denotes the obvious **restriction** of $\sigma$ to $X_1$. A **variable renaming** $\sigma$ is an injective substitution $\sigma : X \to T_{\Sigma}(Y)$ such that $x\sigma \in Y$ for all $x \in X$. Term $t'$ is a **variant** of term $t$ if $t' = t\sigma$ for some variable renaming $\sigma$. The extension of a substitution to terms, and the composition of two substitutions are defined in the usual way.

Given two substitutions $\sigma$ and $\tau$, $\sigma$ is said to be **more general** than $\tau$ if there exists a substitution $\phi$ such that $\sigma\phi = \tau$. Two terms $t$ and $t'$ **unify** if there exists a substitution $\sigma$ such that $t\sigma = t'\sigma$. In this case $\sigma$ is called a **unifier** of $t$ and $t'$. The set of unifiers of any two terms is either empty or it has a most general element (up to variable renaming) called the **most general unifier**. □

**Definition 3 (equations, goals, clauses and programs).** An **equation** is a term of the form $t = t'$, where $t$ and $t'$ are terms of the same sort, say $s$, and '=' $\in \Sigma_{s\cdot s,bool}$ is as in Definition 1. A **goal** $G$ is a multiset of equations $G = \{\!| t_1 = t'_1,\ldots,t_n = t'_n |\!\}$.[3]

A **conditional equation** or **clause** $C$ is of the form $C \equiv (P \to l = r)$, where $P$ is a goal (the **premise**) and $l = r$ is an equation (the **conclusion**) where $l$ is not a variable.[4] Terms $l$ and $r$ are called the **left-** and the **right-hand side** of $C$, respectively (shortly **lhs** and **rhs**). A clause $C$ can also be written as $t_1 = t'_1,\ldots,t_n = t'_n \to l = r$, or simply $l = r$ if $P = \emptyset$. A clause is **left-linear** if its left-hand side is linear; it is **collapsing** if the rhs is a variable which occurs in the lhs. The concepts of substitution application, of variable set, and of variant are extended from terms to equations, goals and clauses in an obvious manner. A **program** $\mathcal{P}$ is a finite set of clauses, $\mathcal{P} = \{C_i\}_{i\in I}$. □

**Definition 4 (narrowing).** Let $\mathcal{P}$ be a program. A goal $G$ **narrows in one step to $G'$ with substitution $\sigma$ using** $C$, written $G \to_{\sigma,C} G'$ if $C \in \mathcal{P} \cup \{=\}$, and one of the following two conditions is satisfied:

[*Base Rule*] $C \in \{=\}$, $G = \{\!| t_1 = t'_1,\ldots,t_n = t'_n |\!\}$ for some $n \ge 1$, $\sigma$ is a most general unifier for $t_i$ and $t'_i$ (for some $1 \le i \le n$), and $G' = \{\!| t_1 = t'_1,\ldots,t_{i-1} = t'_{i-1}, t_{i+1} = t'_{i+1}, t_n = t'_n |\!\}\sigma$.

[*Narrowing Rule*] $C \in \mathcal{P}$ is a clause, and there exist a variant $C' \equiv (P \to l = r)$ of $C$ with $var(C') \cap var(G) = \emptyset$, a non-variable term $t$ and a goal $K$ having a unique occurrence of variable $x_0 \notin var(G) \cup var(C')$, such that:

1. $G = K[x_0 \leftarrow t]$;
2. $t$ and $l$ are unifiable with most general unifier $\sigma'$;

[3] Multisets are represented as lists of items (possibly with duplicates) delimited by '$\{\!|$' and '$|\!\}$'. If $M$ and $M'$ are multisets, $M \uplus M'$ denotes their multiset union. We consider a goal as a *multiset* instead of as a *set* of equations, because this is closer to real implementations of narrowing and of logic programming, where usually duplicates in a goal are not detected.

[4] We do not impose any restriction on variables in clauses (like $var(r) \subseteq var(l)$) because we do not address here problems of completeness.

3. $G' = (K[x_0 \leftarrow r] \uplus P)\sigma'$;
4. $\sigma = \sigma' \restriction var(G)$.

A goal $G$ **narrows to $G'$ with substitution $\sigma$ using program $\mathcal{P}$**, written $G \rightarrow^*_{\sigma,\mathcal{P}} G'$, if there exist a sequence $G = G_0, G_1, \ldots, G_n = G'$ of goals, a sequence $C_1, \ldots, C_n$ of elements of $\mathcal{P} \cup \{=\}$, and a sequence $\sigma_1, \ldots, \sigma_n$ of substitutions such that $G_{i-1} \rightarrow_{C_i,\sigma_i} G_i$ for all $0 < i \leq n$, and $\sigma = \sigma_1\sigma_2 \ldots \sigma_n$. Substitution $\sigma$ is called a **computed answer substitution**. □

Narrowing, as defined above, is intrinsically sequential: each application of the Narrowing Rule rewrites a *single* subterm of the current goal. We introduce now a simple notion of *parallel narrowing*, which allows for the application of a rule to a set of identical subterms of a goal.

**Definition 5 (parallel narrowing).** A goal $G$ **narrows in one parallel step of degree $n$ to $G'$** (with substitution $\sigma$ using $C$), written $G \rightarrow^{||n}_{\sigma,C} G'$ if:

[*Parallel Narrowing Rule*] $C \in \mathcal{P}$ is a clause, $n \geq 1$, and there exist a variant $C' \equiv (P \rightarrow l = r)$ of $C$ with $var(C') \cap var(G) = \emptyset$, a non-variable term $t$ and a goal $K$ having exactly $n$ occurrences of variable $x_0 \notin var(G) \cup var(C')$, such that conditions 1 to 4 of the Narrowing Rule of Definition 4 are satisfied.

We write $G \rightarrow^{||*}_{\sigma,\mathcal{P}} G'$ if $G$ narrows to $G'$ with a narrowing derivation possibly including parallel steps. □

It should be quite evident that the Parallel Narrowing Rule does not increase the computational power of narrowing, if only narrowing derivations ending with the empty goal $\emptyset$ are considered (as it is often the case). More precisely, if $G \rightarrow^{||*}_{\sigma,\mathcal{P}} \emptyset$, then $G \rightarrow^*_{\sigma,\mathcal{P}} \emptyset$; that is, narrowing is able to return any computed answer substitution returned by parallel narrowing.

## 2.2 Jungles and Jungle Rewriting

Jungles are special hypergraphs that are suited to represent (collections of) terms, making explicit the sharing of common subterms. They are essentially equivalent to *directed acyclic graphs*, as discussed in [CMREL 91]. In this section we introduce the category of jungles (see also [HP 91]) and the basic definitions of jungle rewriting, which is a specific graph rewriting formalism defined along the lines of the "double-pushout" approach ([EPS 73, Eh 87]).

**Definition 6 (hypergraphs). A hypergraph $H$ over $\underline{\Sigma}$** is a tuple $H = (V, E, src, trg, m, l)$, where $V$ is a set of **nodes**, $E$ is a set of **hyperedges**, $src$ and $trg : E \rightarrow V^*$ are the **source** and **target** functions, $l : V \rightarrow S$ maps each node to a sort of $\underline{\Sigma}$, and $m : E \rightarrow \Sigma$ maps each edge to an operator of $\underline{\Sigma}$. A **path** in a hypergraph from node $v$ to node $u$ is a sequence of hyperedges $\langle e_1, \ldots, e_n \rangle$ such that either $n = 0$ and $v = u$, or $v \in src(e_1)$, $u \in trg(e_n)$, and for each $1 < i \leq n$, $\exists w \in src(e_n) . w \in trg(e_{n-1})$. A hypergraph is **acyclic** if there are no non-empty paths from one node to itself. A hypergraph is **discrete** if it contains no edges. For a hypergraph $H$, $indegree_H(v)$ (resp. $outdegree_H(v)$) denotes the number of occurrences of a node $v$ in the target strings (resp. source strings) of all edges of $H$. □

In the rest of the paper we will often omit the prefix 'hyper-', calling *hypergraphs* and *hyperedges* simply *graphs* and *edges*. A *jungle* is a hypergraph satisfying some additional conditions which makes it suitable to represent collections of (finite) terms with possibly shared subterms.

**Definition 7 (jungles).** A **jungle over** $\underline{\Sigma}$ is an acyclic graph $J = (V, E, src, trg, m, l)$ over $\underline{\Sigma}$ such that 1) for each node $v \in V$, $outdegree_J(v) \leq 1$, and 2) the labeling of the edges is consistent with both the number and the labeling of the connected nodes, i.e., $\forall e \in E$, $m(e) \in \Sigma_{\mathtt{w},\mathtt{s}} \cup \Gamma_{\mathtt{w},\mathtt{s}} \Leftrightarrow l^*(src(e)) = \mathtt{s} \wedge l^*(trg(e)) = \mathtt{w}$, where $l^*$ is the obvious extension of $l$ to tuples. The **roots** of a jungle are defined as $ROOTS(J) = \{v \in V_J \mid indegree_J(v) = 0\}$. □

*Example 1 (representation of hypergraphs and jungles).* Let $\underline{\Sigma}$ be the following signature with equality, that will be used in all the examples of this paper: $\underline{\Sigma} = (S = \{\mathtt{A}, \mathtt{B}, \mathtt{bool}\}$, $\Sigma = \{\Sigma_{\mathtt{B}\cdot\mathtt{A},\mathtt{A}} = \{\mathtt{g}, \mathtt{h}, \mathtt{f}\}, \Sigma_{\mathtt{A}\cdot\mathtt{A},\mathtt{bool}} = \{=\}, \Sigma_{\mathtt{B}\cdot\mathtt{B},\mathtt{bool}} = \{=\}\}$, $\Gamma = \{\Gamma_{\epsilon,\mathtt{A}} = \{\mathtt{a}, \mathtt{b}\}$, $\Gamma_{\epsilon,\mathtt{B}} = \{\mathtt{d}\}$, $\Gamma_{\mathtt{A}\cdot\mathtt{B},\mathtt{B}} = \{\mathtt{c}\}$, $\Gamma_{\epsilon,\mathtt{bool}} = \{\mathtt{true}\}\}$. In Figure 1 two hypergraphs over $\underline{\Sigma}$ are shown. Edges and nodes are depicted as box and circles, respectively, with the label drawn inside and sometimes (for the need of reference) with a unique name depicted outside.

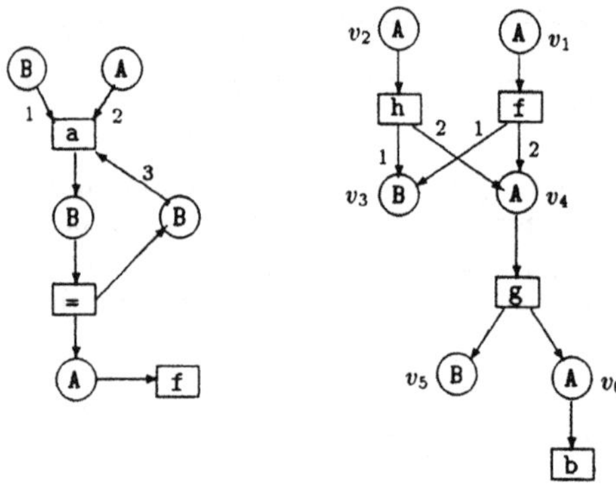


**Fig. 1.** Examples of hypergraphs.

The right graph is also a jungle while the left one is not. Thanks to the properties of jungles, we will use sometimes a linear representation for them (which is borrowed from DACTL [GKSS 88]). For example. the jungle in Figure 1 is represented linearly as $\{v_1 : \mathtt{f}(v_3, v_4 : \mathtt{g}(v_5, v_6 : \mathtt{b})), v_2 : \mathtt{h}(v_3, v_4)\}$ (if necessary, the sort of a node will be indicated by a superscript). Thus a jungle $J$ is regarded as a set of pairs $\{v_1 : t_1, \ldots, v_n : t_n\}$ where the $v_i$'s are the roots of $J$, and the $t_i$ are essentially terms over $\underline{\Sigma}$ where each subterm is labeled by a node of $J$. If a node name appears more than once (like $v_4$ in the example), the associated subterm does not need to be repeated. Such a situation expresses subterm sharing. □

**Definition 8 (categories HGraph$_{\underline{\Sigma}}$ and Jungle$_{\underline{\Sigma}}$).** A **morphism of hypergraphs (over** $\underline{\Sigma}$**)** $f : H_1 \rightarrow H_2$ consists of a pair of functions $f = (f_V : V_1 \rightarrow V_2, f_E : E_1 \rightarrow E_2)$ between edges and nodes respectively, which are compatible with the source and target functions and which are label preserving. A morphism $f = (f_V, f_E)$ is **injective** (resp. **surjective**) iff both $f_V$ and $f_E$ are injective functions (resp. surjective). A morphism $f$ is

**an isomorphism** if it is both injective and surjective. Morphisms of jungles are defined exactly in the same way. Because of the structure of jungles, a morphism is uniquely determined by its behaviour on roots. That is, if $f, g : J \to J'$, $f = g \Leftrightarrow f_V(v) = g_V(v)\ \forall v \in ROOTS(J)$ [HP 91].

The category whose objects are hypergraphs over $\underline{\Sigma}$ and whose arrows are hypergraph morphisms will be denoted by $\mathbf{HGraph}_{\underline{\Sigma}}$. The full subcategory of $\mathbf{HGraph}_{\underline{\Sigma}}$ including all jungles will be called $\mathbf{Jungle}_{\underline{\Sigma}}$. □

We introduce now the basic concepts of the algebraic theory of graph rewriting ([EPS 73, Eh 87]) for the specific case of the category of jungles. In the following we assume to refer to a given, fixed signature $\underline{\Sigma}$.

**Definition 9 (jungle rewrite rules and rewriting systems).** **A jungle (rewrite) rule** $p = (L \xleftarrow{l} K \xrightarrow{r} R)$ is a pair of jungle morphisms $l : K \to L$ and $r : K \to R$, where $l$ is injective. Without loss of generality, throughout the paper we will assume that $K$ is a subjungle of $L$ and that $l$ is the corresponding inclusion. The jungles $L$, $K$, and $R$ are called the **left-hand side** (lhs), the **interface**, and the **right-hand side** (rhs) of $p$, respectively. A **hyperedge replacement** (shortly **HR-**) **jungle rule** is a rule such that $L$ is of the form $\{v_0 : \mathtt{f}(v_1, \ldots, v_n)\}$ for some operator $\mathtt{f}$ of rank $n$ (i.e., $L$ has just one edge with label $\mathtt{f}$ connected to $n+1$ distinct nodes), and $K$ is the discrete jungle obtained from $L$ by removing the unique edge (thus $K$ has $n+1$ nodes). A **(HR-) jungle rewriting system** is a finite set $\mathcal{R}$ of (HR-) jungle rewrite rules. □

The application of a jungle rule $p$ to a jungle $J$ is modelled by a double pushout construction: in order to define it, we need to introduce the notions of *pushout* and of *pushout complement*.

**Definition 10 (pushout and pushout complement).** Given a category $C$ and two arrows $b : K \to B$, $d : K \to D$ of $C$, a triple $\langle H, h : B \to H, c : D \to H \rangle$ is called a **pushout** of $\langle b, d \rangle$ if:

[*Commutativity*] $h \circ b = c \circ d$;
[*Universal Property*] for all objects $H'$ and arrows $h' : B \to H'$ and $c' : D \to H'$, with $h' \circ b = c' \circ d$, there exists a unique arrow $f : H \to H'$ such that $f \circ h = h'$ and $f \circ c = c'$.

In this situation, $H$ is called a **pushout object** of $\langle b, d \rangle$. Moreover, given arrows $b : K \to B$ and $h : B \to H$, a **pushout complement** of $\langle b, h \rangle$ is a triple $\langle D, d : K \to D, c : D \to H \rangle$ such that $\langle H, h, c \rangle$ is a pushout of $b$ and $d$. In this case $D$ is called a **pushout complement object** of $\langle b, h \rangle$. □

**Definition 11 (jungle rewriting).** Given a jungle $J$, a jungle rewrite rule $p = (L \xleftarrow{l} K \xrightarrow{r} R)$, and an **occurrence** (i.e., a jungle morphism) $g : L \to J$, **a direct rewriting from $J$ to $H$ using $p$ (based on $g$)** exists if and only if the diagram in Figure 2 can be constructed, where both squares are required to be pushouts in $\mathbf{Jungle}_{\underline{\Sigma}}$. In this case, the graph $D$ (obtained as pushout complement object of $\langle l, g \rangle$) is called the **context** jungle, and we write $J \Rightarrow_p H$. Given a jungle rewriting system $\mathcal{R}$ we say that a jungle $J$ **rewrites** to a jungle $H$ using $\mathcal{R}$ (written $J \Rightarrow^*_{\mathcal{R}} H$) if there exist a sequence of jungles $J_0 = J, J_1, \ldots, J_n = H$ and a sequence of jungle rules $p_1, \ldots, p_n$ of $\mathcal{R}$ such that $J_i \Rightarrow_{p_{i+1}} J_{i+1}$ for all $0 \le i \le n-1$. □

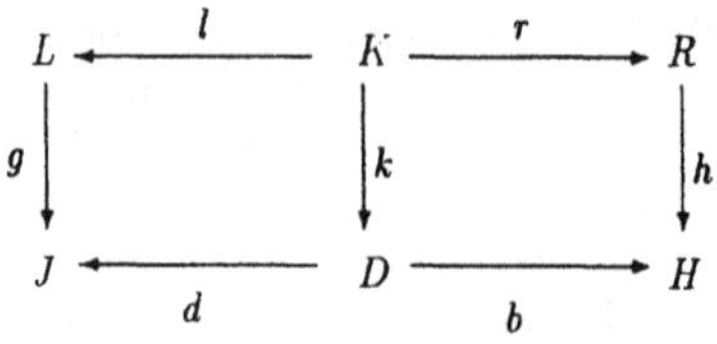


**Fig. 2.** Jungle rewriting via double-pushout construction.

Given a jungle rule and an occurrence, in general the double pushout construction can fail, because the required pushout and pushout complement do not always exist in the category of jungles. However, for the kind of rules that we will consider in this paper a pushout complement always exists, and we assume to choose it in a canonical way.

**Proposition 12 (canonical pushout complement).** *Let $l : K \to L$ be an inclusion of jungles which is surjective on nodes, and let $g : L \to J$ be a jungle morphism. Then there exists at least one pushout complement of $\langle l, g\rangle$. Moreover the* **canonical pushout complement** *of $\langle l, g\rangle$ is the triple $\langle D, k : K \to D, d : D \to J\rangle$ defined as follows. Jungle $D$ is obtained from $J$ by deleting all the edges which are in $g_E(E_L)$ but not in $g_E(l_E(E_K))$; $d$ is the resulting inclusion; and $k$ is the restriction of $g$ to $K$. By construction $d$ is a bijection of nodes.* □

Therefore the application of a jungle rewrite rule to a given occurrence morphism requires just the existence of the right-hand side pushout, and if this exists the result is unique up to isomorphism by a general property of colimits. A theorem characterizing the existence of pushouts in category **Jungle**$_{\Sigma}$ in terms of the existence of most general unifiers is reported in Section 3.

Before closing this section we introduce the notion of *track function* ([HP 91]), which will be helpful later on. Since the inclusion $l$ of any jungle rule we will consider is surjective on nodes, no node will ever be deleted by the application of a rule. Thus we can "track" the evolution of a node via a suitable function.

**Definition 13 (track functions).** Let $p = (L \xleftarrow{l} K \xrightarrow{r} R)$ be a jungle rule such that the inclusion $l : K \to L$ is surjective on nodes. Then the **track function associated with** $p$ is defined as $tr_p = r_V \circ l_V^{-1} : L_V \to R_V$. Similarly, if $J \Rightarrow_p H$ is a direct rewriting as in Figure 2 using a canonical pushout complement (therefore $d : D \to J$ is a bijection on nodes), then the **track function associated with** $J \Rightarrow_p H$ is defined as $tr = b_V \circ d_V^{-1} : J_V \to H_V$. In order to make the track function explicit, sometimes we will denote the above direct rewriting as $J \Rightarrow_{tr,p} H$. □

## 3 Narrowing Calculus and Hyperedge Replacement Jungle Rewriting

In this section we explore the relationship between jungles and terms over a given signature, and we show that hyperedge replacement jungle rewriting faithfully corresponds to narrowing. This result is an extension of the result presented in [CR 93], where it is shown that hyperedge replacement jungle rewriting faithfully models both Term Rewriting Systems and Logic Programming.

If $J$ is a jungle, a term of sort $\mathbf{s}$ can be extracted from each $\mathbf{s}$-labeled node $v$ of $J$ by 'unfolding' the subjungle rooted at $v$. Thus every jungle is considered to represent a multiset of terms (i.e., one term for each root node: clearly, two roots may have the same associated term), and, viceversa, every multiset of terms can be represented by one or more jungles, depending on the degree of sharing of identical subterms.

**Definition 14 (jungles and multisets of terms).** Let $J$ be a jungle over $\underline{\Sigma}$. The **variables** of $J$ are the family of $S$-indexed sets $Var(J) = \{Var_{\mathbf{s}}(J)\}_{\mathbf{s} \in S}$ defined as $Var_{\mathbf{s}}(J) = \{v \in V_J \mid outdegree_J(v) = 0 \wedge l_J(v) = \mathbf{s}\}$. The function $term_J$ associates with each node of $J$ labeled by $\mathbf{s}$ a term in $T_{\underline{\Sigma}_{\mathbf{s}}}(Var(J))$. It is defined inductively as follows:

- $term_J(v) = v$, if $v \in Var_{\mathbf{s}}(J)$,
- $term_J(v) = \mathtt{op}(term_J(v_1), \ldots, term_J(v_n))$, if there exists $e \in E_J$ with $src_J(e) = v$, $trg_J(e) = v_1 \cdot \ldots \cdot v_n$, and $m_J(e) = \mathtt{op} \in \Sigma_{\mathbf{w},\mathbf{s}} \cup \Gamma_{\mathbf{w},\mathbf{s}}$ for some $w \in S^*$.

The multiset of terms represented by jungle $J$ is defined as $TERMS(J) = \{\!|term_J(v) \mid v \in ROOTS(J)|\!\}$. If $T$ is a multiset of terms, we say that jungle $J$ **represents** $T$ iff $T \setminus TERMS(J) = \{\!|v_1, \ldots, v_n|\!\}$, where $v_i$ is a variable occurring in $TERMS(J)$ for all $1 \leq i \leq n$. By $ROOTS_{=}(J)$ we denote the subset of root nodes of $J$ having an outgoing edge labeled by '=' i.e., $ROOTS_{=}(J) = \{v \in ROOTS(J) \mid \exists e \in E_J \,.\, s_J(e) = v \wedge m_J(e) = \text{'='}\}$. $GOAL(J)$ is the goal represented by $J$, that is, the multiset of terms of the form $s = s'$ represented by $J$: $GOAL(J) = \{\!|terms_J(v) \mid v \in ROOTS_{=}(J)|\!\}$. □

*Example 2* . Let us consider the jungle of Figure 1. It contains two variable nodes $v_3$ and $v_5$, both of sort $\mathtt{B}$. The function $term_J$ is defined as $term_J(v_1) = \mathtt{f}(v_3, \mathtt{g}(v_5, \mathtt{b}))$, $term_J(v_2) = \mathtt{h}(v_3, \mathtt{g}(v_5, \mathtt{b}))$, $term_J(v_3) = v_3$, $term_J(v_4) = \mathtt{g}(v_5, \mathtt{b})$, $term_J(v_5) = v_5$, and $term_J(v_6) = \mathtt{b}$. Finally, $TERMS(J) = \{\!|\mathtt{f}(v_3, \mathtt{g}(v_5, \mathtt{b})), \mathtt{h}(v_3, \mathtt{g}(v_5, \mathtt{b}))|\!\}$. □

Many non-isomorphic jungles can represent the same multiset of terms (or the same goal), depending on the degree of sharing. We are interested here in characterizing a representation which manifests minimal sharing (see [HP 91, CR 93]).

**Proposition 15 (variable-collapsed jungles).** *A jungle $J$ is* **variable-collapsed** *iff for any node $v \in V_J$, $indegree_J(v) > 1 \Rightarrow v \in Var(J)$ (that is, only variable nodes can be shared). If $T$ is a multiset of terms, all the variable-collapsed jungles representing $T$ are isomorphic: by $\mathcal{J}(T)$ we denote one of them.* □

Like jungles represent naturally collections of terms, so jungle morphisms correspond to term substitutions.

**Definition 16 (the substitution induced by a morphism and by a track function).** Let $J$ and $J'$ be jungles, and let $h : J \to J'$ be a jungle morphism. The node component of $h$ (i.e., $h_V : V_J \to V_{J'}$) induces a substitution $\sigma_h : Var(J) \to T_{\underline{\Sigma}}(Var(J'))$, defined as $x\sigma_h = term_{J'}(h_V(x))$. In a similar way, also the track functions introduced in Definition 13 are associated with substitutions. □

We are now ready to present the relationship between narrowing and hyperedge replacement jungle rewriting. First we show how a program can be translated into an HR-jungle rewriting system.

**Definition 17 (the HR-jungle rules for equality).** For each sort $\mathbf{s} \in S$, the **(HR-jungle) equality rule** for $\mathbf{s}$ is the rule:

$$HR^{=}_{\mathbf{s}} \equiv \left(L = \{v_0^{\mathtt{bool}} : v_1^{\mathbf{s}} = v_2^{\mathbf{s}}\} \xleftarrow{l} K = \{v_0^{\mathtt{bool}}, v_1^{\mathbf{s}}, v_2^{\mathbf{s}}\} \xrightarrow{r} R = \{v_0^{\mathtt{bool}} : \mathtt{true}, v_3^{\mathbf{s}}\}\right)$$

where $r(v_1^{\mathbf{s}}) = r(v_2^{\mathbf{s}}) = v_3^{\mathbf{s}}$. □

**Definition 18 (the HR-jungle rewriting system representing a program).** Given a clause $C \equiv (t_1 = t'_1, \ldots, t_n = t'_n \rightarrow \mathtt{f}(s_1, \ldots, s_m) = s')$, its **HR-jungle representation** is the HR-jungle rewrite rule $\mathcal{HR}(C) = (L \stackrel{l}{\leftarrow} K \stackrel{r}{\rightarrow} R)$ where

- $L = \{v_0 : \mathtt{f}(v_1, \ldots, v_n)\}$; $K = \{v_0, v_1, \ldots, v_n\}$; $l_V(v_i) = v_i$ for all $0 \leq i \leq n$.
- $R = \mathcal{J}(\{\!|s', s_1, \ldots, s_m, t_1 = t'_1, \ldots, t_n = t'_n|\!\})$, i.e., $R$ is the variable-collapsed jungle representing the rhs and all the arguments of the lhs of $C$, together with all the equations in the premise of $C$, regarded as terms of sort **bool**.
- Morphism $r$ is determined (up to isomorphisms) by the following conditions: 1) $term_R(r(v_0)) = s'$; 2) $term_R(r(v_i)) = s_i$ for all $1 \leq i \leq n$; and 3) $\forall i \in \{0, \ldots, n\} . (r(v_i) \notin ROOTS(R) \vee \exists j \neq i . r(v_i) = r(v_j)) \Rightarrow r(v_i) \in Var(R)$.

The **HR-jungle representation** of a program $\mathcal{P} = \{C_i\}_{i \leq n}$ is the HR-jungle rewriting system $\mathcal{HR}(\mathcal{P}) = \{\mathcal{J}(C_i)\}_{i \leq n} \cup \{HR^{=}_{\mathbf{s}}\}_{\mathbf{s} \in S}$, where $HR^{=}_{\mathbf{s}}$ is as in Definition 17. □

*Example 3 (the HR-jungle rewrite rule representing a clause).* Let $C$ be the following clause over the signature $\Sigma$ of Example 1:

$$C \equiv (y = \mathtt{g}(\mathtt{d}, z) \rightarrow \mathtt{f}(\mathtt{c}(y, x), y) = \mathtt{g}(x, y))$$

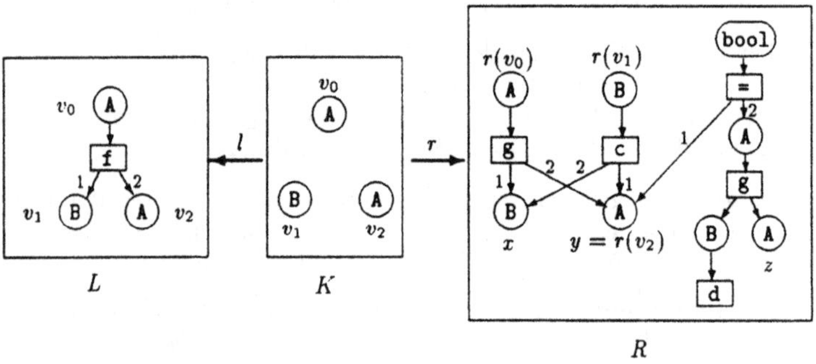


**Fig. 3.** A sample jungle rule.

Then the HR-jungle rewrite rule representing $C$, $\mathcal{HR}(C)$, is the rule shown in Figure 3. This rule can be understood more easily by resorting to a 'normalized' presentation of the clause. If $t_1 = t'_1, \ldots, t_n = t'_n \rightarrow \mathtt{f}(s_1, \ldots, s_m) = s'$ is a clause. its 'normal' form is

$$\mathtt{f}(v_1, \ldots, v_m) = s' \text{ where } t_1 = t'_1, \ldots, t_n = t'_n, v_1 \equiv s_1, \ldots, v_m \equiv s_m$$

with $v_1, \ldots, v_m$ fresh, distinct variables. The HR-jungle representation $(L \stackrel{l}{\leftarrow} K \stackrel{r}{\rightarrow} R)$ of the clause can be considered as a direct translation of this normal form, since $L$ represents exactly $\mathtt{f}(v_1, \ldots, v_n)$, $R$ represents all the equations of the premise, the term $s'$ and the arguments of the lhs of $C$, $s_1, \ldots, s_n$, and finally morphism $r$ forces the identifications $v_i \equiv s_i$. For example, the normal form of the above clause of (which should be compared with the rule in Figure 3) is:

$$\mathtt{f}(v_1, v_2) = \mathtt{g}(x, y) \text{ where } y = \mathtt{g}(\mathtt{d}, z), v_1 \equiv \mathtt{c}(y, x), v_2 \equiv y.$$

The following result states that in the category of jungles, the pushout of two arrows exists iff the associated substitutions have a unifier. For a proof we refer to [CR 93].

**Proposition 19 (pushouts in Jungle$_{\underline{\Sigma}}$ as most general unifiers).**

1. *Let $r : K \to R$ and $d : K \to D$ be two jungle morphisms. Let $\sigma_r : Var(K) \to T_{\underline{\Sigma}}(Var(R))$ and $\sigma_d : Var(K) \to T_{\underline{\Sigma}}(Var(D))$ be the associated substitutions (see Definition 16). Then the pushout of $\langle r, d\rangle$ exists in* **Jungle**$_{\underline{\Sigma}}$ *if and only if there exist two substitutions $\theta : Var(R) \to T_{\underline{\Sigma}}(Y)$ and $\theta' : Var(D) \to T_{\underline{\Sigma}}(Y)$ which 'unify' $\langle \sigma_r, \sigma_d\rangle$, in the sense that $\sigma_r\theta = \sigma_d\theta'$.*
2. *If $\langle g, f\rangle$ is a pushout of $\langle r, d\rangle$ in* **Jungle**$_{\underline{\Sigma}}$ *then $\langle \sigma_g, \sigma_f\rangle$ is a most general unifier of $\langle \sigma_r, \sigma_d\rangle$, i.e., for each pair of substitutions $\langle \theta, \theta'\rangle$ such that $\sigma_r\theta = \sigma_d\theta'$, there exists a $\sigma$ such that $\sigma_g\sigma = \theta$ and $\sigma_f\sigma = \theta'$.* □

We are almost ready to present the main result of this section, stating the faithfulness of the HR-jungle representation of programs, in the sense that every jungle rewriting starting from a jungle $J$ and using the HR-jungle representation of a program $\mathcal{P}$ has the same effect of a narrowing derivation starting from $GOAL(J)$ and using $\mathcal{P}$. This is a consequence of the fact that every direct jungle rewriting performed with the HR-jungle representation of a clause corresponds to one parallel narrowing step, and, similarly, that every direct rewriting via an equality rule corresponds to one application of the Base Rule. Since in general a jungle $J$ may represent other terms besides $GOAL(J)$ (which should be considered as "garbage"), we have to forbid the useless rewriting steps which act on those terms, because they do not correspond to any narrowing step. This motivates the next definition.

**Definition 20 (garbage rewritings).** Let $C$ be a clause, and let $J$ be a jungle. A direct rewriting $J \Rightarrow_{\mathcal{HR}(C)} H$ based on $g$ is a **garbage rewriting** iff there exist no path from the nodes in $ROOTS_{=}(J)$ to the node which is the image via $g$ of the unique root node of the the lhs of $\mathcal{HR}(C)$. □

**Theorem 21 (faithfulness of jungle representation of clauses).**

1. *Let $\mathcal{HR}(C) = (L \xleftarrow{l} K \xrightarrow{r} R)$ be the HR-jungle representation of clause $C$, and let $J \Rightarrow_{tr,\mathcal{HR}(C)} H$ be a non-garbage direct rewriting as in Figure 2, with track function $tr$. Moreover, let $n$ be the number of paths from the nodes in $ROOTS_{=}(J)$ to the image through $g$ of the (unique) root node of $L$ (clearly $n \geq 1$). Then $GOAL(J)$ narrows to $GOAL(H)$ in one parallel step of degree $n$ with substitution $\sigma_{tr} \upharpoonright var(GOAL(J))$ using clause $C$, where substitution $\sigma_{tr}$ is as in Definition 16.*
2. *Let $HR_{\mathbf{s}}^{=}$ be an equality rule, as defined in Definition 17, and suppose that $J \Rightarrow_{tr,HR_{\mathbf{s}}^{=}} H$. Then $GOAL(J) \rightarrow_{\sigma,=} GOAL(H)$ where $\sigma = \sigma_{tr} \upharpoonright var(GOAL(J))$, i.e., $GOAL(J)$ narrows to $GOAL(H)$ in one step via the application of the Base Rule.*
3. *Let $\mathcal{P}$ be a program and $\mathcal{HR}(\mathcal{P})$ be its associated HR-jungle rewriting system. Then from the two points above it follows that if jungle $J$ rewrites to $H$ using $\mathcal{HR}(\mathcal{P})$ (i.e., $J \Rightarrow^{*}_{tr,\mathcal{HR}(\mathcal{P})} H$) without garbage rewritings and $H$ has no edges labeled by '$=$', then $GOAL(J) \rightarrow^{*}_{\sigma,\mathcal{P}} \emptyset$, where $\sigma = \sigma_{tr} \upharpoonright var(GOAL(J))$.* □

From a practical point of view, the last theorem ensures that jungle rewriting using the HR-jungle representation of a program $\mathcal{P}$ can be used as a correct interpreter for narrowing using $\mathcal{P}$ if, as usual, one is interested in computations ending with the empty goal. In fact, in this case every jungle rewriting corresponds to a narrowing derivation computing the same substitution.

# 4 From Hyperedge Replacement to Elementary Rewriting.

We just showed that hyperedge replacement jungle rewriting faithfully models the Narrowing Calculus, essentially at the same level of abstraction. The goal of this and of the next section is to show that jungle rewriting is also suitable for describing narrowing at a lower level of abstraction, corresponding to the basic steps that a real narrowing interpreter performs in order to narrow a given goal. The interpreter we refer to is the LANAM interpreter shortly described in Section 6. It is worth anticipating here that the LANAM actually uses (a suitable encoding of) jungles to represent terms and goals, instead of trees. Therefore the sharing of subterms modelled by jungles is directly implemented in the LANAM.

The LANAM splits a single narrowing step into more elementary steps, like the selection of the rule to be applied, the comparison of the topmost operators of two terms (during a *lazy* unification), and the actual rewriting. Some of these steps concern the control strategy, and will be addressed in Section 5; the others model elementary transformations of the current goal, and will be treated here. Each of such elementary transformations performed by the LANAM on the current goal enjoys some properties which are easily motivated by efficiency considerations. Informally, these properties are:

1. *Constant Time Applicability*: The applicability of an elementary transformation at a given position in the current jungle can be checked in constant time.
2. *Node Persistence*: Any two distinct nodes of the current jungle will never be merged during the computation.
3. *Locality*: The transformation affects only a small, bounded portion of the current jungle.

It is easy to understand why a direct rewriting of a jungle $J$ via the application of an HR-jungle rule (like those introduced in Definition 18) does not correspond in general to an elementary transformation. In fact, none of the above conditions is ensured to hold:

1. By Definition 11 a rule is applicable to a given occurrence if the double pushout of Figure 2 can be constructed; existence of the pushout complement is ensured by Proposition 12, but by Proposition 19 to check for the existence of the right pushout is equivalent to compute the most general unifier of two substitutions, and this is linear in the dimension of the involved substitutions ([PW 78]).
2. *Node persistence* can be formalized by requiring that the track function of a direct rewriting be injective. This is in general not true.
3. As a consequence, in general the extent of the modifications caused by the application of a jungle rule cannot be bounded in space by a constant. In fact, the right pushout can cause the "folding" of subjungles of arbitrary depth. See [CR 93] for a discussion about this topic.

Nevertheless, it is possible to define a syntactic restriction on rules and a new applicability condition that are sufficient to guarantee that a jungle rewriting step is an elementary transformation. This is formalized as follows.

**Definition 22 (elementary rewritings).** A jungle rule $p = (L \xleftarrow{l} K \xrightarrow{r} R)$ is **injective** if $l$ is a bijection on nodes and $r$ is injective; it is **rooted** if $L$ has a unique root. An occurrence $g : L \rightarrow J$ is called **maximal (with respect to $p$)** if for all $v \in Var(L)$, either $g_V(v) \in Var(J)$ or $tr_p(v) \in Var(R)$, and for all $v, v' \in Var(L)$ such that $tr_p(v), tr_p(v') \notin Var(R)$, $g_V(v) \neq g_V(v')$ (see Definition 13).

A direct jungle rewriting $J \Rightarrow_p H$ based on $g : L \rightarrow J$ is **elementary** iff $p$ is rooted and injective, and $g$ is maximal. In this case we write $J \Rightarrow_p^e H$. Similarly, $J \Rightarrow_{\mathcal{R}}^{e*} H$

means that there is a jungle rewriting from $J$ to $H$ using rules of $\mathcal{R}$ where each step is elementary. □

**Proposition 23 (elementary jungle rewritings are "elementary").** *Let $p = (L \stackrel{l}{\leftarrow} K \stackrel{r}{\rightarrow} R)$ be a rooted, injective rule, and let $g : L \rightarrow J$ be an occurrence. Then:*

1. *It can be checked in constant time if $g$ is maximal with respect to $p$. Moreover, if this is true, then $p$ is applicable to $J$, i.e., there exists an $H$ such that $J \Rightarrow^e_p H$. Therefore if one restricts the application of rules to maximal occurrences, applicability can be checked in constant time.*
2. *The track function of an elementary direct rewriting is injective.* □

In the rest of this section we sketch how to transform the jungle rewriting system representing a program $\mathcal{P}$, $\mathcal{HR}(\mathcal{P})$, into an equivalent rewriting system $\mathcal{E}(\mathcal{P})$ whose rules correspond directly to the elementary transformations performed by the LANAM. Exploiting the last proposition, the idea is to put in $\mathcal{E}(\mathcal{P})$ a finite set of rooted, injective rules for each rule $p$ in $\mathcal{HR}(\mathcal{P})$: these rules should simulate $p$ when applied to maximal occurrences. In particular, for each jungle rule in $\mathcal{HR}(\mathcal{P})$ of the form $\mathcal{HR}(C)$ (see Definition 18), we will put in $\mathcal{E}(\mathcal{P})$ the set of all rules "subsumed" by it, while each equality rules will be replaced by a set of injective rules implementing *lazy* unification.

**Definition 24 (subsumption of rules).** Let $p = (L \stackrel{l}{\leftarrow} K \stackrel{r}{\rightarrow} R)$ and $p' = (L' \stackrel{l'}{\leftarrow} K' \stackrel{r'}{\rightarrow} R')$ be two jungle rules. Then $p$ **directly subsumes** $p'$ (written $p \sqsubset p'$) if either

- there exists a $v \in Var(L)$ such that $tr_p(v) \notin Var(R)$ (thus $term_R(tr_p(v)) = \mathtt{f}(t_1, \ldots, t_n)$ for some operator $\mathtt{f}$): $R = R'$; $L'$ ($K'$) is obtained from $L$ ($K$) by adding a single edge labelled by $\mathtt{f}$, having $v$ as source node, and $n$ distinct, new nodes as target; inclusion $l'$ and morphism $r'$ extend $l$ and $r$, respectively, in the obvious way; or
- there exist $v_1, v_2 \in Var(L)$ such that $tr_p(v_1), tr_p(v_2) \notin Var(R)$; $L'$ ($K'$) is obtained from $L$ ($K$) by identifying nodes $v_1$ and $v_2$; and $R'$ is the pushout object of $r$ and the obvious morphism $k : K \rightarrow K'$. Such a pushout exists iff $term_R(tr_p(v_1))$ and $term_R(tr_p(v_2))$ unify.

We denote by $\sqsubseteq$ the transitive and reflexive closure of $\sqsubset$, and we say that $p$ **subsumes** $p'$ if $p \sqsubseteq p'$. If $p$ is a jungle rule, then **Sub**$(p)$ denotes the set of all rules subsumed by $p$. □

Intuitively, if $p \sqsubset p'$, then $p'$ is obtained from $p$ either by copying a single edge of the rhs $R$ to both $K$ an $L$, or by identifying two distinct variable nodes in $L$ and $K$, and the corresponding nodes in $R$, if this is possible. It is easy to check that when both $p$ and $p'$ are applicable to the same jungle, the result is the same. But the relevant fact is that on the one hand there are jungles to which $p$ can be applied and $p'$ cannot; on the other hand, there are jungles to which $p'$ can be applied with a maximal occurrence, while $p$ cannot; and, finally, if $p$ can be applied to a jungle $J$, then there exists a $p'$ subsumed by $p$ which is applicable to $J$ with a maximal occurrence. These facts suggest to include in $\mathcal{E}(\mathcal{P})$ for each clause $C$ the set **Sub**$(\mathcal{HR}(C))$ of all rules subsumed by $\mathcal{HR}(C)$. However, in order to exploit Proposition 23, this is only worthwhile if all the rules in **Sub**$(\mathcal{HR}(C))$ are injective.[5] It is possible to show that given a clause $C$, the set **Sub**$(\mathcal{HR}(C))$ contains a non-injective rule if and only if $C$ is collapsing or non-left-linear. For non-left-linear clauses, we can assume without loss of generality that all the clauses of a program are

[5] All the rules we consider are rooted.

left-linear;[6] for collapsing clauses, we assume instead that their representation as jungle rule makes use of *indirection links*, which are a common implementation technique (they are used also in the LANAM) allowing one to meet the node persistency requirement also in situations where, from a semantical point of view, two nodes should be merged together. An indirection link is representedby a unary, overloaded operator *ind*, which is supposed to be "skipped" whenever it is found by the interpreter: this is equivalent to assume (only for specification purposes) the existence of the following jungle rule which get rid of indirection links.

**Definition 25 (rule for indirection links).** The ***ind*-rule** (for sort $\mathbf{s}$) is the jungle rewrite rule

$$\left(L = \{v_0^{\mathbf{s}} : ind(v_1^{\mathbf{s}})\} \xleftarrow{l} K = \{v_0^{\mathbf{s}}, v_1^{\mathbf{s}}\} \xrightarrow{r} R = \{v_2^{\mathbf{s}}\}\right)$$

where $r(v_0^{\mathbf{s}}) = r(v_1^{\mathbf{s}}) = v_2^{\mathbf{s}}$. □

Indirection links are used as follows in the represenatation of collapsing clauses.

**Definition 26 (the rule with indirection representing a collapsing clause).** Let $C$ be a collapsing clause with variable $x$ as rhs, and let $\mathcal{HR}(C)$ be its jungle representation as in Definition 18. Then the **jungle representation with indirection of** $C$, denoted $\mathcal{HR}_{ind}(C)$, is obtained from $\mathcal{HR}(C)$ by adding in the right-hand side a new node $v$ and an indirection link (i.e., an edge labelled by $ind$) having $v$ as source and $x$ as target. Moreover, morphism $r : K \to R$ is redefined so that the root node $v_0$ of $L$ is mapped to $v$. □

The next lemma formalizes what claimed above about the possibility of simulating direct rewritings via elementary rewritings, using the set of subsumed clauses.

**Lemma 27 (making direct rewritings elementary).** *Suppose that $p \sqsubseteq p'$, $J \Rightarrow_p H$ based on $g$, $J \Rightarrow_{p'} H'$ based on $g'$, and $g$ and $g'$ are consistent.*[7] *Then $H$ is isomorphic to $H'$.*

*Moreover, suppose that $C$ is not collapsing, and that $J \Rightarrow_{\mathcal{HR}(C)} J'$ based on $g$. Then there exists a unique rule $p \in$ $\mathbf{Sub}(\mathcal{HR}(C))$ such that $J \Rightarrow_p^e J'$. As a consequence, $J \Rightarrow_{\mathcal{HR}(C)} J'$ iff $J \Rightarrow^e_{\mathbf{Sub}(\mathcal{HR}(C))} J'$.*

*Finally, if $C$ is a collapsing clause, then the last statement holds by replacing $\mathcal{HR}(C)$ with $\mathcal{HR}_{ind}(C)$.* □

We sketch now how equality rules are handled. Since their role is to perform an atomic unification of terms and we want to model instead *lazy* unification, we replace each equality $HR_{\mathbf{s}}^{=}$ rule with a set of injective rules $E_{\mathbf{s}}^{=}$ modeling the various steps of the unification process. For the sake of space we do not show the rules, but we just describe them informally.

Each rule of $E_{\mathbf{s}}^{=}$ can be applied to a given equation of a goal with a maximal occurrence only if the arguments of the equality operator are of a suitable kind (i.e., variable or term), and they are applicable in mutually exclusive cases. There are therefore four rules in $E_{\mathbf{s}}^{=}$ (because each of the two arguments of '=' can be either a term or a variable), plus one for the case where both arguments are variables *and* they are represented by the same node.

[6] Any non-left-linear clause can be transformed into an equivalent left-linear one by adding suitable equations in the premises.

[7] If $p$ and $p'$ are rooted, $g$ and $g'$ are consistent if $g = g' \circ sub$ with $sub : L \to L'$ being the obvious subsumption morphism.

The rule applicable (with maximal occurrence) when both arguments are proper terms reports success for the comparison of the topmost operators of the arguments, if they are equal, and generates new equality operators for comparing the corresponding subterms. The two rules applicable when either one or the other argument is a variable make a copy the topmost operator of the term argument, and assign it to the variable argument: the subterms of the original term argument become shared. The rule applicable if both arguments are variables introduces an indirection link from one argument to the other. Finally, the rule applicable if both arguments are variables and are the same node simply report success deleting the equality operator.

The next result shows in which sense elementary rewriting using the rules in $E_{\mathbf{s}}^{=}$ is a faithful implementation of the equality rule $HR_{\mathbf{s}}^{=}$, assuming the existence of the *ind*-rule.

**Lemma 28 (faithfulness of rules for lazy unification).** *Let* $\underline{E}_{\mathbf{s}}^{=} = E_{\mathbf{s}}^{=} \cup \{ind\text{-}rule\}$. *Then:*

1. *If* $J \Rightarrow_{tr,HR_{\mathbf{s}}^{=}} H$, *then there exists a jungle* $H'$ *such that* $J \Rightarrow_{tr',\underline{E}_{\mathbf{s}}^{=}}^{e*} H'$, $GOAL(H) = GOAL(H')$, *and* $\sigma_{tr} = \sigma_{tr'}$.
2. *If* $J_0 \Rightarrow_{tr,\underline{E}_{\mathbf{s}}^{=}}^{e*} J_n$ *and* $GOAL(J_n) = \emptyset$, *then for each equality operator in* $J_0$ *having* $v_1$ *and* $v_2$ *as target nodes,* $\sigma_{tr}$ *unifies the terms* $term_{J_0}(v_1)$ *and* $term_{J_0}(v_2)$. □

We are now ready to present the main result of this section, which relates jungle rewriting using the HR-jungle representation of a program with *elementary* rewriting using its injective jungle representation.

**Definition 29 (the injective jungle rewriting system representing a program).** Let $\mathcal{P}$ be a program. Then the **injective jungle representation** of $\mathcal{P}$ is the jungle rewriting system $\mathcal{E}(\mathcal{P})$ containing for each sort $\mathbf{s} \in S$ the rules for lazy unification $E_{\mathbf{s}}^{=}$; for each clause $C$ in $\mathcal{P}$ the set of injective rules subsumed by $\mathcal{HR}(C)$ if $C$ is not collapsing, and by $\mathcal{HR}_{ind}(C)$, if $C$ is collapsing; and the *ind*-rule. Formally, $\mathcal{E}(\mathcal{P}) = \cup\{\mathbf{Sub}(\mathcal{HR}(C)) \mid C \in \mathcal{P}$ and $C$ is not collapsing$\} \bigcup \cup\{\mathbf{Sub}(\mathcal{HR}_{ind}(C)) \mid C \in \mathcal{P}$ and $C$ is collapsing$\} \bigcup \cup_{\mathbf{s} \in S} E_{\mathbf{s}}^{=} \bigcup \{ind\text{-rule}\}$. □

**Theorem 30 (relating HR- and injective representations of a program).** *Let* $\mathcal{HR}(\mathcal{P})$ *be the HR-jungle representation of* $\mathcal{P}$ *(Definition 18) and let* $\mathcal{E}(\mathcal{P})$ *be its injective jungle representation just defined.*

1. *If* $J \Rightarrow_{tr,\mathcal{HR}(\mathcal{P})}^{*} H$, *then there exist a jungle* $H'$ *and a track function* $tr'$ *such that* $J \Rightarrow_{tr',\mathcal{E}(\mathcal{P})}^{e*} H'$, $GOAL(H) = GOAL(H')$, *and* $\sigma_{tr} = \sigma_{tr'}$.
2. *If* $J \Rightarrow_{tr,\mathcal{E}(\mathcal{P})}^{e*} H$ *and* $GOAL(H) = \emptyset$, *then there exists a jungle* $H'$ *and a track function* $tr'$ *such that* $J \Rightarrow_{tr',\mathcal{HR}(\mathcal{P})}^{*} H'$, $GOAL(H') = \emptyset$, *and* $\sigma_{tr} = \sigma_{tr'}$.

□

## 5 Marking and Strategies.

In this section we shortly sketch how the evaluation strategy of a narrowing interpreter can be specified using a slightly modified version of jungle rewriting, which makes use of a suitable marking of nodes. In the evaluation of goals by narrowing we have two sources of nondeterminism, namely (1) where in the goal the next rule will be applied, and (2) which of the rules applicable at a position is chosen. We call the first one **redex-**, the latter one **rule-nondeterminism**. For confluent and strong terminating programs which guarantee unique normal forms the redex nondeterminism is "don't care" for rewriting. To model a narrowing interpreter by jungle rewriting we should be able to describe redex selection strategies. Therefore we extend our notions of jungle and jungle morphism with node markings expressing the state of evaluation in a jungle node. These markings are related to, but different from, the markings used in DACTL ([GKSS 88]). Moreover, they do not enable the specification of arbitrary redex selection strategies, but efficient strategies which can be realized by a compiler for functional logic languages based on graph rewriting are expressible.

**Definition 31 (marked jungles).** A **marked jungle over $\underline{\Sigma}$** is a $\underline{\Sigma}$-jungle $J = (V, E, src, trg, m, l)$ with an extended node labeling $l : V \to \{\bot, *\} \times S$. If $v$ is a node labelled by $(*, \mathbf{s})$ (marked for execution) it could be indicated by $v^*_{\mathbf{s}}$, or simply $v^*$. □

Morphisms of marked jungles are defined in the same way as for jungles, but the label preserving property is weakened: An unmarked node may be mapped on a marked one but not vice versa.

A marked node $v^*$ represents a node where the application of a jungle rule is allowed. Thus marking all nodes in the input goal and in the jungle rewrite rules of $\mathcal{E}(\mathcal{P})$ models exactly jungle rewriting with unmarked jungles as defined in Section 4. To model a redex selection strategy we augment the rules of $\mathcal{E}(\mathcal{P})$ with specific node markings. Additional rules for the propagation of markings in the goal are necessary. As an example, we define a lazy evaluation strategy.

### 5.1 Marked jungle rules for lazy evaluation

Lazy evaluation is based on the assumption that constructors and functions in the specification are strictly separated. Then it is safe to evaluate first the arguments of a function call which are necessary to decide which rule can be applied to the outermost position.

Lazy narrowing reduces a term to its **head normal form (HNF)**. A HNF of a term is a term which cannot be narrowed further on the outermost level (even though subterms may be further narrowable). Thus for a term to become completely narrowed, all subterms must also be narrowed to HNF. We say that an occurrence in a term is **stabilized** if the subterm at this occurrence is in HNF. **Evaluation** of an expression representing a subterm means stabilizing this subterm.

Lazy narrowing applied on goals means that both sides in the goal equations are evaluated simultaneously top-down, until different constructors at corresponding positions occur (and the narrowing fails), or the terms are in normal form and syntactically equal. That means that the computed answer substitution can contain unevaluated subterms where lazy narrowing should recursively be applied in order to complete the computation.

**Definition 32 (marked rules for lazy evaluation).** The set of **marked jungle rules** $\mathcal{M}(\mathcal{P})$ is obtained by marking all rules from $\mathcal{E}(\mathcal{P})$ as follows. For each rule $p = (L \xleftarrow{l} K \xrightarrow{r} R)$ in $\mathcal{E}(\mathcal{P})$ the root node $v_0$ of $L$, its image in $K$, $l^{-1}(v_0)$, its image in $R$, $r \circ l^{-1}(v_0)$, and all root nodes in $ROOTS_=(R)$ are marked with $*$. □

Marking the root nodes ensures a top-down evaluation strategy.

**Definition 33 (lazy direction, rules propagating node markings).** For every pair of jungle rewrite rules $p = (L \xleftarrow{l} K \xrightarrow{r} R)$, $p' = (L' \xleftarrow{l'} K' \xrightarrow{r'} R') \in \mathcal{M}(\mathcal{P})$ such that $p \sqsubset p'$ and $L'$ ($K'$) is obtained from $L$ ($K$) by adding a single edge having $v$ as source node, we call $v$ a **lazy direction for p**. Then $p_p = (L_p \xleftarrow{l_p} K_p \xrightarrow{r_p} R_p)$ is a **rule propagating**

**node markings** for $P$ where $L_p = L, K_p = K$ and $R_p$ is obtained from $R$ by marking $v$ with $*$. The set of these rules is denoted by $\mathcal{P}(\mathcal{P})$. □

Since $p$ is a rule from $\mathcal{M}(\mathcal{P})$, its root nodes are already marked. $L$ contains the already computed part of the pattern. Node $v$ indicates the direction where the evaluation must proceed in order to apply a rule with maximal occurrence at the position where $L$ matches. Therefore $v$ is marked for execution by $p_p$.

The special rules of Section 4 for indirection and for lazy unification are augmented with markings in an obvious way, so that they propagate them top-down in the goal to be evaluated.

### 5.2 Sequentialisation of condition evaluation

Marking all the root nodes representing conditions in the right-hand sides of the rules of $\mathcal{M}(\mathcal{P})$ causes a high degree of parallelization: All conditions of a clause can be evaluated in parallel with the function body. To achieve a sequentialisation of condition evaluation, especially to force the evaluation of all conditions before the function body is constructed, a program transformation is needed. The conditions could be transformed together with the body into a single expression using a special function *if_then*:

Given a clause $C \equiv (t_1 = t_1', \ldots, t_n = t_n' \rightarrow \mathtt{f}(s_1, \ldots, s_m) = s')$, its transformation would be $C' \equiv \mathtt{f}(s_1, \ldots, s_m) = if\_then(t_1 = t_1', \cdots (if\_then(t_n = t_n', s')) \cdots)$ similar to the functional logic language BABEL ([Loo 91]). The function *if_then* is defined by the clause $if\_then(\mathtt{true}, x) = x$. Now the order of condition evaluation can be controlled. The rules in $\mathcal{P}(\mathcal{P})$ generated from the clause for *if_then* would sequentialize the condition evaluation from left to right.

### 5.3 Correctness and Completeness of lazy narrowing

Due to space limitations we will only give a brief and informal overview of correctness and completeness considerations for strategy controlled jungle rewriting.

The correctness of strategy controlled narrowing derivations follows trivially from the fact, that node markings restrict the applicability of jungle rules. Every elementary jungle rewriting step using a rule from $\mathcal{M}(\mathcal{P})$ can be performed in the category of unmarked jungles by the corresponding rule from $\mathcal{E}(\mathcal{P})$. The rules in $\mathcal{P}(\mathcal{P})$ propagate markings, but leave the rewritten goal unchanged. Completeness of strategy controlled narrowing depends on the chosen strategy. For the lazy evaluation strategy it is possible to prove, that every narrowing derivation leading to an empty goal has a corresponding lazy narrowing derivation when the program fulfills the constructor discipline.

**Definition34 (constructor discipline).** A program $P$ fulfills the constructor discipline when for all clauses $C \equiv (t_1 = t_1', \ldots, t_n = t_n' \rightarrow \mathtt{f}(s_1, \ldots, s_m) = s')$ in $P$, all $s_i, i \in \{1, \ldots, m\}$, are data terms and $\mathtt{f} \in \{\Sigma_{\mathbf{W},\mathbf{S}}\}$ is a function. □

Restriction of the marking propagation to lazy directions avoids unnecessary (potential nonterminating) rewriting derivations when the program fulfills an additional uniformity constraint. In [Lo 92] a semantic-preserving transformation into a uniform program is presented.

## 6 Relationship to abstract machine code.

In this section we describe very shortly some features of the LANAM narrowing interpreter, trying to stress the relationship with the jungle rewriting systems presented in the previous sections. A deeper analysis of this relationship is the subject of future work.

Graph reduction is a basic design concept for the implementation of functional programming languages with lazy evaluation. The use of graphs instead of trees supports sharing. All occurrences of a formal parameter are replaced by an unique reference to the graph representing the corresponding argument. Repeated evaluation and unnecessary copying of arguments is prevented.

For functional logic languages lazy evaluation is necessary to avoid many infinite branches in the search tree. Hence graph reduction is used by most implementations for functional logic languages ([Loo 91, Wo 91, Lo 92]).

To describe the transformation from the source language into abstract machine code an intermediate language or formalism is needed, whose level of abstraction is between conditional equations and abstract machine code. In [Lo 92] the language $L_0$ is used, where sharing is expressed by means of a context containing variable bindings. We proposed, alternatively, jungle rewriting as intermediate language, because they have an elegant and abstract operational semantics which supports sharing directly, as shown in the previous sections. Jungle rewriting is flexible enough to express different levels of abstraction, the lowest level being very close to the abstract machine semantics. Usage of jungle rewriting supports correctness proofs and helps to understand the abstract machine instructions. The data structures used by the graph-based abstract machine are shortly presented below, and directly correspond to jungles.

### 6.1 LANAM: a Lazy Abstract NArrowing Machine

The abstract machine LANAM ([Wo 91]) has been designed by the second author for a simulator of the process specification language LOTOS ([EW 92]).

LANAM is an extension of a lazy term rewriting machine LATERM ([Wo 89]) to handle unification and backtracking by concepts adapted from WAM ([War 83]), an abstract machine for PROLOG. Although the LANAM is usable only for first order rules, it has many similarities with the Jump-machine ([Lo 92]) and the stack narrowing machine ([Loo 91]).

Compilation of conditional rewrite rules into LANAM code is based on a transformation of the pattern into decision trees for efficient unification, similar to the technique used for the compilation of LML ([Jos 87]). There is a strong correspondence between this transformation and the rules $\mathcal{P}(\mathcal{P})$ introduced in Definition 33. In fact, the decision tree uses lazy directions at its branches.

### 6.2 Representation of jungles by LANAM

The goal currently narrowed at a certain evaluation state contains already stabilized term nodes on top and unevaluated subterms in lower positions. The stabilized nodes are represented by pairs $(\mathbf{k}, e)$ consisting of a constructor $\mathbf{k}$ and environment pointer $e$ representing the argument nodes. The environment is a frame containing argument pointers $a_1, \ldots, a_n$ following one another in memory. Subterms can be shared, hence the stabilized nodes correspond directly to jungles with an hyperedge pointing to a constructor node $\mathbf{k}$ and argument nodes $a_1, \ldots, a_n$.

Logical (free) variables are represented by empty placeholders in memory, whose content is filled during instantiation by the pair $(\mathbf{k}, e)$ representing the instantiated term. This directly reflects unification performed by a pushout in the category of jungles. Indirection nodes are used when two variables are unified or when the evaluation of a function returns a unbound variable. Section 4 has shown how this is modeled by jungle rule transformations using special rules for lazy unification and indirection.

Unevaluated subterms are represented by closures $(I, e)$, i.e., pairs consisting of the code to evaluate the subterm together with the corresponding environment pointer. The closure technique is an optimization of pure graph reduction which cannot directly be modeled by jungle rewrite rules. It avoids the explicit construction of reducible jungles in memory. The code $I$ directly returns the HNF of the represented subterm. The computed HNF is exactly the same as if jungle rewriting would have been performed.

### 6.3 Unification and backtracking by LANAM

We have only room for a very brief overview of the internals of LANAM. LANAM consists of data structures supporting nested function calls (continuation stack), unification (accumulator, argument and environment stacks) and backtracking (trail and choicepoint stacks). To enable backtracking most of the stacks are realized by linked lists.

LANAM-code of each function consists of four parts:

1. matching code to determine the applicable rewrite rules for given arguments. In terms of jungle rewriting: This code checks whether the pushout on the right-hand side of diagram representing the application of a rule exists. It also performs a part of the pushout construction: The rule variables are bound to argument jungles. This code is based on the construction of a decision tree mentioned above.
2. code for the instantiation of free variables. This code realizes the part of the pushout construction where variables in the input goal are bound by subterms of the left-hand side of a rule.
3. code for the lazy evaluation of the conditions. It computes equality, exactly in the same way as the equality jungle rules introduced in Section 4.
4. code for the right-hand sides of the rewrite rules. This code realizes the remaining part of the pushout construction and activates the evaluation of the constructed jungle.

Jungle rewriting and execution of LANAM code are closely related, especially if we consider the injective rules augmented with markings for the representation of strategies (see Section 5).

## 7 Conclusions and Future Work.

In this paper we showed how jungle rewriting can model narrowing at different levels of abstraction. These correspond to intermediate steps in the compilation of functional logic programs. We described how the jungle rewrite rules representing program clauses can be transformed to model implementation aspects as evaluation strategy, constant time applicability and locality. Correctness and completeness of the most abstract level and the first transformation step was shown.

Our ongoing research concentrates on the formal relationship between jungle rewriting and abstract machine semantics aiming at a correctness prove for a compiler for graph based functional logic languages.

In the paper we just considered acyclic jungles. A generalization of our formalism could permit cyclic jungles as well. This would allow us to model infinite data structures which can be handled by lazy evaluation.

Another interesting extension we are working on is the treatment of higher order functions by jungle rewriting.

## References


[CMREL 91] A. Corradini, U. Montanari, F. Rossi, H. Ehrig, M. Löwe. *Logic Programming and Graph Grammars*. [EKR 91], pp. 221-237.

[CR 93] A. Corradini, F. Rossi. *Hyperedge Replacement Jungle Rewriting for Term Rewriting Systems and Logic Programming*. Theoretical Computer Science, **109**, 1993, pp. 7–48..

[DJ 90] N. Dershowitz, J.-P. Jouannaud. *Rewrite Systems*. Handbook of Theoretical Computer Science (ed. J. van Leeuwen), Vol. B, North Holland, 1990, pp. 243–320.

[EW 92] H. Eertink. D. Wolz. *Symbolic Execution of* LOTOS *Specifications*. IFIP Transactions, Formal Description Techniques V, North-Holland, 1992.

[Eh 87] H. Ehrig. *Tutorial introduction to the algebraic approach of graph-grammars*. [ENRR87], pp. 3-14.

[EKR 91] H. Ehrig, H.-J. Kreowski, G. Rozenberg (Eds.). *Proceedings of the $4^{th}$ International Workshop on Graph-Grammars and Their Application to Computer Science*. LNCS 532, 1991

[ENRR 87] Ehrig, H., Nagl, M., Rozenberg, G., Rosenfeld, A., (Eds.). *Proceedings of the $3^{rd}$ International Workshop on Graph-Grammars and Their Application to Computer Science*, LNCS 291, 1987.

[EPS 73] H. Ehrig, M. Pfender, H.J. Schneider. *Graph-grammars: an algebraic approach*. Proc, IEEE Conf. on Automata and Switching Theory, 1973, pp. 167-180.

[GKSS 88] J.R.W. Glauert, J.R. Kennaway, M.R. Sleep, G.W. Sommer. *Final Specification of DACTL*. Report SYS-C88-11, School of Information Systems, University of East Anglia, Norwich, U.K.

[HP 91] B. Hoffmann, D. Plump. *Implementing Term Rewriting by Jungle Evaluation*. Informatique théorique et Applications/Theoretical Informatics and Applications, **25** (5), 1991, pp. 445–472.

[Jos 87] T. Johnsson. *Compiling Lazy Functional Languages, Part 1*. PhD thesis, Chalmers Tekniska Högskola, Göteborg, Sweden, 1987.

[Ll 87] J.W. Lloyd. *Foundations of Logic Programming*. Springer Verlag, 1984, (2nd Edition 1987).

[Lo 92] H.Lock. *The Implementation of Functional Logic Programming Languages*. PhD thesis, GMD Universität Karlsruhe, 1992.

[Loo 91] R.Loogen. *From Reduction Machines to Narrowing Machines*. TAPSOFT 1991, Springer, LNCS 494, 1991

[Pad 88] P.Padawitz. *Computing in Horn Clause Theories*. Springer EATCS Monographs on Theor. Com. Sci., 1988

[PW 78] M.S. Paterson, M.N. Wegman. *Linear Unification*. Journal of Comp. Syst. Sci., **16** (2), 1978, pp. 251–260.

[War 83] D.H.D.Warren. *An Abstract Prolog Instruction Set*. SRI International, Technical Note 309, October 1983

[Wo 89] D. Wolz, P.Boehm. *Compilation of LOTOS Data Type Specifications*. Proc. of the IFIP TC6-WG 6.1, 9th International Symposium on Protocol Specification, Testing and Verification (ed. E.Brinksma, G.Scollo, C.A.Vissers), 1989

[Wo 91] D. Wolz. *Design of a Compiler for Lazy Pattern Driven Narrowing*. Proceedings of the 7th international workshop on the specification of abstract data types, Springer LNCS 534, 1991.

# Recognizable Sets of Graphs of Bounded Tree-Width

Bruno Courcelle*
Bordeaux-1 University
Laboratoire d'Informatique (associé au CNRS), 351 cours de la Libération.
33405, Talence, France.

Jens Lagergren †
The Royal Institute of Technology
NADA, KTH, S-100 44 Stockholm, Sweden

**Abstract**

We establish that a set of finite graphs of tree-width at most $k$ is recognizable (with respect to the algebra of graphs with an unbounded number of sources) *if and only if* it is recognizable with respect to the algebra of graphs with at most $k$ sources. We obtain a somewhat stronger result for sets of simple finite graphs of tree-width at most $k$.

## 1 Introduction

The notion of a recognizable set of (finite) graphs is essential in some results yielding constructions of context-free graph grammars on one hand (see [4]) and of efficient graph algorithms on the other. In particular, for any fixed $k$, the membership in a recognizable set of graphs of tree-width at most $k$ can be determined in linear time in two different ways: (1) by means of a finite-state tree-automaton operating on a tree encoding a tree-decomposition of width at most $k$ of the given graph (see [4, 6, 3]), or, (2) by means of a "reducing" graph rewriting system, operating directly on the graph and not on a tree-decomposition (see [2]).

In both cases, recognizability plays an essential role. This notion has several essentially equivalent formulations, thoroughly compared in [8, 7, 1]. We only recall the most convenient one for the purposes of the present paper.

Let $m$ be a nonnegative integer and $G$ be a graph. We write $G = H \square K$, where $H$ is a graph with $m$ distinguished vertices that are called its sources, and $K$ is another graph with $m$ sources, if $G$ is the gluing of $H$ and K, resulting from the fusion of the $i$:th sources of $H$ and $K$, for every $i = 1, \ldots, m$. Note that $H$ and $K$ can be viewed as

*Supported by ESPRIT-Basic Research Working group "COMPUGRAPH"
†Supported by the Swedish Natural Science Research Council

two subgraphs of $G$ such that $(H, K)$ is a separation of size $m$ of G, where the vertices common to $H$ and $K$ are the sources of $H$ and $K$. The recognizability of $L$ means that, in order to decide whether a graph $G$ of the form $H\Box K$ belongs to $L$. we need only know to which of finitely many auxilliary pairwise disjoint recognizable sets the graphs $H$ and $K$ do belong. More precisely, these auxilliary sets are the classes of a congruence relative to the operation $\Box$. (This property is called full cutset regularity in Abrahamson and Fellows [1].) An algorithm deciding the membership of $G$ in $L$ can operate as follows: if $G = H\Box K$, and $H$ belongs to some fixed finite set. one replaces it by a (smaller) canonical representative of its class. Such replacements are repeated until one reaches some "irreducible" graph. The original graph $G$ belongs to $L$ if and only if the obtained irreducible graph is a member of a (computable) finite set. This is the basic idea of the reduction algorithm of [2].

The definition of recognizability is based on the existence of a congruence (relative to the operation $\Box$) that has, for each integer $m$, finitely many classes of graphs with $m$ sources. In the present paper, we establish that for a set of graphs of tree-width at most $k$, it suffices to have these facts for the integers $m$ up to $k$ in order to get a recognizable set. This gives an alternative definition of recognizability, equivalent to the first one, but easier to verify. We also establish that for a set of simple graphs of tree-width at most $k$, it suffices to have these facts for $k$ in order to get a recognizable set.

Some extensions and related open problems are discussed in Section 7.

# 2 Graph operations and recognizability

We consider finite and undirected graphs that may have loops and multiple edges. By $V(G)$ we denote the set of vertices of a graph $G$ and by $E(G)$ its set of edges. A graph $G$ has *tree-width at most* $w$ if there is a pair $(X, T)$ such that $T$ is a tree and $X = \{X_t\}_{t\in V(T)}$ a family of subsets of $V(G)$, called *bags*, such that: (1) those nodes $t$ in $T$ whose bags $X_t$ contain a given vertex $v \in V(G)$ induce a subtree of $T$, (2) every pair of adjacent vertices in $G$ share membership in some bag $X_t$, (3) every vertex in $G$ is in some bag $X_t$, and (4) $|X_t| \leq w + 1$ for all $t \in V(T)$. Such a pair $(X, T)$ is called a *tree-decomposition of* $G$ *of width at most* $w$. The smallest $w$ such that $(X, T)$ is a tree-decomposition of width at most $w$ is called the width of $(X, T)$. It is well known that a graph without loops and multiple edges has tree-width at most $k$ *if and only if* it is a partial $k$-tree.

Let $\mathbf{C}$ be a fixed countable set of labels. An *s-graph* can be described as a graph with a possibly empty set of distinguished vertices called sources all with different labels from $\mathbf{C}$. Formally an s-graph (read a sourced graph) is a pair $G = \langle G', f\rangle$ where $G'$ is a graph and $f$ a one-to-one mapping from $C$, a finite subset of $\mathbf{C}$, to $V(G')$. The graph $G'$ is called the underlying graph of $G$. We will call the vertices and edges of $G'$ vertices and edges of $G$, as well. We say that $f(C)$ is the set of sources of $G$. Whenever $c$ is in $C$, $f(c)$ is called the $c$-source of $G$ and $c$ the label of $f(c)$. A vertex that is not a source is called an *internal vertex*. We call $C$ the type of $G$ and denote it by $\tau(G)$. We denote by $\mathbf{G}$ the set of all s-graphs and by $\mathbf{G}(C)$ the set of all s-graphs of type $C$. We identify $\mathbf{G}(\emptyset)$ with the set of all graphs.

By a clique in a graph, we mean an induced subgraph in which any two vertices are linked by at least one edge. Let $G$ and $H$ be two s-graphs. We say that $G$ is an *s-subgraph* of $H$, and we write $G \subseteq H$, if the underlying graph of $G$ is a subgraph of the underlying graph of $H$ and every $c$-source of $G$ is a $c$-source of $H$. Notice, these conditions imply in particular that $\tau(G) \subseteq \tau(H)$. Two s-graphs $G = \langle G', g\rangle$ and $H = \langle H', t\rangle$ are isomorphic if there exists an isomorphism $f$ from $G'$ to $H'$ that preserves source labels, that is, $t = f \circ g$. Of cource, this implies that $G$ and $H$ are of the same type.

We define three types of operations on s-graphs.

*Parallel composition:* Let $G = \langle G', f\rangle \in \mathbf{G}(C)$, $H = \langle H', t\rangle \in \mathbf{G}(C')$, and $K = \langle K', w\rangle \in \mathbf{G}(C \cup C')$. We say that $K$ is the parallel composition of $G$ and $H$ and write $K = G /\!/ H$ *if and only if* $G$ and $H$ are s-subgraphs of $K$ such that:

1. $V(K') = V(G') \cup V(H')$ and $E(K') = E(G') \cup E(H')$; and
2. $E(G') \cap E(H') = \emptyset$, and a vertex $v \in V(K')$ belongs to $V(G') \cap V(H')$ *if and only if* for some $c \in C \cap C'$ the vertex $v$ is the $c$-source of $G$, $H$, and $K$.
3. a vertex $v \in V(K')$ is the $c$-source of $K$ *if and only if* it is the $c$-source of $G$ or $H$.

Note that $/\!/$ is a partial operation. However, if $G /\!/ H$ is not defined, it is possible to find a graph $H'$ isomorphic to $H$ such that $G /\!/ H'$ is defined.

*Source restriction*: Let $C$ be a finite subset of $\mathbf{C}$. We define $\mathbf{rest}_C$ to be the mapping from s-graphs to s-graphs such that $\mathbf{rest}_C(\langle G', f\rangle) = \langle G', f_{|C}\rangle$ where $f_{|C}$ is the restriction of $f$ to $C$. Hence, the sources of $G$ having a label not in $C$ will be vertices in $\mathbf{rest}_C(G)$, but they will be internal vertices not sources. Thus, the type of $\mathbf{rest}_C(G)$ is $\tau(G) \cap C$. Notice, for any s-graph $G = \langle G', f\rangle$, $\mathbf{rest}_\emptyset(G)$ is the underlying graph of $G$, that is $G'$.

*Source renaming*: Let $C$ be a finite subset of $\mathbf{C}$ and $h : C' \to C$ be a bijection. We define $\mathbf{ren}_h : \mathbf{G}(C) \to \mathbf{G}(C')$ to be the mapping such that $\mathbf{ren}_h(\langle G', f\rangle) = \langle G', f \circ h\rangle$.

It is clear that if $G_1$ is isomorphic to $G_2$ and $H_1$ is isomorphic to $H_2$, then $G_1 /\!/ H_1$, $\mathbf{rest}_C(G_1)$, and $\mathbf{ren}_h(G_1)$ are isomorphic to $G_2 /\!/ H_2$, $\mathbf{rest}_C(G_2)$, and $\mathbf{ren}_h(G_2)$, respectively. For every set $L$ of graphs, by $\overline{L}$ we denote the set of graphs that contains a graph $G$ *if and only if* $G$ is isomorphic to some graph in $L$. If $G$ and $H$ are two s-graphs of some type $C$ and $G /\!/ H$ is defined then $\mathbf{rest}_\emptyset(G /\!/ H)$ is a graph without sources. In the introduction, we denoted $\mathbf{rest}_\emptyset(G /\!/ H)$ by $G \Box H$.

Let $L$ be a set of graphs and $C \subseteq \mathbf{C}$. A *$C$-congruence for $L$* is an equivalence relation $\sim$ on $\mathbf{G}(C)$ such that:

(i) any two isomorphic s-graphs in $\mathbf{G}(C)$ are equivalent by $\sim$; and

(ii) for every $G$, $H$, and $K$, all in $\mathbf{G}(C)$, if $G \sim H$, and $G /\!/ K$ and $H /\!/ K$ are defined then

$$\mathbf{rest}_\emptyset(G /\!/ K) \in \overline{L} \textit{ if and only if } \mathbf{rest}_\emptyset(H /\!/ K) \in \overline{L}.$$

An equivalence relation $\sim$ on a set $A$ is ***finite*** *if and only if* it has finitely many equivalence classes. The finiteness condition is equivalent to the existence of a (total) mapping $f$ from $A$ to a finite set $B$ such that the kernel of $f$ is $\sim$ (that is, $a \sim a'$ *if and only if* $f(a) = f(a')$). We say that $L$ is ***recognizable*** *if and only if* for every finite $C \subseteq \mathbf{C}$, there exists a finite $C$-congruence for $\overline{L}$. Note that $L$ is recognizable *if and only if* $\overline{L}$ is recognizable.

**Lemma 2.1** *Let $h : C \to C'$ be a bijection. If $\sim$ is a $C$-congruence for $L$, then the relation $\approx$ on $\mathbf{G}(C')$ defined by*

$$G \approx H \text{ if and only if } \mathbf{ren}_h(G) \sim \mathbf{ren}_h(H)$$

*is a $C'$-congruence for $L$.*

*Proof.* Notice, for all $G, H \in \mathbf{G}(C')$

(i) $G$ is isomorphic to $H$ *if and only if* $\mathbf{ren}_h(G)$ is isomorphic to $\mathbf{ren}_h(H)$; and

(ii) $\mathbf{rest}_\emptyset(G /\!/ H) = \mathbf{rest}_\emptyset(\mathbf{ren}_h(G) /\!/ \mathbf{ren}_h(H))$.

The result follows by straightforward verification. □

Hence, whenever $L$ has a finite $C$-congruence it has a finite $C'$-congruence for each $C'$ of the same cardinality as $C$. When this is the case and the cardinality of $C$ is $k$, we will say that $L$ is $k$-recognizable. So, $L$ is recognizable *if and only if* it is $k$-recognizable for each $k$. Defined in this way recognizability coincides with full cutset regularity as defined in [1]. See section 7 for a discussion of this notion.

# 3 The main theorem

In the next section, we will consider conditions ensuring that for a set of graphs $L$: $L$ is $k'$-recognizable for all $k' \leq k$ *if* $L$ is $k$-recognizable. In section 5, we will go in the other direction. That is, conditions will be considered that ensure that for a set of graphs $L$: $L$ is $k'$-recognizable for all $k' \geq k$ *if* $L$ is $k$-recognizable. In the last section, we will prove that every set of graphs of tree-width at most $k$ satisfies these latter conditions. If the graphs, moreover, are *simple*, that is, without loops and multiple edges, they will also satisfy the former conditions. We obtain the following main result of this paper.

**(Main) Theorem 3.1** *(1) A set of graphs of tree-width at most $k$ is recognizable* if and only if *it is $k'$-recognizable for every $k' \leq k$*

*(2) A set of simple graphs of tree-width at most $k$ is recognizable* if and only if *it is $k$-recognizable.*

We give the proof now, even though, the necessary preliminaries are yet to be established.

*Proof.* (1) Let $L$ be a set of graphs of tree-width at most $k$ that is $k'$-recognizable for every $k' \leq k$. By Proposition (5.2) and (6.2), $L$ is $k+1$-recognizable. Since the argument can be repeated, one gets by induction that $L$ is $k'$-recognizable for every $k' \geq k$.

(2) If $L$ is a set of simple graphs of tree-width at most $k$ and $k$-recognizable, we get by Propositions (4.1) (ii) that $L$ is $k'$-recognizable for every $k' \leq k$. The argument for $k' > k$ is as in (1). Hence, it follows from the initial assumption that $L$ is recognizable. □

# 4 Going to smaller $k$

For every integer $m$, let $\mathbf{G}_m$ denote the set of all s-graphs having at most $m$ loops at each vertex and at most $m+1$ parallel edges between any two vertices. (Hence, $\mathbf{G}_0$ is the set of simple graphs.)

**Proposition 4.1** *Let $L$ be a set of graphs such that at least one of condition (i) and (ii) are satisfied:*

*(i) $L$ is closed under addition and deletion of isolated vertices;*

*(ii) $L \subseteq \mathbf{G}_m$ for some $m$.*

*Then, if $L$ is $k$-recognizable, it is $k'$-recognizable for every $k' \leq k$.*

*Proof.* It is enough to prove that the conditions imply that $L$ is $(k-1)$-recognizable.

Let $C = C' \cup \{c\}$ where $c \notin C'$ and $C'$ has cardinality $k-1$. Let $\sim$ be a finite $C$-congruence for $L$. We shall construct a finite $C'$-congruence $\approx$ for $L$. We will consider two cases, corresponding to the two conditions in the theorem. The second case has two subcases. For each $G \in \mathbf{G}(C')$, let $c_G$ denote an s-graph with no edges and a single vertex which is the $c$-source of $c_G$ but not a vertex of $G$.

*Case(i)* For each $G \in \mathbf{G}(C')$ we let

$$f(G) = [G /\!/ c_G]_\sim .$$

Let $\approx$ be the kernel of $f$. Clearly, $\approx$ is finite since $\sim$ is. So, we only need to prove that $\approx$ is a $C'$-congruence for $L$.

If $G$ and $H$ are two isomorphic s-graphs in $\mathbf{G}(C')$ then $G /\!/ c_G$ is isomorphic to $H /\!/ c_H$. Hence, $G /\!/ c_G \sim H /\!/ c_H$, that is, $f(G) = f(H)$. So $G \approx H$.

Let $G$, $H$, and $K$ be three s-graphs in $\mathbf{G}(C')$ such that $G /\!/ K$ and $H /\!/ K$ are defined, and $G \approx H$, that is, $G /\!/ c_G \sim H /\!/ c_H$. Let $\overline{c}$ denote an s-graph without edges and with a single vertex which is the $c$-source of $\overline{c}$, but not a vertex of any of the s-graphs $G$, $H$, and $K$. Notice, this implies that $G /\!/ c_G$ and $H /\!/ c_H$ are isomorphic to $G /\!/ \overline{c}$ and $H /\!/ \overline{c}$, respectively. Hence,

$$G /\!/ \overline{c} \sim G /\!/ c_G \sim H /\!/ c_H \sim H /\!/ \overline{c}.$$

Clearly,

$$\mathbf{rest}_\emptyset((G /\!/ \overline{c}) /\!/ (K /\!/ \overline{c})) = \mathbf{rest}_\emptyset((G /\!/ K) /\!/ \mathbf{rest}_\emptyset(\overline{c}))$$

and similarly for $H$ instead of $G$. Since $\overline{L}$ is closed under addition and deletion of isolated vertices,

$$\mathbf{rest}_\emptyset(G /\!/ K) \in \overline{L}$$

implies

$$\mathbf{rest}_\emptyset((G /\!/ \overline{c}) /\!/ (K /\!/ \overline{c})) = \mathbf{rest}_\emptyset((G /\!/ K) /\!/ \mathbf{rest}_\emptyset(\overline{c})) \in \overline{L},$$

but $G /\!/ \overline{c} \sim H /\!/ \overline{c}$, hence

$$\mathbf{rest}_\emptyset((H /\!/ K) /\!/ \mathbf{rest}_\emptyset(\overline{c})) = \mathbf{rest}_\emptyset((H /\!/ \overline{c}) /\!/ (K /\!/ \overline{c})) \in \overline{L}$$

so

$$\mathbf{rest}_\emptyset(H /\!/ K) \in \overline{L}.$$

Hence, in this case $\approx$ is a finite $C'$-congruence for $L$.

*Case(ii)* We will use a similar but more complicated construction. For $G \in \mathbf{G}(C')$, we let

$$f(G) = \langle [G /\!/ c_G]_\sim, f'(G) \rangle$$

where $f'(G)$ is the set

$$\overline{\{F \in \mathbf{G}(C') \cap \mathbf{G}_m \mid F \text{ has no internal vertex and } \mathbf{rest}_\emptyset(G /\!/ F) \in \overline{L}\}}.$$

Let $\approx$ be the kernel of $f$. Since $\sim$ is finite and the range of $f'$ is finite, $\approx$ is finite, as well. If $G$ is isomorphic to $H$, we can in almost the same way as in the previous case conclude that $G \approx H$. Let $G$, $H$, and $K$ be three s-graphs in $\mathbf{G}(C')$ such that $G /\!/ K$ and $H /\!/ K$ are defined, and $G \approx H$. Let us assume that $\mathbf{rest}_\emptyset(G /\!/ K) \in \overline{L}$. We need to prove that $\mathbf{rest}_\emptyset(H /\!/ K) \in \overline{L}$.

*Subcase 1:* $K$ has no internal vertex. Since $L \subseteq \mathbf{G}_m$, the s-graph $K$ must be a member of $\mathbf{G}_m$ and of $f'(G)$. Since $G \approx H$, we have $f'(G) = f'(H)$. Hence, $K \in f'(H)$ and $\mathbf{rest}_\emptyset(H /\!/ K) \in \overline{L}$

*Subcase 2:* $K$ has internal vertices. Let $K^c$ be the s-graph in $\mathbf{G}(C')$ obtained by making some internal vertex $x$ of $K$ become the $c$-source. (Notice, since $G /\!/ K$ and $H /\!/ K$ are defined, $x$ is not a vertex in $G$ nor in $H$.) Let $x^c$ denote the s-graph without edges and just one vertex $x$ which is the $c$-source. Obviously,

$$\mathbf{rest}_\emptyset((G /\!/ x^c) /\!/ K^c) = \mathbf{rest}_\emptyset((G /\!/ K)$$

and similarly for $H$ instead of $G$. Hence, from $G /\!/ x^c \sim G /\!/ c_G \sim H /\!/ c_H \sim H /\!/ x^c$ we get that

$$\mathbf{rest}_\emptyset(G /\!/ K) \in \overline{L}$$

implies

$$\mathbf{rest}_\emptyset(H /\!/ K) \in \overline{L}$$

as was to be proved. Hence, in this case, as well, $\approx$ is a finite $C'$-congruence for $L$.
□

# 5 Going to larger $k$ using graph splitting

By $S(G)$ we will denote the set of sources of an s-graph $G$. Let $k$ and $m$ be two non-negative integers, and let $C$ be a finite subset of **C**. An s-graph in $\mathbf{G}(C)$ is $(k,m)$-splittable *if and only if* $k \leq |C|$ and for some s-graphs $G_1, \ldots, G_l$:

$$G = \mathbf{rest}_C(G_1 /\!/ \cdots /\!/ G_l) \tag{1}$$

where for each $i = 1, \ldots, l$ the cardinality of $S(G_i)$ is at most $k$, and

$$|S(G_1) \cup \cdots \cup S(G_l) - S(G)| \leq m.$$

We make some simple observations on the definition given above. First, if $|S(G_i)| < k$ then we can construct a new s-graph $H_i$ such that $|S(H_i)| = k$ by adding $k - |S(G_i)|$ of the sources of $G$ to $G_i$. If $H_i$ is used in the expression (1) instead of $G_i$, the value of the expression is still $G$. Second, if $S(G_i) = S(G_j)$ then we can use the s-graph $G_i /\!/ G_j$ as one graph in the expression expression (1) instead of both $G_i$ and $G_j$ as separate graphs. Third, if for some subset $A$ of cardinality $k$ of $S(G_1) \cup \cdots \cup S(G_l) \cup S(G)$ there is no s-graph $G_i$ such that $S(G_i) = A$, we can add the empty s-graph $\varepsilon$ (that has no vertices, no edges, and no sources) to the expression (1). Hence, if $G$ is $(k,m)$-splittable and $q = |S(G)|$, we can always write $G$ as in (1) with $l = \binom{m+q}{k}$, and $|S(G_i)| = k$ or $G_i = \varepsilon$ for each $G_i$.

Let $l = \binom{m+q}{k}$ and $C_1, \ldots, C_l$ be a fixed enumeration of all the cardinality $k$ subsets of $C \cup \{c_1, \ldots, c_m\}$. We will write

$$G = P(G_1, \ldots, G_l) \tag{2}$$

*if and only if* $k \leq |C|$ and

$$G = \mathbf{rest}_C(G_1 /\!/ \cdots /\!/ G_l)$$

where for each $i = 1, \ldots, l$, $G_i$ is an s-graph with $\tau(G_i) = C_i$ or $G_i = \varepsilon$.

Clearly, $G$ is $(k,m)$-splittable *if and only if* it can be written as in (2).

**Lemma 5.1** *Let $G_1$, $G_2$, and $G_3$ be s-graphs in $\mathbf{G}(C)$, $\mathbf{G}(C')$, and $\mathbf{G}(C'')$, respectively. Then*

$$\mathbf{rest}_\emptyset(\mathbf{rest}_{C''}(G_1 /\!/ G_2) /\!/ G_3) = \mathbf{rest}_\emptyset(G_1 /\!/ \mathbf{rest}_C(G_2 /\!/ G_3)),$$

*that is, whenever one side is defined, so is the other and they are equal.*

*Proof.* Easy verification. The two sides of the equality are both equal to $\mathbf{rest}_\emptyset(G_1 /\!/ G_2 /\!/ G_3)$. □

**Proposition 5.2** *Let $L$ be a set of graphs, $k$ and $m$ be non-negative integers, and $C$ be a subset of* **C** *of cardinality $k+1$, such that, for each $G, H \in \mathbf{G}(C)$, if $\mathbf{rest}_\emptyset(G /\!/ H) \in L$, at least one of $G$ and $H$ is $(k,m)$-splittable. Then, if $L$ is $k$-recognizable, it is $k+1$-recognizable, as well.*

*Proof.* Let $k, m, C$, and $L$ be as in the statement of the proposition with $L$ being $k$-recognizable. For each cardinality $k$ subset $C'$ of $C$, there is a finite $C'$-congruence $\sim_{C'}$ for $L$. For each $G \in \mathbf{G}(C')$, we denote by $[G]$ the equivalence class of $G$ with respect to the $C'$-congruence $\sim_{C'}$. If $k > 0$, we denote by $[\varepsilon]$ the set $\{\varepsilon\}$. We shall construct a finite $C$-congruence $\approx$ for $L$ as the kernel of a mapping $f$ from $\mathbf{G}(C)$ to a finite set $B$. Let $C_1, \ldots, C_l$, and $P$ be associated with $k, m$, and $C$ as in (2) above. For every $G \in \mathbf{G}(C)$, let

$$f(G) = \langle f'(G), f''(G) \rangle$$

where

$$f'(G) = \{\langle [G_1], \ldots, [G_l] \rangle | G = P(G_1, \ldots, G_l)\}$$

and

$$f''(G) = \{\langle [K_1], \ldots, [K_l] \rangle | \mathbf{rest}_\emptyset(G /\!/ P(K_1, \ldots, K_l)) \in \overline{L}\}.$$

Let $\approx$ be the kernel of $f$. Since $\sim$ is finite also $\approx$ is. It is clear that $G \approx H$ if $G$ and $H$ are isomorphic.

Let $G$, $H$, and $K$ be three s-graphs in $\mathbf{G}(C)$ such that $G /\!/ K$ and $H /\!/ K$ are defined, and $G \approx H$, that is, $f(G) = f(H)$. Assume that $\mathbf{rest}_\emptyset(G /\!/ K) \in \overline{L}$. We need to prove that $\mathbf{rest}_\emptyset(H /\!/ K) \in \overline{L}$. There are two cases.

*Case 1:* $G$ is $(k, m)$-splittable. Hence, $G$ can be written $G = P(G_1, \ldots, G_l)$ for some $G_1, \ldots, G_l$. By the definition of $f'$, $\langle [G_1], \ldots, [G_l] \rangle \in f'(G)$. Moreover, since $f'(G) = f'(H)$, $\langle [G_1], \ldots, [G_l] \rangle \in f'(H)$. That is, there are $H_1, \ldots, H_l$ such that $H = P(H_1, \ldots, H_l)$ and $[H_i] = [G_i]$ for each $i = 1, \ldots, l$.

By assumption $\mathbf{rest}_\emptyset(G /\!/ K) = \mathbf{rest}_\emptyset(P(G_1, \ldots, G_l) /\!/ K) \in \overline{L}$. Hence, it is enough to prove that

$$F_i = \mathbf{rest}_\emptyset(P(H_1, \ldots, H_{i-1}, G_i, G_{i+1}, \ldots, G_l)) \in \overline{L}$$

implies

$$F_{i+1} = \mathbf{rest}_\emptyset(P(H_1, \ldots, H_{i-1}, H_i, G_{i+1}, \ldots, G_l)) \in \overline{L}$$

to be able to conclude that $\mathbf{rest}_\emptyset(H /\!/ K) = \mathbf{rest}_\emptyset(P(H_1, \ldots, H_l) /\!/ K) \in \overline{L}$. If $k > 0$ and $G_i = \varepsilon$ then also $H_i = \varepsilon$, and the proof is immediate.

If $k = 0$ or $G_i \neq \varepsilon$, let

$$F^i = H_1 /\!/ \cdots /\!/ H_{i-1} /\!/ G_{i+1} /\!/ \cdots /\!/ G_l$$

and

$$F_i = \mathbf{rest}_\emptyset(\mathbf{rest}_C(G_i /\!/ F^i) /\!/ K).$$

By Lemma (5.1),

$$F_i = \mathbf{rest}_\emptyset(G_i /\!/ \mathbf{rest}_{C'}(F^i /\!/ K))$$

where $C' = \tau(G_i)$ (that is, the set of source labels of $G_i$) and, similarly,

$$F_{i+1} = \mathbf{rest}_\emptyset(H_i /\!/ \mathbf{rest}_{C'}(F^i /\!/ K)).$$

Now, since $G_i \sim H_i$, $F_{i+1} \in \overline{L}$ follows directly from $F_i \in \overline{L}$. This concludes the proof when $G$ is $(k, m)$-splittable.

*Case 2:* $K$ is $(k, m)$-splittable. In this case, $K$ can be written

$$K = P(K_1, \ldots, K_l)$$

for some $K_1, \ldots, K_l$. By the definition of $f''$, $\langle [K_1], \ldots, [K_l] \rangle \in f''(G)$. Moreover, $\langle [K_1], \ldots, [K_l] \rangle \in f''(H)$, since $f''(G) = f''(H)$. That is, there are $M_1, \ldots, M_l$ such that

$$\mathbf{rest}_\emptyset(H /\!/ P(M_1, \ldots, M_l)) \in \mathcal{I} \tag{3}$$

and $[M_i] = [K_i]$ for each $i = 1, \ldots, l$. By the argument used in case 1, it follows from (3) that

$$\mathbf{rest}_\emptyset(H /\!/ K) = \mathbf{rest}_\emptyset(H /\!/ P(K_1, \ldots, K_l)) \in \mathcal{I}$$

as was to be shown. Hence, $\approx$ is a $C$-congruence. □

# 6 Graphs of bounded tree-width are splittable

A tree-decomposition $(X, T)$ of a graph $G'$ is called a rigid tree-decomposition of width $w$ if: $|X_t| = w + 1$ for all $t \in V(T)$ and $|X_t \cap X_s| = w$ for all $(t, s) \in E(T)$.

**Lemma 6.1** *Every graph $G'$ of tree-width at most $w$ and with at least $w + 1$ vertices has a rigid tree-decomposition of width $w$.*

*Proof.* We prove the lemma by induction. Obviously, when $G'$ has exactly $w + 1$ vertices the lemma is true. Assume that it is true for all graphs with less than $n$ vertices and that $G'$ has $n$ vertices. Since $G'$ has tree-width at most $w$ there is a vertex $v$ in $G'$ such that: $v$ has at most $w$ neighbors, call the set of them $A$, and the graph $H$ obtained by making all neighbors of $v$ adjacent to each other and deleting $v$ has tree-width at most $w$. Let $(X, T)$ be a rigid tree-decomposition of $H$. Since $A$ is a clique in $H$ there is some node $r$ in $T$ such that $A \subseteq X_r$. Choose a subset $B$ of $X_r$ such that $A \subseteq B$ and $|B| = w$. Add a new node $r'$ to $T$ and make it adjacent to $r$. Add the set $X_{r'} = B \cup \{v\}$ to $X$. Clearly, $(X, T)$ is a rigid tree-decomposition of width $w$ of $G'$. □

Let $U$ be a subset of the vertex set of a graph $G'$. By $N_{G'}(U)$ we denote the set of neighbors of $U$ in $G'$. By $G' - U$ we denote the graph obtained from $G'$ by removing all the vertices in $U$ together with their incident edges.

**Proposition 6.2** *Let $C$ be a subset of $\mathbf{C}$ of cardinality $w + 1$. Let $G_1$ and $G_2$ be s-graphs in $\mathbf{G}(C)$ such that $\mathbf{rest}_\emptyset(G_1 /\!/ G_2)$ has tree-width at most $w$. Then at least one of $G_1$ and $G_2$ is $(w, w + 1)$-splittable.*

*Proof.* Let $G = G_1 /\!/ G_2$. Let $S = S(G) = S(G_1) = S(G_2) = V(G_1) \cap V(G_2)$. Let $G' = \mathbf{rest}_\emptyset(G)$, $G_1' = \mathbf{rest}_\emptyset(G_1)$, and $G_2' = \mathbf{rest}_\emptyset(G_2)$.

We are looking for a separator $Z$ in either $G_1'$ or $G_2'$ that satisfies certain special conditions. That is, for $i = 1$ or $i = 2$, we shall find a separator $Z$ of cardinality at most $w + 1$ in $G_i'$ such that each connected component $C$ in $G_i' - Z$ satisfies

$$|V(C) \cap S| + |N_G(V(C)) \cap Z| \leq w. \tag{4}$$

The existence of this separator is enough to establish the proposition, since, as easily seen, $|Z| \leq w+1$ together with condition (4) for each connected component in $G_i' - Z$ implies that $G_i$ is $(w, w+1)$-splittable.

Let $(X,T)$ be a rigid tree-decomposition of $G'$. For each edge $(s,t)$ of $T$, let $T_t^s$ denote the component in $T-(t,s)$ that contains $t$. Define $X(s,t)$ as $\cup_{t' \in V(T_t^s)} X_{t'}$ (that is, the set of vertices belonging to the bags of the tree-decomposition corresponding to nodes of $T$ that are "opposite" to $s$ with respect to $t$). Direct each edge in $T$ such that if $s$ is the tail of an edge whose head is $t$ then

$$|(X(s,t) - (X_s \cap X_t)) \cap S| \leq \frac{|S - (X_s \cap X_t)|}{2}.$$

This is always possible, since

$$((X(s,t) - (X_s \cap X_t)) \cap S) \cup ((X(t,s) - (X_s \cap X_t)) \cap S) = S - (X_s \cap X_t).$$

Clearly, in this new directed tree, some vertex $s$ will have indegree 0. Let $C_1, \ldots, C_r$ be the connected components in $G' - X_s$. Let $b(i)$ be a *branch* of $C_i$, that is, a neighbor of $s$ such that

$$V(C_i) \subseteq X(s, b(i)) - (X_s \cap X_{b(i)}).$$

Let $Z_i = X_{b(i)} \cap X_s$ and $Z_i^l = Z_i \cap V(G_l')$ for $l = 1,2$. Since $s$ has indeegre 0, the following holds for each neighbor $t$ of $s$:

$$|(X(s,t) - (X_s \cap X_t)) \cap S| \leq \frac{|S - (X_s \cap X_t)|}{2}.$$

Hence, for each connected component $C_i$ in $G' - X_s$,

$$|V(C_i) \cap S| \leq \frac{|S - Z_i|}{2}. \tag{5}$$

Assume that for some $i$, $|V(C_i) \cap S| + |Z_i^l| \geq w+1$ for $l = 1,2$. Then,

$$2(w+1) \leq 2|V(C_i) \cap S| + |Z_i^1| + |Z_i^2|.$$

But, because of (5) the right hand side of this is at most

$$|S - Z_i| + |Z_i^1| + |Z_i^2|,$$

which is equal to

$$|S| + |Z_i|,$$

since $Z_i^1 \cap Z_i^2 = S \cap Z_i$ and $Z_i^1 \cup Z_i^2 = Z_i$. Since $(X,T)$ is rigid $|Z_i| = w$ holds and, hence, $|S| + |Z_i| = 2w+1$. A contradiction. We conclude that for each $i \in [r]$, either $|V(C_i) \cap S| + |Z_i^1| \leq w$ or $|V(C_i) \cap S| + |Z_i^2| \leq w$.

If for all $i \in [r]$,

$$|V(C_i) \cap S| + |Z_i^1| \leq w \tag{6}$$

then $Z = X_s \cap V(G_1')$ is a separator of the wanted type in $G_1'$. This is so because, each connected component $C$ in $G_1' - Z$ is a subgraph of some connected component $C_i$ in $G' - X_s$ and, hence,

$$N_{G_1'}(V(C)) \cap Z \subseteq N_{G'}(V(C_i)) \cap Z \cap V(G_1') \subseteq Z_i^1$$

and

$$V(C) \cap S \subseteq V(C_i) \cap S.$$

That is, condition (4) is implied by (6). Similarily, if for all $i \in [r]$, $|V(C_i) \cap S| + |Z_i^2| \leq w$, then $Z = X_s \cap V(G_2')$ is a separator of the wanted type in $G_2'$. If neither of these two statements are true then there are some $i$ and $j$, where $i \neq j$, such that:

$$|V(C_i) \cap S| + |Z_i^1| \geq w + 1 \quad \text{(a)},$$

$$|V(C_j) \cap S| + |Z_j^2| \geq w + 1 \quad \text{(b)},$$

$$|V(C_i) \cap S| + |Z_i^2| \leq w \quad \text{(c), and}$$

$$|V(C_j) \cap S| + |Z_j^1| \leq w \quad \text{(d)}.$$

By subtracting the respective sides of (c) and (d) from the sum of the respective sides of (a) and (b), we see that

$$2 \leq |Z_i^1| - |Z_j^1| + |Z_j^2| - |Z_i^2|. \tag{7}$$

First notice, this implies that $Z_i \neq Z_j$, since otherwise $Z_i^1 = Z_j^1$ and $Z_j^2 = Z_i^2$. This and the fact that $(X, T)$ is rigid, that is $|Z_i| = |Z_j| = w$ and $|X_s| = w + 1$, gives that $Z_i \cup Z_j = X_s$ and $|Z_i \cap Z_j| = w - 1$. That is, $|Z_i| - |Z_i \cap Z_j| = |Z_j| - |Z_i \cap Z_j| = 1$. Which implies $|Z_i^1| - |Z_i^1 \cap Z_j^1| \leq 1$ and $|Z_j^2| - |Z_i^2 \cap Z_j^2| \leq 1$. From which we get $|Z_i^1| - |Z_j^1| \leq 1$ and $|Z_j^2| - |Z_i^2| \leq 1$. Assuming that any of these two inequalities is strict, we get

$$|Z_i^1| - |Z_j^1| + |Z_j^2| - |Z_i^2| < 2,$$

a contradiction to (7). We conclude that

$$|Z_i^1| - |Z_j^1| = 1 \tag{8}$$

and

$$|Z_j^2| - |Z_i^2| = 1. \tag{9}$$

Thus, in (7) equality holds.

Second notice, equality holds in (7) *if and only if* equality holds in (a), (b), (c), and (d). Hence, also in (a), (b), (c), and (d) equality holds. We use that equality holds in (a) and (d), and equality (8) to get

$$|V(C_i) \cap S| = |V(C_j) \cap S|. \tag{10}$$

By summing the respective sides of (a) and (b) the following is obtained:

$$|V(C_i) \cap S| + |V(C_j) \cap S| + |Z_i^1| + |Z_j^2| \geq 2(w + 1),$$

that is,

$$|V(C_i) \cap S| + |V(C_j) \cap S| + |Z_i^1 \cup Z_j^2| + |Z_i^1 \cap Z_j^2| \geq 2(w + 1). \tag{11}$$

Since $Z_i^1 \cup Z_j^2 \subseteq X_s$ and $|X_s| = w + 1$, inequality (11) implies

$$|V(C_i) \cap S| + |V(C_j) \cap S| + |Z_i^1 \cap Z_j^2| \geq w + 1. \tag{12}$$

This inequality and $Z_i^1 \cap Z_j^2 \subseteq Z_i^2 \cap S$ give

$$|V(C_i) \cap S| + |V(C_j) \cap S| + |Z_i^2 \cap S| \geq w + 1, \tag{13}$$

and, since $|S| = w + 1$, we conclude that

$$S \subseteq V(C_i) \cup V(C_j) \cup Z_i^2. \tag{14}$$

We will finish the proof by letting $Z = Z_i^2$ and showing that this is a separator of the wanted type in $G'_2$. (It is also possible to show that $Z_j^1$ is a separator of the wanted type in $G'_1$.) Clearly, $|Z_i^2| \leq w + 1$. So, what we need to prove is that each connected component $C$ in $G'_2 - Z_i^2$ satisfies condition (4). If $C$ is a connected component in $G'_2 - Z_i^2$ and $C \cap S \neq \emptyset$, then by (14)

$$V(C) \cap S \subseteq V(C_i) \cup V(C_j).$$

But $C$ intersects at most one of $C_i$ and $C_j$, because, each path from $V(C_i)$ to $V(C_j)$ in $G' - Z_i^2$ contains a vertex in $Z_i$, and $Z_i^2$ is by definition $Z_i \cap V(G'_2)$. So,

$$V(C) \cap S \subseteq V(C_i) \cap S$$

or

$$V(C) \cap S \subseteq V(C_j) \cap S.$$

This together with (10) gives

$$|V(C) \cap S| \leq |V(C_i) \cap S|.$$

From this and (c) we get

$$|V(C) \cap S| + |Z_i^2| \leq w,$$

if we moreover use that $Z = Z_i^2$ we get $|V(C) \cap S| + |N_G(V(C)) \cap Z| \leq w$, as was to be proved. □

# 7 Discussion and open problems

Let us first observe that the results of this paper hold for directed and/or labelled graphs. The constructions and especially the crucial Proposition 6.2 hold for such graphs. They also hold for hypergraphs. In the definition of a tree-decomposition of a hypergraph, one demands that for each hyperedge there is some bag that contains all the vertices incident to the hyperedge. Lemma 4.1 and Proposition 6.2 entail the corresponding results for hypergraphs as follows. Let $H$ be a hypergraph. Let $gr(H)$ be the graph with the same vertex set as the hypergraph $H$, but where two vertices $u$ and $v$ are adjacent *if and only if* there is some hyperedge in $H$ which is incident to both $u$ and $v$. Then $H$ and $gr(H)$ have exactly the same tree- decompositions (see [6]). If $gr(H) = G//G'$ then the vertices of every clique of gr(H) are totally in $G$ or totally in $G'$. It follows that $H$ is $(k,m)$-splittable iff $gr(H)$ is. Hence, the results of Section 6 and whence, the main theorem hold for hypergraphs. We omit further details.

Let us now go back to graphs. We have established that for every set $L$ of simple graphs of tree-width at most $k$ the definition of recognizability can be simplified from "for every $m$, $L$ is $m$-recognizable" to "$L$ is $k$-recognizable" .

In a tree-decomposition of an s-graph, we demand that there is some bag that contains all sources. (The notion of tree-width for an s-graph follows immediately.) As mentioned in the introduction, an algorithm deciding the membership of $G$ in $L$ can operate as follows: if $G = H\Box K$, and $H$ belongs to some fixed finite set, one replaces it by a (smaller) canonical representative of its class. Such replacements are repeated until one reaches some "irreducible" graph. The original graph $G$ belongs to $L$ if and only if the obtained irreducible graph is a member of a (computable) finite set. When the graphs in $L$ are of tree-width at most $k$, it is always possible to find $G = H\Box K$ as above, but where the s-graphs $H$ and $K$ have tree-width at most $k$. So, in some sense, when the graphs in $L$ are of tree-width at most $k$, we are not using the full power of recognizability in this computation. The notion of $m$-recognizability, of a set of graphs $L$, is relative to decompositions of graphs $G$ of the form $G = H\Box K$ where $K$ is a graph with $m$ sources. If in these decompositions the s-graphs $H$ and $K$ are restricted to have tree-width at most $w$ (see the definition of a $C$-congruence in Section 2), then the alternative notion of $(m, w)$-recognizablity is obtained. When $L$ is a set of graphs of tree-width at most $k$, $(k, k)$-recognizablity is exactly what we need to perform the computation described above. It is trivial to see that $m$-recognizability implies $(m, w)$-recognizability for every $w$, but is the converse true? That is, we have the following interesting open question: Given $k$, does there exist $m$ and $w$ such that for each set $L$ of simple graphs of tree-width at most $k$:

if $L$ is $(m, w)$-recognizable, then it is recognizable.

We believe that the answer is yes. In fact, we have the following conjecture.

**Conjecture 7.1** *This property holds for $m = w = k$.*

In the terminology of [1] this conjecture says that for a set of simple graphs of tree-width at most $k$, $k$-cutset regularity is equivalent to full cutset regularity.

Observe that this conjecture is a consequence of the conjecture made in Courcelle [5] saying that in every simple graph $G$ of tree-width at most $k$ (that is, partial $k$-tree), a tree-decomposition can be defined by monadic second-order formulas. Here is an alternative form of this conjecture.

**Conjecture 7.2** *For every $k$, there exists a monadic second-order formulas $g(X_1, \ldots, X_m)$ and $f(x, y, X_1, \ldots, X_m)$ such that, for every simple graph $G$ of tree-width most $k$, there exists sets $X_1, \ldots, X_m$ of vertices and edges satisfying the formula $g$ and for any such $m$-tuple of sets, if one adds all edges linking all pairs of vertices $(x, y)$ such that $f(x, y, X_1, \ldots, X_m)$ holds in $G$, then together with the edges and vertices of $G$, one gets a $k$-tree.*

Let us sketch how 7.1 follows from 7.2 (assuming that the reader knows [4],[7]). Let $L$ be $(k, k)$-recognizable. By the technique of [1, 2, 7] one can prove that $L$ is recognizable with respect to the graph operations of [2] that generate simple graphs of tree-width at most $k$. Hence a finite-state tree-automaton can be constructed which accepts the algebraic expressions over these operations that define graphs belonging to $L$. This automaton can be made to operate on the $k$-tree defined in $G$ by $f(x, y, X_1 \ldots, X_m)$ and can be simulated by a monadic second-order formula by the technique of [4]. It follows that $L$ is monadic second-order definable hence recognizable by the results of [4].

Here are two facts showing that our results do not extend to sets of graphs of unbounded tree-width.

**Fact 7.3** *For every k, there is a set of graphs that is k-recognizable but not (k+1)-recognizable.*

We consider the set of simple graphs consisting of two cliques with same number of vertices sharing a clique with $k+1$ vertices. Since these graphs have no separator with $k$ vertices, they form (quite trivially; we omit details) a $k$-recognizable set. Any two cliques with different number of vertices, among wich $k+1$ are sources, are inequivalent.

**Fact 7.4** *There exists a set of graphs L that is not recognizable but is such that, for each k, the set of graphs in L of tree-width at most k is recognizable (and even finite).*

We take for $L$ the set of cliques with $n$ vertices, augmented with a tail consisting of a path with $n$ vertices, glued to the clique by one of its ends. This set is not recognizable because any two paths of inequal length with one source at one end are inequivalent. (More or less like the language $\{a^n b^n | n > 0\}$ is not regular). For each $k$, only finitely many cliques, whence finitely many graphs in $L$, have tree-width at most $k$.

# References


[1] K. Abrahamson and M. Fellows. Finite automata, bounded tree-width and well-quasiordering. In N. Robertson and P. Seymour, editors, *Graph Structure Theory. Contemporary Mathematics vol. 147*, pages 539–564. AMS, Providence, Rhode Island, 1993.

[2] S. Arnborg, B. Courcelle, A. Proskurowski, and D. Seese. An algebraic theory of graph reduction. Technical Report LaBRI-90-02, Universite de Bordeaux, Jan. 1990. To appear in JACM, extended abstract in Lec. Notes Comp. Sci. 532:70-83. 1991.

[3] S. Arnborg, J. Lagergren, and D. Seese. Problems easy for tree-decomposable graphs. *J. of Algorithms*, 12:308–340, 1991.

[4] B. Courcelle. The monadic second order logic of graphs I: Recognizable sets of finite graphs. *Information and Computation*, 85:12–75, March 1990.

[5] B. Courcelle. The monadic second-order logic of graphs V: On closing the gap between definability and recognizability. *Theoret. Comput. Sci.* 80:153–202, 1991.

[6] B. Courcelle. The monadic second-order logic of graphs III: Tree-decompositions, minors and complexity issues. *Informatique Théorique et Applications*, 26:257–286, 1992.

[7] B. Courcelle. Recognizable sets of graphs: equivalent definitions and closure properties. Technical Report 92-06, Bordeaux-1 University, 1992. To appear in Mathematical Structures in Computer Science.

[8] A. Habel, H.J. Kreowski, and C. Lautemann. A comparison of compatible, finite, and inductive graph properties. *Theoret. Comput. Sci.*, 89:33–62, 1991.

# Canonical derivations for high-level replacement systems

Hartmut Ehrig*
Hans-Jörg Kreowski**
Gabriele Taentzer*

*Technische Universität Berlin, Germany
**Universität Bremen, Germany

## Abstract

Canonical derivations, previously studied for string and graph grammars only, are generalized from graph grammars to high-level replacement systems, short HLR-systems. These systems were recently introduced to provide a common categorical framework for different types of replacement systems on complex objects, including graphs, hypergraphs, structures and algebraic specifications. It turns out that basic results concerning synthesis and analysis of parallel derivation sequences in HLR-systems, obtained in previous papers, can be extended to construct canonical parallel derivation sequences which are optimal w.r.t. leftmost parallelism. The main results show the existence and uniqueness of canonical derivations under weak assumptions for the underlying categories of HLR-systems. These results are specialized to graphs, hypergraphs, Petri nets, algebraic specifications and others by classifying the underlying categories with respect to the assumptions. This leads to interesting new results for most of the corresponding HLR-systems.

## 1. Introduction

Canonical derivations, previously studied for string and graph grammars only (see, e.g., [Kre 76/77]), are generalized from graph grammars to high-level replacement systems recently introduced in [EHKP 91a/b]. They provide a common categorical framework for different types of replacement systems based on categories and pushouts. For example the merging of two objects like the union of sets, the concatenation of strings, the unification of expressions or the gluing of graphs can often be constructed as a pushout in a suitable category. Hence a double pushout of the form as given in definition 2.2.2 can be interpreted in the following way:
The top level forms a production; the vertical morphisms match the production with the bottom level; C remains if L is removed from G; H is obtained if R is added to C; and all together specifies the application of the production to G yielding H. Indeed, direct derivations in many graph and hypergraph grammars, the transformations of Petri nets, the evaluation of expressions and the rule-based modular design of algebraic specifications can be modelled in this way.

If the underlying category has coproducts (which are usually disjoint unions), the parallel production p1+p2 of two productions p1 and p2 can be constructed leading to parallel derivations. Using the compatibility of pushouts and coproducts, each parallel derivation can be

This work is partly supported by the ESPRIT Basic Research Working Group 7183 (COMPUGRAPH).

sequentialized in such a way that only atomic productions are applied. In other words, parallel derivations may speed up the derivation process, but do not increase the generative power. Parallel derivations and their sequentializations can be considered as equivalent. Derivations that are optimal with respect to leftmost parallelism among the equivalent derivations are called canonical. We show that canonical derivations can be constructed under weak conditions using a shift operator on parallel derivations. Moreover, we give sufficient conditions for the uniqueness of canonical derivations.

The results on canonical derivations are shown in section 3 based on the axiomatic approach to analysis, synthesis and independence in section 2. In section 4, we study the particular cases of sequential independence and butterfly independence. Using the parallelism theorem in [EHKP 91a/b] for sequential independence and so-called HLR1-categories (where HLR1 summarizes a certain set of pushout and pullback conditions), we can give particular constructions for analysis, synthesis, shift and canonical derivations. Uniqueness can be shown if some conditions are added to HLR1 yielding HLR1*. Moreover, we show that the results remain true if the conditions are relaxed from HLR1 and HLR1* to HLR0.5 and HLR0.5* respectively provided that sequential independence is replaced by butterfly independence. Finally we classify various categories of graphs, hypergraphs, Petri nets, algebraic specifications and others with respect to the conditions HLR0.5, HLR0.5*, HLR1 and HLR1* in section 5. In particular, we give examples that separate the classes from each other. Because of lack of space, we give proof ideas in most cases only. Actually, parts of the proofs can be done by adaptations of the corresponding proofs for the graph grammar case as given in [Kre 76/77] (and the other ones are given in our long version [EKT 92] of this paper).

## 2. Axiomatic Concepts for Parallelism, Analysis and Synthesis

In this section we review the basic concepts of HLR-systems as given in [EHKP 91a/b], and we present an axiomatic approach for independence, analysis, synthesis and different parallelism properties of parallel derivations generalizing the respective concepts in [Kre 76/77] and [Ehr 79].

### 2.1 General Assumptions

Let **Cat** be a category with a distinguished class **M** of morphisms and binary coproducts which are compatible with **M**, i.e. for all morphisms f, g in **M** also the coproduct, written f+g, is in **M**. Since all categorical colimit and limit constructions are unique up to isomorphism, we consider all constructions up to isomorphism only, but do not mention these isomorphisms explicitly. How the isomorphisms may be handled in the context of the double pushout approach is systematically studied in [CELMR 93] for the graph case and can be extended for the HLR-case.

**2.2 Definition (Derivations and HLR-Systems):** 1. A production $p = (L \leftarrow K \rightarrow R)$ in **CAT** consists of a pair of objects (L,R), called the left- and right-hand side respectively, an object K, called interface, and two morphisms $K \rightarrow L$ and $K \rightarrow R$ belonging to the class **M**.

2. Given a production p as above a direct derivation $G \Rightarrow H$ via p, short $p{:}G \Rightarrow H$ or $G \Rightarrow_p H$, from an object G to an object H in **CAT** is given by two pushout diagrams (1) and (2), called double pushout, with p in the top row and G and H in the bottom corners:

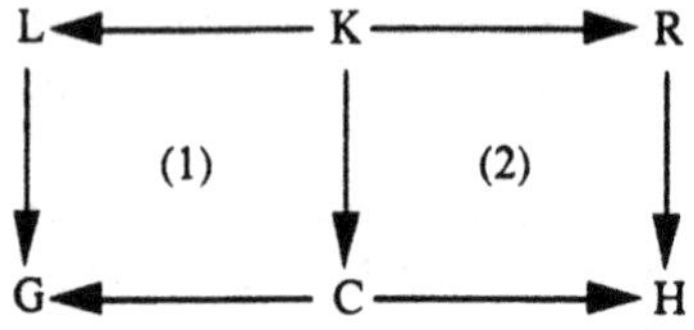


3. A derivation $G \Rightarrow_* H$ via productions (p1,...,pn) from G to H of length n is a sequence of $n \geq 0$ direct derivations $G = G0 \Rightarrow_{p1} G1 \Rightarrow_{p2} \dots \Rightarrow_{pn} Gn = H$. We denote this derivation by $G \Rightarrow_* H$ over **P** if **P** is a set of productions with $p1,\dots,pn \in \mathbf{P}$.

4. An HLR-system H = (S, **P**, **T**) in **CAT** consists of a start object S, a set **P** of productions, and a class **T** of terminal objects in **CAT**. The language L(H) of H is given by the class of all terminal objects in **CAT** derivable from S by **P**, i.e. $L(H) = \{G \in \mathbf{T} \mid S \Rightarrow_* G \text{ over } \mathbf{P}\}$.

**2.3 Definition (Parallel Derivations):** 1. Let $p = (L \leftarrow K \rightarrow R)$ and $p' = (L' \leftarrow K' \rightarrow R')$ be productions. Then the production $(L+L' \leftarrow K+K' \rightarrow R+R')$, where + is the binary coproduct in **CAT**, is called parallel production of p and p' and denoted by p+p'.

2. A direct parallel derivation $G \Rightarrow X$ is a direct derivation via a parallel production p+p'.

3. A parallel derivation s is a derivation sequence via parallel productions. It is called distinct, if all productions occurring in s are pairwise distinct.

Note that the parallel production p+p' is again a production in the sense of definition 2.2.2 because the coproduct of **M**-morphisms is again in **M**. The construction of parallel productions can be iterated leading to productions p1+...+pn $(n \geq 1)$, which are again called parallel productions. If **CAT** has an initial object I with the identity morphism $I \rightarrow I$ in **M**, the zero production $0 = (I \leftarrow I \rightarrow I)$ becomes a unit for parallel productions.

**2.4 Definition (General Independence Relation):** A general independence relation INDEP, short independence, for an HLR-system H is a subclass of all parallel derivations of length 2 in H. If s ∈ INDEP, we say that s is independent.

**2.5 Definition (Analysis and Synthesis):** Let INDEP be a general independence relation for an HLR-system. Let ANAL and SYNT be two binary relations on parallel derivations.

1. ANAL is called analysis if, for each (s',s) ∈ ANAL, s' is a direct parallel derivation of the form $G0 \Rightarrow G2$ via p1+p2 and s is an independent derivation of the form $G0 \Rightarrow G1 \Rightarrow G2$ via (p1,p2). In this case, s is called (p1-)sequentialization of s', and we write s' $ANAL_{p1}$ s.

2. SYNT is called synthesis if the inverse relation $SYNT^{-1} = \{(s',s) \mid (s,s')SYNT\}$ is an analysis. If (s,s') ∈ SYNT, where s is a parallel derivation of the form G0 ⇒ G1 ⇒ G2 via (p1,p2), s' is called (p2-)parallelization of s, and we write s $SYNT_{p2}$ s'.
3. ANAL is called unique (complete) if for each direct parallel derivation s' there is at most (at least) one s with (s',s) ∈ ANAL.
4. SYNT is called unique (complete) if for each independent derivation s:G0 ⇒ G1 ⇒ G2 there is at most (at least) one s' with (s,s') ∈ SYNT.
5. In case of uniqueness, we write $ANAL_{p1}$(s') = s ($SYNT_{p2}$(s) = s') instead of s' $ANAL_{p1}$ s (s $SYNT_{p2}$ s').

Note that the commutativity of p1+p2 implies that for s' there is also at least one independent derivation s2:G0 ⇒ G1' ⇒ G2 via (p2,p1) with s' $ANAL_{p2}$ s2 provided that the analysis is complete. This means that analysis and synthesis lead to relations $ANAL_{p1}$, $ANAL_{p2}$ and $SYNT_{p2}$, which are functions between corresponding sets of derivations in case of unique analysis and unique synthesis. Given an analysis ANAL, there is always a corresponding synthesis defined by the inverse relation. By definition, the inverse relation of a synthesis is an analysis. But one may consider in general two relations that are not inverse to each other.

**2.6 Definition (Associativity of Analysis):** Given an HLR-system with unique analysis, then the analysis is called associative if for each direct parallel derivation s0:G0 ⇒ G3 via p1+p2+p3 with sequentializations s1:G0 ⇒ G1 ⇒ G3 via (p1,p2+p3), and s2:G0 ⇒ G2 ⇒ G3 via (p1+p2,p3), i.e. s1 = $ANAL_{p1}$(s0) and s2 = $ANAL_{(p1+p2)}$(s0), there is a unique direct derivation G1 ⇒ G2 via p2, such that s12:G0 ⇒ G1 ⇒ G2 via (p1,p2), and s23:G1 ⇒ G2 ⇒ G3 via (p2,p3) are sequentializations of s12':G0 ⇒ G2 via p1+p2, and s23':G1 ⇒ G3 via p2+p3 respectively, i.e. s12 = $ANAL_{p1}$(s12') and s23 = $ANAL_{p2}$(s23').

**2.7 Definition (Parallelism Properties):** Let H be an HLR-system.
1. A parallelism structure π = (INDEP,ANAL,SYNT) on H consists of an independence relation INDEP for H, an analysis ANAL and a synthesis SYNT.
2. An HLR-system with parallelism (a πHLR-system for short) is a pair (H,π) consisting of an HLR-system and a parallelism structure π on H.
3. A πHLR-system (H,π) with π = (INDEP,ANAL,SYNT) is called faithful if ANAL and SYNT are complete and if the following properties hold for all derivations s and t via (p1,p2) and s' and t' via p1+p2:

| | |
|---|---|
| (a/s-faithfulness) | If (s,s') ∈ SYNT and (s',t) ∈ ANAL, then s = t. |
| (s/a-faithfulness) | If (s',s) ∈ ANAL and (s,t') ∈ SYNT, then s' = t'. |

Note that faithfulness guarantees a bijective correspondence between independent derivations and direct parallel derivations. The a/s-faithfulness (s/a-faithfulness) requires that the composition of two relations is the identity which does not imply that both relations are functions. But if both properties hold simultaneously then both relations are already bijective functions, which are inverse to each other in addition.

**2.8 Definition (Distributivity of Independence):** Given a πHLR-system (H,π) with π = (INDEP,ANAL,SYNT), then the independence is called distributive, if for all derivation diagrams of the shape

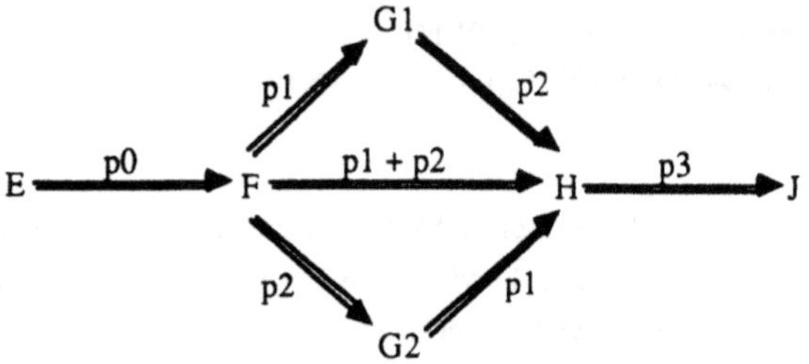


where $F \Rightarrow G1 \Rightarrow H$ via (p1,p2) and $F \Rightarrow G2 \Rightarrow H$ via (p2,p1) are sequentializations of $F \Rightarrow H$ via p1+p2, we have:

1. $F \Rightarrow H \Rightarrow J$ via (p1+p2,p3) is independent if and only if $G1 \Rightarrow H \Rightarrow J$ via (p2,p3) and $G2 \Rightarrow H \Rightarrow J$ via (p1,p3) are independent,
2. $E \Rightarrow F \Rightarrow H$ via (p0,p1+p2) is independent if and only if $E \Rightarrow F \Rightarrow G1$ via (p0,p1) and $E \Rightarrow F \Rightarrow G2$ via (p0,p2) are independent.

## 3. Canonical Derivations Based on Axiomatic Concepts

In this section we introduce a shift operation for parallel derivations as a particular combination of analysis and synthesis defined in the previous section in an axiomatic way for πHLR-systems. This allows to define shift equivalence, canonical derivations and two main results concerning existence and uniqueness of canonical derivations. These constructions and results generalize the corresponding constructions and results of canonical derivations for graph grammars in [Kre 76/77] and [Ehr 79].

**3.1 Definition (Shift):** Let (H,π) be a πHLR-system with π = (INDEP,ANAL,SYNT). Let s: $G \Rightarrow_* G0 \Rightarrow_{p1} G1 \Rightarrow_{p+p2} G3 \Rightarrow_* G'$ be a distinct parallel derivation. Let $G1 \Rightarrow_p G2 \Rightarrow_{p2} G3$ be a sequentialization of $G1 \Rightarrow_{p+p2} G3$. Let $G0 \Rightarrow_{p1} G1 \Rightarrow_p G2$ be independent (w.r.t. INDEP) and $G0 \Rightarrow_{p1+p} G2$ be a parallelization of it.
Then s': $G \Rightarrow_* G0 \Rightarrow_{p1+p} G2 \Rightarrow_{p2} G3 \Rightarrow_* G'$ is a shift of p in s, denoted by s $SHIFT_p$ s'. The set of all shifts of p in s is denoted by $SHIFT_p(s)$.

Note that shift is essentially the composition of an analysis and a synthesis step in a certain context resulting in a derivation where p is applied one step earlier than originally. The distinctness guarantees that there is at most one occurrence of a production p in a derivation s. If p does not occur in s or if p occurs, but either the analysis or the following synthesis is not defined, then $SHIFT_p(s)$ is considered as empty. If (H,π) is faithful and p occurs in s, the analysis step is always defined with a unique result. But the synthesis step may not be defined because the derivation $G0 \Rightarrow_{p1} G1 \Rightarrow_p G2$ is not always independent. In case of independence however the synthesis is unique such that $SHIFT_p(s)$ has at most one element. If this element is s', we write $SHIFT_p(s) = s'$.

**3.2 Definition (Equivalence):** Two distinct parallel derivations s and s' in a πHLR-system are called <u>equivalent</u>, written s ≡ s', if we have s = s' or there are distinct parallel derivations s0,...,sn with s = s0 and s' = sn and productions p1,...,pn such that

(*) $si \in SHIFT_{pi}\, s(i-1)$ or $s(i-1) \in SHIFT_{pi}\,(si)$ $(i = 1,...,n)$.

**3.3 Definition (Canonical Derivations):** Let (H,π) be a πHLR-system.

1. A distinct parallel derivation sequence s is called <u>canonical</u>, short <u>canonical derivation</u>, if $SHIFT_p(s)$ is empty for all atomic productions p.
2. (H,π) has <u>canonical derivations</u> if for each distinct parallel derivation s there is a canonical derivation s' with s ≡ s'.
3. (H,π) has <u>unique canonical derivations</u> if it has canonical derivations and for each pair of canonical derivations s' and s" we have s' = s" provided that s' ≡ s".

Note that, by definition, a shift improves the leftmost parallelism. Hence a canonical derivation is a derivation with optimal leftmost parallelism. In the following theorem we will show that the existence of shift implies the existence of canonical derivations. The existence of unique shift together with additional properties implies unique canonical derivations. In general, unique shift may only define existence but not uniqueness of canonical derivations. If the πHLR-system has an effective construction for unique canonical derivations and an effective equality test, then equivalence of (distinct) derivations s1 and s2 is decidable. In fact, we only have to construct the corresponding unique canonical derivations s1' ≡ s1 and s2' ≡ s2 and to check equality of s1' and s2', because we have s1 ≡ s2 if and only if s1' = s2'. Nonunique canonical derivations have not been studied, but they seem to be useful as derivations with local optimal leftmost parallelism.

**3.4 Theorem (Canonical Derivations)**

Let (H,π) be a πHLR-system with π = (INDEP,ANAL,SYNT). Then we have:

1. (H,π) has canonical derivations.
2. (H,π) has unique canonical derivations if the following properties hold:
   (a) (H,π) is faithful.
   (b) ANAL is associative.
   (c) INDEP is distributive.
   (d) The underlying category **CAT** has an initial object I with $id_I \in M$.

Note that a canonical derivation SHIFT(s) equivalent to s is the result of the nondeterministic procedure that shifts non-zero productions while shiftable ones exist. The proof is based essentially on the corresponding constructions in [Kre 76/77] for graph grammars, where the zero production is used for simplification of the technical reasoning. Hence condition (d) is only assumed for technical reasons, while conditions (a) - (c) are essential, at least for our proof.

**Proof idea:** Although the corresponding proofs in [Kre 76] are given in the context of graph grammars only, sections 3 - 5 of [Kre 76] concerning analysis, synthesis and shift, equivalence of parallel derivations and canonical parallel derivations are independent of graph grammars. They use only corresponding results for graph grammars in sections 1 and 2 of [Kre 76] concerning analysis and synthesis of parallel derivations and unique synthesis of binatural parallel derivations. The results of sections 1 and 2 of [Kre 76] which are used in the main results of sections 3 - 5 of [Kre 76] have been formulated as axioms for HLR-systems in section 2 of this paper.

## 4. Analysis, Synthesis and Canonical Derivations for HLR-Systems

In this section we consider two different notions of independence for direct derivations in HLR-systems. In addition to sequential independence in the sense of [EHKP 91a/b] we introduce the new notion of butterfly independence, which generalizes "schwache Unabhängigkeit" in the sense of [Kre 76/77]. The main advantage of butterfly independence is the fact that we can drop some of the HLR1-properties used in [EHKP 91a/b] to define parallelism, analysis and synthesis, the properties actually needed are called HLR0.5. In fact, both notions of independence lead to πHLR-systems with standard constructions for analysis and synthesis and hence with canonical derivations. The constructed analysis is always unique. Analysis and synthesis turn out to be complete in HLR0.5- and HLR1-categories. Moreover, both notions of independence become equivalent in HLR1-categories. Adding the condition "**M**-pushouts are pullbacks", where the corresponding properties are called HLR0.5* and HLR1* respectively, we obtain uniqueness for synthesis and canonical derivations.
Note, that the notion of HLR1* was introduced in [EP 91b] in order to distinguish the case with uniqueness (HLR1*) from that without uniqueness (HLR1). This distinction was not made in [EHKP 91a/b], where the case with uniqueness was called HLR1 and the notion HLR1* was not introduced at all. Our notion of HLR1* includes two additional weak conditions (d) and (f) in 4.1 below which are not mentioned in [EP 91b].

**4.1 Definition (Different Notions of HLR-Categories):** 1. An <u>HLR0-category</u> is a category **CAT** together with a distinguished class **M** of morphisms.
2. An <u>HLR0.5-category</u> is an HLR0-category with (a) - (c):
(a) Existence of semi-**M**-pushouts, i.e. pushouts (1) in **CAT** of morphisms $m:A \to B$ and $g:A \to C$ with $m \in$ **M** and general $g \in$ **CAT**.

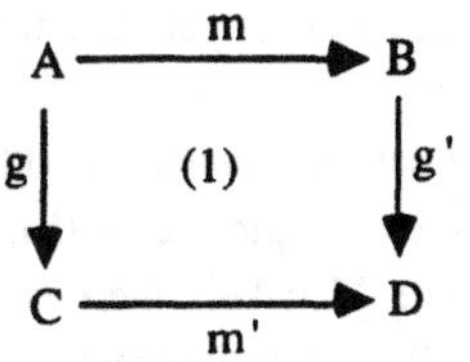


(b) Inheritance of **M**-morphisms under pushouts, i.e. for each pushout (1) as above $m \in \mathbf{M}$ implies $m' \in \mathbf{M}$.

(c) Existence of binary coproducts which are compatible with **M**, i.e. $f, g \in \mathbf{M}$ implies $f+g \in \mathbf{M}$ for the coproduct of morphisms.

3. An HLR0.5*-category is an HLR0.5-category with (d) - (f):

(d) **M** is a class of monomorphisms in **CAT**, i.e. $\mathbf{M} \subseteq$ Monos(**CAT**).

(e) **M**-pushouts are pullbacks, i.e. each pushout (1) as above with $f, g \in \mathbf{M}$ is already a pullback.

(f) Existence of an initial object I in **CAT** with $id_I \in \mathbf{M}$.

4. An HLR1-category is an HLR0.5-category with (g) and (h):

(g) **M**-pushout-pullback decomposition property, i.e. given

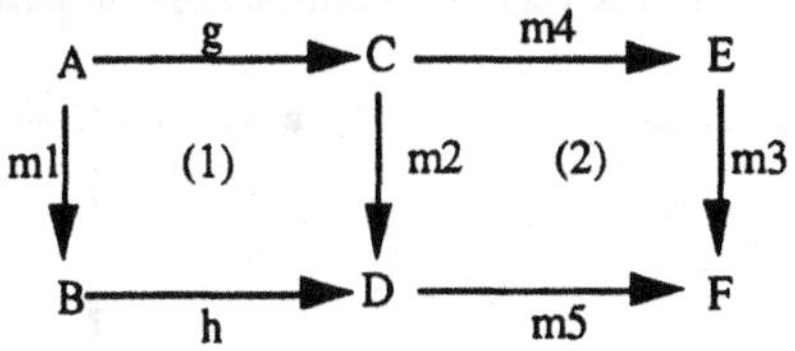


with $(1) \cup (2)$ pushout, (2) pullback, $mi \in \mathbf{M}$ for $i = 1,\ldots,5$ implies that also (1) and (2) are pushouts separately.

(h) Existence and inheritance of **M**-pullbacks, i.e. existence of pullbacks (2) as above with given $m3, m5 \in \mathbf{M}$ such that also $m2, m4 \in \mathbf{M}$.

5. An HLR1*-category is an HLR1-category with conditions (d) - (f) above.

**Remarks, Interpretation and Examples**

1. The main concepts of HLR-systems including productions, derivations, and the generated language can be formulated already in HLR0-categories. HLR0.5- and HLR1-categories are used to obtain particular constructions of analysis, synthesis and canonical derivations while HLR0.5$^*$- and HLR1$^*$-categories lead to uniqueness of these constructions.

2. If **M** = **Mor**, the class of all morphisms in **CAT**, we have the following special meaning:

HLR0: any category,

HLR0.5: existence of pushouts and binary coproducts,

HLR0.5$^*$: HLR0.5 with initial object, all morphisms are monomorphisms and pushouts are pullbacks,

HLR1: HLR0.5 with pullbacks and pushout-pullback decomposition,

HLR1*: HLR0.5* and HLR1.

In fact, the basic category **SETS** of sets and functions with **M** = **Mor** is HLR0.5, but not HLR0.5* and not HLR1 (see counterexample 2.2 in [EK 79]). This means that for all categories **CAT** built over **SETS** with **M** = **Mor** there is a good chance for HLR0.5, but almost no chance for HLR0.5* or HLR1 (see theorem 5.1 below for more details). This implies that HLR-systems, where the morphisms in the productions are allowed to be noninjective, can be handled in the framework of HLR0.5 only. This leads to existence but in general not to uniqueness results.

3. In general there are good chances for **CAT** to be HLR0.5*, HLR1 or HLR1* if **M** is the class $\mathbf{M}_{inj}$ of all injective morphisms or a suitable subclass $\mathbf{M}_{spe}$ of it (see theorem 5.1 below). This leads to existence and in the case HLR0.5* and HLR1* also to uniqueness results (see theorem 5.4 below).

4. In theorem 5.3 we will show that the hierarchy corresponding to the numbering of HLR-categories is proper, while HLR0.5* and HLR1 are incomparable.

**4.2 Definition (Independence and Butterfly Independence):** Given an HLR0-system, a direct derivation G ⇒ H ⇒ X via (p,p') consisting of four pushouts (PO1) - (PO4)

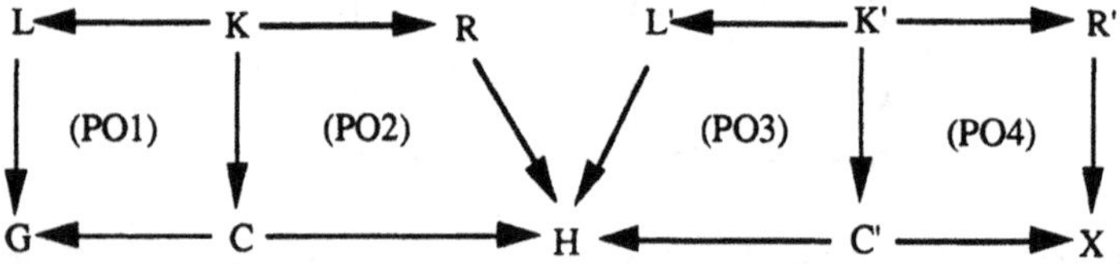


is called

1. <u>sequential independent</u>, or short <u>independent</u>, if there are morphisms L' → C and R → C' such that L' → C → H = L' → H and R → C' → H = R → H.

2. <u>butterfly-independent</u>, or short <u>b-independent</u>, if there are pushouts (1) - (3), called <u>pushout butterfly</u>, such that (1) ∪ (3) = (PO2) and (2) ∪ (3) = (PO3).

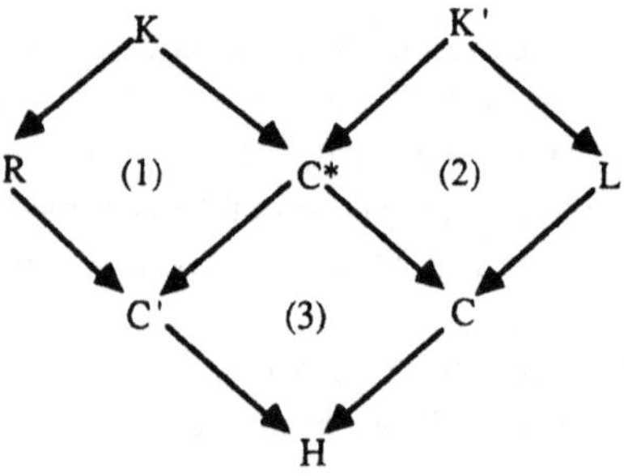

**Remarks, Examples and Interpretation**

1. In the case of graph grammars sequential independence is equivalent to the fact that the intersection of R and L' in H is equal to the intersection of K and K' in H, i.e. the intersection of R and L' consists of gluing points only. Moreover, independence implies butterfly independence in any HLR1-category and the inverse implication is true in any case. But it is open to find an interesting example of independence which is not butterfly independent in an HLR0-category.

2. Note, that in general neither independence nor b-independence are based on unique decompositions. In fact, the morphisms L' → C and R → C' which exist in the case of independence are unique if and only if C → H and C' → H are monomorphisms. They are monomorphisms in HLR0.5-categories with **M** ⊆ Monos. The diagram (3) in the case of b-independence is unique if (3) is a pullback, which is true in HLR0.5$^*$-categories. Hence we have uniqueness of both notions of independence in HLR0.5$^*$-categories and they are equal in HLR1-categories.

3. The condition "**M** ⊆ Monos(**CAT**)" is missing in [EHKP 91a/b], but seems to be necessary in order to show the bijective correspondence in the parallelism theorem. In fact, starting with analysis applied to a parallel derivation we obtain butterflies similar to (1) - (3) above. These butterflies lead to a sequential independent sequence as given above. But in the following synthesis construction we may use different morphisms L' → C and R → C', still satisfying the conditions for independence. Even if $C^*$ is pullback object in (3) this may lead to different morphisms K → $C^*$ and K' → $C^*$,which still satisfy K → $C^*$ → C = K → C and K' → $C^*$ → C' = K' → C'. This, however, will lead to a different morphism K+K' → $C^*$ for the parallel derivation which is the result of the synthesis construction. Hence first analysis and then synthesis in HLR0.5$^*$ and similar in HLR1$^*$ may be different to the identity if we do not require **M** ⊆ Monos(**CAT**). It remains open to find a counterexample for uniqueness of b-independence in HLR1-categories with condition "**M**-pushouts are pullbacks" only. In fact, it is even difficult to find an interesting category with class **M** satisfying "**M**-pushouts are pullbacks" but not "**M** ⊆ Monos(**CAT**)".

4. Example 1.13 in [Kre 76/77] shows for **M** = Mor(**CAT**) that there are two different direct parallel derivations with the same sequentialization G ⇒ H ⇒ X which is butterfly-independent (with two different butterflies leading to two different parallel derivations). Especially this is an example of non-unique independence and non-unique butterfly independence in the case of graph grammars with **M** = **Mor**.

**4.3 Construction (Standard Analysis and Synthesis)**

Let H be an HLR-system in an HLR0-category **CAT**. Let $INDEP_b$ be the butterfly independence and $INDEP_s$ be the sequential independence (which is discussed in part c below). Then H can be extended to the πHLR-system (H,$\pi_b$) with $\pi_b$ = ($INDEP_b$,$ANAL_b$,$SYNT_b$) where $ANAL_b$ and $SYNT_b$ are constructed in part a and part b respectively.

(a) $ANAL_b$: The analysis of a given direct parallel derivation is defined as follows if all used pushouts exist in **CAT** (otherwise it is undefined).

Given a direct parallel derivation s: G ⇒ X via p+p'

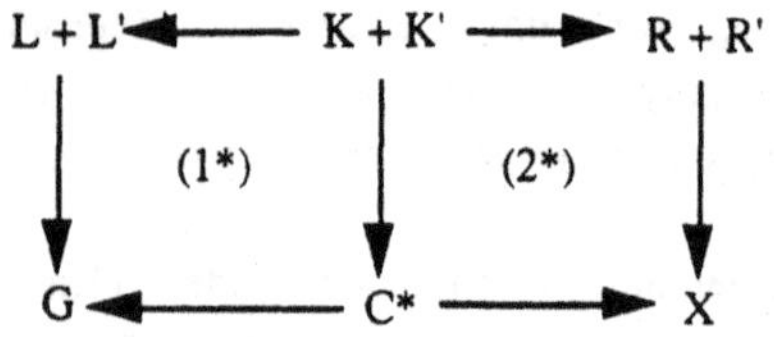


the pushouts ($1^*$) and ($2^*$) can be decomposed into two pushout butterflies using coproduct and pushout properties (the so-called Butterfly-lemma).

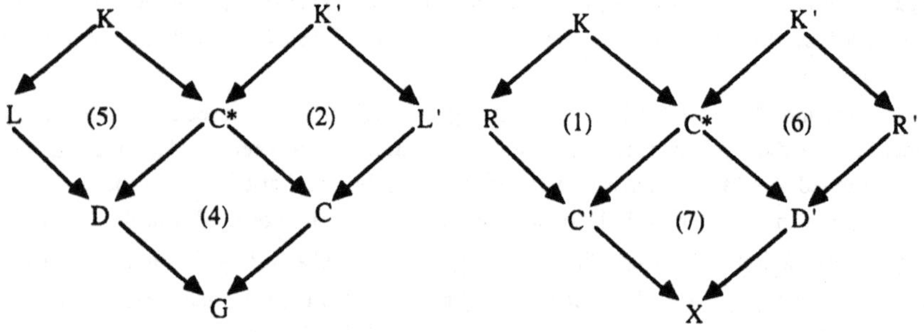


By construction of H in (3) as a pushout, we obtain the pushout butterfly in 4.2 with pushouts (1), (2) and (3) and a derivation G ⇒ H ⇒ X via (p,p') as in 4.2 with (PO1) = (4) ∪ (5), (PO2) = (1) ∪ (3), (PO3) = (2) ∪ (3) and (PO4) = (6) ∪ (7). This derivation is butterfly-independent as shown by the pushout butterfly in 4.2 and is considered as the result of our analysis construction.

(b) $SYNT_b$: The synthesis of a given derivation is defined as follows if all used coproducts and pushouts exist in **CAT** (otherwise it is undefined).

Given a b-independent derivation G ⇒ H ⇒ X via (p, p') as in 4.2 with pushouts (PO1) - (PO4), we have the existence (but in general not uniqueness) of a pushout butterfly as in 4.2.2. Nevertheless, we use the same numbering of pushouts as in part (a) above. Now we are able to decompose (PO1) into pushouts (4) and (5) using $K \to C = K \to C^* \to C$, and to decompose (PO4) into pushouts (6) and (7) using $K' \to C' = K' \to C^* \to C'$. Hence, we have obtained pushouts (1),(2),(4),(5),(6) and (7) as in part (a) leading to two new pushout butterflies (2),(4),(5) and (1),(6),(7). Application of the Butterfly-lemma implies pushouts ($1^*$) and ($2^*$) and hence a direct parallel derivation G ⇒ X via p+p' which is defined to be a result of synthesis. The choice of another pushout butterfly (1'), (2'), (3') may lead to a different result of synthesis. In fact, (1'), (2') and (3') may differ from (1), (2) and (3) in all morphisms except of those which are given in (PO2) and (PO3), especially we may have an object $C^*$ in (3') which is not isomorphic to $C^*$ in (3).

(c) If **CAT** is an HLR1-category, sequential independence and b-independence are equivalent such that $INDEP_s = INDEP_b$. In fact, existence and inheritance of **M**-pullbacks can be used to construct $C^*$ as a pullback of $C \to H$ and $C' \to H$, and the **M**-pushout-pullback decomposition property can be applied twice to obtain the pushout butterfly in 4.2.2. Conversely, b-independence implies independence anyway. Consequently, the HLR1-case allows one to consider the πHLR-system $(H,\pi_b)$ with the variant $\pi_b = (INDEP_s,ANAL_b,SYNT_b)$, i.e. sequential independence instead of b-independence.

By construction, $ANAL_b$ is a unique analysis because all used pushouts are unique if they exist at all. In contrast to that, $SYNT_b$ is not unique in general because there may be various choices of pushout butterflies as pointed out above.

**4.4 Theorem (Existence of Analysis, Synthesis, Shift and Canonical Derivations)**

Let $(H,\pi_b)$ with $\pi_b = (INDEP,ANAL_b,SYNT_b)$ be the πHLR-system in an HLR0.5- or HLR1-category **CAT** as constructed in 4.3 such that $INDEP = INDEP_b$ in case of HLR0.5 and $INDEP = INDEP_b$ or $INDEP = INDEP_s$ in case of HLR1. Then the following hold:

1. $ANAL_b$ and $SYNT_b$ are complete, and $ANAL_b$ is associative.
2. $(H,\pi_b)$ is a/s-faithful and has canonical derivations.

**Proof idea:** First of all let us point out that it suffices to assume an HLR0.5-category with butterfly independence because independent derivations in an HLR1-category are b-independent and vice versa.

1. Completeness of $ANAL_b$ and $SYNT_b$ follows from the constructions in 4.3 (a) and (b) using the properties of HLR0.5-categories in 4.1.2 (a) - (c). Associativity of $ANAL_b$ follows from proof of lemma 1.8 in [Kre 76] which in fact is valid in any HLR0.5-category.

2. The proof of a/s-faithfulness of $(H,\pi_b)$ is based on the fact that the Butterfly-lemma defines a bijective correspondence. Using definition 3.1 $(H,\pi_b)$ induces a shift construction such that we get canonical derivations due to theorem 3.4.1.

**4.5 Theorem (Uniqueness of Synthesis, Shift and Canonical Derivations)**

Let $(H,\pi_b)$ with $\pi_b = (INDEP_s,ANAL_b,SYNT_b)$ be the πHLR-system in an HLR1*-category **CAT** as constructed in 4.3. Let $SHIFT_b$ be the induced shift. Then the following hold:

1. $SYNT_b$ and $SHIFT_b$ are unique, and $INDEP_s$ is distributive (w.r.t. $ANAL_b$).
2. $(H,\pi_b)$ is faithful. and has unique canonical derivations.

Moreover, the first parts of properties 1 and 2 are also satisfied in an HLR0.5*-category with butterfly independence instead of independence.

**Proof idea:**

1. Using the reduction of independence in HLR1-categories to butterfly independence in

HLR0.5-categories, it suffices to show the stated uniqueness for butterfly independence in HLR0.5*-categories which is a consequence of uniqueness of butterfly independence in HLR0.5*-categories (see remark 2 of 4.2). Distributivity of $INDEP_s$ is shown in lemma 2.8 of [Kre 76] and remains valid in HLR1*-categories.

2. We have already completeness of $ANAL_b$ and $SYNT_b$ by 4.4.1 and a/s-faithfulness of $(H,\pi_b)$ by 4.4.2. Moreover, s/a-faithfulness in HLR0.5*-categories follows from the constructions in 4.3 and the possibility of unique reconstructions using the Butterfly-lemma. This implies faithfulness of $(H,\pi_b)$. Uniqueness of canonical derivations for $(H,\pi_b)$ in HLR1*-categories with independence follows from 3.4.2 where properties (a) and (c) are shown already above. For properties (b) and (d) see 4.4.1 and (f) in 4.1.

Note that it is an open problem whether distributivity of butterfly independence is satisfied in HLR0.5*-categories. In that case we also have properties 2 and 4 in HLR0.5*-categories with butterfly independence.

## 5. Applications

In this section we reconsider examples studied in our papers [EHKP 91a], [EP 91b] and [PER 93] and classify them w.r.t. the properties HLR0, HLR0.5, HLR0.5*, HLR1, HLR1* defined in section 4. In particular, we consider the category SET of sets, TOP of topological spaces, GRAPHS of graphs with color preserving morphisms, POGRAPHS of graphs with preordered label sets and color changing morphisms and its subcategory CPOGRAPHS where all labels are completely ordered, HYPGRAPH of hypergraphs and STRUCT of relational structures. Moreover, we regard categories of Petri nets, i.e. P-T-NETS of place tansition nets and AHLNET of algebraic high-level nets combining Petri nets and algebraic specifications where HLR-properties have been studied recently in [PER 93]. The practical relevance of AHL-net transformations is shown in [REP 93]. Finally, three different types of categories of algebraic specifications as presented in [EP 91b] are taken into account. Algebraic specification grammars were recently invented by Parisi-Presicce [Par 91] to describe rule-based modular system design. A first junction between algebraic specifications and graph grammars was made in [EP 91a] and in [EP 93] the relevance of canonoical derivations for algebraic specifications is discussed.

Moreover, we show that we have a proper hierarchy of HLR-properties except of HLR0.5* and HLR1 which are incomparable. Especially we are able to present a new example, algebraic specifications with conservative morphisms as class **M**, which are HLR0.5* but not HLR1. These results together with the main theorems in section 4 allow to analyse which kind of HLR-system has existence or existence and uniqueness of analysis, synthesis, shift and canonical derivations.

### 5.1 Theorem (HLR-Properties of Different Application Categories)

In the following table 1 we summarize the HLR-properties of several example categories with different classes **M** of distinguished morphisms where **Mor**, $\mathbf{M}_{inj}$ and $\mathbf{M}_{spe}$ are the classes of

all morphisms, all injective morphisms, and all special morphisms respectively with:
(1) = injective and color preserving; (2) = injective and strict; (3) = injective and conservative, (4) = $f_{SPEC}$ strict injective and $f_P$, $f_T$ injective where $f_{SPEC}$ is the specification morphism component of AHLNET-morphisms and the components $f_P$ and $f_T$ are functions on places and transitions, resp.

**Remarks**

1. The category **SPEC1** is the category of algebraic specifications and morphisms of type 1, i.e. f#(E1) is derivable from E2 for f:(SIG1, E1) → (SIG2, E2) (see [EM 85]). An injective specification morphism is called
(a) strict, if $f\#^{-1}$(E2) is derivable from E1, and
(b) conservative, if for all SPEC1-algebras A1 there is a SPEC2-algebra A2 s.t. $V_f$(A2) = A1, where SPECi = (SIGi, Ei) (i = 1, 2).
2. In [EP 91b] it was still open whether (**SPEC1**, $M_{strict}$) is HLR1 or HLR1*. In fact it is not even HLR0.5, because strict morphisms are not preserved under pushouts (see [MS 85]). In [MVS 85] our strict morphisms are called conservative. We use conservative in a semantical sense as advocated in [DG 92], where inheritance under pushouts is shown.
3. Additional negative results are mentioned in [EKT 92].

| Category | Name | Reference | Mor | $M_{inj}$ | $M_{spe}$ |
|---|---|---|---|---|---|
| **SETS** | sets | [EHKP 91a] | HLR0.5 | HLR1* | |
| **TOP** | topological spaces | [EHKP 91a] | HLR0.5 | HLR0.5 | |
| **GRAPHS** | graphs | [EHKP 91a] | HLR0.5 | HLR1* | |
| **POGRAPHS** | graphs with p.o.label sets | [EHKP 91a] | HLR0 | HLR0 | |
| **CPOGRAPHS** | graphs with cpo label sets | [EHKP 91a] | HLR0.5 | HLR0.5 | HLR1* (1) |
| **HYP-GRAPHS** | hypergraphs | [EHKP 91a] | HLR0.5 | HLR1* | |
| **P-T-NETS** | Petri nets | [EHKP 91a] | HLR0.5 | HLR1* | |
| **STRUCT** | structures | [EHKP 91a] | HLR0.5 | HLR1 | HLR1* (2) |
| **SPEC1** | algebraic specs type 1 | [EP 91b] | HLR0.5 | HLR0.5 | HLR0 & not HLR0.5 (2)<br>HLR0.5* & not HLR1 (3) |
| **SPEC2** | algebraic specs type 2 | [EP 91b] | HLR0.5 | HLR1 | HLR1* (2) |
| **SPEC3** | algebraic specs type 3 | [EP 91b] | HLR0.5 | HLR1* | |
| **AHLNET** | algeb.high-level nets | [PER 93] | HLR0.5 | HLR1 | HLR1* (4) |

Table 1: HLR-Properties for Different Application Categories

**Proof idea:** In the column **Mor** all categories except of **POGRAPHS** are HLR0.5, because we have existence of pushouts and binary coproducts. **POGRAPHS** fails to have pushouts in general (see [EHKP 91a]). In none of these cases we have HLR0.5* or HLR1, because not all morphisms are monomorphisms and the pushout-pullback decomposition property fails already in **SETS** (see counterexample 2.2 in [EK 79]).
For the columns $\mathbf{M}_{inj}$ and $\mathbf{M}_{spe}$ we refer to the given references. Especially it is shown in our long version [EKT 92] that ($\mathbf{SPEC1}$, $\mathbf{M}_{con}$) is HLR0.5* but not HLR1 where $\mathbf{M}_{con}$ is the class of injective and conservative specification morphisms.

**5.2 Theorem (Hierarchy of HLR-Categories)**
For the HLR-categories defined in 4.1 we have the following hierarchy of classes of categories where I1 - I5 are proper inclusions and HLR0.5$^*$ and HLR1 are not comparable:

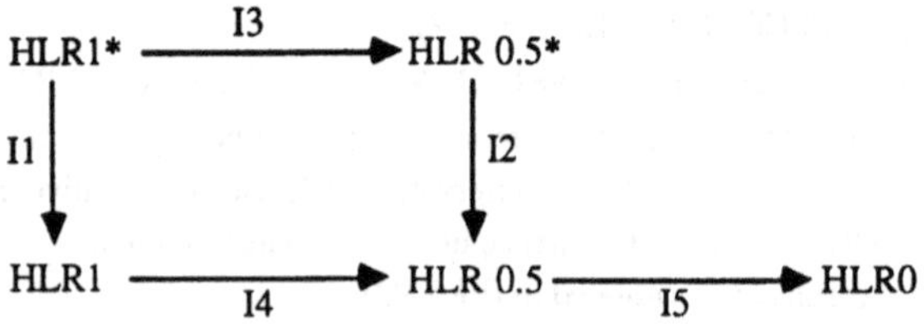


**Proof idea:** By definition in 4.1 we have inclusions I1 - I5. As shown in the proof of 5.1 in [EKT 92] ($\mathbf{SPEC2}$, $\mathbf{M}_{inj}$) is a HLR1-category which is not HLR1$^*$ and also a HLR0.5-category which is not HLR0.5$^*$. Moreover ($\mathbf{SPEC1}$, $\mathbf{M}_{con}$) is a HLR0.5$^*$-category which is not HLR1 and hence also not HLR1$^*$. Hence it is also an example which is HLR0.5 and not HLR1. Finally ($\mathbf{POGRAPHS}$, $\mathbf{M}_{inj}$) is HLR0 but not HLR0.5.

**5.3 Theorem (Existence and Uniqueness of Canonical Derivations in Different Applications of HLR-Systems)**
In the following table 2 we summarize the main results for HLR-systems in the different application categories discussed in 5.1 where the following notation is used:

| | |
|---|---|
| IND: | Independence relation (see 4.2.1) |
| B-IND: | butterfly independence relation (see 4.2.2) |
| E0: | Existence of canonical derivations (see 4.3 with 3.4.1) |
| E: | Existence of complete analysis, synthesis, shift and canonical derivations (see 4.4) |
| E+U: | Existence and uniqueness of analysis, synthesis, shift and canonical derivations (see 4.5) |
| NOT: | Assumptions of theorem 4.3, 4.4 and 4.5 are not satisfied. |

The special morphisms in columns 4 and 5 corresponding to (1) - (3) are

| | |
|---|---|
| (1): | injective and color preserving |
| (2): | injective and strict |
| (3): | injective and conservative |
| (4): | $f_{SPEC}$ strict injective and $f_P$, $f_T$ injective |

**Remark**

In the case (SPEC1, $M_{con}$) we have no results for independence, but existence and uniqueness of analysis, synthesis, shift and parallelism for butterfly independence and existence of canonical derivations (see 4.4).

**Proof:** Direct consequence of theorems 4.3, 4.4, 4.5 and 5.1 where the HLR-properties in table 1 have the following consequences for table 2:
HLR0 implies "NOT" for IND and "E0" for B-IND
HLR0.5 implies "NOT" for IND and "E" for B-IND
HLR0.5* implies "NOT" for IND and "E+U" for B-IND (see remark)
HLR1 implies "E" for IND and B-IND
HLR1* implies "E+U" for IND and B-IND

| | (Mor, B-IND) | ($M_{inj}$, B-IND) | ($M_{inj}$, IND) | ($M_{spe}$, B-IND) | ($M_{spe}$, IND) |
|---|---|---|---|---|---|
| SETS | E | E + U | E + U | | |
| TOP | E | E | NOTE | | |
| GRAPH | E | E + U | + U | | |
| POGRAPHS | E0 | E0 | NOT | | |
| CPOGRAPHS | E | E | NOT | E + U (1) | E + U (1) |
| HYP-GRAPHS | E | E + U | E + U | | E + U (2) |
| P-T-NETS | E | E + U | E + U | | |
| STRUCT | E | E | E | E + U (2) | |
| SPEC1 | E | E | NOT | E0 (2) / E + U* (3) | NOT (2) / NOT (3) |
| SPEC2 | E | E | E | E + U (2) | E + U (2) |
| SPEC3 | E | E + U | E + U | | |
| AHLNET | E | E | E | E + U (4) | E + U (4) |

Table 2: Main Existence and Uniqueness Results

# References


[CELMR 93] Corradini, A.; Ehrig, H.; Löwe, M.; Montanari, U.; Rossi, F.: Abstract graph derivations in the double pushout approach, this volume

[DG 92] Diaconescu, R.; Goguen, J.; Stefaneas, P.: Logical Support for Modularisation, in Proc. ADT/COMPASS Workshop 1991, Springer LNCS 655 (1992)

[EHKP 91a] Ehrig, H.; Habel, A.; Kreowski, H.-J.; Parisi-Presicce, F.: Parallelism and Concurrency in High-Level Replacement Systems, Math. Struct. in Comp. Science 1 (1991), 361-404

[EHKP 91b] Ehrig, H.; Habel, A.; Kreowski, H.-J.; Parisi-Presicce, F.: From Graph Grammars to High-Level Replacement Systems, Proc. 4th Int. Workshop on Graph Grammars and Their Applications to Computer Science, Springer LNCS 532 (1991), 269-291

[Ehr 79] Ehrig, H.: Introduction to the Algebraic Theory of Graph Grammars (A Survey) in: Graph Grammars and Their Application to Computer Science and Biology, Springer LNCS 73, (1979), 1-69

[EK 79] Ehrig, H.; Kreowski, H.-J.: Pushout Properties: An Analysis of Gluing Constructions for Graphs, Math. Nachrichten 91 (1979), 135-149

[EKT 92] Ehrig, H.; Kreowski, H.-J.; Taentzer, G.: Canonical Derivations for High-level Replacement Systems, Computer Science Report no.6/92, University of Bremen, 1992

[EKMRW 81] Ehrig, H.; Kreowski, H.-J.; Maggiolo-Schettini, A.; Rosen, B.; Winkowski, J.: Transformation of Structures: An Algebraic Approach, Math. Syst. Theory 14 (1981), 305-334

[EM 85] Ehrig, H.; Mahr, B.: Fundamentals of Algebraic Specification 1 - Equations and Initial Semantics. EATCS Monographs on Theoretical Computer Science, Vol. 6, Springer (1985)

[EP 91a] H. Ehrig, F. Parisi-Presicce: Algebraic Specification Grammars: A Junction Between Module Specifications and Graph Grammars, Proc. 4th Int. Workshop on Graph Grammars and Application to Computer Science, Springer LNCS 532(1991), 292-310

[EP 91b] H. Ehrig, F. Parisi-Presicce: Nonequivalence of Categories for Equational Algebraic Specifications in View of High-Level-Replacement Systems, Techn. Report No 91-16, TU Berlin, FB 20, 1991

[EP 93] Ehrig, H.; Parisi-Presicce, F.: Interaction between Algebraic Specification Grammars and Modular System Design, Proc. AMAST'93

[Kre 76] Kreowski, H.-J.: Kanonische Ableitungssequenzen für Graphgrammatiken, Techn. Report No. 76-26, TU Berlin, FB 20, 1976

[Kre 77] Kreowski, H.-J.: Manipulation von Graph Transmanipulationen, PhD Thesis, TU Berlin, 1977

[MS 85] Maibaum, T.; Sadler, M.: Axiomatizing Specification Theory, Proc. 3rd ADT-Workshop, Bremen, 1984, Informatik-Fachberichte 116 (1985), 171 - 177

[MVS 85] Maibaum, T.S.E.; Veloso, P.A.S.; Sadler, M.R.: A Theory of Abstract Data Types for Program Development: Bridging the Gap? Proc. TAPSOFT'85, Vol 2, Springer LNCS 186 (1985), 214-230

[Par 91] Parisi-Presicce, F.: Foundations of Rule-Based Design of Modular Systems, TCS 83, No. 1 (1991)

[PER 93] Padberg, J.; Ehrig, H.; Ribeiro, L.: Algebraic High-Level Net Transformation Systems, Techn. Report No. 93 - 12, TU Berlin, FB 20, 1993

[REP 93] Ribeiro, L.; Ehrig, H.; Padberg, J.: Formal Development of Concurrent Systems using Algebraic High-Level Nets and Transformations, Techn. Report No. 93 - 13, TU Berlin, FB 20, 1993, and Proc. SE-Conf. Brazil, 1993

# A Computational Model for Generic Graph Functions

Marc Gemis Jan Paredaens Peter Peelman*

Jan Van den Bussche†

University of Antwerp (UIA)
Department of Mathematics and Computer Science
University of Antwerp (UIA)
Universiteitsplein 1
B-2610 Antwerp, Belgium
Email: {gemis,pareda,peelman,vdbuss}@wins.uia.ac.be.

**Abstract**

The *generic graph machine*, a Turing machine-like computation model for generic graph functions, is introduced. A configuration of this machine consists of a number of machine instances that each are in a state and point to two nodes of a graph. During the execution of a step, the machine instances perform in parallel a local transformation on the graph and are each replaced by a number of other machine instances. It is proved that the generic graph machines express a large and natural class of generic graph functions.

## 1 Introduction

If we represent a database as a (labeled) graph, database manipulations can be modeled as partial functions mapping graphs to graphs [AGP+92]. We call such functions ***graph functions***. However, not all graph functions can be interpreted as data manipulations. Indeed, they should satisfy at least the following two requirements. First, the function should be *computable* in the classical sense, i.e., there must be a Turing machine which, when presented an encoding of a graph as a string over some finite alphabet, produces a similar encoding of the result of the function applied to the graph. Second, the result of the function should depend only on the logical structure of the input graph, and not on the particular encoding used for the computation [AU79]. Nowadays known as *genericity* (for a recent account see [AV91]), this second requirement is equivalent with the requirement that the function preserves graph isomorphisms.

A powerful language for expressing graph functions is the language provided by the GOOD model, a graph-oriented model for object databases [GPVG90a, GPVG90b, GPVdBVG]. Recently, it has been shown [VdBVGAG92] that all generically computable graph functions of interest (in a sense which can be made precise) are expressible in

*Supported by the DPWB under program IT/IF/13.

†Research Assistant of the NFWO.

GOOD. This result gave evidence that the class of GOOD graph functions serves as a yardstick for the "completeness" of data manipulation formalisms.

What was lacking up to now, however, is a Turing machine-like model for expressing GOOD graph functions which works directly on graphs (rather than on strings). It is the purpose of the present paper to introduce such a computation model, called the *generic graph machine* (GGM). A configuration of this machine consists of a number of machine instances that each are in a state and point to two nodes of a graph. During the execution of a step, the machine instances perform in parallel a local transformation on the graph and are each replaced by a number of other machine instances. The high degree of in GGM computations guarantees genericity.

The GGM model differs from other proposed computation models on graphs. *Storage modification machines* [Sch80] are too restricted for our purposes. *On-site* computation models [Lei89] work only as acceptors (not as transducers) and are not computationally complete. *Graph grammars* [EKR90] are typically not designed to express the kind of deterministic graph functions we have in mind in this paper. Closer in spirit to the GGM are the *generic machine* of [AV91] and the *database method scheme* model of [DV91]. A brief discussion of these two models is given in the conclusion of this paper.

This paper is organized as follows. Section 2 presents an informal example of a GGM. Section 3 contains the essential definitions of the GGM model. It also presents a diagrammatic form of the GGM, and a number of examples. In Section 4, the GOOD language is briefly recalled. In Section 5, the basic GGM model is extended with a duplicate elimination operation. In Section 6, it is shown that the GGM-expressible graph functions are precisely those expressible in GOOD. In Section 7, we conclude and indicate further developments of this work.

# 2 An Example of a Generic Graph Machine

A GGM, illustrated in Figure 1, consists of a finite state control and a number of *machine instances* (MIs), and works on a graph. Each MI is in its own state, and has a head and a pointer, both of which point to a node of the graph. Initially, the underlying graph is an arbitrary given input graph, and there is only one MI in a fixed starting state with head and pointer pointing to a particular fixed node. In the figure the graph is enclosed in a rectangle, the MIs are depicted by gray circles, labeled by their state, and the head and pointer of each MI is indicated by a dotted line.

In one transition of the GGM, each MI can—depending on its state—do the following:

1. change state, or die;
2. perform a simple action on the graph, such as the addition or deletion of a node or edge;
3. move its head and pointer to other nodes, possibly splitting up into several new MIs.

The actions of the different MIs are performed in parallel.

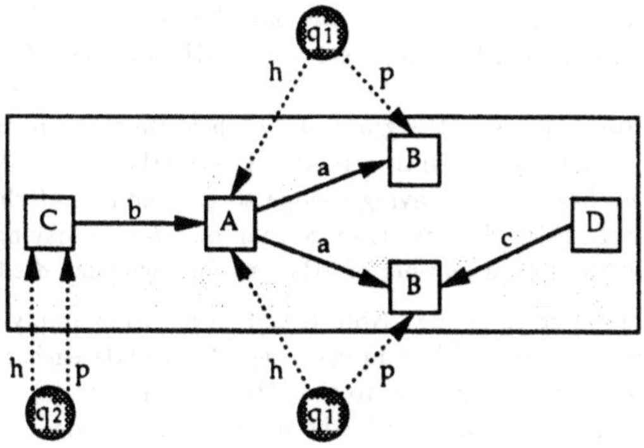


Figure 1: Introductory GGM. The circles denote MIs.

We now present an introductory example of a GGM. The GGM of Figure 2 creates for every triple ($A$-node, $d$-edge, $B$-node), a $C$-node with an outgoing $a$-edge to the $A$-node and an outgoing $b$-edge to the $B$-node. It also deletes the $d$-edge.

$$\begin{aligned}
\delta(q_0) &= (?A, q_1 : q_7)\\
\delta(q_1) &= (P := H, q_2)\\
\delta(q_2) &= (?(d, B), q_3 : q_7)\\
\delta(q_3) &= (+(b^{-1}, C), q_4)\\
\delta(q_4) &= (+a, q_5)\\
\delta(q_5) &= (?(b, B), q_6 : q_7)\\
\delta(q_6) &= (-d^{-1}, q_7)
\end{aligned}$$

Figure 2: An example of a GGM.

The GGM starts with one MI, in state $q_0$ and with head and pointer on a unique, special node $A_0$, that is assumed to be present in the graph. Because of the value of $\delta(q_0)$, this MI splits up into a set of new MIs, one for each $A$-node in the graph. All the new MIs are in state $q_1$, and each one has its head pointing to a different $A$-node. The pointer is copied from the old MI, so it still points to the $A_0$-node, for each of the new MIs. It is possible that there are no $A$-nodes in the graph. In this case the initial MI simply changes state to $q_7$.

So, there are two possibilities:

- there are a number of MIs in state $q_1$ (and these are the only MIs of the GGM). In this case all these MIs execute in parallel $\delta(q_1)$, which copies the head to the pointer (and changes state to $q_2$). The result is that for each MI the head and pointer are on the same $A$-node;

- there is one MI in state $q_7$ (and it is the only MI of the GGM). Since $\delta(q_7)$ is not defined, the MI dies, and because there are no MIs left, the GGM halts.

Hence, if there are $A$-nodes in the graph, for each of them, there is one MI, in state $q_2$, and with head and pointer pointing to it. These MIs execute $\delta(q_2)$, which says: if possible, move the head via a leaving $d$-edge to a $B$-node and change state to $q_3$; otherwise, change state to $q_7$. If there is more than one $d$-edge leaving the $A$-node to a $B$-node, the MI splits up into a number of MIs, one for each such $d$-edge.

Again all the MIs in state $q_7$ die. Note that this does not imply that the machine stops, because there may still be MIs in state $q_3$. These MIs each create one $C$-node and one $b$-edge pointing from that node to the MIs head (a $B$-node). Furthermore, they change state to $q_4$ and move their head to the newly created node.

The MIs in state $q_4$ each create an $a$-edge from their head to their pointer and change to state $q_5$.

Next, the MIs in state $q_5$ move their head via the $b$-edge back to the $B$-node and get into state $q_6$. Note that this head movement is always possible (the $b$-edge was created before), so no MI will change to state $q_7$.

Finally, the MIs are in state $q_6$ and remove the $d$-edge between their pointer and their head.

# 3 Definition of Generic Graph Machines

Basically, we assume the existence of the following sets: $U$ of *nodes*; $NL$ of *node labels*; and $EL$ of *edge labels*. For technical accuracy the symbols "$H$" and "$P$" are assumed not to be in $EL$.

A *graph* is a triple $G = (N, E, \lambda)$ with:

- $N$, the set of nodes, a finite subset of $U$;
- $E$, the set of labeled edges, a finite subset of $N \times EL \times N$;
- $\lambda$, the node labeling function, a total function from $N$ to $NL$. We can extend $\lambda$ to $E$ by defining $\lambda(e) = a$, for every $e = (m, a, n)$ in $E$.

A *Generic Graph Machine* (GGM) is a four-tuple $(Q, q_0, A_0, \delta)$ with:

- $Q$ a finite set of states;
- $q_0 \in Q$, the initial state;
- $A_0 \in NL$, the initial node label;
- $\delta : Q \rightarrow \Theta$, the transition function, a partial function, with:

$$\begin{aligned}\Theta = \{\, & (0,p), (=,p:r), (H := P, p), (P := H, p), \\ & (+a,p), (+a^{-1},p), (-a,p), (-a^{-1},p), \\ & (+(a,A),p), (+(a^{-1},A),p), (-P,p), \\ & (?A,p:r), (?(a,A),p:r), (?(a^{-1},A),p:r) \mid a \in EL, A \in NL, p, r \in Q \,\}.\end{aligned}$$

A *configuration* of a GGM is a pair $(G, C)$, with:

- $G = (N, E, \lambda)$ a graph;
- $C \subseteq \{ (q, head, pointer) \mid q \in Q, head, pointer \in N \}$.

Each element of $C$ is called a *machine-instance* (MI) of the configuration. We call a configuration where $C$ is a singleton, a *one-MI configuration*.

We now describe the effect of one step of a GGM on a graph. So, we specify the transition from a configuration $Conf = (G, C)$ to a configuration $Conf' = (G', C')$. We will do this in two steps. In Subsection 3.1, we state the effect of a GGM on a one-MI configuration. In Subsection 3.2, we generalize to configurations with more than one MI.

## 3.1 Effect of a GGM on a One-MI Configuration

Let $Conf = (G, C)$ be a one-MI configuration of a GGM, with $G = (N, E, \lambda)$ and $C = \{ M \}$ with $M = (q, head, pointer)$. Depending on $\delta(q)$, the GGM transforms $Conf$ into a configuration $Conf' = (G', C')$, in the following way:

$\delta(q)$ is undefined: $G$ remains unchanged and $M$ is removed from $C$ (we say it *dies*). Formally, $G' = G$ and $C'$ is empty.

$\delta(q) = (0, p)$: is used to change $M$ from state $q$ to state $p$ without transforming $G$. Formally, $G' = G$ and $C' = \{ (p, head, pointer) \}$.

$\delta(q) = (=, p : r)$: $M$ changes to state $p$ if its head and pointer are equal; else it changes to state $r$. Formally, $G' = G$; if $head = pointer$ then $C'$ is $\{ (p, head, head) \}$, otherwise $C'$ is $\{ (r, head, pointer) \}$.

$\delta(q) = (P := H, p)$, resp. $(H := P, p)$: $M$ copies its head to its pointer, resp. its pointer to its head. Formally, $G' = G$; $C'$ is $\{ (p, head, head) \}$, resp. $\{ (p, pointer, pointer) \}$.

$\delta(q) = (+a, p)$, resp. $(+a^{-1}, p)$: $M$ adds to $G$ an $a$-edge between its head and its pointer. Formally, $G' = (N, E', \lambda)$, with $E' = E \cup \{ (head, a, pointer) \}$. resp. $E' = E \cup \{ (pointer, a, head) \}$; $C' = \{ (p, head, pointer) \}$.

$\delta(q) = (-a, p)$, resp. $(-a^{-1}, p)$: $M$ deletes from $G$ the $a$-edge between its head and its pointer. Formally, $G' = (N, E', \lambda)$, with $E' = E \setminus \{ (head, a, pointer) \}$. resp. $E' = E \setminus \{ (pointer, a, head) \}$; $C' = \{ (p, head, pointer) \}$.

$\delta(q) = (+(a, A), p)$, resp. $(+(a^{-1}, A), p)$: $M$ adds to $G$ a new $A$-node and an $a$-edge between the new node and its head. $M$ also changes its head to the new node. Formally, $G' = (N', E', \lambda')$ with $N' = N \cup \{ n \}$ for some $n$ in $U \setminus N$; $\lambda'(m) = \lambda(m)$ for all $m$ in $N$; $\lambda'(n) = A$; $E' = E \cup \{ (head, a, n) \}$, resp. $E' = E \cup \{ (n, a, head) \}$; and $C' = \{ (p, n, pointer) \}$.

$\delta(q) = (-P, p)$: $M$ deletes from $G$ its pointer node (and the edges connected to it). Concretely, $pointer$ is deleted[1] from $U$; $G' = (N', E', \lambda')$ with $N' = N \setminus \{ pointer \}$ and

[1] To be entirely correct, it would be necessary to augment configurations with an additional subset $V$ of $U$ of deleted nodes. $V'$ would then equal $V \cup \{pointer\}$, and in the case $\delta(q) = (+(a, A, p))$ we would require that $n \in U \setminus (N \cup V)$.

$E' = E \cap (N' \times EL \times N')$; $\lambda'(m) = \lambda(m)$ for all $m \in N'$; and $C'$ is $\{(p, head, head)\}$ if $head \neq pointer$, and the empty set otherwise.

$\delta(q) = (?A, p : r)$: is used to locate all $A$-nodes of the graph. $M$ splits up into several MIs, one for each $A$-node. The state of each such new MI is $p$, its head is an $A$-node, and its pointer is the old pointer. If there are no $A$-nodes in the graph, $M$ simply changes to state $r$. Formally, $G' = G$. Let $D = \{(p, head', pointer) \mid \lambda(head') = A\}$; if $D$ is empty, then $C' = \{(r, head, pointer)\}$, otherwise $C' = D$.

$\delta(q) = (?(a, A), p : r)$, resp. $(?(a^{-1}, A), p : r)$: is used to navigate through the graph. $M$ splits into several new MIs: a new MI is created for each $A$-node that is connected by an $a$-edge to the head of $M$. The state of each such new MI is $p$, its head is an $A$-node, and its pointer is the old pointer. If the former procedure does not create any MIs, $M$ changes to state $r$. Formally, $G' = G$. Let $D = \{(p, head', pointer) \mid \lambda(head') = A \wedge (head, a, head') \in E$, resp. $(head', a, head) \in E\}$; if $D$ is empty, then $C' = \{(r, head, pointer)\}$, otherwise $C' = D$.

## 3.2 Effect of a GGM on a General Configuration

Let $Conf = (G, C)$ be a configuration of a GGM with $G = (N, E, \lambda)$ and $C = \{M_1, \ldots, M_k\}$. Let $(G_i', C_i')$ be the configuration obtained by applying the GGM to $(G, \{M_i\})$. For each $i$, $G_i' = (N_i', E_i')$. We will assume that the nodes added by different MIs are different; so, for $1 \leq i < j \leq k$, $(N_i \setminus N)$ and $(N_j \setminus N)$ are disjoint.

The GGM transforms $Conf$ into a configuration $Conf' = (G', C')$, with:

- $G' = (N', E', \lambda')$, with:
  - $N' = \bigcap_{i=1}^{k} N_i' \cup \bigcup_{i=1}^{k} (N_i' \setminus N)$,
  - $E' = (\bigcap_{i=1}^{k} E_i' \cup \bigcup_{i=1}^{k} (E_i' \setminus E)) \cap (N' \times EL \times N')$,
  - $\lambda'(n) = \lambda(n)$, if $n \in \bigcap_{i=1}^{k} (N_i' \cap N)$, and $\lambda'(n) = \lambda_i'(n)$, if $n \in (N_i' \setminus N)$;
- $C' = \{(q, head, pointer) \in \bigcup_{i=1}^{k} C_i' \mid q \in Q \wedge head, pointer \in N'\}$.

We see from the above definition of $C'$ that an MI dies not only if $\delta$ is not defined on its state (as in the previous subsection), but also if its head or pointer no longer belong to the graph. Hence, many MIs of a configuration can be killed by a $-P$-operation. As a matter of fact, MIs $(q_1, 1, 2)$ and $(q_1, 2, 1)$ kill each other if $\delta(q_1) = (-P, q_2)$.

The operation of a GGM can be visualized as follows. Each MI takes into account its state. The $\delta$-function indicates the action to be taken on the actual graph by the MI. First all the MIs check in parallel if they can perform the action. Then, again in parallel, they adjust the graph and move their head and pointer (if required). Finally all the MIs whose head and pointer are no longer in the graph, are deleted. This kind of behavior is known as *systolical* behavior. Note that from the above definition follows that this behavior is unambiguous.

For example, if $\delta(q_1) = (?(a, B), q_3 : q_4)$ and $\delta(q_2) = (-a, q_5)$, and the GGM is in configuration $(G, \{(q_1, 1, 1), (q_2, 1, 2)\})$, then the next configuration is $(G', \{(q_3, 2, 1), (q_5, 1, 2)\})$ ($G$ and $G'$ are given in Figure 3).

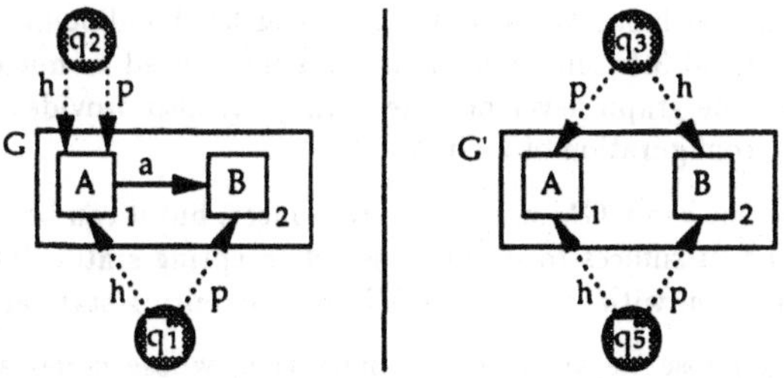


Figure 3: The systolical behavior of a GGM.

Also note that since the second component of a configuration is a set, two identical MIs merge into one MI. Clearly two MIs are identical iff their state as well as their head and their pointer are equal. For example, if $\delta(q_1) = (0, q_4)$, $\delta(q_2) = (-P, q_4)$ and $\delta(q_3) = (?A, q_4 : q_5)$, and the GGM is in configuration $(G, \{(q_1, 1, 1), (q_2, 1, 2), (q_3, 2, 1)\})$, then the next configuration is $(G', \{(q_4, 1, 1)\})$ ($G$ and $G'$ are given in Figure 4).

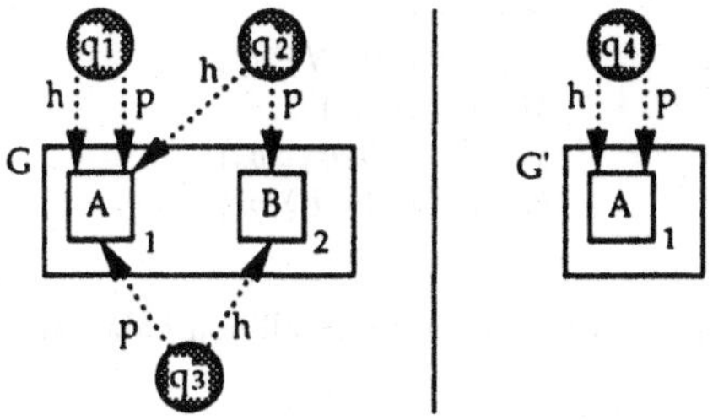


Figure 4: The merging of MIs.

## 3.3 Expressing Graph Functions with GGMs

In this subsection we describe how GGMs express graph functions. The transition from configuration *Conf* to configuration *Conf'*, as defined in Subsection 3.2 is denoted by $Conf \Longrightarrow Conf'$, and $\stackrel{+}{\Longrightarrow}$ denotes the transitive closure of this relation.

Let $\mathcal{F}$ be a graph function, and let $M = (Q, q_0, A_0, \delta)$ be a GGM such that $A_0$ does not occur in the graphs in the domain or range of $\mathcal{F}$. $M$ expresses $\mathcal{F}$ iff for each graph $G$ in the domain of $\mathcal{F}$ holds:

- let $G_1$ be the graph consisting of $G$ and an additional $A_0$-labeled node $n$;
- let $G_1'$ be the graph consisting of $\mathcal{F}(G)$ and $n$;
- then $(G_1, \{(q_0, n, n)\}) \stackrel{+}{\Longrightarrow} (G_1', \emptyset)$.

We also say that $M$ *transforms* $G$ into $\mathcal{F}(G)$.

So, a GGM starts with one MI, in state $q_0$, with as head and pointer the $A_0$-node, and ends when there are no MIs left. The $A_0$-node is introduced to make sure that during the transformation the graph never becomes empty. It also provides a uniform way for handling the start configuration of a GGM.

As defined up to now, a GGM works as a transducer, but it can be used as an acceptor as well. To this end, it suffices to specify a set of accepting states. Those graphs which generate a configuration with at least one MI in an accepting state are accepted.

For example, suppose we are given a binary tree, whose nodes are labeled $N$ and whose edges are labeled $e$. Possibly as a subtask of a more complicated task, we must determine whether the tree is unbalanced, i.e., whether there are two paths from the root to a leaf with different length. This can be done using the program fragment of Figure 5. Assume that upon entrance of the program, exactly one MI is present, in state $q_0$ and with head and pointer on the root. This MI initiates a traversal of the tree, following every path by splitting in each node. Each MI that arrives in a leaf creates a $B$-node. Thus, if an MI arrives in a leaf and there is already a $B$-node, the tree is non-balanced. Since it takes two steps to test for the existence of a $B$-node and to create it, a 0-operation must be added to synchronize the parallel traversal.

$$
\begin{aligned}
&A = \{ q_A \} \\
&\delta(q_0) = (?(e, N), q_1 : q_2) \\
&\delta(q_1) = (0, q_0) \\
&\delta(q_2) = (?B, q_A : q_3) \\
&\delta(q_3) = (+(b, B), q_4)
\end{aligned}
$$

Figure 5: A GGM accepting all non-balanced trees.

### 3.4 GGM Diagrams

It is often convenient to represent a GGM diagrammatically. Given a GGM $M$, the associated diagram is a flowchart graph whose nodes represent those states of $M$ on which the transition function $\delta$ is defined. Recall that for a state $q$, $\delta(q)$ is of the form $(\theta, p)$ or $(\theta, p : r)$. The diagram node representing $q$ is then labeled $\theta$, and has, depending on the form of $\delta(q)$, either one outgoing edge to the node representing $p$, or two outgoing edges, one labeled Y to the node representing $p$, and one labeled N to the node representing $r$. In case $\delta$ is not defined on $p$ or $r$, the outgoing edge is a loose exit edge not connected to any target node. Finally, there is a loose entry edge to the node representing the initial state $q_0$. An example is shown in Figure 6: the diagram in this figure represents the GGM of Figure 5.

Two GGM diagrams can be composed in the well-known way, by connecting the first diagram's exit edges to the second diagram's entry edge. This technique will be frequently used in Section 6.

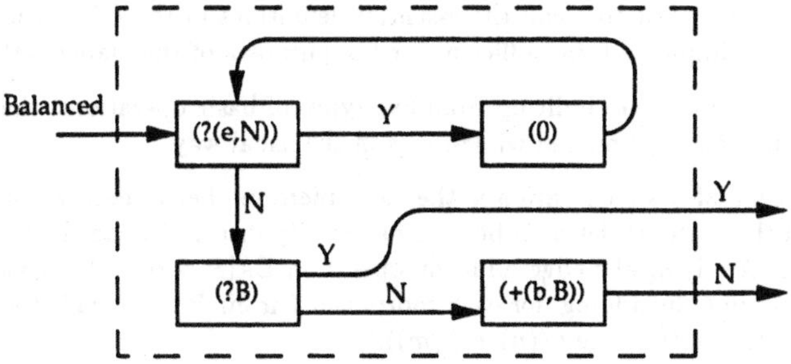


Figure 6: Diagram of GGM **Balanced**

## 3.5 Examples of GGMs

In this subsection we present in a less detailed way two more examples of GGMs. The GGM of the first example connects every node of a directed, acyclic graph (DAG) with all the roots of the DAG from which it can be reached. The input to this GGM is a DAG with $N$-labeled nodes and $e$-labeled edges. The first part is the localization of the root nodes. The characteristic of a root node is that it has no incoming edges. This can easily be detected by testing in every $N$-node whether or not an inverse $e$-edge can be followed to another $N$-node. The test is made such that there remains an MI with head and pointer on every root node.

All these MIs keep there pointer fixed and traverse the DAG with their head (as in the example of Figure 5). By using an edge addition, we link all nodes that can be reached from a particular root node to that root node. An MI dies when it reaches a leaf node.

Another example is a GGM that tests whether or not a graph is a tree. The input to this GGM is a arbitrary graph with $N$-labeled nodes and $e$-labeled edges. The first part of the program localizes a unique root node. In a way similar to the start of the previous program, we tag every root by a *Root*-node.

Up to now we only detected the graphs that have no roots. We detect graphs containing several root nodes by placing an MI with head and pointer on every *Root*-node, moving the head of these MIs to all other *Root*-nodes, and then testing if there are MIs with head and pointer different.

Now the graph is a tree iff starting from the root, every node can be reached by a unique path. This is tested by traversing the graph from the root on and tagging every visited node. If a visited node is already tagged, if a node has more than one tag, or if there nodes without tags, the graph is not a tree.

# 4 GOOD

As already mentioned in the introduction, it is the goal of this paper to introduce a Turing machine-like computation model for the GOOD graph functions. Therefore, it is appropriate at this point to recall the essential definitions of the GOOD language. We will do this in a simplified form sufficient for the purposes of this paper [vR92].

Programs in GOOD are built up from five types of basic operations; the language is closed off under composition and while-loops in the usual way.

The first type of basic operation is the *edge addition*. Let $G$ be a graph, $P$ another graph (called the *pattern*), let $n, m$ be (not necessarily distinct) nodes in $P$, and let $e$ be an edge label. Applying the edge addition operation $\mathrm{EA}[P, n, m, e]$ to $G$ yields a graph $G'$, obtained from $G$ by adding, for each morphism $f$ from $P$ to $G$ (called a *matching* of the pattern $P$ in $G$), the edge $(f(n), e, f(m))$.

The *node addition* is defined as follows. Let $n_1, \ldots, n_k$ be nodes in $P$, let $e_1, \ldots, e_k$ be edge labels, and let $K$ be a node label. Applying the node addition operation

$$\mathrm{NA}[P, n_1, \ldots, n_k, e_1, \ldots, e_k, K]$$

to $G$ yields a graph $G'$, obtained from $G$ by adding, for each matching $f$ of $P$ in $G$, a $K$-labeled new node with an $e_i$-labeled edge to $f(n_i)$ for each $i = 1, \ldots, k$, provided such a node does not yet exist in $G'$.

The *edge deletion* is defined as follows. Let $(n, e, m)$ be an edge in $P$. Applying the edge deletion operation $\mathrm{ED}[P, n, e, m]$ to $G$ yields a graph $G'$, obtained from $G$ by deleting, for each matching $f$ of $P$ in $G$, the edge $(f(n), e, f(m))$.

The *node deletion* is defined as follows. Let $n$ be a node in $P$. Applying the node deletion operation $\mathrm{ND}[P, n]$ to $G$ yields a graph $G'$, obtained from $G$ by deleting, for each matching $f$ of $P$ in $G$, the node $f(n)$.

The fifth type of basic operation, the *abstraction*, is different in spirit from the other four and is used for duplicate elimination. Let $n$ be a node in $P$, let $e, e'$ be edge labels, and let $K$ be a node label. We can define the following equivalence relation among the nodes of $G$:

$$m \equiv_e m' \Leftrightarrow \{m'' \mid \text{edge } (m, e, m'') \text{ in } G\} = \{m'' \mid \text{edge } (m', e, m'') \text{ in } G\}.$$

We say in this case that $m$ and $m'$ are *duplicates w.r.t. e*. Denote the restriction of $\equiv_e$ to those nodes that equal $f(n)$ for some matching $f$ of $P$ in $G$ by $\equiv_e |_P$. Then applying the abstraction operation $\mathrm{AB}[P, n, e, e', K]$ to $G$ yields a graph $G'$, obtained from $G$ by adding, for each equivalence class $Z$ w.r.t. $\equiv_e |_P$, a $K$-labeled new node with an $e'$-labeled edge to each $m \in Z$, provided such a node does not yet exist in $G'$.

Finally, GOOD programs can now be defined as follows. Each basic operation is a program. The composition $Q_1; Q_2$ of two programs is a program. If $Q$ is a program and $K$ is a node label, then the while-loop $\mathrm{Loop}[K, Q]$ is a program. The semantics of basic operations was defined above. The semantics of composition is the obvious one, with the extra provision that a node which is deleted by a node deletion operation in $Q_1$ cannot be recreated by a node addition or abstraction operation in $Q_2$. The semantics of the while-loop is "repeat $Q$ as long as there is a $K$-labeled node in the graph."

# 5 Duplicate Elimination

The GOOD model includes the abstraction operation, which groups each equivalence class of duplicate nodes together with a single new node which can then serve as a unique representative for the class. The basic GGM model as defined up to now cannot provide this functionality; in fact, it is equivalent to GOOD without abstraction (as can be seen from the proofs which will be presented in Section 6). Therefore, in this section, we extend GGMs with duplicate elimination.

To this end, we specify a number of states which we call "red." Each time some MI is in a red state, duplicate elimination if performed on the current configuration before moving to the next configuration. The actual construction by which duplicates are eliminated is more explicit and global than the one provided by the abstraction operation, and is formally defined next.

Let $(G, C)$ be a GGM configuration, and let $n_1, n_2$ be two nodes of $G$. We say that $n_1$ and $n_2$ are *duplicates* if the transposition of $n_1$ and $n_2$, which can be applied to $G$ and $C$ in the canonical way, leaves $G$ and $C$ invariant. Hence, $n_1$ and $n_2$ are logically interchangeable within the configuration.

The duplicate elimination now transforms $(G, C)$ into a new one $(G^{\text{d.e.}}, C^{\text{d.e.}})$ as follows. Call a node in $G$ *red* if it is under the head of a MI in a red state. Let $\mathcal{Z}$ be the partition of $N$ in equivalence classes according to the following equivalence relation:

- If $n_1, n_2$ are red, then $n_1$ and $n_2$ are equivalent if $n_1$ and $n_2$ are duplicates;
- If $n$ is not red, then $n$ is only equivalent to itself.

For each non-singleton equivalence class $Z$, we take a new representative node $n_Z \in U \backslash N$. We extend this to singleton equivalence classes $Z = \{n\}$ by putting $n_Z = n$. Now in the new configuration, each representative replaces its equivalence class. Formally, we have:

- $N^{\text{d.e.}} = \bigcup_{Z \in \mathcal{Z}} \{n_Z\}$;
- $\lambda(n_Z) = \lambda(n)$ for $n \in Z$;
- $E^{\text{d.e.}} = \bigcup_{Z, Z' \in \mathcal{Z}} \{(n_Z, e, n_{Z'}) \mid \exists (n, e, n') \text{ in } G, n \in Z, n' \in Z'\}$;
- $C^{\text{d.e.}} = \{(q, n_Z, n_{Z'}) \mid \exists (q, n, n') \in C, n \in Z, n' \in Z'\}$.

# 6 Equivalence of GGM and GOOD

We are now ready to show that the GGM model is equivalent to the GOOD model. The significance of this result was already pointed out in the introduction. It shows that every graph function expressible by a GGM is generically computable, and moreover, that the GGM model is complete in this respect.

## 6.1 Simulation of GOOD by GGM

**Theorem.** *For every GOOD-program $Q$, there exists a GGM $M$ such that for all graphs $G$ and $G'$ holds that $Q$ transforms $G$ into $G'$ iff $M$ transforms $G$ into $G'$.*

**Proof.** First, observe that it is sufficient to construct for each of the five types of basic GOOD operations an equivalent GGM diagram. Indeed, composition of programs can then be simulated by composing the associated diagrams; a while-loop Loop$[K, Q]$ can be simulated by augmenting the transition table of the diagram associated to $Q$ with an entry of the form $\delta(q) = (?K, p : r)$.

The GGM that simulates a GOOD-operation has three major parts: The first part tags all matchings of the pattern of the GOOD-operation. The second part of the GGM performs the actual operation, using the tags constructed in the first part. The last part is needed to remove auxiliary nodes and edges, that were used in the previous parts. We will now describe the first two parts in more detail:

**Part 1: Pattern matchings**

The tagging[2] of the matchings of a pattern $P$ involves three steps.

The first step is to tag all combinations of nodes of $P$. In this way, at least all matchings of the pattern are indicated. The tagging of the node-combinations of $P$ is done in an inductive way on the nodes $A_1, \ldots, A_n$ of $P$. Every combination of nodes $A_1, \ldots, A_n$ in the graph is marked by adding an $M_n$-node, that is linked to every $A_i$ by an $i$-edge (Diagrams $N_{1,0}$, $N_{i,0}$ and $N_{i,j}$ of Figure 7).

First an $M_1$-node is created for every $A_1$-node. Then, by induction on $i$, for every pair of nodes $M_{i-1}$ and $A_i$ in the graph, a new node $M_i$ is created. The new node is linked to the $A_i$-node and, via node $M_{i-1}$, to nodes $A_1, \ldots, A_{i-1}$.

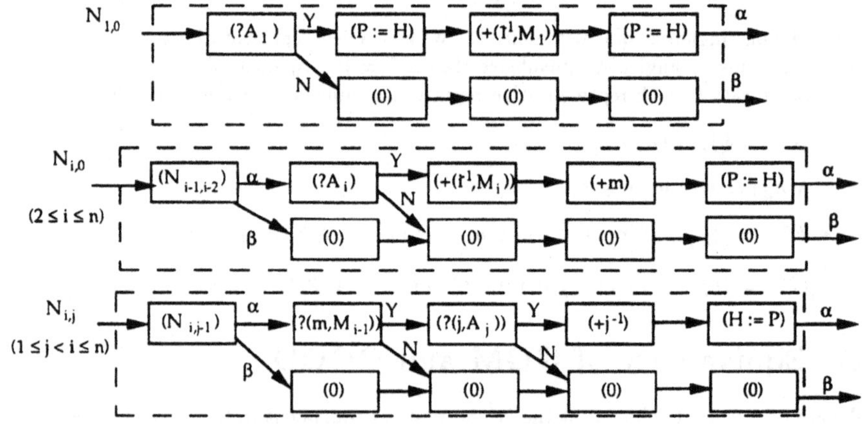


Figure 7: Diagram for node-combinations.

For every $M_n$-node there is one MI of the GGM. However, many of the tagged combinations are not matchings of $P$, because they lack some of the edges $e_1, \ldots, e_m$ ($e_i =$

[2]We assume that the labels used for the tagging are not yet used in the graph.

$(A_{i_1}, a_i, A_{i_2})$) of $P$. The next part of the GGM (Figure 8) isolates the MIs that point to correct matchings of $P$ from the other ones.

The occurrence of the edges of $P$ is checked sequentially. Note that the MIs pointing to correct matchings of $P$ stay on the upper path of the diagram. To link this diagram with the previous one, we can take $N_{n,n-1} = E_0$.

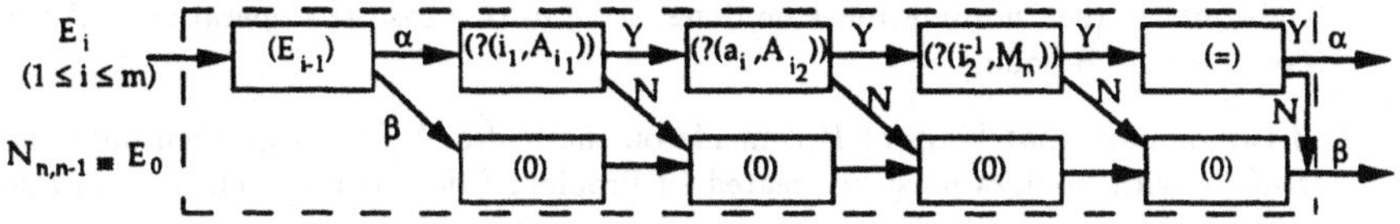


Figure 8: Diagram for the existence of edge $e_i$.

The correct matchings of $P$ are now distinguished from the others by creating a $P$-labeled node for every MI of the GGM that is at the upper path of the diagram, and by linking these $P$-nodes to their corresponding $A_i$-nodes (in a way similar to diagrams $N_{i,0}$ and $N_{i,j}$).

**Part 2: Operations**

**Edge Deletion** The diagram of Figure 9 is used to simulate an edge deletion ED$[P, A_{i_1}, a_i, A_{i_2}]$. The diagram starts with MIs with as head and pointer $P$-nodes, one for each matching of $P$. First, the head and the pointer are changed to the $A_{i_1}$-node of the matching. Then, the head is changed (via the $P$-node, to stay in the same matching) to the $A_{i_2}$-node. Finally, the $a_i$-edge between pointer and head is removed.

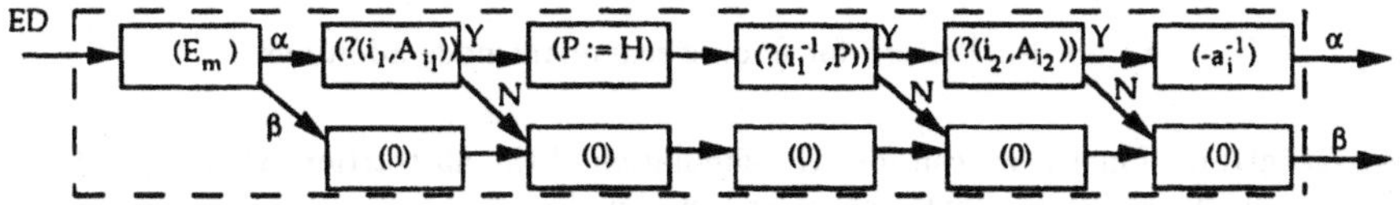


Figure 9: Diagram for an edge deletion.

**Edge Addition** The diagram for an edge addition is similar to the one for an edge deletion. Of course, in the last step an edge is added instead of deleted. Note that by definition of a graph, an edge with a certain label can only occur once between two nodes, so if the node that must be added is already present, the operation will have no effect.

**Node Deletion** A node deletion ND$[P, A_i]$ is simulated by moving the head (which is a $P$-node) to its corresponding $A_i$-node, copying the head to the pointer, moving back the head to the original $P$-node, and deleting the pointer node.

**Node Addition** The next operation that must be simulated is the node addition

$$\mathrm{NA}[P, A_{i_1}, \ldots, A_{i_k}, b_1, \ldots, b_k, B].$$

We cannot generate a $B$-node for each $P$-node, since this gives rise to the following two problems:

1. if for a matching of $P$ in the graph, there is already a $B$-node with $b_i$-edges to the right nodes, then nothing must be done, since in this case the operation will not create a new B-node.

2. if two or more matchings of $P$ coincide on the nodes $A_{i_1}, \ldots, A_{i_k}$, then only one $B$-node with $b_i$-edges must be created (if Problem 1 does not occur), since in this case the operation will create only one new B-node.

The first problem can be solved by checking the graph for the pattern of Figure 10 and performing a node deletion of the $P$-node of that pattern. To solve problem 2, we take nodes $A_{i_1} \ldots A_{i_k}$ as the first $k$ nodes in the search for the pattern matchings. So, a $M_k$-node indicates those nodes that will be connected to a new node.

Instead of generating a $B$-node for every $P$-node, we follow the path of $m$-edges from the $P$-node to the $M_k$-node, thereby eliminating redundant matchings. We have thus created precisely those $B$-nodes that will be created by the node addition operation.

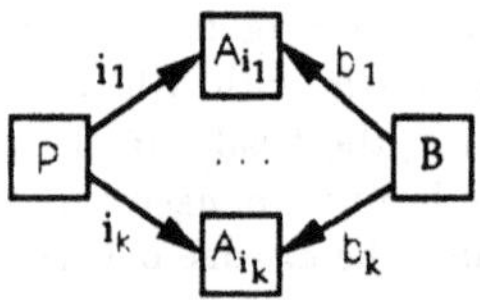


Figure 10: Pattern for finding redundant node additions.

**Abstraction** Finally, we consider the simulation of the abstraction $\mathrm{AB}[P, A_i, a, b, B]$. For simplicity, we assume that $a$-edges end in $C$-nodes.

We first put a $\neg a$-edge between every pair of $(A, C)$-nodes that are not connected with an $a$-edge. The next step is to determine the equivalence relation of the $A$-nodes on their outgoing $a$-edges. This is done by adding an $e$-edge between every pair of $A$-nodes and then deleting the $e$-edges between non-equivalent $A$-nodes. Two $A$-nodes are non-equivalent if there is an $(a, \neg a^{-1})$-path or an $(\neg a, a^{-1})$-path between them. Now we create for every $A$-node in the pattern, a $B'$-node with a $b$-edge between them. If there is a $(b, e)$-path between a $B'$ and an $A$-node, we add a $b$-edge between the $B'$ and $A$-node. Now we can perform duplicate elimination on the $B'$-nodes.

The duplicate elimination is performed with $B'$-nodes rather than directly with $B$-nodes, in order to avoid the creation of unneeded $B$-nodes (a situation similar to problem 1 of the node addition simulation above). It now remains to remove the $B'$-nodes that indicate existing abstractions, before copying them to $B$-nodes.

## 6.2 Simulation of GGM by GOOD

**Theorem.** *For every GGM $M$, there exists a GOOD-program $Q$ such that for all graphs $G$ and $G'$ holds that $M$ transforms $G$ into $G'$ iff $Q$ transforms $G$ into $G'$.*

**Proof.** The GOOD-program that simulates a GGM is one big while-loop. In each pass of the loop, one transformation step of the GGM is simulated. Due to the parallelism of the GGM, it is necessary to first tag all nodes involved in the current transformation, before executing the actual operations.

Since a GGM describes a transition between configurations, we extent the graph $G$ such that it represents a configuration. Therefore we add a node with label $Q$ for every state $q$ of $M$, a node with label $A_0$, and a node with label $MI$, for every MI. Each $MI$-node has one $q$-edge to a state-node, and one $h$-edge and $p$-edge to represent the head and the pointer. So, initially there is one $MI$-node, with $q$-edge to node $Q_0$, and $h$- and $p$-edge to node $A_0$.

The condition to stay in the loop of the simulation is the existence of $MI$-nodes. In the first part of the body, for each $MI$-node a set of $MI'$-nodes is created. These nodes represent the successors of the $MI$-node in the next configuration. The $MI'$-nodes also have additional edges to mark the nodes and edges that have to be deleted in the second part of the loop-body.

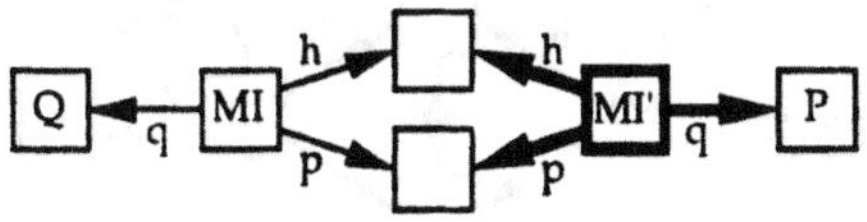


Figure 11: Creation of $MI'$-nodes for $\delta(q) = (0, p)$.

The node addition of Figure 11 is used to create the $MI'$-nodes for the operation $\delta(q) = (0, p)$. In fact, Figure 11 represents a whole family of node additions, obtained by letting unlabeled nodes stand for any node label occurring in the $\delta$-function.

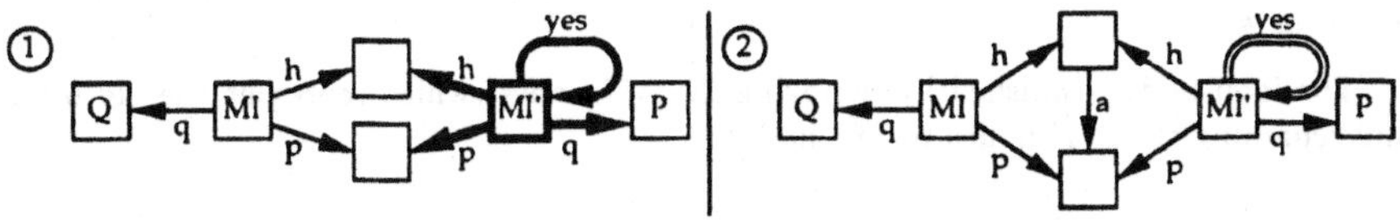


Figure 12: Creation of $MI'$-nodes for $\delta(q) = (+a, p)$.

To avoid problems when sequencing a parallel edge addition and edge deletion, we must add an edge only if it is not yet present, and we must delete it only if it is present

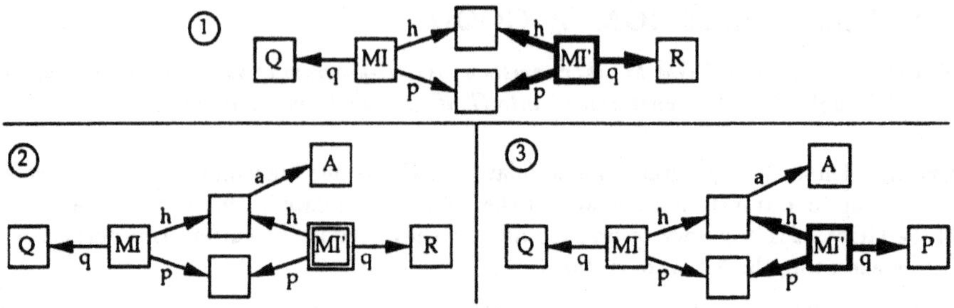


Figure 13: Creation of $MI'$-nodes for $\delta(q) = (?(a, A), p : r)$.

in the graph. The fact that the operation is allowed is indicated by a *yes*-edge. The construction of the *yes*-edge is shown in Figure 12, on the example of $\delta(q) = (+a, p)$.

For the creation of the $MI'$-nodes for $\delta(q) = (?(a, A), p : r)$, we need to get in state $r$ if no $a$-edge to an $A$-node is found. The simulation for this operation is given in Figure 13.

All other operations can be simulated in a way similar to these three examples.

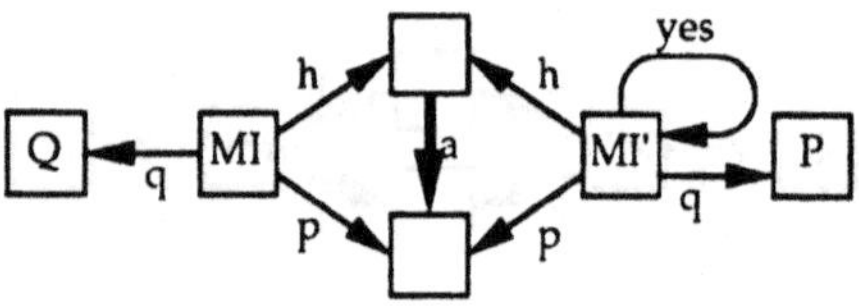


Figure 14: Execution of $\delta(q) = (+a, p)$

The second part of the loop-body does the actual additions and deletions of nodes and edges. The edge addition of Figure 14 is used for performing the actual addition of the $a$-edge for $\delta(q) = (+a, p)$. Note that the addition is done only for those parts of the graph where the *yes*-edge is present. The other additions and deletions are performed in a similar way.

The loop-body is finished by removing the $MI$-nodes (which represent the old MIs) and renaming the $MI'$-nodes to $MI$-nodes.

# 7 Concluding Remarks

Two computation models that are very close in spirit to the GGM model presented here are the *generic machine* (GM) of [AV91] and the *database method scheme* model (DMS) of [DV91].

GMs are designed for the computation of generic functions on relational structures; the GM model is a conservative extension of the basic Turing machine model. One might at first suspect that our GGM model is nothing but the restriction of the GM model to graph structures. This is not the case. GGMs are specifically tuned to work on graphs; for example, they lack a Turing machine tape. Furthermore, GMs focus on domain-preserving functions: specialized to graphs, GMs would not be able to add new nodes to a graph. Nevertheless, the two models are, as already mentioned, very close in spirit.

A DMS consists of a number of straight-line procedures which execute in parallel on a database of objects which is organized as a graph. From this description, the similarity to the GGM model is obvious. We conjecture that DMSs can be simulated by GGMs with only constant overhead in space and time, and vice versa.

For expository reasons, we have deliberately avoided a discussion of the complexity of GGM computations. It can be shown that GGMs can be simulated by ordinary Turing machines with polynomial overhead in combined space-time resources, and vice versa. Due to the parallelism involved, combined space-time complexity seems to be the natural measure of resource consumption in connection with GGM computations.

We plan to present a detailed comparison between GGMs, GMs, and DMSs, as well as an extensive discussion of complexity issues, in a forthcoming report.

# References


[AV91] S. Abiteboul and V. Vianu. Generic computation and its complexity. In *Proceedings 23rd ACM Symposium on Theory of Computing*, pages 209–219, 1991.

[AU79] A.V. Aho and J.D. Ullman. Universality of data retrieval languages. In *Proceedings of the ACM Symposium on Principles of Programming Languages*, pages 110–120, 1979.

[AGP+92] M. Andries, M. Gemis, J. Paredaens, I. Thyssens, and J. Van den Bussche. Concepts for graph-oriented object manipulation. In A. Pirotte, C. Delobel, and G. Gottlob, editors, *Advances in Database Technology—EDBT'92*, volume 580 of *Lecture Notes in Computer Science*, pages 21–38. Springer-Verlag, 1992.

[DV91] K. Denninghoff and V. Vianu. The power of methods with parallel semantics. In *Proceedings 17th International Conference on Very Large Data Bases*, pages 221–232, 1991.

[EKR90] H. Ehrig, H.-J. Kreowski, and G. Rozenberg, editors. *Graph-Grammars and Their Application to Computer Science*, volume 532 of *Lecture Notes in Computer Science*. Springer-Verlag, 1990.

[GPVdBVG] M. Gyssens, J. Paredaens, Jan Van den Bussche, and Dirk Van Gucht. A graph-oriented object database model. To appear in *IEEE Transactions on Knowledge and Data Engineering*.

[GPVG90a] M. Gyssens, J. Paredaens, and D. Van Gucht. A graph-oriented object database model. In *Proceedings of the Ninth ACM Symposium on Principles of Database Systems*, pages 417–424. ACM Press, 1990.

[GPVG90b] M. Gyssens, J. Paredaens, and D. Van Gucht. A graph-oriented object database model for database end-user interfaces. In H. Garcia-Molina and H.V. Jagadish, editors, *Proceedings of the 1990 ACM SIGMOD International Conference on Management of Data*, number 19:2 in SIGMOD Record, pages 24–33. ACM Press, 1990.

[Lei89] D. Leivant. Descriptive characterizations of computational complexity. *Journal of Computer and System Sciences*, 39(1):51–83, 1989.

[Sch80] A. Schönhage. Storage modification machines. *SIAM J. Comput.*, 9(3):490–508, 1980.

[VdBVGAG92] J. Van den Bussche, D. Van Gucht, M. Andries, and M. Gyssens. On the completeness of object-creating query languages. In *Proceedings 33rd Symposium on Foundations of Computer Science*, pages 372–379. IEEE Computer Society Press, 1992.

[vR92] J. van Rossum. Master's thesis, Technical University of Eindhoven, 1992. (in Dutch).

# Graphs and Designing

Ewa Grabska
*Instytut Informatyki UJ*
*ul. Nawojki 11, 30-072 Kraków, Poland*

**ABSTRACT.** In this paper the idea of defining graphical models or pictures by specifying their logical structures (graphs) and the possible ways of realization of such structures (realization schemes) is presented. The fact that the structure of the object is independent of its realization may be useful in the process of designing.

## 1. Introduction

This paper should be seen as a part of thorough study of graphical modeling. Construction of a graphical model is based on a selection of modeling primitives and a collection of modeling procedures for their instantiation and manipulation [19].

The object of our study will be the graphical models themselves in terms of a precise formalism which will serve as the conceptual basis for designing graphical models, helping us to understand how they are created.

It is obvious that for graphical models information processing concerns the manipulation of high dimensional structures, first of all graphs, rather than linear structure. But in our opinion the graph methods of processing, representation and generation of patterns applied today allow to introduce certain innovations.

The innovation proposed in this paper is so called *realization scheme*. It is a kind of mapping which assigns a set of graphical models of an object to a graph representing the structure of this object. In other words, we present the idea of defining graphical models by specifying their logical structures (graphs) and the possible ways of realization of such structures (realization schemes). The realization schemes contain information about basic geometrical shapes which the models consist of and transformations that the shapes must undergo. They reflect also certain design criteria. Thanks to this approach one graph can have several different graphical models associated with it through different realization schemes.

Most of graphical models have their logical structure rather complex and consequently graphs describing them have to be large. Graph grammars are commonly applied to generate such structures. In literature there are numerous syntactical approaches to pattern generation (see: e.g., [3, 4, 5, 6, 7, 8, 11, 12, 17, 24, 25]). The notion of realization scheme makes it possible to define a new grammar approach based on attributed graph grammars. In this approach the proposed realization scheme can be substituted for the classical domain of semantics (see: [2, 20]) and then semantics rules allow us to construct an attribution from a given realization scheme for a generated graph.

It should be pointed out that attributed graph grammars are not considered here. We concentrate our attention only on some components of such grammars, i.e., on the notion of realization scheme and special graphs which are useful in graphical modeling.

The paper is organized as follows. Section 2 contains preliminaries. In Section 3 we deal with spatial patterns which allow to define the concept of realization (graphical model) in Section 5. Section 4 contains the definition of composition graph (CP-graph), i.e., such graph whose nodes are equipped with two kinds of labels. Section 5 is the fundamental part of the paper. The notions of realization scheme and graphical model are introduced and some examples of applications are given. The presentation of a certain property of realization schemes, which is useful in the process of designing ends this section. Finally, in Section 6 we present Conclusions.

# 2. Preliminaries

In this section we recall some notations and notions which are necessary in the sequel.

(2.1) For a set $A$, $\mathcal{P}(A)$ denotes the *power set of* $A$ and $\#A$ the *cardinality of* $A$.

(2.2) Let $\beta$ be a function defined on $A$ and $A' \subseteq A$. The *restriction of* $\beta$ *to* $A'$ is denoted by $\beta \mid A'$.

(2.3) By a *graph* we understand a system $g = (V, E, s, t)$ where $V$ and $E$ are finite sets called the *set of nodes* and the *set of edges*, respectively, and $s, t : E \rightarrow V$ are two mappings assigning to each edge $e \in E$ elements $s(e)$ and $t(e)$ called the *source* and the *target*, respectively.
Given a graph $g$, let us denote by $V_g, E_g, s_g$ and $t_g$ the set of nodes, the set of edges, the source and target functions, respectively.

(2.4) Let $A_1, ..., A_n$ be finite sets and let $g$ be a graph.
Any function $\xi : V_g \rightarrow A_1 \times \cdots \times A_n$ where $n \geq 1$ will be called a *node-labelling function*.
The system $(g, \xi)$ will be called the *labelled graph over* $A_1, ..., A_n$.

(2.5) $\mathbb{R}$ denotes the set of real numbers and $\mathbb{R}^n$ the Euclidean space of dimension $n$, for some $n \geq 1$. The Euclidean space of dimension $n$, equipped with the ordinary distance function $\rho : \mathbb{R}^n \times \mathbb{R}^n \rightarrow \mathbb{R}$ is denoted by $(\mathbb{R}^n, \rho)$ or, briefly, by $\mathbb{R}^n$.

(2.6) Let $F$ be a set of transformations of $\mathbb{R}^n$ into $\mathbb{R}^n$. We shall say that the sets $A$ and $B$ are *conjugate with respect to* $F$ if there exists $f \in F$ such that $f(A) = B$. The set $F$ will be called the set of *admissible transformations*.

Various sets of admissible transformations may be considered, e.g., the sets of *isometries*, the set of *similarity transformations*, or the set of *affine transformations*.

# 3. Spatial Patterns

In many design applications, graphical models are constructed with instances (transformed copies) of the geometrical shapes that are defined in a *prototypical* set. The basic geometrical shapes for the graphical model depend on the type of model under consideration. But, there exist properties which are common for most of prototypical sets.

**Definition 3.1** Let $F$ be a set of admissible transformations of $\mathbb{R}^n$ into $\mathbb{R}^n$.

(3.1) A *part* in $\mathbb{R}^n$ is a bounded subset in $\mathbb{R}^n$,
(3.2) A set $S$ of parts is *prototypical* over $F$ if $S$ does not contain a pair of parts $A$ and $B$ that are conjugate with respect to $F$. □

It is easy to note that for each set of parts a prototypical set can be constructed. Elements of the prototypical set will be called *prototypical parts* .

**Remark 3.1** Since it is not known in advance where instances of prototypical parts (transformed copies) are to be placed in the *world coordinate system* , i.e., the coordinates referenced by a user, the definition of a prototypical part must be stated in an independent coordinate system, which is then transformed to world coordinate. This independent reference is called the *master coordinate system* and is defined for each prototypical part.

In creating a graphical model it is vital to arrange its components (transformed copies of prototypical parts). Basically, the components are arranged in such a way as to achieve a design goal. The goal can be of technical or aesthetic nature.
On the other hand, a graphical model can be seen as the union of transformed prototypical parts, i.e., as the subset of $\mathbb{R}^n$. This representation of the graphical model is called the *spatial pattern*. To define spatial patterns (cf. [23]) we shall need the notion of *spatial structure*.

**Definition 3.2** Let $F$ be a set of admissible transformations of $\mathbb{R}^n$ into $\mathbb{R}^n$. Let $S$ be a prototypical set over $F$. A subset $Q \subset S \times F$ is said to be a *spatial structure in* $\mathbb{R}^n$. □

In other words, a spatial structure is defined as a set of ordered pairs, the first element of which is a prototypical part, and the second a transformation that specifies the orientation of this prototypical part with respect to the world coordinate system.

**Definition 3.3** Let $F$ be a set of admissible transformations of $\mathbb{R}^n$ into $\mathbb{R}^n$. Let $S$ be a prototypical set over $F$. Let $Q \subset S \times F$ be a spatial structure in $\mathbb{R}^n$. Let $I_s : \mathbb{R}^n \to \mathbb{R}^n$ $(s \in S)$ be the transformation of the master coordinate system for the prototypical part $s$ into world coordinates. The *spatial pattern* for $Q$ is denoted by $Z(Q)$ and is defined as follows:

$$Z(Q) = \bigcup_{(s,f)\in Q} f \circ I_s(s)$$ □

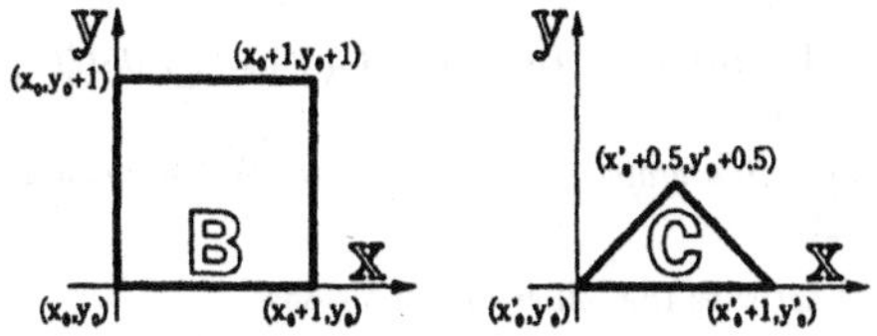

Fig.1: The set of prototypical parts

**Example 3.1** Let us consider the set $F$ of similarity transformations of $\mathbb{R}^2$ into $\mathbb{R}^2$ and the set of geometric figures consisting of the square $B$ and the triangle $C$ given by Fig.1, and Fig.2 presenting a part of the Pythagoras Tree. Let the translation with vector $(x, y)$, rotation through $\alpha$ relative to the coordinate origin and scaling according to the value $v$ be denoted by $Tr(x, y)$, $Rt(\alpha)$ and $Sc(v)$, respectively. It is easy to see that the square and the triangle shown in Fig.1 can be elements of

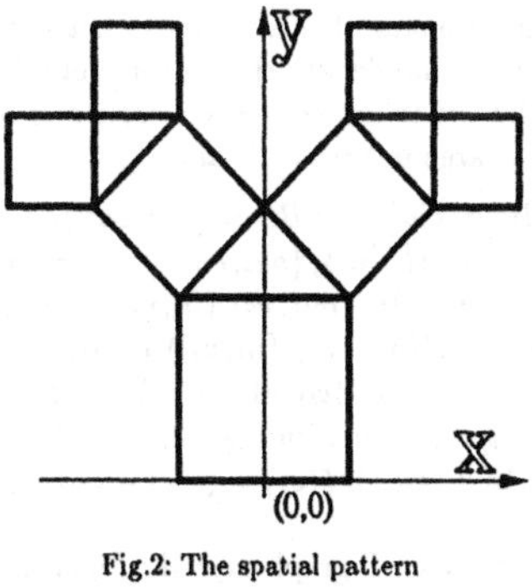

Fig.2: The spatial pattern

a prototypical set $S$ over the set $F$ of similarities and the part of the Pythagoras Tree is a spatial pattern corresponding to the following spatial structure:

$$
\begin{aligned}
Q = \{ & (B, f_B{}^1[= Tr(-0.5,0)]), && (C, f_C{}^1[= f_B{}^1 \circ Tr(0,1)]), && (B, f_l{}^1[= f_C{}^1 \circ f_{RS}]), \\
& (B, f_r{}^1[= f_C{}^1 \circ f_{TRS}]), && (C, f_l{}^2[= f_l{}^1 \circ Tr(0,1)]), && (C, f_r{}^2[= f_r{}^1 \circ Tr(0,1)]), \\
& (B, f_l{}^2 \circ f_{RS}), && (B, f_l{}^2 \circ f_{TRS}), && (B, f_r{}^2 \circ f_{RS}), \\
& (B, f_r{}^2 \circ f_{TRS})\}
\end{aligned}
$$

where $f_{RS} = Rt(45°) \circ Sc(0.5\sqrt{2})$, and $f_{TRS} = Tr(0.5, 0.5) \circ Rt(-45°) \circ Sc(0.5\sqrt{2})$. ◇

We shall see later that for Pythagoras Tree there exist recursive formulas allowing us to define transformations of its spatial structure at an arbitrary level of detail.

# 4. Composition Graphs

In this section we shall define graphs which will represent relationships between component parts of spatial patterns.

In our definition of graphs the pattern theory developed by Grenander [9] is used as a helpful framework. Grenander drew his inspiration from Wittgenstein's Tractatus Logico-Philosophicus [1]. In the following, let $A$, $BI$, and $BO$ be fixed alphabets of *labels of prototypical parts, in-bonds,* and *out-bonds,* respectively. Let $W$ be a nonempty finite subset of $A \times \mathcal{P}(BI) \times \mathcal{P}(BO)$, and $\alpha, \beta_I, \beta_O$ be defined by $w = (\alpha(w), \beta_I(w), \beta_O(w))$, $w \in W$.

Let $V$ be a set of nodes and $\xi : V \to W$ be a node-labelling function. Put

$$B_I(V) = \bigcup_{v \in V} \beta_I(\xi(v)) \times \{v\} \quad \text{and} \quad B_O(V) = \bigcup_{v \in V} \beta_O(\xi(v)) \times \{v\} \tag{4.1}$$

**Definition 4.1** By a *composition graph (CP-graph)* $c$ over $W$ we mean a labelled graph $c = ((V, E, s, t), \xi)$ such that

1. $E \subseteq B_O(V) \times B_I(V)$ satisfies the following conditions:
   - for each $(b_O, v) \in B_O(V)$ there exists at most one $(b_I, u) \in B_I(V)$ such that $((b_O, v), (b_I, u)) \in E$,
   - for each $(b_I, u) \in B_I(V)$ there exists at most one $(b_O, v) \in B_O(V)$ such that $((b_O, v), (b_I, u)) \in E$,
   - for each $((b_O, v), (b_I, u)) \in E$ $v \neq u$, and
2. for each $e = ((b_O, v), (b_I, u)) \in E$ $s(e) = v$ and $t(e) = u$. □

In the following we shall fix a universe $Z$ of nodes with a common labelling and shall consider CP-graphs which can be built from nodes of this universe. Let $\Gamma = (\gamma(z) : z \in Z)$ be a nonempty family of elements of $W$. By $L(W, \Gamma)$ we denote the universe of CP-graphs $c = ((V, E, s, t), \xi)$ with $V \subseteq Z$ and $\xi = \gamma \mid V$. It is worthwhile to notice that concrete sets of CP-graphs will be regarded as subsets of $L(W, \Gamma)$ for sufficiently large but fixed $W$ and $\Gamma$.

**Example 4.1** Let $A = \{a, b\}$, $BI = \{in\}$ and $BO = \{out, out'\}$.
Let $W = \{(a, \{in\}, \{out\}), (a, \{in\}, \emptyset), (b, \{in\}, \{out, out'\})\}$, $V = \{v_1, v_2, v_3, v_4, v_5, v_6, v_7, v_8, v_9, v_{10}\}$,
$E = \{((out, v_1), (in, v_2)), ((out, v_2), (in, v_4)), ((out', v_2), (in, v_3)), ((out, v_4), (in, v_6)), ((out, v_3), (in, v_5)),$
$((out', v_5), (in, v_7)), ((out, v_5), (in, v_8)), ((out', v_6), (in, v_9)), ((out, v_6), (in, v_{10}))\}$,
and the source and the target of each edge $((b_O, v), (b_I, u))$ is defined as $v$ and $v'$, respectively.
Let $\xi : V \to W$ be a node-labelling function such that: $\xi(v_1) = \xi(v_3) = \xi(v_4) = (a, \{in\}, \{out\})$,
$\xi(v_2) = \xi(v_5) = \xi(v_6) = (b, \{in\}, \{out, out'\})$, $\xi(v_7) = \xi(v_8) = \xi(v_9) = \xi(v_{10}) = (a, \{in\}, \emptyset)$.

It is obvious that $((V, E, s, t), \xi)$ is a CP-graph over $W$. Fig.3 shows the CP-graph with in-bonds and out-bonds represented by the graphic symbols  and , respectively, and edges by lines from out-bonds of their sources to in-bonds of their targets. ◇

**Remark 4.1** The introduced above compositions graphs (CP-graphs) are a modification of nets presented in [10, 14]. When giving a systematic introduction into CP-graphs we revive some of the ideas from the early seventies, proposed by Feder [5]. Feder has defined so-called plex-structures. Each symbol of a plex-structure may have an arbitrary number of attaching points for connecting to other symbols. A symbol of N-attaching points is called an N-attaching-point entity (NAPE). Structures

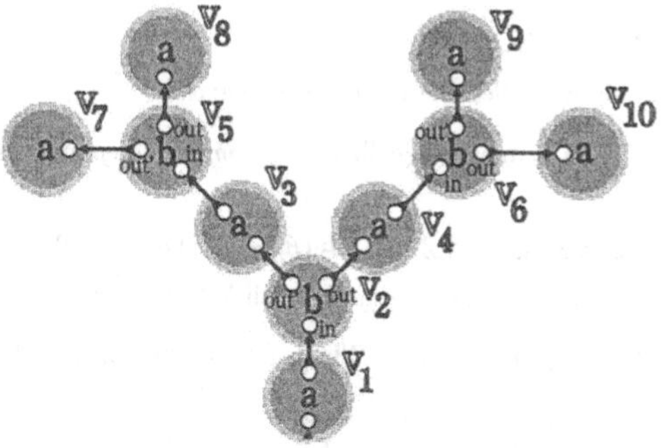


Fig.3: The CP-graph defined in Example 4.1

formed by interconnecting NAPEs are plex structures. Feder has proposed them as representations for realizations in $I\!R^2$. An extension of Feder's plex grammar is the three-dimensional plex grammar presented in [13, 16]. Both rules of 2D-plex and 3D-plex grammars are described by means of the string notation.
When considering CP-graphs it is worthwhile also to refer them to hypergraphs (see: e.g., [18, 21, 22]). The basic units of hypergraphs are hyperedges, i.e., objects, which may have many "tentacles" to grip at their source and target nodes rather than only two "arms" as the edges of ordinary graphs. Comparing CP-graphs with hypergraphs one can say that CP-graphs are composed of "hypernodes". However, in our approach to design graphical models CP-graphs as descriptions of relationships between components parts of the models are more useful than hypergraphs.

## 5. Realization Schemes

On the one hand, we are interested in spatial patterns, which are sets of points. On the other hand, we suggest the use of CP-graph structures to represent the patterns. The notion of a realization scheme allows to bridge the gap between CP-graphs and spatial patterns, and as it has been already mentioned it is most interesting for us to study it.
Let $\Gamma = (\gamma(z) : z \in Z)$ be a nonempty family of elements of a nonempty finite set $W$. Let $L(W, \Gamma)$ be the universe of CP-graphs $c = ((V, E, s, t,), \xi)$ with $V \subseteq Z$ and $\xi = \gamma \mid V$.
In the following, let

$$B(Z) = B_I(Z) \cup B_O(Z) \tag{5.2}$$

The set $B(Z)$ is the set of all in-bonds and out-bonds for the nodes of all CP-graphs of $L(W, \Gamma)$ (see: (4.1)).
For a given universe $L(W, \Gamma)$ of CP-graphs a realization scheme consists of a set of admissible transformations, a set of prototypical parts in $I\!R^n$ over this set of transformations, four mappings: *prototype* and *fragment assignments*, a *fitting function*, and a predicate of *applicability*. The prototype assignment assigns prototypical parts to nodes of which CP-graphs of $L(W, \Gamma)$ can be built. The fragment assignment assigns fragments or points of the prototypical parts to bonds of the nodes corresponding

to these parts. The fitting function determines elements of the set of admissible transformations that specify transformations of prototypical parts according to design criteria defined by means of the applicability.

**Definition 5.1** By a *realization scheme for* $L(W,\Gamma)$ *in* $\mathbb{R}^n$ we mean $D = (F, S, \omega_1, \omega_2, \chi, \varphi)$ where

1. $F$ is a set of admissible transformations $f : \mathbb{R}^n \to \mathbb{R}^n$,
2. $S$ is a prototypical set over $F$,
3. $\omega_1 : Z \to S$ is a mapping, called the *prototype assignment,*
4. $\omega_2 : B(Z) \to \mathcal{P}(\mathbb{R}^n)$ is a mapping, called the *fragment assignment*, satisfying the following conditions:
   - $\omega_2(b, z) \subseteq \omega_1(z)$ for each $z \in Z$ and $b \in (\beta_I(\gamma(z)) \cup \beta_O(\gamma_O(z)))$, and
   - $\gamma(z) = \gamma(z')$ implies $\omega_2(b, z) = \omega_2(b, z')$ for all $b \in (\beta_I(\gamma(z)) \cup \beta_O(\gamma_O(z)))$,
5. $\chi : \bigcup_{i\geq 1}(F \times B_Z)^i \times \bigcup_{j\geq 1}(B_Z)^j \supset K \to \mathcal{P}(\bigcup_{k\geq 1} F^k)$ is a mapping, called the *fitting function* ,
6. $\varphi$ is a predicate, called the *applicability* , such that $\varphi((a,d), f) =$ true for each $(a, d) \in K$ and $f \in \chi(a, d)$. □

The meaning of the fitting function is that given labels (in-bonds or/and out-bonds) of nodes of $Z$ and admissible transformations corresponding to these nodes this function specifies a possible set of transformations which can be assigned to the potential adjacent nodes and determined by means of their labels.
To explain the notions introduced in Def.5.1 we shall give the following examples.

**Example 5.1** Let us return to Example 3.1 and Example 4.1.
Let $\Gamma = (\gamma(z) : z \in Z)$ be a sufficiently large family of elements of $W$. Let us assume that $F$ contains all similarities and the prototypical set $S$ consists of the square $B$ and the triangle $C$ defined in Example 3.1 (see: Fig.1).
Let $\omega_1(z) = B$ (square) for all $z \in Z$ with $\alpha(\gamma(z)) = a$
and $\omega_1(z) = C$ (triangle) for all $z \in Z$ with $\alpha(\gamma(z)) = b$.
Denote the line segment from the point $p_i$ to $p_j$ by $[p_i, p_j]$.

Let $\omega_2(in, z) = [(x_0, y_0), (x_0+1, y_0)]$ for all $z \in Z$ with $\gamma(z) \in \{(a, \{in\}, \{out\}), (a, \{in\}, \emptyset)\}$
and $\omega_2(out, z) = [(x_0, y_0+1), (x_0+1, y_0+1)]$ for all $z \in Z$ with $\gamma(z) = (a, \{in\}, \{out\})$.
Let $\omega_2(in, z) = [(x_0', y_0'), (x_0'+1, y_0')]$
and $\omega_2(out', z) = [(x_0', y_0'), (x_0'+0.5, y_0'+0.5)]$
and $\omega_2(out, z) = [(x_0'+0.5, y_0'+0.5), (x_0'+1, y_0')]$ for $z \in Z$ with $\gamma(z) = (b, \{in\}, \{out, out'\})$.

Let us assume that the applicability says that there exists coincidence of the appropriate pairs of sides of the neighbouring polygonals (see: Fig.2 and Fig.3). We have

1. $\chi((f, (out, z)), (in, z')) = \{f_1, f_2\}$ where
   - $f_1 = f \circ Tr(0, 1)$
   - $f_2 = f \circ Tr(0, 1) \circ Rt(180°)$

   for all $z, z' \in Z$ such that $\gamma(z) = (a, \{in\}, \{out\})$ and $\gamma(z) = (b, \{in\}, \{out, out'\})$.
2. $\chi((f, (out, z)), (in, z')) = \{f_3, f_4\}$ and $\chi((f, (out', z)), (in, z')) = \{f_5, f_6\}$ where

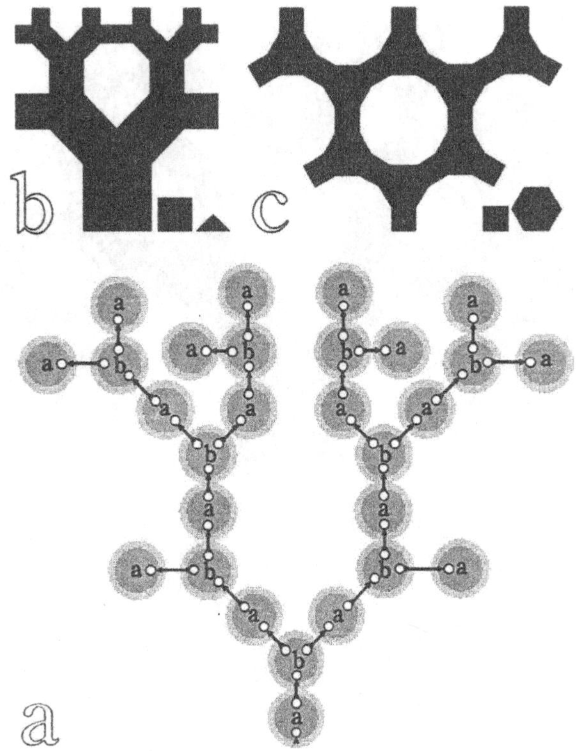


Fig.4: The two graphical models and their CP-graph

- $f_3 = f \circ Tr(0.5, 0.5) \circ Rt(-45°) \circ Sc(0.5\sqrt{(2)})$
- $f_4 = f \circ Tr(1, 0) \circ Rt(135°) \circ Sc(0.5\sqrt{(2)})$
- $f_5 = f \circ Rt(45°) \circ Sc(0.5\sqrt{(2)})$
- $f_6 = f \circ Tr(0.5, 0.5) \circ Rt(-135°) \circ Sc(0.5\sqrt{(2)})$

for all $z, z' \in Z$ such that $\gamma(z) = (t, \{in\}, \{out, out'\})$ and $\alpha(\gamma(z')) = a$
and
$h \in \chi((f, (b, z)), (b', z'))$ iff $h(I_{\omega_1(z')}(\omega_2(b', z'))) = f(I_{\omega_1(z)}(\omega_2(b, z))$ where
$I_{\omega_1(z)}$ and $I_{\omega_1(z')}$ are the transformations of the master coordinate systems into world coordinates for the prototypical parts $\omega_1(z)$ and $\omega_1(z')$, respectively. ◇

The part of the Pythagoras Tree and the Pythagoras Tree shown in Fig.4b and Fig.5a, respectively, are graphical models described by means of the realization scheme $D$ defined above.
The structures of the parts of the Pythagoras Tree and the honey-comb shown in Fig.4b and Fig.4c, respectively, are represented by the same CP-graph (Fig.4a). But, these models are defined by two realization schemes with different prototypical parts which are shown in the bottom right hand

Fig.5: The two graphical models with the same graph structure

corners in Fig.4b and Fig.4c. For an appropriately large graph, applying different realization schemes we can obtain graphical models which are completely different from one another. For instance it is difficult to find any similarity between the Pythagoras Tree and the honey-comb shown in Fig.5a and Fig.5b, respectively. And yet, these models have structures described by the same graph.
It is necessary to notice that the fitting function $\chi$ and the predicate $\varphi$ of the realization scheme $D$ defined in Example 5.1 are of a special simplified form. Now, we shall outline one of realization schemes in which $\chi$ and $\varphi$ are more complex.

**Example 5.2** Let us assume that we would like to design the case (see: Fig.6a) whose overal dimensions, i.e., width, depth, and height are denoted by $\alpha$, $\beta$, and $\alpha_1$, respectively. Additionally, the ratio of the depth $\beta_1$ of the high case part to the depth $\beta$ equals to $d$ and the height $\alpha_2$ of the low case part is such that $0 < \alpha_2 < \alpha_1$.
The designer would like to make his/her final decision about the proportions of the case not before he/she can see potential solutions in the form of perspective views on the monitor screen. When the designer wants to change proportions of the case he/she can change at least one of the following numbers $\alpha$, $\beta$, $\alpha_1$, $d$, and $\alpha_2$ himself/herself or can use for this purpose an appropriate random number generator.
The designer needs a graphical model of the case in three dimensional space. The model is composed of rectangles (see: Fig.6b) and its structure can be described by means of the CP-graph shown in Fig.6c.
Let us outline the realization scheme $D'$ for this model.
Let $W=\{(a,\emptyset,\{out_1,out_2,out_3,out_4,out_5,out_6\}),(a,\{in_1,in_2\},\{out_1,out_2\}),(a,\{in_1\},\{out_1,out_2\}),$ $(a,\{in_1,in_2,in_3\},\{out_1\}),(a,\{in_1,in_2,in_3\},\emptyset)\}$.
Let the family $\Gamma=(\gamma(v): v\in V=\{v_1,v_2,v_3,v_4,v_5,v_6,v_7,v_8\})$, where

$$\gamma(v_1)=(a,\emptyset,\{out_1,out_2,out_3,out_4,out_5,out_6\}),$$
$$\gamma(v_2)=\gamma(v_7)=(a,\{in_1,in_2\},\{out_1,out_2\}),$$
$$\gamma(v_3)=\gamma(v_8)=(a,\{in_1\},\{out_1,out_2\}).$$
$$\gamma(v_4)=(a,\{in_1,in_2,in_3\},\{out_1\}),$$
$$\gamma(v_5)=\gamma(v_6)=(a,\{in_1,in_2,in_3\},\emptyset),$$

Let $F$ be a set of admissible transformations $f: \mathbb{R}^3 \to \mathbb{R}^3$ containing all scalings (non-uniform), translations, rotations about the positive z-axis and rotations about the positive x-axis.
Denote by $Sc(s_x,s_y,s_z)$, $Rt_z(\Theta)$, $Rt_x(\Theta)$ and $Tr(t_x,t_y,t_z)$ the scaling with parameters $s_x,s_y,s_z$, the rotation about the positive z-axis through an angle $\Theta$, the rotation about the positive x-axis through an angle $\Theta$ and the translation with parameters $t_x,t_y,t_z$, respectively.
Let the set $S$ contain only one square $A$ in the xz-plane shown in Fig.6d.
Let $\omega_1(v_i)=A$ (square) where i = 1,...,8.
The values of the fragment assignment $\omega_2$ are sides or fragments of sides of the square $A$. For example, $\omega_2$ defined for bonds of the node $v_1$ is defined as follows:

$$\omega_2(out_1,v_1)=[(0,0,1),(0,0,0)],$$
$$\omega_2(out_3,v_1)=[(d,0,0),(1,0,0)],$$
$$\omega_2(out_5,v_1)=[(1,0,1),(d,0,1)],$$
$$\omega_2(out_2,v_1)=[(0,0,0),(d,0,0)],$$
$$\omega_2(out_4,v_1)=[(1,0,0),(1,0,1)],$$
$$\omega_2(out_6,v_1)=[(d,0,1),(0,0,1)].$$

Let sizes and the position of the bottom $p_1$ of the case are determined by the transformation $f$. Let us define the value of the function $\chi$ for the the sizes and the positions of four faces of the case $p_2, p_3, p_7$ and $p_8$.
We have $\chi(((f,(out_1,v_1)),(f,(out_2,v_1)),(f,(out_4,v_1)),(f,(out_5,v_1))),$ $((in_1,v_2),(in_1,v_3),(in_1,v_7),(in_1,v_8),(out_1,v_2),(out_1,v_3),(in_2,v_2)))=(h_1,h_2,h_3,h_4)$ where

- $f=Sc(\beta,0,\alpha)$ and $\alpha,\beta\in\mathbb{R}$,
- $h_1=Rt_x(-90°)\circ Sc(d\beta,0,\alpha 1)$,

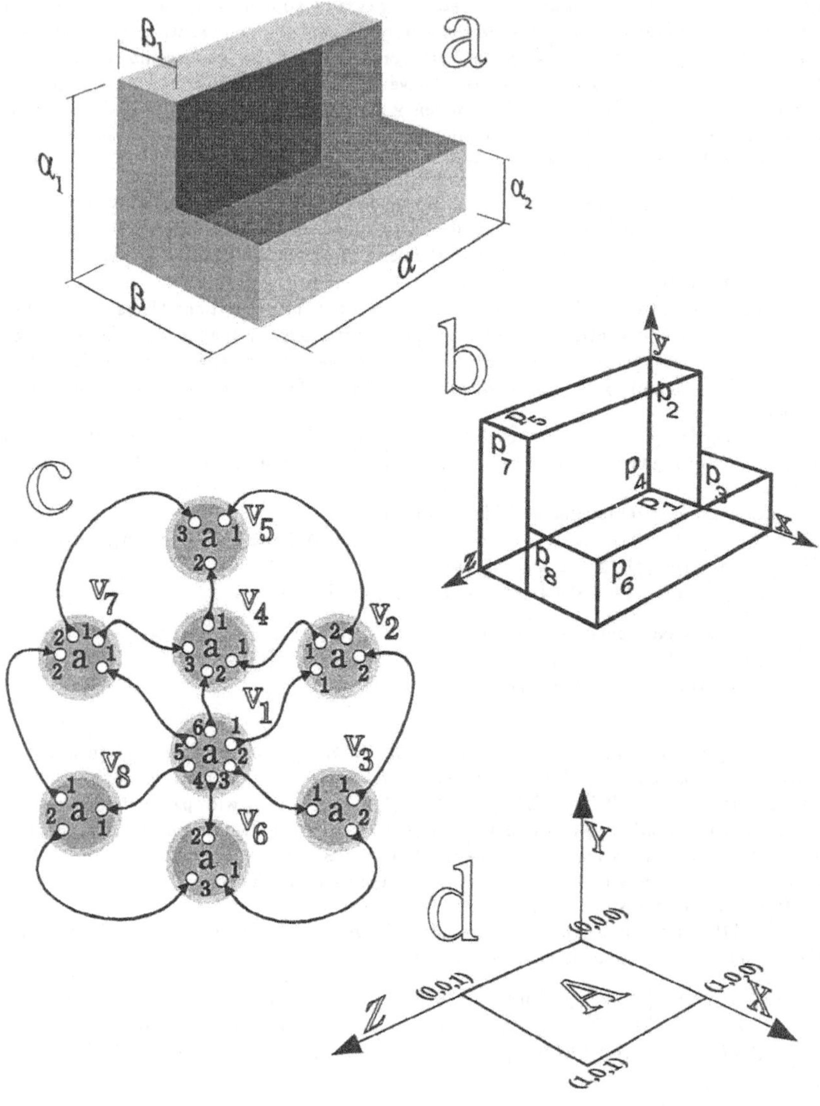


Fig.6: Designing the case

- $h_2 = Tr((d\beta, 0, 0) \circ Rt_x(-90°) \circ Sc((d-1)\beta, 0, \alpha 2)$,
- $h_3 = Tr(0, 0, \alpha) \circ h_1$, and
- $h_4 = Tr(0, 0, \alpha) \circ h_2$

and $\varphi((((f, (out_1, v_1)), (f, (out_2, v_1)), (f, (out_4, v_1)), (f, (out_5, v_1))),$
$((in_1, v_2), (in_1, v_3), (in_1, v_7), (in_1, v_8), (out_1, v_2), (out_1, v_3), (in_2, v_2), (h_1, h_2, h_3, h_4))$ = true iff

- the pairs of the rectangles $p_2$, $p_3$ and $p_7$, $p_8$ form lateral faces which are perpendicular to the bottom $p_1$, and
- the side of $p_2$ parallel to the y-axis is longer than the side of $p_3$ parallel to this axis.

Let $D = (F, S, \omega_1, \omega_2, \chi, \varphi)$ be realization scheme for $L(W, \Gamma)$. In general a fitting function $\chi$ can be specified for a concrete CP-graph $c \in L(W, \Gamma)$. This is done by finding an assignment of addmissible transformations to nodes of $c$ such that for each two sequences of the nodes of $c$ if the first sequence with transformations assigned to the nodes and the second sequence without transformations constitute an argument of $\chi$ then transformations assigned to the nodes of the second sequence define a value of $\chi$ for this argument.
To describe it in a formal way we shall introduce the notion of an attributed CP-graph, then some operators defined on the domain of the fitting function, and at the end the notion of a compatible CP-graph with a given realization scheme.

**Definition 5.2** Let $F$ be a set of admissible transformations for $L(W, \Gamma)$. Each pair $g = (c, \delta)$, where $c \in L(W, \zeta)$ and $\delta : V_c \to F$, is called an *attributed* composition graph over $W$. □

Let $D = (F, S, \omega_1, \omega_2, \chi, \varphi)$ be a realization scheme for $L(W, \Gamma)$ and $c \in L(W, \Gamma)$.
Let $(a, d) \in \bigcup_{i \geq 1} (F \times B_Z)^i \times \bigcup_{j \geq 1} B_Z^j$, where

- $a = (a_{i_1 j_1}, a_{i_2 j_2}, ..., a_{i_n j_n})$ where $a_{i_k j_k} = (f_{i_k}, (b_{i_k j_k}, z_{i_k}))$ $1 \leq k \leq n$, $z_{i_k} \in V_c$ and $b_{i_k j_k} \in \beta_I(\gamma(z_{i_k})) \cup \beta_O(\gamma(z_{i_k}))$,
- $d = (d_{i_{n+1} j_{n+1}}, ..., d_{i_{n+m} j_{n+m}})$ where $d_{i_k j_k} = (b_{i_k j_k}, z_{i_k})$ $n+1 \leq k \leq n+m$.

Let us define the following operators:

$P_n(a_{i_1 j_1}, ..., a_{i_{k+n} j_{k+n}}) = (a_{i_1 j_1}, ..., a_{i_n j_n})$, i.e., the operator $P_n$ gives n-th first elements of a given sequence,

$Q_n(a_{i_1 j_1}, ..., a_{i_{k+n} j_{k+n}}) = (a_{i_{1+n} j_{1+n}}, ..., a_{i_{k+n} j_{k+n}})$, i.e., the operator $Q_n$ removes n-th first elements of a given sequence,

$T(a_{i_1 j_1}, a_{i_2 j_2}, ..., a_{i_n j_n}) = (d_{i_1 j_1}, ..., d_{i_n j_n})$, i.e., the operator $T$ removes transformations from all elements of a given sequence,

$U(a_{i_1 j_1}, a_{i_2 j_2}, ..., a_{i_n j_n}) = (f_{i_1}, ..., f_{i_n})$, i.e., the operator $U$ gives transformations for all elements of a given sequence.

**Definition 5.3** Given a realization scheme $D = (F, S, \omega_1, \omega_2, \chi, \varphi)$ for $\zeta$ in $\mathbb{R}^n$, an attributed CP-graph $g = (c, \delta)$ with $c \in L(W, \zeta)$ is said to be *compatible* with $D$ if for each $(a_{i_1 j_1}, a_{i_2 j_2}, ..., a_{i_n j_n})$, where $a_{i_k j_k} = (\delta(z_{i_k}), (b_{i_k j_k}, z_{i_k}))$ $1 \leq k \leq n$ such that there exists $m < n$ such that $(P_m(a_{i_1 j_1}, a_{i_2 j_2}, ..., a_{i_n j_n}), T \circ Q_m(a_{i_1 j_1}, a_{i_2 j_2}, ..., a_{i_n j_n})) \in K$, where $K$ is a domain of the function $\chi$, we have $U \circ Q_m(a_{i_1 j_1}, ..., a_{i_n j_n}) \in \chi(P_m(a_{i_1 j_1}, ..., a_{i_n j_n}), T \circ Q_m(a_{i_1 j_1}, ..., a_{i_n j_n}))$. □

To explain the compability of CP-graphs with $D$ we shall give an example.

**Example 5.3** Let $c$ be a CP-graph defined in Example 4.1. Let us return to Example 5.1 and let us consider the realization scheme given in this example.
Let the function $\delta : V_c \rightarrow F$ be defined as follows:

$\delta(v_1) = Tr(-0.5, 0)$, $\delta(v_2) = \delta(v_1) \circ Tr(0,1)$,
$\delta(v_3) = \delta(v_2) \circ Rt(45°) \circ Sc(0.5\sqrt{(2)})$, $\delta(v_4) = \delta(v_2) \circ Tr(0.5, 0.5) \circ Rt(-45°) \circ Sc(0.5\sqrt{(2)})$,
$\delta(v_5) = \delta(v_3) \circ Tr(0,1)$, $\delta(v_6) = \delta(v_4) \circ Tr(0,1)$,
$\delta(v_7) = \delta(v_5) \circ Rt(45°) \circ Sc(0.5\sqrt{(2)})$, $\delta(v_8) = \delta(v_5) \circ Tr(0.5, 0.5) \circ Rt(-45°) \circ Sc(0.5\sqrt{(2)})$,
$\delta(v_9) = \delta(v_6) \circ Rt(45°) \circ Sc(0.5\sqrt{(2)})$, $\delta(v_{10}) = \delta(v_6) \circ Tr(0.5, 0.5) \circ Rt(-45°) \circ Sc(0.5\sqrt{(2)})$.

The CP-graph $g = (c, \delta)$ is compatible with $D$. ◇

**Definition 5.4** Given a realization scheme $D = (F, S, \omega_1, \omega_2, \chi, \varphi)$ for $\zeta$ in $\mathbb{R}^n$ and an attributed CP-graph $g = (c, \delta)$ which is compatible with $D$, by a ***realization*** of $g$ according to $D$ we mean

$$r_D(g) = Z(\bigcup_{(v \in V_c)} (\omega_1(v), \delta(v)) = \bigcup_{(v \in V_c)} (\delta(v) \circ I_{\omega_1(v)})(\omega_1(v))$$

where $I_{\omega_1(v)}$ is the transformation of the master coordinate system for the prototypical part $\omega_1(v)$ into world coordinates system. □

**Example 5.4** Let $g$ be an attributed graph given in Example 5.3. The spatial pattern presented in Example 3.1 and shown in Fig.2 is a realization of $g$ which is compatible with $D$ defined in Example 5.1. ◇

At the end of this section we shall consider a certain relation between simplified realization schemes that can be useful in the process of designing. First, we shall introduce the following definition:

**Definition 5.5** Let $D = (F, S, \omega_1, \omega_2, \chi, \varphi)$ be a realization scheme for $L(W, \Gamma)$.
The realization scheme $D$ is *simple* if its fitting function $\chi$ is defined on the set $(F \times B_Z) \times B_Z$ and $\varphi(((f, (b, z)), (b', z')), h) =$ true iff $h(I_{\omega_1(z')}(\omega_2(b', z'))) = f(I_{\omega_1(z)}(\omega_2(b, z)))$. □

In other words, in a simple realization scheme its applicability predicate says that each value $h$ of the fitting function $\chi$ is determined in such a way that there exists coincidence the part being the copy of the value of fragment assignment $\omega_2$ transformed by $h$ with the part which is the copy of the value of $\omega_2$ transformed by $f$ being the transformation of the argument of $\chi$ defined for $h$.
Let us note that the realization scheme $D$ given in Example 5.1 and defined for the Pythagoras Tree is simple.

**Definition 5.6** Let $D = (F, S, \omega_1, \omega_2, \chi, \varphi)$ be realization scheme for $L(W, \Gamma)$. By a *constrained part* of $\omega_1(z)$ we mean

$$cst(\omega_1(z)) = \bigcup_{(b \in B_z)} \omega_2(b, z)$$

The complement of $cst(\omega_1(z))$ in $\omega_1(z)$ is called a *changeable part* of $\omega_1(z)$ and is denoted by $chp(\omega_1(z))$. □

For each $z \in Z$ we have

$$\omega_1(z) = cst(\omega_1(z)) \cup chp(\omega_1(z)) \tag{5.3}$$

It is worthwhile to note that bonds are arguments of fragment assignment $\omega_2$ whose values are arguments of predicate of $\varphi$. If $chp(\omega_1(z)) = \emptyset$ then any change in $\omega_1(z)$ can influence the values of $\varphi$.

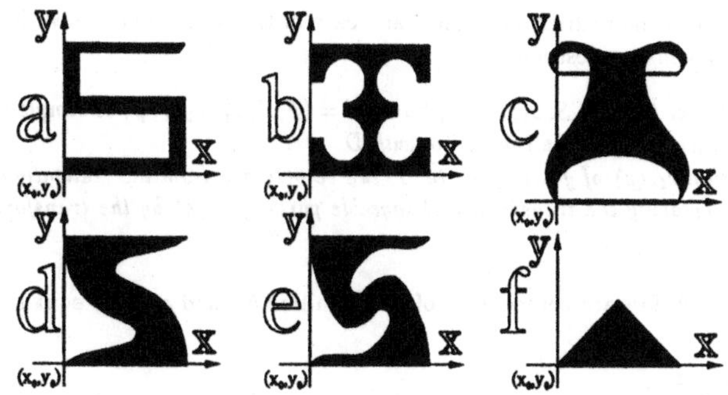


Fig.7: The prototypical parts of $D_1$, $D_2$, $D_3$, $D_4$, and $D_5$

**Example 5.5** Let us return to the realization scheme $D$ defined for the Pythagoras Tree in Example 5.1. The prototypical part being the triangle is completely determined by its constrained part.
Let us consider the three realization schemes $D_1, D_2$ and $D_3$ which differ from the scheme $D$ only in one element of their prototypical sets. Figures 7a,b, and c present the prototypical parts which belong to the alphabets of $D_1, D_2$ and $D_3$, respectively, and correspond to the square of the scheme $D$. Let assume that the prototypical parts are defined in the same master coordinate system as the square in the figure 1 and the two line segments $[(x_0, y_0), (x_0 + 1, y_0)]$ and $[(x_0, y_0 + 1), (x_0 + 1, y_0 + 1)]$ which define the constrained part of the square in $D$ are contained in each prototypical part presented in Fig.7.
Let us consider the next two realization schemes which differ from $D$ only in shapes of prototypical parts. The prototypical parts for $D_4$ and $D_5$ replacing the square of the scheme $D$ are shown in Fig.7d and e, respectively. Let assume that the constrained part of this square is contained in each of them and that the second common prototypical part for the systems (Fig.7f) has its constrained part equals to the triangle of $D$. The realizations compatible with $D_1, D_2, D_3, D_4$ and $D_5$ are shown in Fig.8. ◇

Fig.8: The realizations compatible with $D_1, D_2, D_3, D_4$ and $D_5$

The relations between the realization in Fig.2 and each of the realizations shown in Fig.8 are summarized in the following proposition:

**Proposition 5.1** *Let $D = (F, S, \omega_1, \omega_2, \chi, \varphi)$ and $D' = (F, S', \omega_1', \omega_2, \chi, \varphi)$ be simple realization schemes for $L(W, \Gamma)$. Let $g = (c, \delta)$ be compatible with $D$.*
*Then the realization $r_{D'}(g)$ of $g$ according to $D'$ can be obtained from the realization $r_D(g)$ of $g$ according to $D$ by replacing the transformed changeable parts of $r_D(g)$ by the transformed changeable parts of $r_{D'}(g)$.*
Proof:
The proof is a straightforward consequence of the definition 5.5 and and the equality 5.3. ∎

# 6. Conclusions

The notion of realization scheme introduced in this paper supplies formal basis for implementation of design tools that offer the user much freedom in creating of graphical models. The fact that the structure of the object is independent of its realization may be useful in the process of designing. Due to it we can design an object applying one of two strategies: one that consists in changing of connections between nodes of the graph while prototypical parts remain unchanged, or the other changing prototypical parts or transformations while the graph remains unchanged.
In this paper a realization is defined for an attributed composition graph $g = (c, \delta)$ which is compatible with the given realization scheme $D$. There does not exist a general procedure which constructs an attribution $\delta$ from a given composition graph $c$ and a given realization scheme $D$ such that the attributed composition graph $g = (c, \delta)$ is compatible with $D$. The method of finding an attribution depends on the type of graphical model under consideration similarly as the method of finding a solution of the equation $a_0 + a_1x + ... + a_nx^n = 0$ depends on the degree $n$.
As it has been considered, the notion of realization scheme makes it possible to define a new approach to attributed grammars and can be treated as an extension of the classical semantic domain.
In a realization scheme only geometrical attributes of graphical models have been mentioned. In our future work we intend to develop the realization scheme so that other attributes of models, e.g., color, texture and surface pattern will be also taken into consideration.

**Acknowledgement.** The author wish to thank you prof. Józef Winkowski for several useful suggestions.

# References

[1] Wittgenstein L.: Tractatus Logico-Philosophicus, Sixth Edition, London, 1955.
[2] Knuth D. E: Semantics of context-free languages, Mathematical Systems Theory 2, 127-145, 1968.
[3] Pfaltz J. L, and Rosenfeld A.: Web Grammars, Proc.Int.Joint Conf. Art. Intelligence. , 1969
[4] Shaw A. C.: A Formal Description Schema as a Basis for Picture Processing Systems, Inf. Contr.14, 9-52, 1969.
[5] Feder J.: Plex Languages, Inform. Sci.3, 225-241, 1971.
[6] Fu K. S.: Syntactic Method in Pattern Recognition, Academic Press, New York, 1974.
[7] Gips J.: Shape Grammars and Their Uses, Artificial Perception, Shape Generation and Computer Aesthetic, Birkhäuser Verlag, Basel/Stuttgart, 1975.
[8] Stiny G.: Pictoral and Formal Aspects of Shapes and Shapes Grammars, Basel, Switherland, Birkhäuser-Verlag, 1976.
[9] Grenander U.: Regular Structures, Springer-Verlag, 1981.
[10] Grabska E.: Pattern Synthesis by Means of Graph Theory, Ph.D. Thesis, Institute of Mathematics, Jagiellonian University, 1982.

[11] Maurer H. A., Rozenberg G., and Welzl E.: Using String Languages to Describe Picture Languages, Information and Control 54, 155-185, 1982.

[12] Bunke H.: Graph Grammars as a Generative Tool in Image Understanding, Lecture Notes in Computer Science 153, Springer Verlag,Berlin, 8-19, 1983.

[13] Lin W. C., and Fu K. S.: A Syntactic Approach to 3D Object Representation, IEEE Trans. Patt. Anal. Mach. Intell., vol. PAMI-6, no.3, 351-364, 1984.

[14] Göttler H., and Grabska E.: Attributed Graph-Grammars and Composition Nets, Proceedings WG'85, H.Noltemeier (Ed.), Trauner Verlag, 119-130, 1985.

[15] Hearn D., and Baker M. P.: Computer Graphics, Prentice-Hall,Inc., Englewood Cliffs, New Jersey, 1986.

[16] Lin W. C., and Fu K. S.: A Syntactic Approach to Three-Dimensional Object Recognition, IEEE Trans. Syst., Man, Cybern., vol. SMC-16, no.3, 405-422, 1986.

[17] Prusinkiewicz P.: Graphical Applications of L-Systems, Proc. of Graphics Interface'86 - Vision Interface '86, 247-253, 1986.

[18] Bauderon M., and Courcelle B.: Graph Expressions and Graph Rewritings, Mathematical Systems Theory 20, 83-127, 1987.

[19] Mäntylä M.: An Introduction to Solid Modeling, Computer Science Press, Rockville, Maryland, 1988.

[20] Engelfriet J, and Heyker L. M.: The Term-Generating Power of Context-Free Hypergraph Grammars and Attribute Grammars, Report No 17/89, Department of Computer Science, Leiden University, 1989.

[21] Habel A.: Hyperdge Replacement: Grammars and Languages, Ph.D. Thesis, Bremen, 1989.

[22] Habel A., and Kreowski H. J.: Collages and Patterns Generated by Hyperedge Replacement, Report No 15/90. Bremen University, 1990.

[23] Woodbury R. F., Carlson C. N., and Heisserman J. A.: Geometric Design Spaces, Intelligent CAD II, IFIP, 337-353, 1990.

[24] Grabska E.; Design Spaces and Aesthetic Measure, EnvironmentalDesign,Aesthetic Quality and Information Technology; Focus Symposium Proceedings, Baden-Baden, Germany, Oksala T. (ed.), 25-32, 1992.

[25] Grabska E., and Hliniak G.: Structural Aspects of CP-graph Languages, ZN UJ MXCI Prace Informatyczne, z5, 1993.

# ESM systems and the composition of their computations

D. Janssens
Dept. of Computer science
Free University of Brussels (VUB)

**Abstract.** ESM systems are graph rewriting systems where productions are morphisms in a suitable category, ESM. The way graphs are transformed in ESM systems is essentially the same as in actor grammars, which were introduced in [JR]. It is demonstrated that a rewriting step corresponds to a (single) pushout construction, as in the approach from [L]. Rewriting processes in ESM systems are represented by computation structures, and it is shown that communication of rewriting processes corresponds to a gluing operation on computation structures. In the last section we briefly sketch how one may develop a semantics for ESM systems, based on computation structures, that is compositional w.r.t. the union of ESM systems.

## Introduction

Actor systems were introduced in [H] and [HB] as a model of concurrent computation, and actor grammars were introduced in [JR] as a formal model for actor systems, aimed at providing a formal tool for the implementation of actor systems on a parallel computer. Hence in this framework a graph rewriting system is viewed as a program and its rewriting processes are the runs or computations of that program.

In the first part of the paper the relation between actor grammars and the single pushout approach to graph rewriting of [L] is clarified. To this aim a category ESM is considered, where objects are graphs and morphisms represent graph transformations. A graph rewriting system is a set of such ESM morphisms – hence the name ESM systems. The actor grammars of [JR] and [JLR] are ESM systems that are equipped with a set of initial graphs, and where productions are of a special form. Our main interest is in developing the theory of general ESM systems, including those that cannot be directly interpreted as a formal description of an actor system; therefore we prefer the name ESM systems.

In the second part of the paper the composition of rewriting processes in an ESM system is considered. A representation of rewriting processes, called computation structures, is introduced and it is demonstrated that communication between computations corresponds to a gluing operation on computation structures. Then a set of $P$–valid computation structures is associated with an ESM system $P$: the set of computation structures that describe a rewriting process of $P$. However it turns out that the semantics obtained in this way is not compositional with respect to the union of ESM sytems. In the last section it is briefly explained how one can get a compositional semantics: the notion of a computation structure is replaced by that of a conditional computation. A conditional computation consists of a computation structure $C$ and a condition $cnd$; $(C, cnd)$ is $P$–valid if $C$ may occur in combination with another computation structure $C'$ such that $C'$ satisfies the condition. Each conditional computation of a composed system $P_1 \cup P_2$ is then obtained as the composition of a conditional computation of $P_1$ and a conditional computation of $P_2$.

# 1 Preliminaries

In this section we recall some basic terminology to be used in the paper.

(1) For a set $A$, $Id_A$ denotes the identity relation on $A$. For a relation $R \subseteq A \times A$, $R^*$ denotes the reflexive and transitive closure of $R$. We will often write $R^*_A$ to stress that the closure is taken with respect to the set $A$. The relation $R$ is *antisymmetric* if $(x, y) \in R$, $(y, x) \in R$ implies that $x = y$. For sets $A$ and $B$, the difference of $A$ and $B$ is denoted by $A - B$. The union of disjoint sets $A$ and $B$ is often denoted by $A \oplus B$. For disjoint sets $A$, $B$ and functions $f_1 : A \to C$ and $f_2 : B \to C$, the common extension of $f_1$ and $f_2$ to a function from $A \oplus B$ into $C$ is denoted by $f_1 \oplus f_2$.

(2) Let $\Sigma$ and $\Delta$ be sets. A $(\Sigma, \Delta)$*-graph* is a system $g = (V, E, \phi)$ where $V$ is a finite set (called the *set of nodes of* $g$), $E \subseteq V \times \Delta \times V$ (called the *set of edges* of $g$), and $\phi$ is a function from $V$ into $\Sigma$ (called the *node-labeling function of* $g$). For a $(\Sigma, \Delta)$-graph $g$, its components are denoted by $V_g$, $E_g$ and $\phi_g$, respectively.

(3) Let $g$ and $h$ be $(\Sigma, \Delta)$-graphs, let $f : V_g \to V_h$ be an injective function and let $R \subseteq V_g \times \Delta \times V_g$. Then the set $\{(f(x), \delta, f(y)) \mid (x, \delta, y) \in R\}$ is denoted by $f(R)$. We use a similar notation in the case where $R \subseteq V_g \times V_g$ or $R \subseteq (V_g \times \Delta) \times (V_g \times \Delta)$, and for the inverse relation $f^{-1}$. $f(g)$ denotes the graph $(f(V_g), f(E_g), \phi_h \circ f)$.

(4) A *graph morphism from* $g$ *into* $h$ is an injective function $f : V_g \to V_h$ such that $\phi_g = \phi_h \circ f$ and $f(E_g) \subseteq E_h$. $f$ is a *graph isomorphism* if its inverse is a graph morphism from $h$ into $g$. $Id^{gr}_g$ denotes the identical graph morphism on $g$.

(5) Let $g$ and $h$ be $(\Sigma, \Delta)$-graphs. Then $h$ is a *subgraph of* $g$ if $V_h \subseteq V_g$, $E_h \subseteq E_g$ and $\phi_h$ is the restriction of $\phi_g$ to $V_h$. For a subset $A$ of $V_g$, the *subgraph of* $g$ *induced by* $A$ is the graph $(A, E_g \cap (A \times \Delta \times A), \phi')$, where $\phi'$ is the restriction of $\phi_g$ to $A$. The graphs $g$ and $h$ are *disjoint* if $V_g \cap V_h = \emptyset$. For disjoint graphs $g$ and $h$, $g \oplus h$ denotes the graph $(V_g \oplus V_h, E_g \oplus E_h, \phi_g \oplus \phi_h)$.

Throughout this paper we assume that $\Sigma$ and $\Delta$ denote arbitrary but fixed alphabets.

# 2 ESM morphisms and ESM systems

In this section we first recall some basic notions of actor grammars. The definition of actor productions we use is the one from [J], however the relations $Emb$ and $Iden$ are denoted

$R^S$ and $R^t$, respectively. In [JR] and [JLR], additional restrictions were imposed on the productions to ensure that the application of a production to a graph corresponds to an event in an actor system. However these restrictions are irrelevant for the purpose of this paper.

**Definition 2.1** *An actor production is a 4-tuple $(g, h, R^s, R^t)$ where $g$ and $h$ are nonempty $(\Sigma, \Delta)$-graphs, $R^s \subseteq (V_g \times \Delta) \times (V_h \times \Delta)$, and $R^t \subseteq V_g \times V_h$.*

An actor grammar is a set of actor productions equipped with a set of $(\Sigma, \Delta)$-graphs which are to be used as initial graphs of rewriting processes.

**Definition 2.2** *An actor grammar is a pair $(P, Init)$ where $P$ is a set of actor productions and Init is a set of $(\Sigma, \Delta)$-graphs.*

We now define how actor productions are used to rewrite graphs. The definition has an operational flavor: it specifies how, given an actor production $\pi$, a graph $k$ and an occurrence $g'$ of the left-hand side of $\pi$ in $k$, one may construct the graph $m$ that results from applying $\pi$ to $g'$. We first consider sequential rewriting steps: steps consisting of the replacement of a single occurrence of the left-hand side of a production. Concurrent steps are considered later.

**Definition 2.3** *Let $\pi = (g, h, R^s, R^t)$ be an actor production. Let $k$ be a $(\Sigma, \Delta)$-graph and let $i : g \to k$ be a graph morphism. Then the graph obtained by the application of $\pi$ to $k$ via $i$ is the graph $m$ obtained as follows.*

*(1) Let $g' = i(g)$, and let $k'$ be the subgraph of $k$ induced by $V_k - V_{g'}$. Let $j : h \to h'$ and $i' : k' \to m'$ be isomorphisms such that $h'$ and $m'$ are disjoint.*

*(2) Let $\tilde{m} = h' \oplus m'$ (consider $j$ and $i'$ as graph morphisms into $\tilde{m}$). Then $m$ is the graph obtained from $\tilde{m}$ by adding, for each $(x, \delta, y) \in E_k$, the following edges to $E_{\tilde{m}}$.*

*(2.1) If $x, y \in V_{g'}$, then add all edges $(j(u), \mu, j(w))$ such that $((i^{-1}(x), \delta), (u, \mu)) \in R^s$ and $(i^{-1}(y), w) \in R^t$.*

*(2.2) If $x \in V_{k'}$ and $y \in V_{g'}$, then add all edges $(i'(x), \delta, j(w))$ such that $(i^{-1}(y), w) \in R^t$.*

*(2.3) If $x \in V_{g'}$ and $y \in V_{k'}$, then add all edges $(j(u), \mu, i'(y))$ such that $((i^{-1}(x), \delta), (u, \mu)) \in R^s$.*

*(2.4) If $x, y \in V_{k'}$, then add nothing.*

The definition is illustrated in Figure 2.1.

Obviously, (2) of Definition 3.2 specifies the embedding mechanism of ESM systems: it says how $R^s$ and $R^t$ are used to infer edges of $m$ from an arbitrary edge $e = (x, \delta, y)$ of $k$. If $x, y \in V_g$, (case 2.1), then such an edge $e'$ is obtained as follows:

(i) select a pair in $R^s$ such that its first component corresponds (via $i$) to the "source part" $(x, \delta)$ of $e$; let $(u, \mu)$ be the second component of this pair,

(ii) select a pair in $R^t$ such that its first component corresponds (via $i$) to the "target part" $y$ of $e$; let $w$ be the second component of this pair,

(iii) let $e'$ be the edge corresponding (via $j$) to $(u, \mu, w)$.

Hence, informally speaking, the relations $R^s$ and $R^t$ describe the way source parts and target parts of edges are transferred from $k$ to $m$. The edge label $\delta$ of an edge $(x, \delta, y)$ is considered part of the source part; thus the mechanism does not treat sources and targets of edges in a symmetric way. Cases (2.2) through (2.4) are the obvious counterparts of (2.1) where $x$ and/or $y$ are not replaced: then the source part and/or the target part remains unchanged.

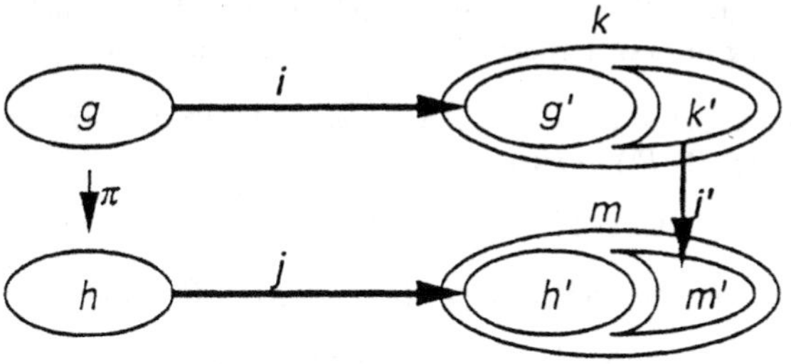


Figure 2.1

**Remark 2.1** *The graph $m$ is determined only up to an isomorphism by $\pi$, $k$ and $i$. It is, of course, uniquely determined if one also specifies $i'$ and $j$. In that case we say that $m$ is the graph obtained by the application of $\pi$ to $k$ via $(i, i', j)$.*

Since we are interested in the use of graph rewriting as a model of concurrent computation, we have to allow the concurrent application of productions to a graph: we allow that several nonoverlapping occurrences of left-hand sides of productions are replaced simultaneously. Although Definition 2.3 can be adapted in a straightforward way to cover concurrent rewriting, it is well-known that one may also describe a concurrent step by first constructing the disjoint union of several productions, and then applying the composed production. This is also the approach taken in [JR] and [JLR]. The concurrent composition of productions is defined as follows.

**Definition 2.4** *Let $\pi_1 = (g_1, h_1, R_1^s, R_1^t)$ and $\pi_2 = (g_2, h_2, R_2^s, R_2^t)$ be actor productions such that $g_1$ and $g_2$ are disjoint, and $h_1$ and $h_2$ are disjoint. The concurrent composition of $\pi_1$ and $\pi_2$ is the actor production $(g_1 \oplus g_2, h_1 \oplus h_2, R_1^s \oplus R_2^s, R_1^t \oplus R_2^t)$.*

Observe that edges connecting concurrently replaced subgraphs of $k$ may correspond to edges of $m$ according to (2.3) of Definition 2.3. Also, Definition 2.4 may be extended in the obvious way to allow the concurrent composition of non-disjoint actor productions: before one constructs their composition, $\pi_1$ and $\pi_2$ may be replaced by isomorphic, but disjoint copies.

In the remaining part of this section it is demonstrated that the graph grammars considered in this paper fit into the single pushout approach to graph rewriting (see, e.g.,

[L]). To this aim a category ESM (for extended structure morphism) is defined, where the objects are $(\Sigma,\Delta)$-graphs and the arrows represent the transformations between $(\Sigma,\Delta)$-graphs obtained by rewriting them. An actor production may be viewed as an ESM morphism, and the application of a production to a graph corresponds to the construction of a pushout in ESM.

For $(\Sigma,\Delta)$-graphs $g$ and $h$, an ESM morphism $R$ from $g$ into $h$ consists of three relations, $R=(R^c,R^s,R^t)$, where $R^s$ and $R^t$ play the same role as the last two components of an actor production (i.e., they describe the transfer of source and target parts of edges). It it is required that $R$ is structure-preserving in the sense that $h$ contains all edges that are transferred (in the way described in (2) of Definition (2.3)) from $g$ via $R^s$ and $R^t$. The relation $R^c$ describes the causal relationship between nodes of $g$ and nodes of $h$ introduced by the rewriting process considered. The notion is formally defined as follows.

**Definition 2.5** *Let $g$ and $h$ be $(\Sigma,\Delta)$-graphs. An ESM morphism from $g$ into $h$ is a 3-tuple $R=(R^c,R^s,R^t)$ of relations such that*

*(1) $R^c, R^t \subseteq V_g \times V_h$ and $R^s \subseteq (V_g\times\Delta)\times(V_h\times\Delta)$,*

*(2) $R^t\subseteq R^c$ and, for each $\delta,\mu\in\Delta$, $((x,\delta),(y,\mu))\in R^s$ implies $(x,y)\in R^c$,*

*(3) $\{(u,\mu,w)\in V_h\times\Delta\times V_h \mid$ there exists an edge $(x,\delta,y)\in E_g$ such that $((x,\delta),(u,\mu))\in R^s$ and $(y,w)\in R^t\}\subseteq E_h$.*

Note that $R_\emptyset=(\emptyset,\emptyset,\emptyset)$ is an ESM morphism from $g$ into $h$. For a $(\Sigma,\Delta)$-graph $g$, the 3-tuple $(Id_{V_g}, Id_{V_g\times\Delta}, Id_{V_g})$ is an ESM morphism from $g$ into $g$. It is called the *identical ESM-morphism on* $g$, and it is denoted by $Id_g^{ESM}$. For ESM morphisms $R_1=(R_1^c,R_1^s,R_1^t): g\to h$ and $R_2=(R_2^c,R_2^s,R_2^t): h\to z$, the *composition of* $R_1$ *and* $R_2$, denoted by $R_1\circ R_2$, is the ESM-morphism $(R_2^c\circ R_1^c, R_2^s\circ R_1^s, R_2^t\circ R_1^t)$ from $g$ into $z$. It is easily verified that the $(\Sigma,\Delta)$-graphs together with ESM morphisms form a category, ESM.

**Remark 2.2** *For $(\Sigma,\Delta)$-graphs $g,h$ and a graph morphism $f: g\to h$, the 3-tuple $(f,\overline{f},f)$, where $\overline{f}=\{((x,\delta),(f(x),\delta))\mid x\in V_g\}$, is an ESM morphism from $g$ into $h$. Observe that $Id_g^{gr}$ corresponds to $Id_g^{ESM}$. Hence ESM morphisms are an extension of the usual morphisms of graph structures in two ways: they are based on relations (rather than functions), and edge labels need not be preserved.*

For an ESM morphism $R=(R^c,R^s,R^t)$ from $g$ into $h$ and $E\subseteq E_g$, the set

$$\{(u,\mu,w)\in V_h\times\Delta\times V_h \mid \text{there exists an edge } (x,\delta,y)\in E \text{ such that } ((x,\delta),(u,\mu))\in R^s \text{ and } (y,w)\in R^t\}$$

is denoted by $R(E)$. Hence (3) of Definition 2.5 may be expressed as: $R(E_g)\subseteq E_h$.

The notion of an ESM morphism is closely related to the notion of a structured transformation, investigated in [JR] and [JLR]. In fact, for an ESM morphism $R=(R^c,R^s,R^t)$ from $g$ into $h$, the 4-tuple $(g,h,R^s,R^t)$ is a structured transformation.

We now consider the relationship between ESM morphisms and actor productions. On the one hand, it is clear that, for each ESM morphism $R: g\to h$, the 4-tuple $(g,h,R^s,R^t)$

is an actor production. On the other hand, for an actor production $\pi = (g, h, R^s, R^t)$, let $R_\pi = (R^c, R^s, R^t)$, where $R^c = V_g \times V_h$, and let $\overline{h} = (V_h, E_h \cup R_\pi(E_g), \phi_h)$. Then $R_\pi : g \to \overline{h}$ is an ESM morphism and $\overline{\pi} = (g, \overline{h}, R^s, R^t)$ is an actor production. Moreover, $\overline{\pi}$ may be applied to a graph $k$ via a graph morphism $i$ if and only if $\pi$ may be applied to $k$ via $i$, and the resulting graph is the same in both cases. In this sense one may replace each production $\pi$ of an actor grammar $G$ by the corresponding production $\overline{\pi}$ without changing the way graphs are rewritten. Since $\overline{\pi}$ is uniquely determined by $R_\pi, g$ and $\overline{h}$, one may view the set of productions of $G$ as a set of ESM morphisms (such as $R_\pi$). Observe that an ESM morphism $R : g \to h$ used in this way is of a restricted form: its first component $R^c$ expresses the causal relationship between nodes of $g$ and nodes of $h$, and since $R$ describes an *atomic* action of $G$, each node of $h$ is causally dependent of each node of $g$. Formally one has the following.

**Definition 2.6** *A primitive ESM morphism is an ESM morphism $R : g \to h$ such that $g, h$ are nonempty and $R^c = V_g \times V_h$.*

In this paper we are interested in the set of all rewriting processes of a system, rather than only those that have their initial graph in a fixed set. This leads to the following definition of a graph rewriting system.

**Definition 2.7** *An ESM system is a set of primitive ESM morphisms.*

The elements of an ESM system are often called *productions*. The following result states that a graph rewriting step in an ESM system (and hence, in an actor grammar) corresponds to a single pushout construction in the category ESM. The proof is a straightforward but somewhat tedious verification; we omit it in this extended abstract.

**Theorem 2.1** *Let $R : g \to h$ be an ESM morphism and let $\pi = (g, h, R^s, R^t)$ be the corresponding actor production. Let $k$ and $m$ be $(\Sigma, \Delta)$-graphs and let $i : g \to k$, $i' : k' \to m'$ and $j : h \to h'$ be graph morphisms such that $m$ is obtained by the application of $\pi$ to $k$ via $(i, i', j)$. Let $Q_\pi = (Q^c, Q^s, Q^t)$, where*

$$\begin{aligned}
Q^c &= \{(i(x), j(y)) \mid (x, y) \in R^c\} \cup \{(x, i'(x)) \mid x \in V_{k'}\}, \\
Q^s &= \{((i(x), \delta), (j(y), \mu)) \mid ((x, \delta), (y, \mu)) \in R^s\} \cup \{((x, \delta), (i'(x), \delta)) \mid x \in V_{k'}\}, \\
Q^t &= \{(i(x), j(y)) \mid (x, y) \in R^t\} \cup \{(x, i'(x)) \mid x \in V_{k'}\}.
\end{aligned}$$

*Then $Q_\pi$ is an ESM morphism from $k$ into $m$ and the following diagram is a pushout in ESM.*

$$\begin{array}{ccc}
g & \xrightarrow{R} & h \\
i\downarrow & & \downarrow j \\
k & \xrightarrow{Q_\pi} & m
\end{array}$$

Observe that the building of rewriting processes of an ESM system $P$ corresponds to building larger ESM morphisms from the primitive ESM morphisms in $P$; there are three ways to do this.

(1) Application to a graph: the pushout construction of Theorem 2.1.

(2) Concurrent composition: the obvious counterpart of Definition 2.4, combining two ESM morphisms $R_1 : g_1 \rightarrow h_1$, $R_2 : g_2 \rightarrow h_2$ into an ESM morphism $R : g_1 \oplus g_2 \rightarrow h_1 \oplus h_2$.

(3) Sequential composition: combining $R_1 : g_1 \rightarrow g_2$, $R_2 : g_2 \rightarrow g_3$ into $R_2 \circ R_1 : g_1 \rightarrow g_3$.

# 3 Computation structures and their composition

In this section a concrete representation of a rewriting process or computation of an ESM system is introduced; it is called a computation structure. The set of computation structures corresponding to an ESM system is defined, and it is demonstrated that interactions between rewriting processes can be described by a gluing operation on the corresponding computation structures. The three operations mentioned at the end of Section 2 (application, concurrent composition, sequential composition) correspond to special cases of this gluing operation.

The way a rewriting process of an ESM system is described by a computation structure is somewhat similar to the way runs of a Petri net are described by Petri net processes. More specifically, a computation structure consists of a $(\Sigma, \Delta)$-graph $g = (V, E, \phi)$ and an ESM morphism $(R^c, R^s, R^t)$ from $g$ into $g$. The nodes of $g$ represent the nodes of the graphs occurring in the rewriting process – they may be compared with the places of a Petri net process. The relation $R^c$ describes the causal relationship between these nodes – the information represented by $R^c$ may be compared with the information in the edges of a Petri net process. The relations $R^s$ and $R^t$ describe how source and target parts of edges are transfered in the rewriting steps. $E$ is the set of all edges of the graphs occurring in the rewriting process. Evidently, $R^s$, $R^t$ and $E$ have no counterpart in Petri net processes, and on the other hand the transitions of a Petri net process have no counterpart in a computation structure. The role of transitions is however comparable to that of a $P$-covering, introduced in Definition 3.2. The notion of a computation structure is formally defined as follows.

**Definition 3.1** *A computation structure is a 4-tuple $(V, E, \phi, R)$ such that*

*(1) $(V, E, \phi)$ is a $(\Sigma, \Delta)$-graph,*

*(2) $R : (V, E, \phi) \rightarrow (V, E, \phi)$ is an ESM morphism,*

*(3) $R^c, R^s$ and $R^t$ are reflexive, transitive and antisymmetric, and*

*(4) for each $((x, \delta), (y, \mu)) \in R^s$, $x = y$ implies $\delta = \mu$.*

**Example 3.1** Let $\Sigma = \{a\}$, $\Delta = \{\alpha, \beta\}$, and let $(V, E, \phi)$ be the graph depicted in Figure 3.1 (a). We have omitted node labels in Figure 3.1 (a); evidently all nodes have label $a$. Let $R = (R^c, R^s, R^t)$, where $R^c, R^s, R^t$ are the transitive and reflexive closure of the relations $(\{1,2\} \times \{4,5\}) \cup (\{3,5\} \times \{6,7\})$, $\{((1,\beta),(5,\beta)), ((5,\beta),(6,\alpha))\}$ and $\{(2,5)\}$, respectively. Then $C = (V, E, \phi, R)$ is a computation structure. $C$ may be interpreted as a description of a rewriting process in the following way. Let $g_1, h_1, g_2, h_2, k$ be the graphs depicted in Figure 3.1 (b) (we assume again that all nodes are labelled by $a$). Let

$$V_1 = V_{g_1} \oplus V_{h_1}, V_2 = V_{g_2} \oplus V_{h_2}, R_1^c = V_{g_1} \times V_{h_1}, R_2^c = V_{g_2} \times V_{h_2},$$

$$R_1^s = \{((r,\beta),(u,\beta))\}^{\bullet}_{V_1}, R_2^s = \{((v,\beta),(x,\alpha))\}^{\bullet}_{V_2}, R_1^t = \{(s,u)\}^{\bullet}_{V_1}, R_2^t = (\emptyset)^{\bullet}_{V_2}.$$

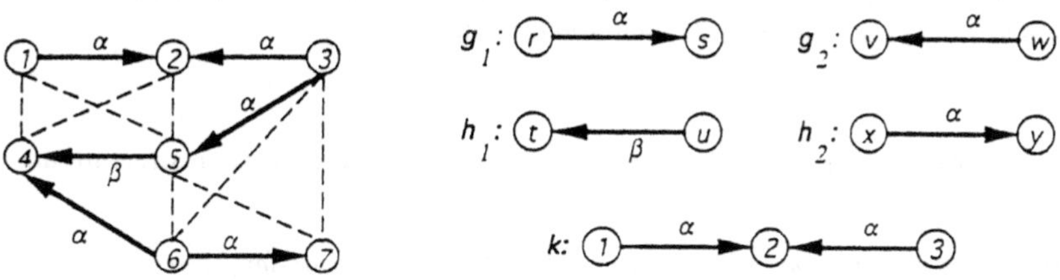


(a) (b)

Figure 3.1

Then $\pi_1 = (R_1^c, R_1^s, R_1^t) : g_1 \to h_1$ and $\pi_2 = (R_2^c, R_2^s, R_2^t) : g_2 \to h_2$ are ESM morphisms, and $\{\pi_1, \pi_2\}$ is an ESM system. $C$ represents the rewriting process in which first the nodes 1,2 of $k$ are replaced by 4,5 using production $\pi_1$, and then nodes 3,5 are replaced by 6,7 using $\pi_2$. Observe that the initial graph of the rewriting process is the subgraph of $C$ (or rather, the $(\Sigma, \Delta)$-graph consisting of the first three components of $C$) induced by the nodes 1,2,3; i.e. the nodes that are minimal with respect to the causal relation $R^c$. The graph resulting from the rewriting process is the subgraph induced by the nodes 4,6,7, which are maximal with respect to $R^c$, and the intermediate graph obtained after the first step is the subgraph induced by the nodes 4,5,3. In general the intermediate graphs correspond to subgraphs induced by cuts of $R^c$.

**Remark 3.1** *For a $(\Sigma, \Delta)$-graph $g = (V, E, \phi)$, the 4-tuple $(V, E, \phi, Id_g^{ESM})$ is a computation structure. It is called the identical computation structure on $g$. The 4-tuple $(\emptyset, \emptyset, \emptyset, R_\emptyset)$ is also a computation structure. It is called the trivial computation structure.*

For a computation structure $C$, its components are denoted by $V_C, E_C, \phi_C$ and $R_C$. The set of minimal and maximal nodes of $C$ with respect to $R_C^c$ are denoted by $Min(C)$ and $Max(C)$, respectively. Computation structures will be compared using *CS morphisms*, which are structure–preserving injective functions on nodes.

**Definition 3.2** *For computation structures $C_1 = (V_1, E_1, \phi_1, R_1)$ and $C_2 = (V_2, E_2, \phi_2, R_2)$, a CS morphism from $C_1$ into $C_2$ is an injective function $f : V_1 \to V_2$ such that $\phi_2 \circ f = \phi_1$, $f(E_1) \subseteq E_2$, $f(R_1^c) \subseteq R_2^c$, $f(R_1^s) \subseteq R_2^s$ and $f(R_1^t) \subseteq R_2^t$.*

Obviously, for a computation structure $C$, the identical relation on $V_C$ is a CS morphism from $C$ into $C$. The computation structures together with CS morphisms form a category CS. It follows from Remark 3.1 that the $(\Sigma, \Delta)$-graphs together with graph morphisms form a subcategory of CS.

One may now associate a set of computation structures with an ESM system $P$. Informally, a computation structure $C$ corresponds to a rewriting process of $P$ if $C$ can be covered by occurrences of productions of $P$ in such a way that these occurrences together induce the relations $R^c$, $R^s$ and $R^t$ between the nodes of $C$. Moreover, each edge of $E_C$ either belongs to the initial graph (i.e., it is only incident to nodes of $Min(C)$), or it is an edge created by one of the occurrences, or it is an edge obtained via $R^s_C$ and $R^t_C$ from another edge. We assume that $P$ is such that, for each $\pi : g \longrightarrow h$ in $P$, $g$ and $h$ are disjoint. Then $\pi$ may be represented by the computation structure

$$C_\pi = (V_g \oplus V_h, E_h \cup \pi(E_g), \phi_g \oplus \phi_h, \pi).$$

The notions of a $P$-covering, a valid $P$-covering and a $P$-valid computation structure are defined as follows.

**Definition 3.3** *Let $P$ be an ESM system and let $C$ be a computation structure. A $P$-covering of $C$ is a collection $(f_i, \pi_i)_{i\in I}$ of pairs such that*

*(1) For each $i \in I$, $\pi_i \in P$ and $f_i : C_{\pi_i} \rightarrow C$ is a CS morphism.*

*(2) For each $i \in I$, let $C_{\pi_i} = (V_i, E_i, \phi_i, R_i)$. Then for each $x \in V_C$, there are at most two indices $i \in I$ such that $x \in f_i(V_i)$, and if $x \in f_i(V_i) \cap f_j(V_j)$ where $i \neq j$, then either $x \in f_i(Max(C_{\pi_i})) \cap f_j(Min(C_{\pi_j}))$ or $x \in f_i(Min(C_{\pi_i})) \cap f_j(Max(C_{\pi_j}))$.*

**Definition 3.4** *Let $P$ be an ESM system, let $C$ be a computation structure and let $(f_i, \pi_i)_{i\in I}$ be a $P$-covering of $C$. Then $(f_i, \pi_i)_{i\in I}$ is a valid $P$-covering of $C$ if the following holds. Let $E_{min} = E_C \cap (Min(C) \times \Delta \times Min(C))$ and, for each $i \in I$, $C_{\pi_i} = (V_i, E_i, \phi_i, R_i)$ and $E_{i,Max} = E_i \cap (Max(C_{\pi_i}) \times \Delta \times Max(C_{\pi_i}))$. Then*

$$R_C = \left(\bigcup_{i\in I} f_i(R_i)\right)^*_{V_C},$$

*where the transitive and reflexive closure is taken componentwise (for $R^c_C, R^s_C, R^t_C$ separately), and*

$$E_C = R_C\left(E_{min} \cup \bigcup_{i\in I} f_i(E_{i,Max})\right).$$

**Definition 3.5** *Let $P$ be an ESM system and let $C$ be a computation structure. Then $C$ is $P$-valid if there exists a valid $P$-covering of $C$.*

We will now introduce a composition operation for computation structures. This composition models the way computations (rewriting processes) interact or communicate. Consider computation structures $C_1, C_2$ and assume that they describe rewriting processes that communicate: each of the processes may use output of the other process as its input. If the rewriting process corresponds to a computation structure $C$, then its input and output correspond to the subgraphs of $C$ induced by $Min(C)$ and $Max(C)$, respectively. Hence the combined rewriting process corresponds to the computation structure obtained from

$C_1$ and $C_2$ by gluing them together, identifying nodes of $Min(C_1)$ with nodes of $Max(C_2)$ and/or vice versa. This gluing operation is formally defined as a pushout construction in CS. Evidently the gluing of $C_1$ and $C_2$ can only be constructed if one specifies which nodes are to be identified: one has to specify in detail how the two rewriting processes interact. Formally one has the following.

**Definition 3.6** *Let $C_1 = (V_1, E_1, \phi_1, R_1), C_2 = (V_2, E_2, \phi_2, R_2)$ be computation structures. A $(C_1, C_2)$-interaction is a 3-tuple $(C_{int}, d_1, d_2)$ where $C_{int} = (V_{int}, E_{int}, \phi_{int}, R_{int})$ is a computation structure and $d_1 : C_{int} \to C_1$, $d_2 : C_{int} \to C_2$ are CS-morphisms such that*

*(1) for each $x \in V_{int}$, either $d_1(x) \in Min(C_1)$ and $d_2(x) \in Max(C_2)$, or $d_1(x) \in Min(C_2)$ and $d_2(x) \in Max(C_1)$, and*

*(2) The relation $(d_1^{-1}(R_1^c) \cup d_2^{-1}(R_2^c))^*_{V_{int}}$ is antisymmetric.*

The condition (1) of Definition 3.6 expresses the fact that nodes of $V_{int}$ may be interpreted as input from $C_2$ into $C_1$ or vice versa. Condition (2) ensures that the pushout object of $d_1$ and $d_2$ exist. Furthermore, since condition (1) implies that $R^c_{int}$ is the identical ESM morphism, one may consider $C_{int}$ as a $(\Sigma, \Delta)$-graph (see Remark 3.1). The composition of computation structures may now be defined as follows.

**Definition 3.7** *Let $C_1, C_2$ be computation structures and let $int = (C_{int}, d_1, d_2)$ be a $(C_1, C_2)$-interaction. The composition of $C_1$ and $C_2$ over int is the set*

$$C_1 \,\Box_{int}\, C_2 = \{(C_{12}, c_1, c_2) \mid \text{the diagram } \begin{array}{ccc} C_{int} & \xrightarrow{d_2} & C_2 \\ d_1\downarrow & & \downarrow c_2 \\ C_1 & \xrightarrow{c_1} & C_{12} \end{array} \text{ is a pushout in } CS\}.$$

We will often write $C_{12} \in C_1 \,\Box_{int}\, C_2$ instead of $(C_{12}, c_1, c_2) \in C_1 \,\Box_{int}\, C_2$. For given $C_1$, $C_2$ and $int$, the computation structure $C_{12}$ is unique up to isomorphism. There exists a concrete construction of $C_{12}$, $c_1$ and $c_2$, which is a straightforward extension of the well-known gluing construction for sets.

The three operations for ESM morphisms mentioned at the end of Section 2 correspond to special cases of the composition operation for computation structures. Indeed, one may represent an ESM morphism $R : g \to h$ by a computation structure $C_R$ (we assume that g and h are disjoint):

$$C_R = (V_g \oplus V_h, E_h \cup R(E_g), \phi_g \oplus \phi_h, R).$$

Then one has the following:

- Application: let $g, h, R, k, m, Q, i$ be as in Theorem 2.1, and assume that $g, h$ are disjoint and $k, m$ are disjoint. Let $C_g$ and $C_k$ be the identical computation structures on $g$ and $k$, respectively. Then $Id_g^{gr}$ and $i$ may be considered as CS morphisms from $C_g$ into $C_R$ and from $C_g$ into $C_k$, respectively. Moreover, $int = (C_g, Id_g^{gr}, i)$ is a $(C_R, C_k)$-interaction and $C_Q \in C_R \,\Box_{int}\, C_k$.

- Sequential composition: the sequential composition of ESM morphisms $R : g \rightarrow h$ and $T : h \rightarrow z$ corresponds to the composition of $C_R$ and $C_T$ where the interaction consists of the identical computation structure $C_h$ on $h$ and inclusion maps from $C_h$ into $C_R$ and $C_T$ (nodes of $C_h$ are maximal in $C_R$ and minimal in $C_T$).

- Concurrent composition: the concurrent composition of ESM morphisms corresponds to a situation where the interaction is empty, i.e., the trivial computation structure and two empty CS morphisms.

# 4 Conditional computations

For an ESM system $P$, a $P$-valid computation structure represents a computation (i.e. a rewriting process) of $P$. Hence one may use the set of all $P$-valid computation structures as a formal semantics of the system $P$. However this approach has the disadvantage that it is not compositional with respect to the union of ESM systems. To illustrate this, consider ESM systems $P_1$, $P_2$ and let $P = P_1 \cup P_2$. Let $\pi_1, \pi_2 \in P_1$, $\nu_1, \nu_2 \in P_2$ and consider the computation structure $C$ of Figure 4.1 (a). It is assumed that $C$ is $P$-valid; the rectangles in Figure 4.1 (a) represent a valid $P$-covering of $C$. In a compositional system one would like to view a computation of the composed system as a composition of computations of the component systems; hence in this case one would like to view $C$ as a composition of $C_1$ and $C_2$, where $C_1$ and $C_2$ correspond to the parts of $C$ realized using productions from $P_1$ and $P_2$, respectively. $C_1$ and $C_2$ are depicted in Figure 4.1 (b) and (c).

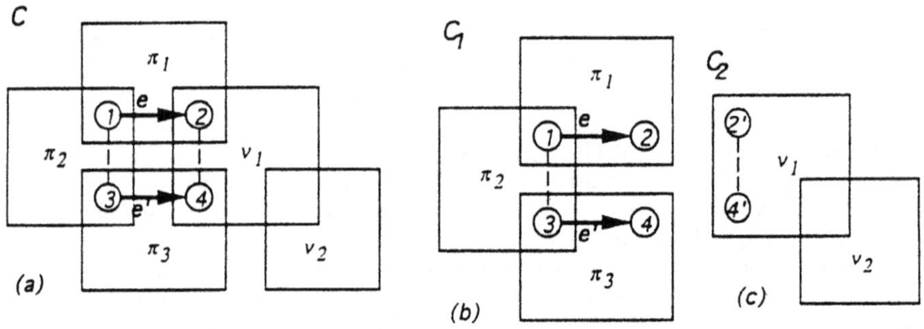

Figure 4.1

Now assume that the edge $e'$ of $C$ is obtained from the edge $e$ by $\pi_2^s$ and $\nu_1^t$, as indicated by the dotted lines. Then the covering of Figure 4.1 (b) is not $P_1$-valid because it does not satisfy the second part of Definition 3.4. Hence it may not belong to the proposed semantics of $P_1$. However if $C_1$ is composed with $C_2$ over an interaction *int* that identifies the nodes 2,2' and 4,4', then the result is $P$-valid.

In the rest of this section it is sketched how the notion of a computation structure may be extended to the notion of a conditional computation; this yields a semantics that has the desired compositional property. In this new approach, the inclusion morphism $c_1 : C_1 \rightarrow C$ is called a $C_1$-*context*, and the fact that $C_1$ becomes valid in this context is represented in the semantics of $P_1$. The basic idea is that the effect of a context on a computation structure $C$ can be described by a *C-condition*. It turns out that, to find out whether a computation structure $C$ becomes valid in a context $ctx : C \to C'$, it is sufficient to know which edges are induced by $C'$ between minimal nodes of $ctx(C)$, and which causal relationships and edge transfer relations are induced by $C'$ between minimal and maximal nodes of $ctx(C)$. Hence the information needed about $ctx$ consists of a pair $(E, R)$, where $E$ is a set of edges on $Min(C)$ and $R$ is an ESM morphism. Such pairs will be used as conditions imposed on $C$-contexts. Formally, one has the following.

**Definition 4.1** *Let $C$ be a computation structure.*

*(1) A $C$-context is a CS morphism from $C$ into a computation structure $C'$.*

*(2) A $C$-condition is a pair $(E, R)$ such that $E \subseteq Min(C) \times \Delta \times Min(C)$ and $R : (Max(C), \emptyset, \phi_{Max}) \to (Min(C), \emptyset, \phi_{Min})$ is an ESM-morphism, where $\phi_{Min}, \phi_{Max}$ are the restrictions of $\phi_C$ to $Min(C)$, $Max(C)$, respectively.*

*(3) Let $ctx : C \to C'$ be a $C$-context and let $cnd = (E_{cnd}, R_{cnd})$ be a $C$-condition. Then $ctx$ satisfies $cnd$ if $ctx(E_{cnd}) \subseteq E_{C'}$, $ctx(R^c_{cnd}) \subseteq R^c_{C'}$, $ctx(R^s_{cnd}) \subseteq R^s_{C'}$ and $ctx(R^t_{cnd}) \subseteq R^t_{C'}$.*

*(4) Let $cnd_1 = (E_1, R_1)$ and $cnd_2 = (E_2, R_2)$ be $C$-conditions. Then $cnd_1$ is a strengthening of $cnd_2$ if $E_2 \subseteq E_1$ and $R_2 \subseteq R_1$ (componentwise).*

For computation structures $C_1$, $C_2$, a $(C_1, C_2)$-interaction $int$, and $C \in C_1 \square_{int} C_2$, the CS morphism $c_1 : C_1 \to C$ is a $C_1$-context and $c_2 : C_2 \to C$ is a $C_2$-context. For a computation structure $C$ and two $C$-conditions $cnd_1$ and $cnd_2$, if $cnd_1$ is a strengthening of $cnd_2$ then $cnd_1$ is at least as restrictive as $cnd_2$ when viewed as a condition on $C$-contexts.

Each $C$-context $ctx : C \to C'$ induces a $C$-condition in the following way.

**Definition 4.2** *Let $ctx : C \to C'$ be a $C$-context. The $C$-condition induced by $ctx$ is the pair $(E, R)$ where*

$$E = ctx^{-1}(E_{C'} \cap ctx(Min(C)) \times \Delta \times ctx(Min(C))),$$

$$R_c = ctx^{-1}(R^c_{C'} \cap ctx(Max(C)) \times ctx(Min(C))),$$

$$R_s = ctx^{-1}(R^s_{C'} \cap (ctx(Max(C)) \times \Delta) \times (ctx(Min(C)) \times \Delta),$$

$$R_t = ctx^{-1}(R^t_{C'} \cap ctx(Max(C)) \times ctx(Min(C))).$$

Observe that $ctx : C \to C'$ satisfies a $C$-condition $cnd$ if and only if the $C$-condition induced by $ctx$ is a strengthening of $cnd$. If $ctx : C \to C'$ is a $C$-context, then one may view the substructure $\tilde{C}$ of $C'$ induced by the image of $C$ under $ctx$ as a copy of $C$ that is enriched (in the sense that edges and relations are added) according to $C'$. If $ctx$ satisfies a $C$-condition $cnd = (E_{cnd}, R_{cnd})$, then $\tilde{C}$ contains as a substructure a copy of $C$ that is enriched according to $(E_{cnd}, R_{cnd})$. This enriched version of $C$ is denoted by $C \square cnd$; formally:

$$C \square cnd = (V_C, R(E_C \cup E_{cnd}), \phi_C, R) \text{ where } R = (R_C \cup R_{cnd})^*.$$

One may now describe the behaviour of an ESM system by a set of pairs $(C, cnd)$, where $C$ is a computation structure and $cnd$ is a $C$-condition. Informally, $C$ is a description of a rewriting process that may occur in any context that satifies $cnd$; in this sense $cnd$ specifies those contexts in which $C$ becomes valid. If $(C, cnd)$ is valid in a context $ctx : C \to C'$, then $C$ will contain only information (edges, relations) that is created in the rewriting process corresponding to $C$. The information in $cnd$, on the other hand, is to be created by the part of $C'$ outside $ctx(C)$. Hence edges incident with nodes of $Min(C)$, and in particular the edges of the initial graph of the rewriting process corresponding to $C$, are specified in $cnd$: $C$ can only occur in a context that provides these edges. Formally, one has the following.

**Definition 4.3** *A conditional computation is a pair $(C, cnd)$ where $C$ is a computation structure such that, for each $(x, \delta, y) \in E_C$, $x \notin Min(C)$ and $y \notin Min(C)$, and cnd is a $C$-condition.*

Observe that a conditional computation $(C, cnd)$ where $cnd = (E_{cnd}, R_\emptyset)$ may be interpreted as a "stand-alone" computation: the only condition that a $C$-context must satisfy is that it supplies the initial graph.

One can now give the counterpart of Definition 3.4 for conditional computations (instead of computation structures).

**Definition 4.4** *Let $P$ be an ESM system and let $(C, cnd)$ be a conditional computation. A valid $P$-covering of $(C, cnd)$ is a $P$-covering $(f_i, \pi_i)_{i\in I}$ of $C \,\square\, cnd$ such that the following holds. Let, for each $i \in I$, $C_{\pi_i} = (V_i, E_i, \phi_i, R_i)$ and $E_{i,Max} = E_i \cap (Max(C_{\pi_i}) \times \Delta \times Max(C_{\pi_i}))$. Then*

$$R_C = \left(\bigcup_{i\in I} f_i(R_i)\right)^*_{V_C} \quad \text{and } E_C = R_C\left(\bigcup_{i\in I} f_i(E_{i,Max})\right).$$

The notion of $P$-validity for conditional computations is defined as follows.

**Definition 4.5** *Let $P$ be an ESM system and let $(C, cnd)$ be a conditional computation. $(C, cnd)$ is $P$-valid if there exists a valid $P$-covering of $(C, cnd)$.*

The set of all $P$-valid conditional computations is denoted by $Ccomp(P)$. In the last part of this section it is demonstrated that the composition of computation structures can be extended into a composition operation on conditional computations, and that the set $Ccomp(P)$ of conditional computations of a composed ESM system $P = P_1 \cup P_2$ is the set of all conditional computations that can be obtained by composing a $P_1$-valid and a $P_2$-valid conditional computation.

Informally, the composition of two conditional computations $(C_1, cnd_1)$ and $(C_2, cnd_2)$ over a $(C_1, C_2)$-interaction $int$ is the set of all conditional computations $(C, cnd)$ such that $C \in C_1 \,\square_{int}\, C_2$ and $cnd$ is a $C$-condition such that each $C$-context $ctx_3 : C \to C_3$ satisfying $cnd$ corresponds, on the one hand, to a $C_1$-context $ctx_1$ that satisfies $cnd_1$, and on the other hand to a $C_2$-context $ctx_2$ that satisfies $cnd_2$. The situation is illustrated by

Figure 4.2; $ctx_1$ is the restriction of $ctx_3$ to the image of $C_1$ in $C$, and $ctx_2$ is the restriction of $ctx_3$ to the image of $C_2$ in $C$.

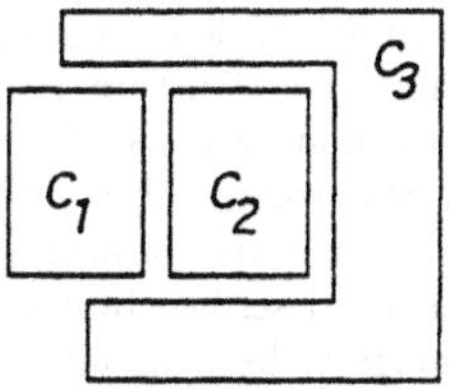


Figure 4.2

Formally, to a computation structure $C \in C_1 \Box_{int} C_2$ and a $C$–condition $cnd$ corresponds a $C_1$–context $\bar{c}_1 = j \circ c_1$, where $j$ is the inclusion morphism from $C$ into $C \Box cnd$. The $C_1$–condition induced by $\bar{c}_1$ is called the $C_1$–condition *induced by* $(C_2, int, cnd)$. Evidently, the $C_2$–condition induced by $(C_1, int, cnd)$ is obtained in a similar way. The composition of conditional computations may now be defined as follows.

**Definition 4.6** *Let* $(C_1, cnd_1)$, $(C_2, cnd_2)$ *be conditional computations and let int be a* $(C_1, C_2)$*–interaction. The composition of* $(C_1, cnd_1)$ *and* $(C_2, cnd_2)$ *over int is the set*

$(C_1, cnd_1) \Box_{int} (C_2, cnd_2) =$
$\{(C, cnd) \mid C \in C_1 \Box_{int} C_2,$ *the* $C_1$*-condition induced by* $(C_2, int, cnd)$ *is a strengthening of* $cnd_1$ *and the* $C_2$*-condition induced by* $(C_1, int, cnd)$ *is a strengthening of* $cnd_2$ $\}$.

It turns out that a semantics for ESM systems based on conditional computations has the desired compositional property.

**Theorem 4.1** *For ESM systems* $P_1, P_2$,

$$Ccomp(P_1 \cup P_2) = \{(C, cnd) \mid \textit{there exist } (C_1, cnd_1) \in Ccomp(P_1), (C_2, cnd_2) \in Ccomp(P_2) \textit{ and a } (C_1, C_2)\textit{-interaction int such that } (C, cnd) \in (C_1, cnd_1) \Box_{int} (C_2, cnd_2)\}.$$

The set $Ccomp(P)$ may be obtained in an inductive way, using the composition of conditional computations. from the productions of $P$ and the identical computation structures. Productions and identical computation structures are then represented as conditional computations in the following way.

(1) Let $\pi : g \rightarrow h$ be a production of $P$. We assume that $g$ and $h$ are disjoint. The conditional computation corresponding to $\pi$ is the pair $(C_\pi, cnd_\pi)$ where $C_\pi = (V_g \oplus V_h, E_h, \phi_1 \oplus \phi_2, \pi)$ and $cnd_\pi = (E_g, R_\emptyset)$.

(2) The identical computation structure on $g$ corresponds to the conditional computation $(C_g, cnd_g)$, where $C_g = (V_g, \emptyset, \phi_g, Id_g^{ESM})$ and $cnd_g = (E_g, R_\emptyset)$.

**Theorem 4.2** *Let* $P$ *be an ESM system and let* $X$ *be the smallest set of conditional computations such that* $X$ *contains* $P$, $X$ *contains all identical computation structures and, for each* $(C_1, cnd_1), (C_2, cnd_2) \in X$ *and each* $(C_1, C_2)$*-interaction int,* $X$ *contains the set* $(C_1, cnd_1) \Box_{int} (C_2, cnd_2)$. *Then* $X = Ccomp(P)$.

# 5 Conclusion

In this paper we have introduced two ways to describe graph rewriting processes in ESM systems. On the one hand, a computation structure yields a detailed description of a rewriting process, including its intermediate graphs. On the other hand an ESM morphism yields a more abstract description, where only the relationships between the nodes of the initial graph and the resulting graph are considered. One has a similar situation in [JLR] and [J]: there computations are described by pairs $(g, C)$, where $g$ is the initial graph and $C$ is a graph containing roughly the same information as a computation structure together with a $P$-covering. The external effect of a computation $(g, C)$ is described by a structured transformation $\Gamma(g, C)$ that can be constructed from $(g, C)$. In a similar way one may construct an ESM morphism from a computation structure (by considering the substructure induced by the minimal and maximal nodes). In fact, applying this construction to conditional computations leads to a more abstract semantics for ESM systems, consisting of pairs $(M, cnd)$ where $M$ describes the external effect of a computation. It turns out that the counterparts of Theorems 4.1 and 4.2 still hold for this more abstract semantics. We conclude that the notion of a computation structure provides a useful tool for the development of a theory of ESM systems, and it seems worthwile to investigate its use for other issues, such as the comparison or the refinement of ESM systems.

## References

[H] C. Hewitt, Viewing Control Structures as Patterns of Passing Messages, *J. Artificial Intel.*, **8** (1977), 323-364.

[HB] C. Hewitt and H. Baker, Laws for Communicating Parallel Processes, *Proc. IFIP 77*, Toronto, 1977, 987-992.

[J] D. Janssens, Equivalence of Computations in Actor Grammars, *Theoretical Computer Science*, **109** (1993), 145-180.

[JLR] D. Janssens, M.Lens and G. Rozenberg, Computation Graphs for Actor Grammars, *Journal of Computer and System Sciences*, **46** (1993), 60-90.

[JR] D. Janssens and G. Rozenberg, Actor Grammars, *Math. Systems Theory*, **22** (1989), 75-107.

[L] M. Löwe, Algebraic Approach to Single–Pushout Graph Transformation, *Theoretical Computer Science*, **109** (1993), 181-224.

# Relational Structures and Their Partial Morphisms in View of Single Pushout Rewriting

Yasuo KAWAHARA[1] and Yoshihiro MIZOGUCHI[2]

[1] Research Institute of Fundamental Information Science, Kyushu University 33, Fukuoka 812, Japan

[2] Deptartment of Control Engineering and Science, Kyushu Institute of Technology, Iizuka 820, Japan

**Abstract.** In this paper we present a basic notion of relational structures which includes simple graphs, labelled graphs and hypergraphs, and introduce a notion of partial morphisms between them. An existence theorem of pushouts in the category of relational structures and their partial morphisms is proved under a certian functorial condition, and it enables us to discuss single pushout rewritings of relational structures.

## 1 Introduction

The algebraic theory of graph grammar, motivated from the study of graph grammars, was initiated by Ehrig and his group [6] in the early 70's. Since then the idea to define graph transformations with so-called double pushout derivations has been extensively developed and applied to various fields [1, 3, 4, 5] of computer science. In 1984 Raoult [14] proposed another idea for graph transformation, so-called single pushout derivations [10, 16], making use of a notion of partial morphisms of term graphs. For example, Raoult [14] showed that the Church-Rosser property and a few properties of critical pairs of production rules if an involved category of graphs and their partial morphisms has pushouts. Of course we cannot generally expect the existence of pushouts in categories of graphs and their partial morphisms.

Thus it is important to present which categories of graphs and their partial morphisms are convenient to studying graph transformations with single pushouts. The paper shows some concrete categories of relational structures and partial morphisms have pushouts and consequently the single pushout approach can be applied in these categories as in [14] and [13]. In the section 2 we introduce a basic notion of relational structures which includes simple graphs [13], labelled graphs [13] and hypergraphs [9], and a few properties on the category of relational structures and their (total) morphisms are stated. Here we should mention that relational structures have been initially studied as foundations of graph transformations by Ehrig et al. [2] and the notion of relational structures treated in this paper is a special case of their original notion. In the section 3 we define a notion of partial morphisms of relational structures to show an existence thoerem (Cf. Theorem 9) of pushouts. This enables us to more generally formalize single pushout rewritings of relational structures. The section 4 is devoted to state that the category of relational structures together with partial morphisms is isomorphic to a category of partial morphisms in the sense of [9, 15].

## 2 Relational Structures

In this section we introduce a basic notion of relational structures which includes simple graphs [13], labelled graphs [13] and hypergraphs [9], and a few properties on the category of relational structures are studied. In what follows we freely use relational calculus [7, 8], namely theory of binary relations on sets.

Let **Set** be the category of sets and (total) functions. A relation of a set $A$ to a set $B$, denoted by $\alpha : A \rightharpoonup B$, is a subset of the cartesian product $A \times B$ of $A$ and $B$, that is, $\alpha \subseteq A \times B$. Usually a function $f : A \to B$ is regarded as a relation by its graph $\{(x, y) \in A \times B | y = f(x)\}$. Thus we identify the identity function $\mathrm{id}_A : A \to A$ of a set $A$ with the diagonal relation $\{(x, x) \in A \times A | a \in A\}$. For two relations $\alpha, \alpha' : A \rightharpoonup B$ a notation $\alpha \sqsubseteq \alpha'$ means that $\alpha$ is a subset of $\alpha'$ as subsets of $A \times B$. In other words, $\sqsubseteq$ denotes the usual inclusion relation of subsets. The composite relation $\beta\alpha$(or, $\beta \cdot \alpha$) of a relation $\alpha : A \rightharpoonup B$ followed by a relation $\beta : B \rightharpoonup C$ is a relation such that

$$(x, z) \in \beta\alpha \iff \exists y \in B : (x, y) \in \alpha \text{ and } (y, z) \in \beta.$$

The inverse $\alpha^\sharp : B \rightharpoonup A$ of a relation $\alpha : A \rightharpoonup B$ is defined by $\alpha^\sharp = \{(y, x) \in B \times A | (x, y) \in \alpha\}$. It is easy to see that a relation $\alpha : A \rightharpoonup B$ is a function if and only if $\alpha\alpha^\sharp \sqsubseteq \mathrm{id}_B$ (the univalency) and $\mathrm{id}_A \sqsubseteq \alpha^\sharp\alpha$. A partial function $f : A \to B$ is a relation with $ff^\sharp \sqsubseteq \mathrm{id}_B$ (univalent). The domain relation $\mathrm{d}(\alpha) : A \rightharpoonup A$ of a relation $\alpha : A \rightharpoonup B$ is defined by $\mathrm{d}(\alpha) = \alpha^\sharp\alpha \sqcap \mathrm{id}_A$, that is,

$$\mathrm{d}(\alpha) = \{(x, x') \in A \times A | x = x' \text{ and } \exists y \in B : (x, y) \in \alpha\}.$$

In general, relations in **Set** are not morphisms(functions) in **Set**, but relations will be used to define a notion of relational structures below. For more details on theory of relations see for example [7, 8, 12, 13].

Let $\mathcal{C}$ be an (abstract) category. A *frame* $< \omega : S \to R, \lambda : T \to R >$, on which relational structures are defined, consists of three functors $S, T, R : \mathcal{C} \to$ **Set** and two natural transformations $\omega : S \to R$, $\lambda : T \to R$.

**Definition 1.** A *relational structure* $< a, \alpha >$ over a frame $< \omega : S \to R, \lambda : T \to R >$ is a pair of an object $a$ of $\mathcal{C}$ and a relation $\alpha : Sa \rightharpoonup Ta$ such that $\lambda_a\alpha \sqsubseteq \omega_a$. A morphism $f$ from a relational structure $< a, \alpha >$ into a relational structure $< b, \beta >$, denoted by $f :< a, \alpha > \to < b, \beta >$, is a morphism $f : a \to b$ in $\mathcal{C}$ satisfying $T(f)\alpha \sqsubseteq \beta S(f)$. □

$$\begin{array}{ccc} Sa & \xrightarrow{\alpha} & Ta \\ \searrow & & \swarrow \\ \omega_a & Ra & \lambda_a \end{array} \qquad \begin{array}{ccc} Sa & \xrightarrow{S(f)} & Sb \\ \alpha\downarrow & & \downarrow\beta \\ Ta & \xrightarrow[T(f)]{} & Tb \end{array}$$

Let $f :< a, \alpha > \to < b, \beta >$ and $g :< b, \beta > \to < c, \gamma >$ be morphisms of relational structures over a frame $< \omega : S \to R, \lambda : T \to R >$. As $T(f)\alpha \sqsubseteq \beta S(f)$ and $T(g)\beta \sqsubseteq \gamma S(g)$ we have

$$T(gf)\alpha = T(g)T(f)\alpha \sqsubseteq T(g)\beta T(f) \sqsubseteq \gamma S(g)S(f) = \gamma S(gf),$$

which shows that the composite $gf :< a, \alpha > \to < c, \gamma >$ is a morphism of relational structures. Also it is clear that $\mathrm{id}_a :< a, \alpha > \to < a, \alpha >$ is a morphism of relational structures, where $\mathrm{id}_a$ is the identity morphism of $a$. Thus relational structures and their morphisms over $< \omega : S \to R, \lambda : T \to R >$ form a category $\mathcal{C} < \omega : S \to R, \lambda : T \to R >$, called the category of relational structures and their morphisms.

When a functor $R : \mathcal{C} \to \mathbf{Set}$ is a constant functor into a singleton set $1(= \{0\})$, that is, $Ra = 1$ for any object $a$ of C, the condition $\lambda_a \alpha \sqsubseteq \omega_a$ is void. (Note that when $Ra = 1$ the graph of a function $\omega_a$ is the greatest relation $Sa \times 1$ of $Sa$ into 1.) In this case we write $\mathcal{C} < S, T >$ for $\mathcal{C} < \omega : S \to R, \lambda : T \to R >$.

Now some examples of relational structures are given in the following:

*Example 1.* **(Simple graphs)** A (simple) graph $< A, \alpha >$ is a pair of a set $A$ and a relation $\alpha : A \rightharpoonup A$. A vertex of $< A, \alpha >$ is an element of $A$, and there is only one directed edge from an vertex $x \in A$ to a vertex $y \in A$ if and only if $(x, y) \in \alpha$. In this sense a graph $< A, \alpha >$ does not admit multiple edges. A graph homomorphism $f$ from a graph $< A, \alpha >$ into a graph $< B, \beta >$ is a function $f : A \to B$ such that $f\alpha \sqsubseteq \beta f$.

$$\begin{array}{ccc} A & \xrightarrow{f} & B \\ {\scriptstyle\alpha}\downarrow & & \downarrow{\scriptstyle\beta} \\ A & \xrightarrow[f]{} & B \end{array}$$

Because of the definition of the composition of relations and the univalency of functions, the condition $f\alpha \sqsubseteq \beta f$ is equivalent a natural condition that $(x, y) \in \alpha$ implies $(f(x), f(y)) \in \beta$, that is, $f$ preserves edges. Here is a concrete example of graphs. Let $A = \{x, y, z\}$, $B = \{x', y'\}$, $\alpha = \{(x, y), (x, z)\} : A \rightharpoonup A$ and $\beta = \{(x', y')\} : B \rightharpoonup B$. Then two graphs $< A, \alpha >$ and $< B, \beta >$ are illustrated by the following figure.

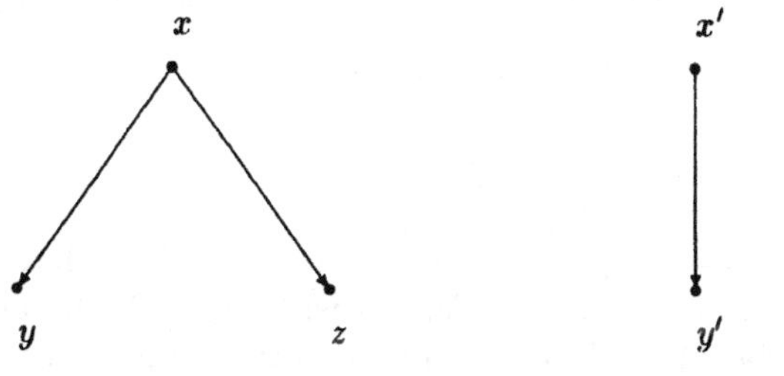


$< A, \alpha >$ $< B, \beta >$

A function $f : A \to B$ with $f(x) = x'$ and $f(y) = f(z) = y'$ is a morphism from $< A, \alpha >$ to $< B, \beta >$, but a function $f' : A \to B$ with $f'(x) = x'$, $f'(y) = x'$ and $f'(z) = y'$ is not a morphism. Finally it is obvious that the category of graphs and graph homomorphisms defined above is identical with the category **Set** $< Id, Id >$ of relational structures. □

*Example 2.* **(Labelled graphs)** Let $\Sigma$ be a set of labels. A $\Sigma$-labelled graph $< A, \alpha >$ is a pair of a set $A$ and a family $\alpha = \{\alpha_\sigma : A \rightharpoonup A \mid \sigma \in \Sigma\}$ of relations. A $\Sigma$-labelled graph homomorphism $f$ from a $\Sigma$-labelled graph $< A, \alpha >$ into a $\Sigma$-labelled graph $< B, \beta >$ is a function $f : A \to B$ such that $f\alpha_\sigma \sqsubseteq \beta_\sigma f$ for any $\sigma \in \Sigma$.

$$\begin{array}{ccc} A & \xrightarrow{f} & B \\ {\scriptstyle \alpha_\sigma}\downarrow & & \downarrow{\scriptstyle \beta_\sigma} \\ A & \xrightarrow[f]{} & B \end{array}$$

A $\Sigma$-copower functor $\Sigma \cdot (-) : \mathbf{Set} \to \mathbf{Set}$ is a functor which assigns to each set $A$ a coproduct(disjoint union) $\Sigma \cdot A = \coprod_{\sigma \in \Sigma} A$ of $\Sigma$ copies of $A$ and to each function $f : A \to B$ a function $\Sigma \cdot f : \Sigma \cdot A \to \Sigma \cdot B$ mapping $\sigma$-th component $A$ into $\sigma$-th component $B$ by $f : A \to B$ for any $\sigma \in \Sigma$. Then it is easy to check that the category of $\Sigma$-labelled graphs and $\Sigma$-labelled graph homomorphisms is equal to the category **Set** $< \Sigma \cdot (-), Id >$ of relational structures. □

*Example 3.* **(Hypergraphs)** Let $\Lambda$ be a set of function symbols with an arity function **arity** $: \Lambda \to \mathbf{N}$, where $\mathbf{N}$ is the set of all positive integers. For a set $A$ we denote by $A^+$ the set of all nonempty tuples of members of $A$ and by $\mathbf{size}_A : A^+ \to \mathbf{N}$ the size function of tuples. A hypergraph $< A, \alpha >$ over $\Lambda$ is a pair of a set $A$ and a relation $\alpha : \Lambda \rightharpoonup A^+$ such that $\mathbf{size}_A \alpha \sqsubseteq \mathbf{arity}$.

$$\begin{array}{ccc} \Lambda & \xrightarrow{\alpha} & A^+ \\ \searrow & & \swarrow \\ \mathbf{arity} & \mathbf{N} & \mathbf{size}_A \end{array}$$

Note that $\alpha$ plays a role of the set of hyperedges and the condition $\mathbf{size}_A \alpha \sqsubseteq \mathbf{arity}$ is equivalent to a condition that $(\omega, x) \in \alpha$ for $\omega \in \Lambda$ and $x \in A^+$ implies $\mathbf{arity}(\omega) = \mathbf{size}_A(x)$. For a function $f : A \to B$ a function $f^+ : A^+ \to B^+$ is defined by a trivial way [14]. Thus we have a functor $(-)^+ : \mathbf{Set} \to \mathbf{Set}$, called nonempty Kleene functor. A homomorphism $f$ from a hypergraph $< A, \alpha >$ into a hypergraph $< B, \beta >$ is a function $f : A \to B$ such that $f^+ \alpha \sqsubseteq \beta$.

$$\begin{array}{ccc} \Lambda & \xrightarrow{\mathrm{id}_\Lambda} & \Lambda \\ {\scriptstyle \alpha}\downarrow & & \downarrow{\scriptstyle \beta} \\ A^+ & \xrightarrow[f^+]{} & B^+ \end{array}$$

Let $\mathbf{const}_\Lambda : \mathbf{Set} \to \mathbf{Set}$ be the constant functor with values $\Lambda$ and $\mathrm{id}_\Lambda$, and $\mathbf{const}_N : \mathbf{Set} \to \mathbf{Set}$ the constant functor with values $N$ and $\mathrm{id}_N$. Then the arity function $\mathbf{arity} : \Lambda \to \mathbf{N}$ induces a natural transformation $\mathbf{arity}\hat{} : \mathbf{const}_\Lambda \to \mathbf{const}_N$. Also it is easy to see that the size function $\mathbf{size}_A : A^+ \to N$ gives a natural transformation $\mathbf{size} : (-)^+ \to \mathbf{const}_N$. At last the category of hypergraphs over $A$ is identical with the category $\mathbf{Set} < \mathbf{arity}\hat{} : \mathbf{const}_A \to \mathbf{const}_N, \mathbf{size} : (-)^+ \to \mathbf{const}_N >$ of relational structures. □

For a set $A$ we denote by $A^*$ the set of all finite strings of $A$ including the empty string $\varepsilon$. For a function $f : A \to B$ a function $f^* : A^* \to B^*$ is defined by $f^*(\varepsilon) = \varepsilon$ and $f^*(aw) = f(a)f^*(w)$ for $a \in A$ and $w \in A^*$. The Kleene functor $(-)^* : \mathbf{Set} \to \mathbf{Set}$ is a functor which assigns to each set $A$ a set $A^*$ and to each function $f : A \to B$ a function $f^* : A^* \to B^*$.

*Example 4.* Let $Id : \mathbf{Set} \to \mathbf{Set}$ be the identity functor of $\mathbf{Set}$ and $(-)^* : \mathbf{Set} \to \mathbf{Set}$ the Kleene functor. A relational structure $< A, \alpha >$ of $\mathbf{Set} < Id, (-)^* >$ such that $\alpha : A \to A^*$ is a function can be regarded as a graph in the sense of Raoult [14]. Let $< B, \beta >$ be another such relational structure and $f :< A, \alpha > \to < B, \beta >$ a morphism of $\mathbf{Set} < Id, (-)^* >$.

$$\begin{array}{ccc} A & \xrightarrow{f} & B \\ {\scriptstyle\alpha}\downarrow & & \downarrow{\scriptstyle\beta} \\ A^* & \xrightarrow[f^*]{} & B^* \end{array}$$

It is trivial that a condition $f^*\alpha \sqsubseteq \beta f$ is the same as a condition $f^*\alpha = \beta f$ for graph morphisms in [14]. Thus the category of graphs defined by Raoult is a subcategory of $\mathbf{Set} < Id, (-)^* >$. □

Throughout the rest of this paper we will assume that a fixed frame $< \omega : S \to R, \lambda : T \to R >$ is given and we will write $\mathcal{C} < \omega, \lambda >$ for $< \omega : S \to R, \lambda : T \to R >$. The forgetful functor $V : \mathcal{C} < \omega, \lambda > \to \mathcal{C}$ is defiend by $V < a, \alpha >= a$ for a relational structure $< a, \alpha >$ of $\mathcal{C} < \omega, \lambda >$ and $Vf = f$ for a morphism $f :< a, \alpha > \to < b, \beta >$ of relational structures. Then the following theorem is concerned with (co)limit creation [11] of the forgetful functor.

**Theorem 2.**

(a) *The forgetful functor* $V : \mathcal{C} < \omega, \lambda > \to \mathcal{C}$ *weakly creates colimits.*
(b) *If* $R : \mathcal{C} \to \mathbf{Set}$ *preserves limits, then* $V : \mathcal{C} < \omega, \lambda > \to \mathcal{C}$ *weakly creates limits.*

Sketch of Proof. Let $\mathcal{X}$ be a small category and a functor $J : \mathcal{X} \to \mathcal{C} < \omega, \lambda >$ map a morphism $u : x \to y$ of $\mathcal{X}$ to a morphism $Ju :< Jx, \alpha_x > \to < Jy, \alpha_y >$.
(a) Let $(a, \tau)$ be a colimit of $VJ : \mathcal{X} \to \mathcal{C}$. Define a relation $\alpha : Sa \to Ta$ by

$$\alpha = \bigsqcup_{x \in Obj(\mathcal{X})} T(\tau_x)\alpha_x S(\tau_x)^\sharp$$

and a natural trasformation $\sigma : J \to \mathbf{const}_{<a,\alpha>}$ by $\sigma_x = \tau_x :< Jx, \alpha_x > \to < a, \alpha >$ for each object $x$ of $\mathcal{X}$. (Remark that we write $S(\tau_x)^\sharp$ for $(S(\tau_x))^\sharp$.) Then it is a routine to see that $(< a, \alpha >, \sigma)$ is a unique colimit of $J$ with $V < a, \alpha >= a$

and $V\tau = \sigma$.
(b) Let $(a, \tau)$ be a limit of $VJ : \mathcal{X} \to \mathcal{C}$. Define a relation $\alpha : Sa \rightharpoonup Ta$ by

$$\alpha = \sqcap_{x \in Obj(\mathcal{X})} T(\tau_x)^\sharp \alpha_x S(\tau_x)$$

and a natural trasformation $\sigma$ : $\mathbf{const}_{<a,\alpha>} \to J$ by $\sigma_x = \tau_x :< a, \alpha > \to < Jx, \alpha_x >$ for each object $x$ of $\mathcal{X}$. Note that $\lambda_a \alpha \sqsubseteq \omega_a$ follows from

$$\sqcap_{x \in Obj(\mathcal{X})} T(\tau_x)^\sharp T(\tau_x) = \mathrm{id}_{Ra}$$

since $T$ transforms $(a, \tau)$ to a limit $(Ta, T\tau)$ of $TVJ$. Then it is also easy to check that $(< a, \alpha >, \sigma)$ is a unique limit of $J$ with $V < a, \alpha >= a$ and $V\tau = \sigma$.
□

The following is an immediate corollary of the last theorem and Theorem 17 and 18 in the Appendix.

**Corollary 3.**

(a) *If $\mathcal{C}$ has colimits, then $\mathcal{C} < \omega, \lambda >$ has colimits.*
(b) *If $\mathcal{C}$ has limits and $R : \mathcal{C} \to$ **Set** preserves limits, then $\mathcal{C} < \omega, \lambda >$ has limits.*
□

## 3 Partial Morphisms

In this section we introduce a notion of partial morphisms of relational structures to discuss single-pushout rewritings of relational structures and show an existence theorem (Cf. Theorem 9) of pushouts, which enables us to more generally formalize rewritings of relational structures.

**Definition 4.** A class $\mathcal{M}$ of monomorphisms of a category $\mathcal{C}$ is said to be *admissible* [15, 9] if it satisfies the following conditions:

(a) $\mathcal{M}$ contains all isomorphisms of $\mathcal{C}$,
(b) $\mathcal{M}$ is closed under composition,
(c) For a morphism $f : d \to b$ and a monomorphism $n : e \to b$ in $\mathcal{M}$ there is a pullback

$$\begin{array}{ccc} p & \xrightarrow{g'} & e \\ {\scriptstyle n'}\downarrow & (*) & \downarrow{\scriptstyle n} \\ d & \xrightarrow[f]{} & b \end{array}$$

such that $n' : p \to d$ is in $\mathcal{M}$. □

It is clear that the class $Mon(\mathcal{C})$ of all monomorphisms of $\mathcal{C}$ is admissible. For example, the class $Mon(\mathbf{Set})$ of all monomorphisms of the category **Set** is admissible. (Note that a monomorphism of **Set** is an injective function.)

Let $\mathcal{M}$ be an admissible class of monomorphisms of $\mathcal{C}$. A *span* $(m, f)$ from $a$ into $b$ in $\mathcal{C}$ (with respect to $\mathcal{M}$), denoted by $(m, f) : a \to b$, is a pair of a monomorphism $m : d \to a$ in $\mathcal{M}$ and a morphism $f : d \to b$ of $\mathcal{C}$. A span $(m, f) : a \to b$ is equivalent to a span $(m', f') : a \to b$ if there is an isomorphism $u$ with $fu = f'$ and $mu = m'$. Trivially this relation is an equivalence relation on the collection of all spans from $a$ into $b$. A *partial morphism* from $a$ into $b$ in $\mathcal{C}$ (with respect to $\mathcal{M}$) is an equivalence class $[m, f]$ of a span $(m, f) : a \to b$ by the equivalence relation. For two partial morphisms $[m, f] : a \to b$ and $[n, g] : b \to c$ a pullback $(*)$ as above can be constructed (since $\mathcal{M}$ is admissible) and the composite of $[m, f]$ followed by $[n, g]$ is defined to be a partial morphism $[mn', gf'] : a \to c$. It is easy to show the well-defindness and the associativity of the composition of partial morphisms. A morphism $f : a \to b$ of $\mathcal{C}$ is identified with a partial morphism $[\mathrm{id}_a, f]$.

Thus all objects of $\mathcal{C}$ and all partial morphsims in $\mathcal{C}$ with respect to $\mathcal{M}$ form a category $P(\mathcal{C}, \mathcal{M})$ [9, 15], called the category of partial morphsims in $\mathcal{C}$ with respect to $\mathcal{M}$. It is obvious that the category $P(\mathbf{Set}, Mon(\mathbf{Set}))$ is identical with the category **Pfn** of sets and (ordinary) partial functions.

**Definition 5.** A functor $S : \mathcal{C} \to \mathbf{Set}$ is *admissible* with respect to $\mathcal{M}$ if it transforms monomorphisms of $\mathcal{M}$ into injective functions and preserves inverse images of monomorphisms of $\mathcal{M}$ (namely, it transforms any pullback $(*)$ as above with $n : e \to b$ in $\mathcal{M}$ into a pullback in $\mathbf{Set}$). $\square$

Notice that every functor $S : \mathcal{C} \to \mathbf{Set}$ preverving pullbacks is admissible with respect to any admissible class of monomorphisms. For a morphism $f : a \to b$ of $\mathcal{C}$ is a monomorphism if and only if the following square is a pullback in $\mathcal{C}$.

$$\begin{array}{ccc} a & \xrightarrow{\mathrm{id}_a} & a \\ {\scriptstyle \mathrm{id}_a}\downarrow & & \downarrow{\scriptstyle f} \\ a & \xrightarrow[f]{} & b \end{array}$$

It is easily seen that if both squares

$$\begin{array}{ccc} A & \xrightarrow{f} & B \\ {\scriptstyle g}\downarrow & & \downarrow{\scriptstyle h} \\ C & \xrightarrow[k]{} & D \end{array} \qquad \begin{array}{ccc} A' & \xrightarrow{f'} & B' \\ {\scriptstyle g'}\downarrow & & \downarrow{\scriptstyle h'} \\ C' & \xrightarrow[k']{} & D' \end{array}$$

are pullbacks in **Set**, then both squares

$$\begin{array}{ccc} A \times A' & \xrightarrow{f \times f'} & B \times B \\ {\scriptstyle g \times g'}\downarrow & & \downarrow{\scriptstyle h \times h'} \\ C \times C' & \xrightarrow[k \times k']{} & D \times D' \end{array} \qquad \begin{array}{ccc} A + A' & \xrightarrow{f + f'} & B + B' \\ {\scriptstyle g + g'}\downarrow & & \downarrow{\scriptstyle h + h'} \\ C + C' & \xrightarrow[k + k']{} & D + D' \end{array}$$

are also pullbacks in **Set**. Therefore both products and copruducts of admissible functors are also admissible. Trivially the identity functor $Id : \mathbf{Set} \to \mathbf{Set}$, a copower functor $\Sigma \cdot (-) : \mathbf{Set} \to \mathbf{Set}$, (nonempty) Kleene functor $(-)^+ : \mathbf{Set} \to \mathbf{Set}$, and a constant functor $\mathbf{const}_A : \mathbf{Set} \to \mathbf{Set}$ are admissible with respect to $Mon(\mathbf{Set})$.

**Lemma 6.** *Let $\mathcal{M}$ be an admissible class of monomorphisms of a category $\mathcal{C}$.*

(a) *An admissible functor $S : \mathcal{C} \to \mathbf{Set}$ with respect to $\mathcal{M}$ can be extended to a functor $S^{\cdot} : P(\mathcal{C}, M) \to \mathbf{Pfn}$.*
(b) *If $S, T : \mathcal{C} \to \mathbf{Set}$ are admissible functors with respect to $\mathcal{M}$ and $\omega : S \to T$ is a natural transformation, then*

$$\omega_b S^{\cdot}(f) \sqsubseteq T^{\cdot}(f)\omega_a \text{ and } \omega_b S^{\cdot}(f) S^{\cdot}(f)^{\sharp} = T^{\cdot}(f)\omega_a S^{\cdot}(f)^{\sharp}$$

*hold for every partial morphism $f : a \to b$ in $P(\mathcal{C}, M)$.*

Proof. (a) For a partial morphism $f = [m, f'] : a \to b$ with a morphism $f' : d \to b$ of $\mathcal{C}$ and a monomorphism $m : d \to a$ in $\mathcal{M}$ we define a partial function $S^{\cdot}(f) = S(f')S(m)^{\sharp} : Sa \to Sb$, where $S(m)^{\sharp}$ is the inverse relation of $S(m)$. (Note that $S(m)^{\sharp}$ is in fact a partial function defined on the image of $S(m)$ since $S(m)$ is injective). It is clear that $S^{\cdot}(f) = S(f)$ for a morphism $f$ of $\mathcal{C}$. (Recall that $[\mathrm{id}_a, f] = f$.) From the admissiblity of $S : \mathcal{C} \to \mathbf{Set}$ it immediately follows that $S^{\cdot}$ preverves the composition, that is, $S^{\cdot}(gf) = S^{\cdot}(g)S^{\cdot}(f)$ for partial morphisms $f : a \to b$ and $g : b \to c$.
(b) Let $f = [m, f'] : a \to b$ be a partial morphism with a morphism $f' : d \to b$ of $\mathcal{C}$ and a monomorphism $m : d \to a$ in $\mathcal{M}$. Then

$$\begin{aligned}
\omega_b S^{\cdot}(f) &= \omega_b S(f')S(m)^{\sharp} \\
&= T(f')\omega_d S(m)^{\sharp} \\
&= T(f')T(m)^{\sharp}T(m)\omega_d S(m)^{\sharp} \\
&= T(f')T(m)^{\sharp}\omega_a S(m)S(m)^{\sharp} \\
&\sqsubseteq T(f')T(m)^{\sharp}\omega_a \\
&= T^{\cdot}(f)\omega_a .
\end{aligned}$$

The latter equality follows similarly. □

In what follows we assume that there is given a frame $< \omega : S \to R, \lambda : T \to R >$ in which functors $S, T, R : \mathcal{C} \to \mathbf{Set}$ are admissible with respect to an admissible class $\mathcal{M}$ of monomorphisms of $\mathcal{C}$.

**Definition 7.** Let $< a, \alpha >$ and $< b, \beta >$ be relational structures over $< \omega, \lambda >$. A partial morphism $f^{\cdot} :< a, \alpha > \to < b, \beta >$ of relational structures (with respect to $\mathcal{M}$) is a partial morphism $f : a \to b$ of $P(\mathcal{C}, M)$ such that $T^{\cdot}(f)\alpha \mathrm{d}(S^{\cdot}(f)) \sqsubseteq \beta S^{\cdot}(f)$. □

Let $f :< a, \alpha > \to < b, \beta >$ and $g :< b, \beta > \to < c, \gamma >$ be partial morphisms of structures over $< \omega, \lambda >$. Since $T^{\cdot}(f)\alpha\mathrm{d}(S^{\cdot}(f)) \sqsubseteq \beta S^{\cdot}(f)$ and $T^{\cdot}(g)\beta\mathrm{d}(S^{\cdot}(g)) \sqsubseteq \gamma S^{\cdot}(g)$, we have

$$\begin{aligned}
T^{\cdot}(gf)\alpha\mathrm{d}(S^{\cdot}(gf)) &= T^{\cdot}(g)T^{\cdot}(f)\alpha\mathrm{d}(S^{\cdot}(g)S^{\cdot}(f)) \\
&= T^{\cdot}(g)T^{\cdot}(f)\alpha\mathrm{d}(S^{\cdot}(f))\mathrm{d}(S^{\cdot}(g)S^{\cdot}(f)) \\
&\sqsubseteq T^{\cdot}(g)\beta S^{\cdot}(f)\mathrm{d}(S^{\cdot}(g)S^{\cdot}(f)) \\
&= T^{\cdot}(g)\beta\mathrm{d}(S^{\cdot}(g))S^{\cdot}(f) \\
&\sqsubseteq \gamma S^{\cdot}(g)S^{\cdot}(f) \\
&= \gamma S^{\cdot}(gf).
\end{aligned}$$

Hence the composite of partial morphisms is also a partial morphism, so relational structures and partial morphisms over $<\omega,\lambda>$ form a category $P(\mathcal{C},\mathcal{M})<\omega,\lambda>$, called the category of relational structures and partial morphisms (with respect to $\mathcal{M}$). When $\mathcal{C}=\mathbf{Set}$ and $\mathcal{M}=Mon(\mathbf{Set})$ we have $\mathbf{Pfn}=P(\mathbf{Set},Mon(\mathbf{Set}))$, so $P(\mathbf{Set},Mon(\mathbf{Set}))<\omega,\lambda>$ will be written $\mathbf{Pfn}<\omega,\lambda>$.

**Lemma 8.** *Let $x:B\to X$ and $y:C\to X$ be partial functions between sets and a square*

$$\begin{array}{ccc} A & \xrightarrow{f} & B \\ {\scriptstyle g}\downarrow & & \downarrow{\scriptstyle h} \\ C & \xrightarrow[k]{} & D \end{array}$$

*a pushout in* $\mathbf{Pfn}$. *If a (total) function $z:D\to X$ satisfies $zh\sqsubseteq x$ and $zk\sqsubseteq y$, then $z=xh^\sharp\sqcup yk^\sharp$.*

Proof. First we show that $z'=xh^\sharp\sqcup yk^\sharp$ is a partial function, that is, $z'z'^\sharp\sqsubseteq \mathrm{id}_X$. But $h^\sharp h\sqsubseteq h^\sharp z^\sharp zh\sqsubseteq x^\sharp x$ because of the totality $\mathrm{id}_D\sqsubseteq z^\sharp z$ of $z$, and similarly $k^\sharp k\sqsubseteq y^\sharp y$ and $k^\sharp h\sqsubseteq y^\sharp x$. Hence we have

$$\begin{aligned} z'z'^\sharp &= xh^\sharp hx^\sharp\sqcup yk^\sharp hx^\sharp\sqcup xh^\sharp ky^\sharp\sqcup yk^\sharp ky^\sharp \\ &\sqsubseteq xx^\sharp xx^\sharp\sqcup yy^\sharp xx^\sharp\sqcup xx^\sharp yy^\sharp\sqcup yy^\sharp yy^\sharp \\ &\sqsubseteq \mathrm{id}_X \quad (\text{ by } xx^\sharp\sqsubseteq\mathrm{id}_X \text{ and } yy^\sharp\sqsubseteq\mathrm{id}_X\ ) \end{aligned}$$

On the other hand $hh^\sharp\sqcup kk^\sharp=\mathrm{id}_D$ since the above square is a pushout in $\mathbf{Pfn}$, so $z=zhh^\sharp\sqcup zkk^\sharp\sqsubseteq xh^\sharp\sqcup yk^\sharp=z'$. Therefore $z'\sqsubseteq z'z^\sharp z\sqsubseteq z'z'^\sharp z\sqsubseteq z$, which proves $z'=z$. □

**Theorem 9.** *Assume that the extension $S^\cdot:P(\mathcal{C},\mathcal{M})\to\mathbf{Pfn}$ of $S$ preserves pushouts. If $<b,\beta>$ and $<c,\gamma>$ are relational structures over $<\omega,\lambda>$ and if a square*

$$\begin{array}{ccc} a & \xrightarrow{f} & b \\ {\scriptstyle g}\downarrow & (1) & \downarrow{\scriptstyle h} \\ c & \xrightarrow[k]{} & d \end{array}$$

*is a pushout in $P(\mathcal{C},\mathcal{M})$, then $h:<b,\beta>\to<d,\delta>$ and $k:<c,\gamma>\to<d,\delta>$ are partial morphisms of relational structures, where $\delta=T^\cdot(h)\beta S^\cdot(h)^\sharp\sqcup T^\cdot(k)\gamma S^\cdot(k)^\sharp$. Moreover, if $h':<b,\beta>\to<d',\delta'>$ and $k':<c,\gamma>\to<d',\delta'>$ are partial morphisms of relational structures satisfying $h'f=k'g$, then there exists a unique partial morphism $t:<d,\delta>\to<d',\delta'>$ of relational structures such that $h'=th$ and $k'=tk$.*

Proof. First we must see that $<d,\delta>$ is a relational structure, that is, $\lambda_d\delta\sqsubseteq\omega_d$. A square

$$\begin{array}{ccc} Sa & \xrightarrow{S^\cdot(f)} & Sb \\ {\scriptstyle S^\cdot(g)}\downarrow & & \downarrow{\scriptstyle S^\cdot(h)} \\ Sc & \xrightarrow[S^\cdot(k)]{} & Sd \end{array}$$

is a pushout in **Pfn** by the assumption, and $\omega_d S^{\cdot}(h) \sqsubseteq R^{\cdot}(h)\omega_b$ and $\omega_d S^{\cdot}(k) \sqsubseteq R^{\cdot}(k)\omega_c$ follow from Lemma 6(b). Remark that $\omega_d$ is a (total) function. Hence $\omega_d = R^{\cdot}(h)\omega_b S^{\cdot}(h)^\sharp \sqcup R^{\cdot}(k)\omega_c S^{\cdot}(k)^\sharp$ by the last lemma and

$$\begin{aligned}\lambda_d\delta &= \lambda_d T^{\cdot}(h)\beta S^{\cdot}(h)^\sharp \sqcup \lambda_d T^{\cdot}(k)\gamma S^{\cdot}(k)^\sharp \\ &\sqsubseteq R^{\cdot}(h)\lambda_b\beta S^{\cdot}(h)^\sharp \sqcup R^{\cdot}(k)\lambda_c\gamma S^{\cdot}(k)^\sharp \\ &\sqsubseteq R^{\cdot}(h)\omega_b S^{\cdot}(h)^\sharp \sqcup R^{\cdot}(k)\omega_c S^{\cdot}(k)^\sharp \\ &= \omega_d.\end{aligned}$$

Next we see that $h :< b,\beta > \rightarrow < d,\delta >$ and $k :< b,\beta > \rightarrow < d,\delta >$ are partial morphisms of relational structures. It simply follows from

$$\begin{aligned}T^{\cdot}(h)\beta \mathrm{d}(S^{\cdot}(h)) &\sqsubseteq T^{\cdot}(h)\beta S^{\cdot}(h)^\sharp S^{\cdot}(h) \quad (\mathrm{d}(S^{\cdot}(h)) = S^{\cdot}(h)^\sharp S^{\cdot}(h) \sqcap \mathrm{id}_{Sb}) \\ &\sqsubseteq \delta S^{\cdot}(h) \quad (\delta = T^{\cdot}(h)\beta S^{\cdot}(h)^\sharp \sqcup T^{\cdot}(k)\gamma S^{\cdot}(k)^\sharp).\end{aligned}$$

Finally assume that $h' :< b,\beta > \rightarrow < d',\delta' >$ and $k' :< c,\gamma > \rightarrow < d',\delta' >$ are partial morphisms satisfying $h'f = k'g$. Then we have $T^{\cdot}(h')\beta \mathrm{d}(S^{\cdot}(h')) \sqsubseteq \delta' S^{\cdot}(h')$ and $T^{\cdot}(k')\gamma \mathrm{d}(S^{\cdot}(k')) \sqsubseteq \delta' S^{\cdot}(k')$. As (1) is a pushout in $P(\mathcal{C},\mathcal{M})$, there exists a unique partial function $t : d \rightarrow d'$ such that $h' = th$ and $k' = tk$. It suffices to prove that $T^{\cdot}(t)\delta \mathrm{d}(S^{\cdot}(t)) \sqsubseteq \delta' S^{\cdot}(t)$. But it follows from

$$\begin{aligned}&T^{\cdot}(t)\delta \mathrm{d}(S^{\cdot}(t)) \\ &\sqsubseteq T^{\cdot}(t)[T^{\cdot}(h)\beta S^{\cdot}(h)^\sharp \sqcup T^{\cdot}(k)^\sharp \gamma S^{\cdot}(k)^\sharp] S^{\cdot}(t)^\sharp S^{\cdot}(t) \quad (\mathrm{d}(S^{\cdot}(t)) \sqsubseteq S^{\cdot}(t)^\sharp S^{\cdot}(t)) \\ &= [T^{\cdot}(t)T^{\cdot}(h)\beta S^{\cdot}(h)^\sharp S^{\cdot}(t)^\sharp \sqcup T^{\cdot}(t)T^{\cdot}(k)\gamma S^{\cdot}(k)^\sharp S^{\cdot}(t)^\sharp] S^{\cdot}(t) \\ &= [T^{\cdot}(h')\beta S^{\cdot}(h')^\sharp \sqcup T^{\cdot}(k')\gamma S^{\cdot}(k')^\sharp] S^{\cdot}(t) \\ &= [T^{\cdot}(h')\beta \mathrm{d}(S^{\cdot}(h')) S^{\cdot}(h')^\sharp \sqcup T^{\cdot}(k')\gamma \mathrm{d}(S^{\cdot}(k')) S^{\cdot}(k')^\sharp] S^{\cdot}(t) \\ &\sqsubseteq [\delta' S^{\cdot}(h') S^{\cdot}(h')^\sharp \sqcup \delta' S^{\cdot}(k') S^{\cdot}(k')^\sharp] S^{\cdot}(t) \\ &\sqsubseteq [\delta' \sqcup \delta'] S^{\cdot}(t) \\ &= \delta' S^{\cdot}(t).\end{aligned}$$

This completes the proof. □

Note that a relational structure $< d,\delta >$ in the above theorem is unique up to isomorphisms. The following is exactly a corollary of the last theorem.

**Corollary 10.** *If the extended functor* $S^{\cdot} : P(\mathcal{C},\mathcal{M}) \rightarrow$ **Pfn** *of* $S$ *preserves pushouts, then the category* $P(\mathcal{C},\mathcal{M}) < \omega,\lambda >$ *of relational structures and partial morphisms has pushouts.* □

The identity functor $Id$ : **Pfn** → **Pfn**, a copower functor $\Sigma\cdot(-)$ : **Pfn** → **Pfn**, and a constant functor $\mathbf{const}_A$ : **Pfn** → **Pfn** preserve pushouts. Hence we obtain the following

**Corollary 11.** *The categories* **Pfn** $< Id, Id >$, **Pfn** $< \Sigma\cdot(-), Id >$ *and* **Pfn** $<$ $\widehat{\mathbf{arity}}, \mathbf{size} >$ *have pushouts.* □

Now we will formalize the single pushout rewritings of relational structures defined in this paper.

A *production rule* $p$ is a triple of two relational structures $< a, \alpha >$, $< b, \beta >$ and a partial morphism $f : a \to b$. (Note that $f$ need not to be a partial morphism of relational structures.) A *matching* to $p$ is a morphism $g :< a, \alpha > \to < x, \xi >$ of relational structures. If a square

$$\begin{array}{ccc} a & \xrightarrow{f} & b \\ {\scriptstyle g}\downarrow & & \downarrow{\scriptstyle h} \\ x & \xrightarrow[k]{} & y \end{array}$$

is a pushout in $P(\mathcal{C}, \mathcal{M})$, we define $\eta = T^{\cdot}(h)\beta S^{\cdot}(h)^{\sharp} \sqcup T^{\cdot}(k)[\xi - T(g)\alpha S(g)^{\sharp}]S^{\cdot}(k)^{\sharp}$. (Notations $\sqcup$ and $-$ denote the set union and the set subtraction, respectively.) The graph $< y, \eta >$ is the reduced graph after applying a production rule $p$ along a matching $g$, and denoted by $< x, \xi > \Rightarrow_{p/g} < y, \eta >$. A square

$$\begin{array}{ccc} < a, \alpha > & \xrightarrow{f} & < b, \beta > \\ {\scriptstyle g}\downarrow & & \downarrow{\scriptstyle h} \\ < x, \xi > & \xrightarrow[k]{} & < y, \eta > \end{array}$$

is called the rewriting square for a production rule $p$ along a matching $g$. (Note that rewriting squares are not necessarily pushouts in $P(\mathcal{C}, \mathcal{M}) < \omega, \lambda >$.)

To indicate that the above definition of rewritings of relational structures is valid, we will show that $< y, \eta >$ is in fact a relational structure (that is, $\lambda_y \eta \sqsubseteq \omega_y$):

$$\begin{aligned} \lambda_y \eta &= \lambda_y T^{\cdot}(h)\beta S^{\cdot}(h)^{\sharp} \sqcup \lambda_y T^{\cdot}(k)[\xi - T(g)\alpha S(g)^{\sharp}]S^{\cdot}(k)^{\sharp} \\ &\sqsubseteq R^{\cdot}(h)\lambda_b \beta S^{\cdot}(h)^{\sharp} \sqcup R^{\cdot}(k)\lambda_x[\xi - T(g)\alpha S(g)^{\sharp}]S^{\cdot}(k)^{\sharp} \quad (6(b)) \\ &\sqsubseteq R^{\cdot}(h)\omega_b S^{\cdot}(h)^{\sharp} \sqcup R^{\cdot}(k)\omega_x S^{\cdot}(k)^{\sharp} \quad (\lambda_b\beta \sqsubseteq \omega_b,\ \lambda_x\xi \sqsubseteq \omega_x) \\ &= \omega_y S^{\cdot}(h)S^{\cdot}(h)^{\sharp} \sqcup \omega_y S^{\cdot}(k)S^{\cdot}(k)^{\sharp} \quad (6(b)) \\ &\sqsubseteq \omega_y \sqcup \omega_y \\ &= \omega_y \end{aligned}$$

In the above computation of relations the naturality of $\lambda$ and $\omega$ and the univalency (namely, $S(h)S(h)^{\sharp} \sqsubseteq \mathrm{id}_{Sy}$ and $S(k)S(k)^{\sharp} \sqsubseteq \mathrm{id}_{Sy}$) of functions have been applied.

With this framework of rewritings of relational structures we generalise the main results in [13], which will be stated below without proof (Cf. Theorem 3.6 and 3.10 in [13]):

**Theorem 12.** *Let* $p_i = (< a_i, \alpha_i >, < b_i, \beta_i >, f_i : a_i \to b_i)$ *be a production rule,* $g_i :< a_i, \alpha_i > \to < x, \xi >$ *a matching to* $p_i$ *and* $< x, \xi > \Rightarrow_{p_i/g_i} < y_i, \eta_i >$ *for* $i = 0, 1$. *If* $f_i :< a_i, \alpha_i > \to < b_i, \beta_i >$ *is a partial morphism of graphs* $(i = 0, 1)$ *and* $\mathrm{d}(g_0^{\sharp}) \sqcap \mathrm{d}(g_1^{\sharp}) \sqsubseteq \mathrm{d}(k_0) \sqcap \mathrm{d}(k_1)$, *then there exist matchings* $g_i' :< a, \alpha_i > \to < y_{1-i}, \eta_{1-i} >$ $(i = 0, 1)$ *and a graph* $< y, \eta >$ *such that* $< y_{1-i}, \eta_{1-i} > \Rightarrow_{p_i/g_i'} < y, \eta >$ $(i = 0, 1)$ $\square$.

**Theorem 13.** *A rewriting system of relational structures is confluent if and only if every critical pair in it is confluent.*

## 4 A Comparison Theorem

This section is devoted to state that the category $P(\mathcal{C}, \mathcal{M}) < \omega, \lambda >$ of relational structures together with partial morphisms is isomorphic to a category of partial morphisms in the sense of [9, 15].

**Proposition 14.** *A morphism $m :< d, \delta > \rightarrow < a, \alpha >$ in $\mathcal{C} < \omega, \lambda >$ is a monomorphism if and only if $m : d \rightarrow a$ is a monomorphism in $\mathcal{C}$.*

Proof. It is immediate that $m :< d, \delta > \rightarrow < a, \alpha >$ is a monomorphism if $m : d \rightarrow a$ is a monomorphism. We have to show its converse. Assume that $m :< d, \delta > \rightarrow < a, \alpha >$ is a monomorphism and two morphisms $x, y : c \rightarrow d$ in $\mathcal{C}$ satisfies $mx = my$. Define $\gamma : Sc \rightarrow Tc$ to be the empty relation. Then $x, y :< c, \gamma > \rightarrow < d, \delta >$ are morphisms of relational structures and so $x = y$ since $m :< d, \delta > \rightarrow < a, \alpha >$ is a monomorphism. □

A monomorphism $m :< d, \delta > \rightarrow < a, \alpha >$ in $\mathcal{C} < \omega, \lambda >$ is called *full* with respect to $\mathcal{M}$ if $m : d \rightarrow a$ is in $\mathcal{M}$ and $\delta = T(m)^\sharp \alpha S(m)$.

**Lemma 15.** *The class $F(\mathcal{M})$ of all full monomorphisms in $\mathcal{C} < \omega, \lambda >$ with respect to $\mathcal{M}$ is admissible.*

Proof. Trivially $F(\mathcal{M})$ contains all isomorphisms of $\mathcal{C} < \omega, \lambda >$ and is closed under compsition. It suffices only to show that for a morphism $h :< b, \beta > \rightarrow < a, \alpha >$ of $\mathcal{C} < \omega, \lambda >$ and a full monomorphism $m :< d, \delta > \rightarrow < a, \alpha >$ in $F(\mathcal{M})$ there is a pullback

$$\begin{array}{ccc} < e, \varepsilon > & \xrightarrow{n} & < b, \beta > \\ {\scriptstyle k}\downarrow & & \downarrow{\scriptstyle h} \\ < d, \delta > & \xrightarrow[m]{} & < a, \alpha > \end{array}$$

with $n :< e, \varepsilon > \rightarrow < b, \beta >$ in $F(\mathcal{M})$. As $\mathcal{M}$ is admissible there is a pullback

$$\begin{array}{ccc} e & \xrightarrow{n} & b \\ {\scriptstyle k}\downarrow & (**) & \downarrow{\scriptstyle h} \\ d & \xrightarrow[m]{} & a \end{array}$$

with $n : e \rightarrow b$ in $\mathcal{M}$. Define $\varepsilon = T(n)^\sharp \beta S(n)$. Then it is a routine work to check that the square (**) is a pullback in $\mathcal{C} < \omega, \lambda >$. (Recall that $R(n)^\sharp R(n) = \mathrm{id}_{Re}$ and $T(k)T(n)^\sharp = T(m)^\sharp T(h)$ because $R, T : \mathcal{C} \rightarrow \mathbf{Set}$ are admissible w.r.t. $\mathcal{M}$.) By the definition $n :< e, \varepsilon > \rightarrow < b, \beta >$ is clearly in $F(\mathcal{M})$. This complets the proof. □

**Theorem 16.** *The category $P(\mathcal{C},\mathcal{M}) < \omega,\lambda >$ of relational structures together with partial morphisms between them is isomorphic to the category $P(\mathcal{C} < \omega,\lambda >, F(\mathcal{M}))$ of partial morphisms in $\mathcal{C} < \omega,\lambda >$ with respect to the class $F(\mathcal{M})$ of all full monomorphisms, that is,*

$$P(\mathcal{C},\mathcal{M}) < \omega,\lambda > \cong P(\mathcal{C} < \omega,\lambda >, F(\mathcal{M})).$$

Proof. Fisrt we construct a functor $\iota : P(\mathcal{C},\mathcal{M}) < \omega,\lambda > \to P(\mathcal{C} < \omega,\lambda >, F(\mathcal{M}))$. Let $f :< a,\alpha > \to < b,\beta >$ be a morphism of $P(\mathcal{C},\mathcal{M}) < \omega,\lambda >$. By the definition a partial morphism $f : a \to b$ in $\mathcal{C}$ is represented as $f = [m, f']$, where $m : d \to a$ is a monomorphism in $\mathcal{M}$ and $f' : d \to b$ is a morphism of $\mathcal{C}$. Define a relation $\delta : Sd \to Td$ by $\delta = T(m)^\sharp \alpha S(m)$. Then $m :< d,\delta > \to < a,\alpha >$ is clearly in $F(\mathcal{M})$. Also, since $\mathrm{d}(S^{\cdot}(f)) = S(m)S(m)^\sharp$, $S(m)^\sharp S(m) = \mathrm{id}_{Sd}$ and $T^{\cdot}(f)\alpha\mathrm{d}(S^{\cdot}(f)) \sqsubseteq \beta S(f^{\cdot})$, we obtain $T(f')\delta \sqsubseteq \beta S^{\cdot}(f')$. Thus a functor $\iota : Pfn(\mathcal{C}) < \omega,\lambda > \to Pfn(\mathcal{C} < \omega,\lambda >, F(\mathcal{M}))$ is defined by $\iota[m : d \to a, f' : d \to b] = [m :< d,\delta > \to < a,\alpha >, f' :< d,\delta > \to < b,\beta >]$. Obviously $\iota$ is an embedding. To prove that $\iota$ is an isomorphism of categories it suffices to see that if a morphism of $P(\mathcal{C} < \omega,\lambda >, F(\mathcal{M}))$ is a morphism of $P(\mathcal{C},\mathcal{M}) < \omega,\lambda >$. Now assume that $f = [m :< d,\delta > \to < a,\alpha >, f' :< d,\delta > \to < b,\beta >]$ is a morphism of $P(\mathcal{C} < \omega,\lambda >, F(\mathcal{M}))$. Then we have

$$\begin{aligned} T^{\cdot}(f)\alpha\mathrm{d}(S^{\cdot}(f)) &= T(f')T(m)^\sharp \alpha S(m)S(m)^\sharp \\ &= T(f')\delta S(m)^\sharp \\ &\sqsubseteq \beta S(f')S(m)^\sharp \\ &= \beta S^{\cdot}(f), \end{aligned}$$

which completes the proof. □

## References


1. V. Claus, H. Ehrig and G. Rozenberg (Eds.), *Graph-Grammars and Their Application to Còmputer Science and Biology*, Lecture Notes in Computer Science **73**(1979).
2. H. Ehrig, H.-J. Kreowski, A. Maggiolo-Schettini, B. K. Rosen and J. Winkowski, *Transformations of structures: an algebraic approach*, Math. Systems Theory **14**(1981), 305–334.
3. H. Ehrig, H.-J. Kreowski and G. Rozenberg (Eds.), *Graph-Grammars and Their Application to Computer Science*, Lecture Notes in Computer Science **153**(1982).
4. H. Ehrig, H.-J. Kreowski and G. Rozenberg (Eds.), *Graph-Grammars and Their Application to Computer Science*, Lecture Notes in Computer Science **532**(1991).
5. H. Ehrig, M. Nagl, G. Rozenberg and A. Rosenfeld (Eds.), *Graph-Grammars and Their Application to Computer Science*, Lecture Notes in Computer Science **291**(1987).
6. H. Ehrig, H. Pfender and H.J. Schneider, *Graph grammars: an algebraic approach*, Proc. 14th Ann. Conf. on Switching and Automata Theory (1973), 167–180.
7. Y. Kawahara, *Pushout-complements and basic concepts of grammars in toposes*, Theoret. Comput. Sci. **77**(1990), 267–289.
8. Y. Kawahara and Y. Mizoguchi, *Categorical assertion semantics in topoi*, Advances in Software Science and Technology, **4**(1992), 137–150.

9. R. Kennaway, *Graph rewriting in some categories of partial morphisms*, Lecture Notes in Computer Science **532**(1991), 490–504.
10. M. Löwe and H. Ehrig, *Algebraic approach to graph transformation based on single pushout derivations*, Lecture Notes in Computer Science **484**(1991), 338–353.
11. S. Mac Lane, *Categories for the Working Mathematician*, (Springer-Verlag, 1972).
12. Y. Mizoguchi, *A graph structure over the category of sets and partial functions*, to appear in Cahiers de topologie et géométorie différentielle catégoriques.
13. Y. Mizoguchi and Y. Kawahara, *Relational graph rewritings*, to appear in TCS.
14. J. C. Raoult, *On graph rewritings*, Theoret. Comput. Sci. **32**(1984), 1–24.
15. E. Robinson and G. Rosolini, *Categories of partial maps*, Inf. Comp. **79**(1988), 95–130.
16. P.M. van den Broek, *Algebraic graph rewtiting using a single pushouts*, Lecture Notes in Computer Science **493**(1991), 90–102.

# 5 Appendix

In this appendix we recall some basic notions of functors, natural transformations, and (co)limits creation on category theory [11].

A *category* $\mathcal{C}$ consists of a family $Obj(\mathcal{C})$ of objects and sets $\mathcal{C}(a,b)$ of morphisms from $a$ into $b$ for each pair $(a,b)$ of objects, as follows:

(a) a morphism $gf$(or, $g\cdot f$) $\in \mathcal{C}(a,c)$, the composite of $f$ followed by $g$, is defined for morphisms $f\in\mathcal{C}(a,b)$ and $g\in\mathcal{C}(b,c)$,
(b) the associativity $h(gf)=(hg)f$ holds for morphisms $f\in\mathcal{C}(a,b)$, $g\in\mathcal{C}(b,c)$ and $h\in\mathcal{C}(c,d)$,
(c) for each object $a$ of $\mathcal{C}$ there is a morphism $\mathrm{id}_a\in\mathcal{C}(a,a)$, the identity morphism of $a$, such that $f\mathrm{id}_a=\mathrm{id}_b f=f$ for $f\in\mathcal{C}(a,b)$.

A morphism $f\in\mathcal{C}(a,b)$ is also denoted by $f:a\to b$, and $a$ and $b$ are called the domain and the codomain of $f$, respectively.

There are many examples of categories [11], for example, the category **Set** of sets and (total) functions, the category **Pfn** of sets and partial functions, and the category **Rel** of sets and (binary) relations.

Let $\mathcal{C}$ and $\mathcal{D}$ be categories. A *functor* $F:\mathcal{C}\to\mathcal{D}$ consists of an object function $F$, which assigns to each object $a$ of $\mathcal{C}$ an object $Fa$ of $\mathcal{D}$ and a morphism function (also written $F$), which assigns to each morphism $f:a\to b$ of $\mathcal{C}$ a morphism $F(f):Fa\to Fb$ of $\mathcal{D}$ in such a way that $F(\mathrm{id}_a)=\mathrm{id}_{Fa}$ for each object $a$ of $\mathcal{C}$ and $F(gf)=F(g)F(f)$ whenever the composite $gf$ is defined in $\mathcal{C}$. Every category $\mathcal{C}$ has the identity functor $Id:\mathcal{C}\to\mathcal{C}$ such that $Id(a)=a$ for each object $a$ of $\mathcal{C}$ and $Id(f)=f$ for each morphism $f$ of $\mathcal{C}$. For functors $F:\mathcal{C}\to\mathcal{D}$ and $G:\mathcal{D}\to\mathcal{E}$ the composite functor $GF:\mathcal{C}\to\mathcal{E}$ is defined by $GF(a)=G(F(a))$ for each object $a$ of $\mathcal{C}$ and $GF(f)=G(F(f))$ for each morphism $f$ of $\mathcal{C}$.

Let $d$ be a fixed object of a category $\mathcal{D}$. Then a trivial functor $\mathbf{const}_d:\mathcal{C}\to\mathcal{D}$, called the *constant functor* of $\mathcal{C}$ with a value $d$, is defined by $\mathbf{const}_d(a)=d$ for each object $a$ of $\mathcal{C}$ and $\mathbf{const}_d(f) = \mathrm{id}_d$ for each morphism $f$ of $\mathcal{C}$.

Let $F : \mathcal{C} \to \mathcal{D}$ and $F' : \mathcal{C} \to \mathcal{D}$ be functors. A *natural transformation* $\lambda : F \to F'$ consists of a function which assigns to each object $a$ of $\mathcal{C}$ a morphism $\lambda_a : Fa \to Ga$ of $\mathcal{D}$ such that the following square

$$\begin{array}{ccc} Fa & \xrightarrow{F(f)} & Fb \\ {\scriptstyle \lambda_a}\downarrow & & \downarrow{\scriptstyle \lambda_b} \\ F'a & \xrightarrow[F'(f)]{} & F'b \end{array}$$

commutes (i.e. $F'(f)\lambda_a = \lambda_b F(f)$) for $f : a \to b$ of $\mathcal{C}$.

Let $\mathbf{const}_d, \mathbf{const}_{d'} : \mathcal{C} \to \mathcal{D}$ be constant functors with objects $d, d'$ of $\mathcal{D}$. Every morphism $h : d \to d'$ of $\mathcal{D}$ defines a natural transformation $\hat{h} : \mathbf{const}_d \to \mathbf{const}_{d'}$ by $\hat{h}_a = h$ for each object $a$ of $\mathcal{C}$.

Let $J : \mathcal{X} \to \mathcal{C}$ be a functor and $a$ an object of $\mathcal{C}$. For a natural transformation $\lambda : \mathbf{const}_a \to J$ and a morphism $f : b \to a$ of $\mathcal{C}$ a natural transformation $\lambda * f : \mathbf{const}_b \to J$ is defined by $(\lambda * u)_x = \lambda_x f$ for each object $x$ of $\mathcal{X}$.

A *limit* of a functor $J : \mathcal{X} \to \mathcal{C}$ is a pair $(a, \lambda)$ of an object $a$ of $\mathcal{C}$ and a natural transformation $\lambda : \mathbf{const}_a \to J$ such that for each natural transformation $\lambda' : \mathbf{const}_b \to J$ there is a unique morphism $f : b \to a$ with $\lambda * f = \lambda'$. A limit $(a, \lambda)$ of a functor $J$ is unique up to isomorphism. A category $\mathcal{C}$ has limits if every functor $J : \mathcal{X} \to \mathcal{C}$ from a small category $\mathcal{X}$ (in which the family $Obj(\mathcal{X})$ of all objects of $\mathcal{X}$ forms a set) has a limit.

Let $J : \mathcal{X} \to \mathcal{C}$ and $F : \mathcal{C} \to \mathcal{D}$ be functors and $\lambda : \mathbf{const}_a \to J$ a natural transformation. Then we can define a natural transformation $F\lambda : \mathbf{const}_{Fa} \to FJ$ by $(F\lambda)_x = F(\lambda_x)$ for each object $x$ of $\mathcal{X}$.

A functor $F : \mathcal{C} \to \mathcal{D}$ *preserves* a limit $(a, \lambda)$ of a functor $J : \mathcal{X} \to \mathcal{C}$ if $(Fa, F\lambda)$ is also a limit of the compoite $FJ$. A functor $F : \mathcal{C} \to \mathcal{D}$ preserves limits if $F$ preserves all limits of functors from small categories.

A functor $F : \mathcal{C} \to \mathcal{D}$ *weakly creates* limits for a functor $J : \mathcal{X} \to \mathcal{C}$ if for every limit $(d, \tau)$ of the composite $FJ : \mathcal{X} \to \mathcal{D}$ there is exactly one limit $(a, \lambda)$ of $J$ with $Fa = d$ and $F\lambda = \tau$. (Remark that this definition of limit creation is slightly different from that in [11, p. 108].) A functor $F : \mathcal{C} \to \mathcal{D}$ weakly creates limits if $F$ weakly creates all limits of functors from small categories.

The following is a standard theorem [11] indicating a relationship between creation and preservation of limits.

**Theorem 17.** *If a functor $F : \mathcal{C} \to \mathcal{D}$ weakly creates limits and $\mathcal{D}$ has limits, then $\mathcal{C}$ has limits and $F : \mathcal{C} \to \mathcal{D}$ preserves limits.*

Proof. The first conclusion directly follows from the definition of limit creation. We will see only the latter conclusion. Assume that $(a, \lambda)$ is a limit of a functor $J : \mathcal{X} \rightarrow \mathcal{C}$ from a small category $\mathcal{X}$. As $\mathcal{D}$ has all small limits, there is a limit $(d, \tau)$ of $FJ$. Since $F : \mathcal{C} \rightarrow \mathcal{D}$ creates limits there is a unique limit $(b, \sigma)$ of $J$ with $Fb = d$ and $F\sigma = \tau$. But limits are unique up to isomorphism, so there is an isomorphism $f : a \rightarrow b$ with $\lambda = \sigma * f$. Hence $F\lambda = F(\sigma * f) = F\sigma * Ff = \tau * Ff$,so $(Fa, F\lambda)$ is a limit of $FJ$. This completes that $F$ preserves limits. □

All concepts of category theory have their dual [11]. The following are dual notions concerned with limits stated above.

Let $J : \mathcal{X} \rightarrow \mathcal{C}$ be a functor and $a$ an object of $\mathcal{C}$. For a natural transformation $\lambda : J \rightarrow \mathbf{const}_a$ and a morphism $f : a \rightarrow b$ of $\mathcal{C}$ a natural transformation $f * \lambda : J \rightarrow \mathbf{const}_b$ is defined by $(f * \lambda)_x = f\lambda_x$ for each object $x$ of $\mathcal{X}$.

A *colimit* of a functor $J : \mathcal{X} \rightarrow \mathcal{C}$ is a pair $(a, \lambda)$ of an object $a$ of $\mathcal{C}$ and a natural transformation $\lambda : J \rightarrow \mathbf{const}_a$ such that for each natural transformation $\lambda' : J \rightarrow \mathbf{const}_b$ there is a unique morphism $f : a \rightarrow b$ with $f * \lambda = \lambda'$. A colimit $(a, \lambda)$ of a functor $J$ is unique up to isomorphism. A category $\mathcal{C}$ has colimits if every functor $J : \mathcal{X} \rightarrow \mathcal{C}$ from a small category $\mathcal{X}$ has a colimit.

Let $J : \mathcal{X} \rightarrow \mathcal{C}$ and $F : \mathcal{C} \rightarrow \mathcal{D}$ be functors and $\lambda : J \rightarrow \mathbf{const}_a$ a natural transformation. Then we can define a natural transformation $F\lambda : FJ \rightarrow \mathbf{const}_{Fa}$ by $(F\lambda)_x = F(\lambda_x)$ for each object $x$ of $\mathcal{X}$.

A functor $F : \mathcal{C} \rightarrow \mathcal{D}$ *preserves* a colimit $(a, \lambda)$ of a functor $J : \mathcal{X} \rightarrow \mathcal{C}$ if $(Fa, F\lambda)$ is also a colimit of the compoite $FJ$. A functor $F : \mathcal{C} \rightarrow \mathcal{D}$ preserves colimits if $F$ preserves all colimits of functors from small categories.

A functor $F : \mathcal{C} \rightarrow \mathcal{D}$ *weakly creates* colimits for a functor $J : \mathcal{X} \rightarrow \mathcal{C}$ if for every colimit $(d, \tau)$ of the composite $FJ : \mathcal{X} \rightarrow \mathcal{D}$ there is exactly one pair $(a, \lambda)$ of an object $a$ of $\mathcal{C}$ and a natural transformation $\lambda : J \rightarrow \mathbf{const}_a$ with $Fa = d$ and $F\lambda = \tau$, and if. moreover, this pair $(a, \lambda)$ is a colimit of $J$. A functor $F : \mathcal{C} \rightarrow \mathcal{D}$ weakly creates colimits if $F$ weakly creates all colimits of functors from small categories.

The following is a dual of Theorem 5.1.

**Theorem 18.** *If a functor $F : \mathcal{C} \rightarrow \mathcal{D}$ weakly creates colimits and $\mathcal{D}$ has colimits, then $\mathcal{C}$ has colimits and $F : \mathcal{C} \rightarrow \mathcal{D}$ preserves colimits.* □

# Single Pushout Transformations of Equationally Defined Graph Structures with Applications to Actor Systems *

Martin Korff

Computer Science Department, Technical University of Berlin,
D-1000 Berlin 10, Franklinstr. 27/28, Germany
martin@cs.tu-berlin.de

**Abstract.** This work has practically been motivated by an approach of modelling actor systems using algebraic graph grammars. It turned out that essential requirements on graph structures modelling computational states could nicely be expressed as conditional equations.
These and other examples lead then to a general investigation of single pushout transformations within categories of equationally defined graph structures i.e., certain algebras satisfying a set of given equations, and partial morphisms. Fundamentally we characterize pushouts in these equationally defined categories as the corresponding pushouts in the supercategory of graph structures without equations if and only if the pushout object already satisfies the given equations. For labeled graph structures this characterization can be inherited from the unlabeled case, but only for a restricted class of equations. For a special kind of so-called local equations in particular, interesting graph transformation results carry over to the new setting. The use and the effects of such equations are illustrated and discussed for corresponding graph grammar modellings of a client/server problem considered as an actor system.

**Keywords:** Algebraic Graph Transformations, Equationally Defined Graph Structures, Actor Systems

## 1 Introduction

The algebraic approach to graph transformation originates in [EPS73, Ehr79] and was extended in [Löw90, LE91]. According to the definition of a direct transformation as a double or single pushout in a suitable category the original and the latter variants are also called double and single pushout approaches respectively ([EKL91]). The practical use of the algebraic approach was shown by a number of applications to several fields in Computer Science and related areas e.g., functional and logical programming, production systems, semantical networks, actor systems, library and information systems, shared term rewriting, etc. [Pad82, Kor91, CR93, EK80, KW87, HP91]. On the other hand the algebraic graph grammar approach is distinguished by a comparably large number of theoretical results concerning e.g., embedded, parallel, concurrent, amalgamated, and distributed derivations [Ehr79, BEHL87, Hab92, BFH87, Löw90]. In [Löw90] the single pushout approach was introduced not only for graphs, but also hypergraphs, and similar graph-like structures subsumed by the notion of a graph structure. Transformations are defined as single pushouts in categories of graph structures i.e., algebras, and partial homomorphisms induced by certain algebraic signatures called graph structure signatures. Recently the theoretical approach for the transformation of graphs structures, graph structures combined with simple labels or algebraic attributes, relational structures, Petri nets, algebraic specifications, etc., was unified by the notion of a High Level Replacement System. Rewriting steps are defined as double or single pushouts in an axiomatic framework of categories ([EHKP92]).

* This work has been partly supported by the ESPRIT Working Group 7183 "Computing by Graph Transformation (COMPUGRAPH II)"

This provided the starting point for the present work. Graph structure signatures naturally allow the formulation of conditional equations. Transformations are defined as single pushouts in the resulting full subcategory of graph structures and partial morphisms satisfying these equations. But these categories do not have pushouts in general: This means that rules cannot be applied at arbitrary matches. Due to this fact the new approach is not covered by the currently existing framework of High-Level Replacement systems in the single pushout approach ([EL93]). Thus, graph transformation results require an investigation of its own.

Fundamentally we prove that pushouts in these categories are characterized by the corresponding pushouts in the supercategory of graph structures without equations if and only if the pushout object already satisfies the given equations. For labeled graph structures this characterization can be inherited from the case without labels, but only for a restricted class of so-called bleachable equations. Since this new approach includes the case without equations, in general the corresponding theory becomes weaker. However for a restricted class of so-called local equations, interesting graph transformation results carry over to the new setting. Due to its local effect, the restriction to local equations guarantees the existence of coproducts. This allows a general construction of parallel rules and weakened forms of independency notions such that the Church-Rosser Theorem and the synthesis of parallel derivations are guaranteed. Moreover there are results concerning derived and sequentially composed rules.

This work has practically been motivated by an approach of modelling actor systems using algebraic graph grammars. They, in turn, were stimulated by a previous graph grammar model of [JR89] similar to the NLC-approach [JR80]. Essential requirements on graph structures modelling computational states could nicely be expressed as conditional equations w.r.t. an appropriate graph structure signature.

The actor paradigm of computation is especially suitable as a model of highly concurrent systems ([Cli81, Agh88]). Asynchronous message passing provides the only means of communication between actors. Moreover there are some locality laws restricting the creation of new actors and messages. A client/server problem dealing with the interconnection of workstations and printers is considered as an actor system. Two examples are provided which are essentially based on a graph grammar model in [Kor93] enriched by different types of conditional equations. Due to the above characterization the operational effect of such equations is advantageously used to model basic system properties: rule applications are restricted to those cases where the graph obtained via the intuitive gluing construction in [Löw90] satisfies the equations.

The paper starts with an introduction to actor systems and the modelling of a computer network system using equationally defined graph structures. A review of basic notions and results from [Löw90] is contained in chapter 3. Based on an elaborated investigation in [Kor92], chapter 4 contains a presentation of the main notions and results concerning the transformation of equationally defined graph structures. Moreover chapter 4 includes a formal version of the introductory examples. In the conclusion the main facts are summarized and future research topics are addressed.

Finally I would like to thank Hartmut Ehrig and Andrea Corradini for discussions and valuable comments concerning topics of this work.

## 2 Actor Systems as an Example for the Transformation of Equationally Defined Graph Structures

The actor paradigm of computation has been designed as a model for massively concurrent systems ([Cli81, Agh88, JR89]). A number of computational agents, called actors, may communicate via asynchronous message passing only. The arrival of each message takes an arbitrary but finite amount of time. Each actor has its own private storage[2] which particularly includes a buffer for arriving messages and a list of actor addresses, the actor acquaintance list. The processing of a message by its target actor is called an event. The participants of an event are the actor's acquaintances and the actor addresses contained in the message (message acquaintances). As

[2] In fact, the actor paradigm allows (slightly) different models. We summarize the most typical settings.

a result of an event new actors may be created, new messages may be sent, and the actor's acquaintance list may be updated. Only participants and newly created actors may become the target of a newly sent message or be referred to in the updated actor's acquaintance list.

In [JR89], further continued in [JLR90, JR91] actor systems were modelled by so-called actor grammars. Structured transformations, on which actor grammars are essentially based, primarily focus on *specificational* aspects. Rather than directly defining the transformation result for an arbitrary given graph in some functional manner, the approach inductively relates specific graphs and their transformation results by rules and a number of composing operations hereon. Contrastingly, using algebraic graph transformations emphasizes on an *operational* understanding of graph productions where rules implement local changes *within* some surrounding context. In [Kor93] a algebraic graph grammar model for actor systems is discussed. But consistency conditions which were essential in [JR80] could not be expressed in the mere single pushout approach by [Löw90]. The use of conditional equations lead to an integrated categorical framework.

In the graph grammar models events are viewed as atomic in time. Thus each snapshot of an ongoing actor computation shows a system state between events. These states are modelled by graphs. Each application of a graph rule models an event. In the examples below this basic idea is concretized and applied to a simple client/server system. The first example is based on unlabelled, the second on labelled graph structures.

**Example 2.0.1** (*Marking Print Jobs*) Each of a number of workstations may send jobs to printers. Each printer depends on a scheduling process selecting jobs to be printed next. The crucial point is to ensure that, at any time, there is at most one job selected. The specification of this system as an actor system is straightforward: workstations • and printers □ are considered as actors modeled as vertices. Jobs are messages modelled as edges targeted (no source) to its target actor (vertex). Whenever a job has been selected by the scheduling process this is modelled by a specific edge, called marker, targeting (no source) to a job. The graph $G2$ in the sample transformation below illustrates a sample situation where four workstations with three jobs waiting for transmission are connected to only one printer where one job is already marked for printing. There are three types of events: **send**, **mark**, and **print** events. Each event is described by one of the rules below, outlined by round corner boxes. Items on the left hand side and on the right hand side which are not preserved specify the deletion and addition of items in an application respectively (preserved items are indicated by numbers; morphisms are indicated by thicker lines).

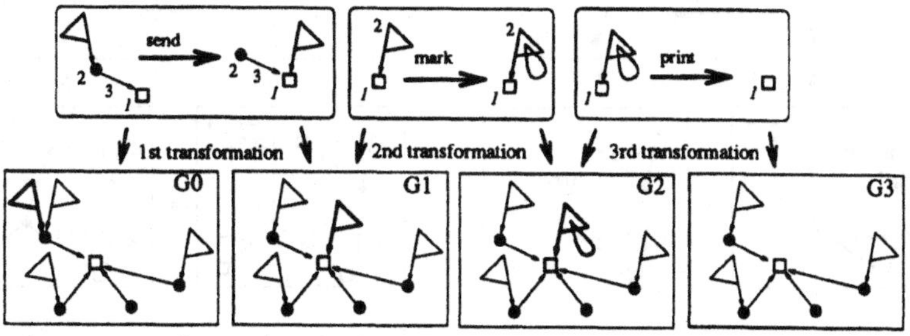


In order to ensure that a selected job is always printed before the scheduler selects a second, we propose three alternative equations demanding a consistency property of all graph structures concerned:

(1) **for all** $m, m' \in \mathit{Marker}$ : $\qquad m = m'$

(2) **for all** $m, m' \in \mathit{Marker}$ : $\mathit{prt}(\mathit{job}(m)) = \mathit{prt}(\mathit{job}(m')) \rightarrow \quad m = m'$

(3) **for all** $m, m' \in \mathit{Marker}$ : $\mathit{prt}(\mathit{job}(m)) = \mathit{prt}(\mathit{job}(m')) \rightarrow \mathit{job}(m) = \mathit{job}(m')$

The first is most rigorous since all markers are required to be identical: at most one marker is allowed in the system represented by a graph. The second contrastingly allows the existence of two markers in a graph only if they mark jobs belonging to different printers. The third is even more liberal since multiple markers may accumulate on the very same job. The effect of such equations is to restrict the transformations as described above to those where all graphs are consistent. All equations hold for all left and right hand sides graphs of the three rules. Moreover they are satisfied by the graphs $G0, G1, G2$, and $G3$ above as well as by $G1', G2'$, and $G3'$ below. The graphs $H1, H2$, and $H3$ are the results of the corresponding applications of the **marking** rule disregarding all equations. Requiring (exclusively) the first, second, or third equation turns graphs $H1, H2, H3$, graphs $H2, H3$, and graph $H3$ respectively into inconsistent ones such that the rule cannot be applied then.

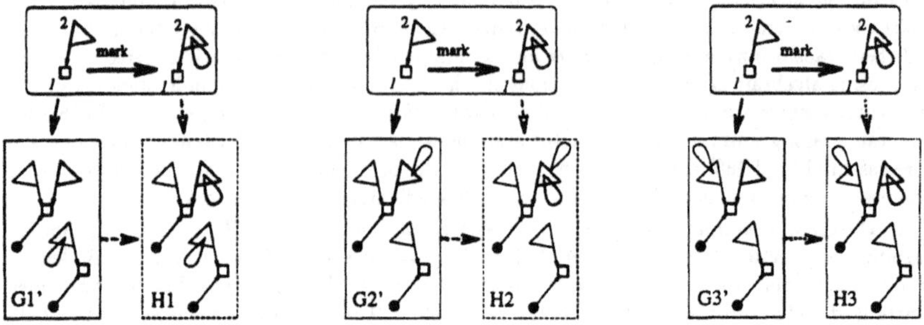


**Example 2.0.2** (*Adding New Printers*) The previous system is slightly extended. Each workstation shall be able to ***uniquely*** address one of a number of printers. This can be achieved by labelling the edges representing corresponding lines between each workstation and each of its printer acquaintances. Uniqueness is formalized by imposing the conditional equation requiring that two edges must be equal whenever they have the same source source (workstation) and the same label (printer name):

**for all** $e, e' \in Lines : s(e) = s(e')$ and $l(e) = l(e')$ implies $e = e'$

The figure below shows a situation where a message approaches a workstation containing a name and a reference to a new printer. The corresponding event of making this printer with name "laser" accessible to the workstation is modelled by the rule **add-laser**. The left and the right side show situations where **add-laser** can and cannot be applied respectively.

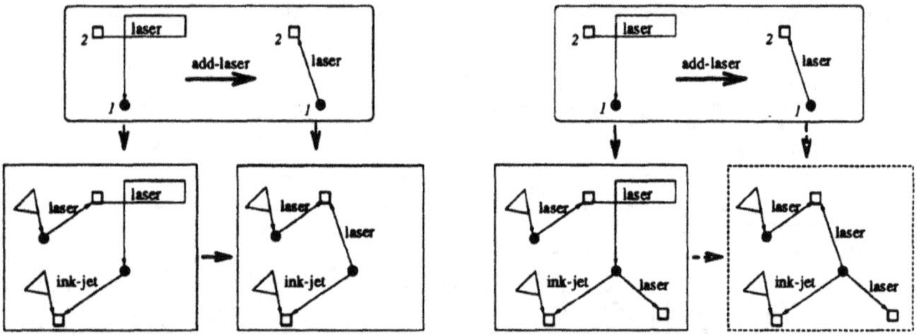


## 3 Short Review of Basic Notions and Results

Rather than for 'classical' graphs, the single pushout approach in [Löw90] has been formulated for a more general class of algebras (see [EM85]), so-called graph structures. This notion is motivated by the fact that a category of algebras and partial homomorphisms is cocomplete if and only if the algebras are graph structures. Covering a wide range of graph-like structures, basic notions from the double pushout graph transformation approach [Ehr79], carry over to the framework of single pushout transformations.

A *signature* $Sig = (S, OP)$ consists of a set of sorts and a family of sets $(OP_{w,s})_{w \in S^*, s \in S}$ of operation symbols. For $op \in OP_{w,s}$, we also write $op : w \to s$. A *Sig-algebra* $A$ is an $S$-indexed family $(A_s)_{s \in S}$ of carrier sets together with an $OP$-indexed family of mappings $(op^A)_{op \in OP}$ such that $op^A : A_{s_1} \times \ldots \times A_{s_n} \to A_s$ if $op \in OP_{s_1 \ldots s_n, s}$. A *homomorphism* $f : A \to B$ between two $Sig$-algebras $A$ and $B$ is a sort-indexed family of total mappings $f = (f_s : A_s \to B_s)_{s \in S}$ such that $op^B(f(x)) = f(op^A(x))$. A *partial homomorphism* $g : A \to B$ between $Sig$-algebras $A$ and $B$ is a homomorphism $g! : dom(g) \to B$ from a subalgebra $dom(g) \subseteq A$ to $B$.

The category of all $Sig$-algebras and all partial homomorphisms between them is denoted by $\boldsymbol{Sig^P}$. $\boldsymbol{Sig}$ denotes the subcategory of $\boldsymbol{Sig^P}$ with the same class of objects and all total homomorphisms.

The basic definition in [Löw90] is that of a graph structure signature motivated by the following Theorem 3.0.4 which has been proven in [Löw90] too.

**Definition 3.0.3** (*Graph Structure Signature*) A *graph structure signature* $GSig$ is a signature which contains unary operator symbols only. $GSig$ is called *hierarchical* if there is no infinite sequence $op_1 : s_1 \to s_2, op_2 : s_2 \to s_3, \ldots, op_i : s_i \to s_{i+1}, op_{i+1} : s_{i+1} \to s_{i+2}, \ldots$ of operator symbols. All $GSig$-algebras are called *graph structures*. •

**Theorem 3.0.4** (*Cocompleteness of* $\boldsymbol{Sig^P}$) The category $\boldsymbol{Sig^P}$ is finitely cocomplete if and only if $Sig$ is a graph structure signature. •

The *initial object* $I$ w.r.t. to some graph structure signature is the empty "graph". The *coproduct* of two graph structures is (induced by) the disjoint union of their carriers. The explicit construction of pushouts is provided by the following proposition 3.0.5 which has been proven in [Löw90].

**Proposition 3.0.5** (*Pushouts in Graph Structures*) Let $GSig$ be a graph structure signature and $f : A \to B, g : A \to C \in \boldsymbol{GSig^P}$. The pushout of $f$ and $g$ in $\boldsymbol{GSig^P}$, i.e. $(D, f^* : C \to D, g^* : B \to D)$, can be constructed in four steps: [3]

1. Construction of the *gluing object*: Let $\underline{A}$ be the greatest subalgebra of $A$ satisfying (i) $\underline{A} \subseteq dom(f) \cap dom(g)$ and (ii) $x \in \underline{A}, y \in A$ with $f(x) = f(y)$ or $g(x) = g(y)$ implies $y \in \underline{A}$.

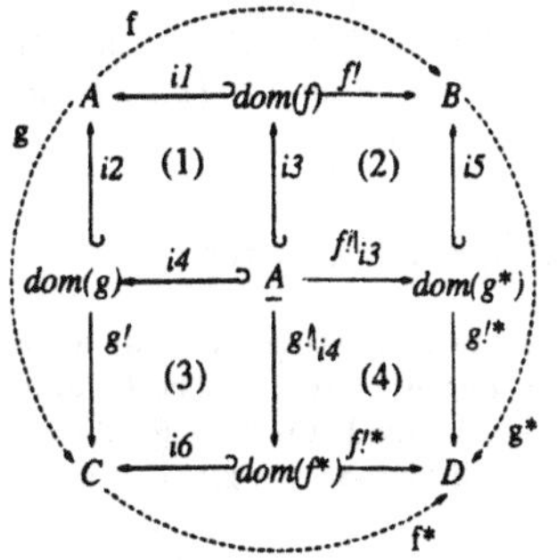


2. Construction of the *definedness area* of $f^*$ and $g^*$: Let $dom(g^*)$ be the greatest subalgebra of $B$ whose carriers are contained in $B - f(A - \underline{A})$. Symmetrically, $dom(f^*)$ is the greatest subalgebra in $C - g(A - \underline{A})$.
3. *Gluing*: Let $D = (dom(g^*) \uplus dom(f^*))_{/\equiv}$, where $\equiv$ is the least equivalence relation containing $\sim$, where $x \sim y$ if there is $z \in \underline{A}$ with $f(z) = x$ and $g(z) = y$.
4. *Pushout morphisms*: $f^*$ is defined for all $x \in dom(f^*)$ by $f^*(x) = [x]_\equiv$. The morphism $g^*$ is defined symmetrically.

•

The notion of a *graph structure* is justified by the observation that graphs, hypergraphs and other graph-like structures can be seen as algebras w.r.t. appropriate graph structure signatures.

*Directed graphs*, for instance, are algebras with respect to the following signature:

| | |
|---|---|
| **Sig** GRAPH = | I.e., each graph $G$ consists of a set of vertices $G_V$, a set of edges $G_E$, |
| **Sorts** $V, E$ | and two unary mappings $s^G, t^G : E_G \to V_G$ which provide source |
| **Opns** $s, t : E \to V$ | and target vertices for each edge. |

[3] The explicit representation of the definedness areas of partial homomorphisms by inclusion morphisms allows to visualize the whole situation in $GSig$. Note that the squares (1), (2), and (3) commute and (4) is a pushout in $GSig$ due to proposition 3.0.5 (3. and 4.).

Transformation rules and matches of rules are described by partial homomorphisms. The intuition of a rule $r : L \rightarrow R$ is that all objects in $L - dom(r)$ shall be deleted, all objects in $R - r(L)$ shall be added and $dom(r)$ resp. $r(L)$ provides the gluing context. Defining a transformation as a pushout in $\boldsymbol{GSig}^P$ captures this intuitive behaviour of deletion, gluing, and addition of graphical elements in a very compact and precise mathematical way as shown in the following proposition 3.0.5.

**Definition 3.0.6** (*Rule, Match, Transformation*) Let GSig be a graph structure signature. A *rule* $r : L \rightarrow R$ is a partial morphism in $\boldsymbol{GSig}^P$. A *match* for $r$ in an object $G \in \boldsymbol{GSig}^P$ is a total morphism $m : L \rightarrow G$. The *direct transformation* of $G$ with the rule $r$ at a match $m$ is the pushout of $m$ and $r$ in $\boldsymbol{GSig}^P$. The corresponding pushout object is called *derived graph*. Transformations are sequences of direct transformations.

Note that the transformation with $r$ deletes destructively, due to proposition 3.0.5 (2.), in the sense that all objects in $G$ pointing (for example by source or target mappings) to deleted objects are deleted themselves; if two objects $x \in L - dom(r)$ and $y \in dom(r)$ in a rule's left hand side are identified by a match, the effect of deletion is dominant.

Using the categorical formulation of definitions — which is the fundamental idea of high level replacement systems — most of the theoretical results concerning e.g. embedded, parallel, concurrent, and amalgamated transformations carried over to transformation systems defined in related categories. As examples for such categories we like to mention $\boldsymbol{RGSig}^P$, $\boldsymbol{AGSig}^P$, and $\boldsymbol{LGSig}^P$. In $\boldsymbol{RGSig}^P$ consisting of relational graph structures graphs are considered as relational structures in the sense of [EKMS+81] but homomorphisms are allowed to be partial. In $\boldsymbol{AGSig}^P$, graph structures are attributed by arbitrary algebras; homomorphisms are composed from a partial graph structure morphism and a total homomorphism on the algebra part. In the category of graph structures with labels $\boldsymbol{LGSig}^P$, distinguished label sorts are fixly interpreted as labels and homomorphisms are required to be identities on label sorted components.

Due to the great practical importance of labeled graph structures, also for the current investigations, we briefly give the precise definition.

**Definition 3.0.7** (*Labeled Graph Structure Signature*) A *labeled* graph structure signature $LSig = (LS, LOP, LabS, \Sigma)$ is a graph structure signature $(LS, LOP)$ together with non-empty label alphabets $\Sigma = (\Sigma_s)_{s \in LabS}$ for *label sorts* $LabS \subseteq \{s \in LS | \neg\exists op : s \rightarrow s' \in LOP\}$.

$\boldsymbol{LSig}^P$ denotes the category of all $(LS, LOP)$-algebras $A$ with $A_s = \Sigma_s$ for all $s \in LabS$ and all partial $(LS, LOP)$-homomorphisms $h$ between these with $h_s = id_s$ for all $s \in LabS$. *Labeled graph structures* are algebras in $\boldsymbol{LSig}^P$.

## 4 Transformations of Equationally Defined Graph Structures

Viewing graphs and similar structures as algebras w.r.t. an algebraic signature naturally allows to add equations in the sense of [EM85] — as they have been used in section 2. The following notions and results from algebraic specifications with equations shall provide the basis for the new approach where the notion of a graph structure signature is replaced by that of a graph structure specification.

A *specification* $Spec = (Sig, E)$ consists of a signature $Sig = (S, OP)$ and a set $E$ of *conditional equations*. Each $e = (X, Pre, Conc)$ in $E$ consists of a family $X = (X_s)_{s \in S}$ of pairwise disjoint variable sets $X_s$, a family of equations $Pre = (t_i, t'_i)_{i \in I}$, the *premises*, and a distinguished equation $Conc = (t, t')$, the *conclusion*, such that each $(t, t')$ is a pair of terms $t, t' \in T_{OP,s}(X)$ having the same sort. Usually we will write $e$ as

$$\forall x_{11}, \ldots, x_{1n} \in sort_1, \cdots, x_{m1}, \ldots, x_{mo} \in sort_m : t_1 = t'_1, \ldots, t_k = t'_k \rightarrow t = t'.$$

$e$ is *satisfied* by a $Sig$-algebra $A$, written $A \models e$, if for each assignment $ass_A = X \rightarrow A$ it holds that $t^A = t'^A$ whenever $t_i^A = t_i'^A$ for all $i \in I$. Here $.^A : T_{OP}(X) \rightarrow A$ denotes the unique extension of $ass_A$ to $T_{OP}(X)$ (see [EM85]).

A *Spec-algebra* $A$ is a $Sig$-algebra which additionally *satisfies* all equations $e \in E$, written $A \models E$.

The category of all *Spec*-algebras and all partial homomorphisms between them is denoted by $\boldsymbol{Spec}^P$. $\boldsymbol{Spec}$ denotes the subcategory of $\boldsymbol{Spec}^P$ with the same class of objects and all total homomorphisms.

Let $\cdot_{/E} : \boldsymbol{Sig} \to \boldsymbol{Spec}$ be the free (left adjoint) functor w.r.t. the inclusion $\subseteq : \boldsymbol{Spec} \hookrightarrow \boldsymbol{Sig}$. It uniquely extends the free construction $A \mapsto (A_{/E}, \boldsymbol{nat} : A \to A_{/E})$ where $A_{/E}$ is $A$ factored by the congruence relation induced by the equations $E$; the morphism $\boldsymbol{nat} : A \mapsto A_{/E}$ satisfies the universal property that for all $f : A \to B$ with $B \models E$, there is a unique morphism $f^* : A_{/E} \to B$ such that $f = f^* \circ \boldsymbol{nat}$. Since $\boldsymbol{Spec}$ is a subcategory of $\boldsymbol{Spec}^P$ with the same object-class, the total and surjective morphism $\boldsymbol{nat}$ is inherited by $\boldsymbol{Spec}^P$, although in general it may not have the universal property there.

### 4.1 Colimits

The notion of a graph structure specification induces a subcategory of graph structures satisfying the given set of equations. In particular this subcategory does not have all coproducts and pushouts. In this section they are characterized. For a certain class of so-called local equations, the existence of arbitrary coproducts is guaranteed.

**Definition 4.1.1** (*Graph Structure Specification*) A graph structure specification is a specification $GSpec = (GSig, E)$ where $GSig$ is a graph structure signature. •

**Proposition 4.1.2** (*Closure Properties*) Let $GSpec$ be a graph structure specification. Then $\boldsymbol{GSpec}^P$ is closed w.r.t. subalgebras, homomorphic images, homomorphic preimages of monomorphisms (injective morphisms), and epi-mono-factorizations. •

Although for a graph structure signature $GSig$, the category $\boldsymbol{GSig}^P$ has all pushouts, this does not generally hold for graph structure specifications. The following example will demonstrate this. It is based on the graph structure signature *Vertices* which allows its graph structures to have vertices only. The equations which are added in the graph structure specification *OneVertex* require them to have at most one vertex. So the graphs $A = \emptyset, B, C$, and thus the graph morphisms $f : A \to B$ and $g : A \to C$ are in $OneVertex^P$. But there is no pushout of $f$ and $g$ in $\boldsymbol{OneVertex}^P$ according to the following arguments. Note that $D = B + C$ is the pushout in $\boldsymbol{Vertices}^P$ according to proposition 3.0.5. From this we naturally obtain $g'_{/E}$ and $f'_{/E}$ using the functor $\cdot_{/E}$ mentioned above. The properties of $\cdot_{/E}$ provide that $g'_{/E}$ and $f'_{/E}$ form the pushout of $f$ and $g$ in $\boldsymbol{OneVertex}$. However, since we allow morphisms to be partial they cannot be pushout of $f$ and $g$ in $\boldsymbol{OneVertex}^P$: By replacing $f'_{/E}$ by $\overline{f} = \emptyset$ we obtain a second commutative diagram for which there is no morphism $u : D_{/E} \to D_{/E}$ such that $g'_{/E} = u \circ g'_{/E}$ and $\overline{f} = u \circ f'_{/E}$. A symmetric argument provides that all potential pushout morphisms must be defined on $dom(f'_{/E})$ and $dom(g'_{/E})$ respectively. Hence there is no other candidate for a pushout, and thus no pushout at all.

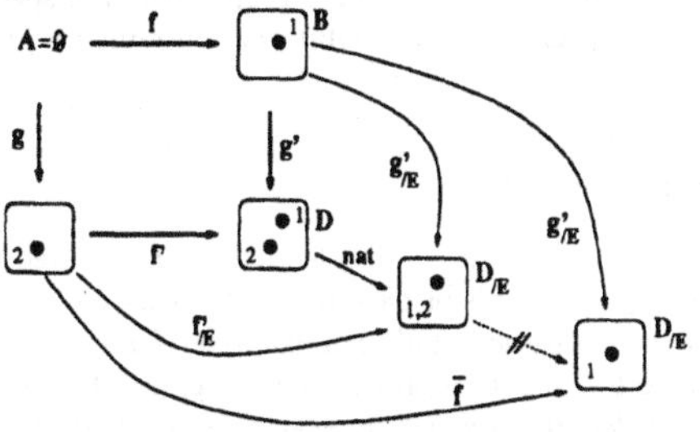


Sig *Vertices* =
Sorts *Vertices*

Sig *OneVertex* = *Vertices* +
Eqns $\forall x, y \in$ Vertex : $x = y$

The arguments of this example carry over to a general characterization of pushouts in $\boldsymbol{GSpec}^P$ (see [Kor92]): In order to construct the pushout of two morphisms $f$ and $g$ in $\boldsymbol{GSpec}^P$ we first construct their pushout $g'$ and $f'$ in $\boldsymbol{GSpec}^P$ with pushout object $D$. If now $D \models E$, $g'$ and $f'$ are already the desired pushout morphisms; otherwise there is no pushout of $f$ and $g$ in $\boldsymbol{GSpec}^P$. In fact, the restriction to *hierarchical* graph structure specifications $GSpec = (GSig, E)$ is essential. See [Kor92] for a corresponding counterexample concerning non-hierarchical graph structures.

**Theorem 4.1.3** (*Pushouts in hierarchical* $\boldsymbol{GSpec^P}$) Let $GSpec = (GSig, E)$ be a hierarchical graph structure specification. Given two arrows $f$ and $g$ in $\boldsymbol{GSpec^P}$, their pushout in $\boldsymbol{GSpec^P}$ exists if and only if the pushout object $D$ of $f$ and $g$ in category $\boldsymbol{GSig^P}$ satisfies the equations in $E$: in this case the pushout object is $D$ itself. ∙

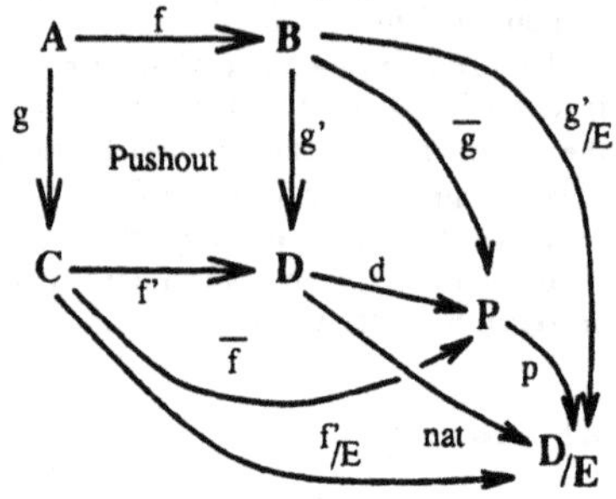


*Proofsketch:* The proof consists of the following parts.

If: We observe that each $GSpec$-algebra and all partial $GSpec$-homomorphisms are $GSig$-algebras and $GSig$-homomorphisms respectively, because $\boldsymbol{GSpec^P}$ is a *full* subcategory of $\boldsymbol{GSig^P}$. Let $C \rightarrow D \leftarrow B$ be the pushout in $\boldsymbol{GSig^P}$ of a pair of $GSpec$-homomorphisms $C \leftarrow A \rightarrow B$ where $D$ is a $GSpec$-algebra. Commutativity is preserved and the universal pushout property w.r.t. morphisms $B \rightarrow X \leftarrow C$ in $\boldsymbol{GSig^P}$ holds also in $\boldsymbol{GSpec^P}$ because only $GSpec$-algebras $X$ need to be considered.

Only if: It can be shown that, if there is a pushout of $f$ and $g$ in $\boldsymbol{GSpec^P}$, then this must be (isomorphic to) the pushout in $\boldsymbol{GSig^P}$ on which the equations $E$ have been imposed i.e., if $(f, g)$ has pushout $B \xrightarrow{\bar{g}} P \xleftarrow{\bar{f}} C$ in $\boldsymbol{GSpec^P}$ then $B \xrightarrow{g'_{/E}} D_{/E} \xleftarrow{f'_{/E}} C$ is pushout of $(f, g)$ too.
But $D_{/E}$ cannot be the pushout object in $\boldsymbol{GSpec^P}$ if $D_{/E} \neq D$.
Thus, there is no pushout of $(f, g)$ in $\boldsymbol{GSpec^P}$ whenever $D \not\models E$.

□

The empty graph structure is initial in $\boldsymbol{GSpec^P}$. Based on this observation there is a characterization of coproducts similar to theorem 4.1.3. The category $\boldsymbol{GSpec^P}$ does not have arbitrary coproducts This is shown by the example above too (simply forget about $f$ and $g$). However, coproducts of graph specifications generally exist if equations are restricted to local equations — and many practically interesting equations are local (see section 4.4).

**Definition 4.1.4** (*Local Equation*) An equation $e = (X, Pre, Conc) \in E$ is called *local* if (1.) there are no variables besides those in the conclusion i.e., $X = Vars(Conc)$ and (2.) *either* $|X| = 1$ *or* there is at least one (non-label sorted)[4] equation $eq \in Pre$ such that $Vars(eq) = X$, where $Vars(p)$ denotes the set of all variables in an equation $p$. ∙

**Theorem 4.1.5** (*Coproducts in* $\boldsymbol{GSpec^P}$) Let $GSpec = (GSig, E)$ be a graph structure specification where $E$ consists of local equations only. Then $\boldsymbol{GSpec^P}$ has arbitrary coproducts which coincide with coproducts in $\boldsymbol{GSig^P}$. ∙

## 4.2 Graph Structure Specifications with Labels

As in the case without equations ([Löw90]), all notions from the 'unlabeled' case can be lifted to 'labeled' ones. Since labels are globally fixed and homomorphisms have to be identities on labels, this allows one to formulate equations which cannot be satisfied at all. For example, if $a, b \in \Sigma$ are two arbitrary distinct labels then the equation "$a = b$" cannot be satisfied by any graph structure. The essential difference between unlabeled and labeled graph structure specifications stems from the fact that the category of labeled graph structure specifications does not allow morphisms to identify items with different labels. Thus, $\cdot_{/E} : \boldsymbol{LGSig} \rightarrow \boldsymbol{LGSpec}$ is no longer a free functor

[4] see below

w.r.t. the inclusion $\subseteq$: $\boldsymbol{LGSpec} \hookrightarrow \boldsymbol{LGSig}$ which particularly implies that the proof of theorem 4.1.3 cannot be lifted to the labeled case. However, in [Kor92] for a particular class of labeled equations, called *bleachable* equations, pushouts in the category $\boldsymbol{LGSpec}$ are characterized in the way of theorem 4.1.3. This result was achieved by proving that for bleachable equations it is possible to transform a labeled graph structure specification into an unlabeled, categorically equivalent one. The syntactical definition of bleachable equations in [Kor92] is highly technical. Therefore we take its semantic characterization as the definition here. See section 4.4 for examples.

**Definition 4.2.1** (*Bleachable equation*) Let $LSpec = (LSig, E)$ be a labeled graph structure specification with labeled graph structure signature $LSig = (LS, LOP, \Sigma, LabS)$. Then a conditional equation $e = (X, Pre, Conc)$ with $Pre = (t_i, t_i')_{i \in I}$, and $Conc = (t, t')$ is called *bleachable* if for all $A \in \boldsymbol{LSig}^P$ and all assignments $ass : X \to A$ it holds that $t_i^A = t_i'^A$ for all $i \in I$ implies that all labels which can be obtained by applying the same sequence of operations to both elements $t^A$ and $t'^A$ are equal. ▪

Consider a pushout construction in $\boldsymbol{LSig}^P$ with pushout object $D$ as shown in the picture of theorem 4.1.3. The above condition ensures then that whenever the premise of a bleachable equation evaluates in $D$ to true — which requires the evaluation of the conclusion in $D$ also to be true — the morphism $nat : D \to D_{/E}$ uniquely existing in $\boldsymbol{LSpec}$ by definition does not identify any labels. Then the proof from theorem 4.1.3 can be reused, obtaining a result analogous to the non-labeled case.

### 4.3 Graph Transformation Results

In this section we introduce some results concerning the theory of transformation of $\boldsymbol{GSpec}$ algebras, based on those in [Löw90] These results are due to the characterization of pushouts in theorem 4.1.3, the closure properties of $\boldsymbol{GSpec}^P$ in proposition 4.1.2 and general properties of categories. All proofs of theorems and propositions below can be found in [Kor92].

In the following let $GSpec = (GSig, E)$ be a fixed graph structure specification, and let $r : L \to R$ and $r' : L' \to R'$ be two rules.

Inspired by high level replacement systems, the definition of rules, matches, and transformations is carried over from section 3.

**Definition 4.3.1** (*Rule, Match, Transformation*) Let GSpec be a graph structure specification. A *rule* $r : L \to R$ is a partial morphism in $\boldsymbol{GSpec}^P$. A *match* for $r$ in an object $G \in \boldsymbol{GSpec}^P$ is a total morphism $m : L \to G$. The *direct transformation* of $G$ with the rule $r$ at a match $m$ is the pushout of $m$ and $r$ in $\boldsymbol{GSpec}^P$. The corresponding pushout object is called *derived graph*. Transformations are sequences of direct transformations. ▪

**Definition 4.3.2** (*Redex*) If a graph structure $G$ can be transformed by a $r$ at $m : L \to G$ in $\boldsymbol{GSpec}^P$ this is denoted by $G \overset{r,m}{\Longrightarrow} r_m\langle G\rangle$. In this case $m$ is called a *redex*. ▪

***Rule-Derived Rules.*** The Rule-Derived Rule Theorem states that the effect of a transformation is of local nature only. Each 'two-level' application of a rule can equivalently be done in one step. Moreover, general applications of rules can be replaced by applications of minimal rules restricted to injective matches.

**Definition 4.3.3** (*Rule-Derived Rule*) Let $G \overset{r,m}{\Longrightarrow} r_m\langle G\rangle$ be a transformation. Then the transformation morphism $r_m : G \to r_m\langle G\rangle$ is called a *rule-derived rule*. It is *minimal* if the redex $m$ is surjective. ▪

**Proposition 4.3.4** (*Rule-Derived Rule*) Let $rd : G \to H$ be a rule-derived rule w.r.t. $r$. Then there is a direct transformation $K \overset{r,m'}{\Longrightarrow} M$ whenever there is a direct transformation $K \overset{rd,n}{\Longrightarrow} M$.

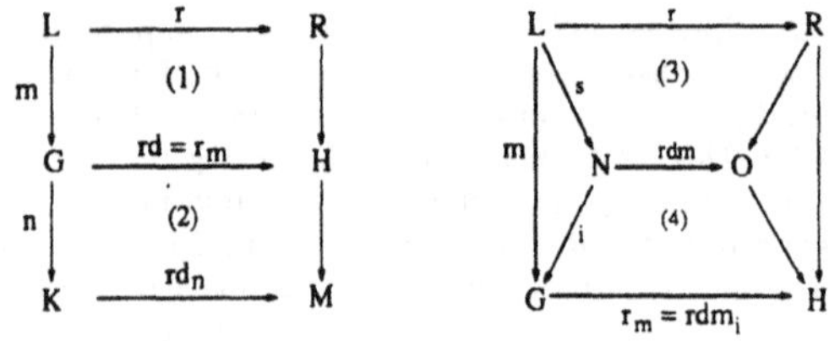


Moreover for each direct transformation $G \stackrel{rm}{\Longrightarrow} H$, there is a *minimal rule-derived rule* $rdm : N \rightarrow O$ i.e., there is a total surjective match $s : L \rightarrow N$, with injective match $i : N \rightarrow G$ such that $r_m = rdm_i$.

*Proofsketch:* (1) and (2) is a composed pushout. $N = m(L) \models E$, thus $i$ is mono, implying that $O \rightarrow H$ is mono too. $H \models E$ implies $O \models E$. □

*Local Church Rosser, Parallelism.* The Local Church Rosser Theorem states that, whenever a sequence of rule applications is independent, the resulting graph can also be obtained if the order of rule applications is switched. The Parallelism Theorem states that such transformations are equivalent to an application of the corresponding parallel rule. The 'classical' overlapping in gluing-items only is no longer sufficient for ensuring that one of the parallely competing rules can still be applied to the result of the other one. The reason for this is that generally the effect of conditional equations is not only of local nature.

**Definition 4.3.5** (*Parallel Independence, Parallel Rule*) Two matches $m : L \rightarrow G$ and $m' : L' \rightarrow G$ are *unconnected* if $G$ consists of two disjoint parts, each of which containing an image of a match i.e., $G = G_m + G_{m'}$, $m(L) \subseteq G_m$, and $m'(L') \subseteq G_{m'}$.

The *parallel rule* $r + r' : L + L' \rightarrow R + R'$ is constructed from $L + L'$ and $R + R'$, the coproducts of the left and right hand sides respectively, and $r + r'$, the mediating morphism from $L + L'$ to $R + R'$.

**Theorem 4.3.6** (*Local Church-Rosser/Parallelism*) Let $E$ consist of local equations only. Then, given two unconnected redices $m : L \rightarrow G$ and $m' : L' \rightarrow G$, there are redices $\overline{m'} = r_m \circ m' : L' \rightarrow r_m\langle G\rangle$ and $\overline{m} = r'_{m'} \circ m : L \rightarrow r'_{m'}\langle G\rangle$ such that $r'_{\overline{m'}}\langle r_m\langle G\rangle\rangle = r + r'_{m+m'}\langle G\rangle = r_{\overline{m}}\langle r'_{m'}\langle G\rangle\rangle$

*Proofsketch:* Unconnected redices are parallel independent in the sense of [Löw90] and thus the graph $X = r + r'_{m+m'}\langle G\rangle$ can be computed in $GSig^P$. By definition 4.3.5 $G$ there are of two subgraphs which are independently replaced by applying rules $r$ and $r'$. Hence $r_m\langle G\rangle \models E$ and $r'_{m'}\langle G\rangle \models E$ implies $X \models E$. □

Counterexamples showing that this Theorem generally does not hold if redices are not required to be unconnected are given in [Kor92].

*Sequential Composition.* The effect of one rule applied after another can be summarized by a third rule, the sequential composition of both. There, the construction depends on the intersection of both rules in the intermediate graph structure. This is captured by the notion of a sequential relation.

**Definition 4.3.7** (*Sequential Independence, Sequential Composition*) Given a direct transformation $G \stackrel{rm}{\Longrightarrow} H$, graph morphisms $m_r : R \rightarrow H$ and $m' : L' \rightarrow H$ are *s(equentially) independent* if there is a redex $v : L' \rightarrow G$ such that $m' = r_m \circ v$.

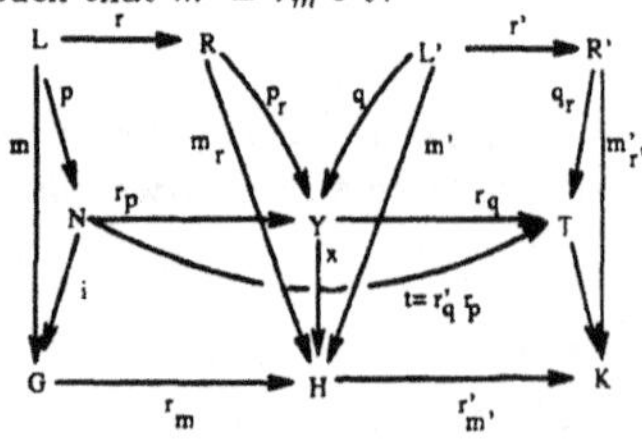

The derived rule $r'_{m'} \circ r_m$ is a ***sequential composition*** of $r$ and $r'$ if $m_r$ and $m'$ are jointly surjective. •

**Proposition 4.3.8** (*Sequential Independence*) Two direct transformations $G \overset{r_m}{\Longrightarrow} H$ and $H \overset{r'_{m'}}{\Longrightarrow} K$ in $\boldsymbol{Spec^P}$ are s-independent iff $m'(L') \cap m_r(R) \subseteq m_r \circ r(glue(r, m))$.

If they are s-independent, then there is a redex $z$ for the parallel rule $r + r'$ such that $K$ can be derived from $G$ by applying $r + r'$ at $z$ i.e., $K = (r + r')_z\langle G\rangle$. •

*Proofsketch:* The Proposition holds in $\boldsymbol{Sig^P}$ ([Löw90]) and that $G, H, K \models E$. □

**Theorem 4.3.9** (*Sequential Composition*) For direct transformation $G \overset{r_m}{\Longrightarrow} H$ and $H \overset{r'_{m'}}{\Longrightarrow} K$ there is a sequential composition $t : N \to T$ of $r$ and $r'$ and a match $i : N \to G$ such that $t$ transforms $G$ to $K$ at $i$ i.e., $K = t_i\langle G\rangle$. •

*Proofsketch:* $G, H, K \models E$, $N = m(L) \cup r_m^{-1} \circ n(M) \subseteq G$. And since $Alg^P(Spec)$ is closed w.r.t. subalgebras $N \models E$. $i : N \to G$ is injective, so is $x : Y \to H$. Moreover $x$ is total. Thus, $Y \models E$ due to $H \models E$. Similarly we conclude $tt : T \to K$ is total and injective, thus $T \models E$. □

### 4.4 Formal version of the actor system examples

In the first example (2.0.1) graph structures were used to describe different systems containing the following types of objects: workstations, printers, lines from workstations to printers (the actor acquaintances), print jobs at workstations, print jobs at printers, and markers for selected print jobs. These objects are formally considered as elements of corresponding graph structure carriers as specified by the sorts in the graph structure specification $Ex1$ below.

**Spec** Ex1 =
**Sorts** $WS, Printer, Line, WJob, PJob, Marker$
**Opns** $s : Line \to WS$
$t : Line \to Printer$
$ws : WJob \to WS$
$prt : PJob \to Printer$
$job : Marker \to PJob$
**Eqns** $\forall m, m' \in Marker : \quad m = m'$
$\forall m, m' \in Marker : \quad prt(job(m)) = prt(job(m')) \to \quad m = m'$
$\forall m, m' \in Marker : \quad prt(job(m)) = prt(job(m')) \to job(m) = job(m')$

Correspondingly the operation symbols of $Ex1$ must be interpreted as follows: for each line connecting a workstation $w$ and a printer $p$, there are two operations $s$(ource) and $t$(arget) yielding $w$ and $p$ respectively (i.e. elements of a $Line$-carrier is a 'normal' edge). For each job $j$ which is present at a workstation $w$, $ws(j)$ returns $w$. Analogously, for each job $j$ which is present at a printer $p$, $prt(j)$ returns $p$. Finally, $job(m)$ describes the job to which a marker $m$ has been attached.

The left and right hand sides of all rules **send, mark** and **print** as well as $G0, G1, G2$, and $G3$ and $G1', G2'$, and $G3'$ are graph structures in $\boldsymbol{Ex1^P}$ — for any of the above equations — since there is at most one marker in each of them. For assigning $m$ and $m'$ to the only two markers in each of the graphs $H1, H2$, and $H3$ we obtain that the first equation is never satisfied, that the second equation is satisfied for $H1$ only, and that the third equation is satisfied for $H2$ and $H3$ only. By theorem 4.1.3 and definition 4.3.1 the **mark**ing rule can then only be applied if the resulting graph satisfies the required equation. The **send**ing and **print**ing rules can always be applied since each of these equation is satisfied for a (potential) derived graph structure if and only if it is for the original graph structure.

Both the second and the third equation are *local* since the variable set only contains variables which are in the conclusion and these variables are also contained in the (unique equation in the) premise. The latter condition is not satisfied by the first equation since there is (no equation in the) premise at all. Hence the first equation is ***non-local.*** For the second and third equation, jobs at different printers may always be marked in parallel provided that each of the corresponding jobs can be marked sequentially.

As an example for another practically interesting kind of equations take the following, which requires a distinguished sort *Switch* to be included in the above graph structure specification: **for all** $s \in Switch,\ p, p' \in Printer : p = p'$. Its effect is that at most one printer is allowed in the system whenever there is a *Switch*-element. Removing these elements 'disables' the equation: if there is no assignment of variables to elements of the algebra the equation becomes satisfied by definition.

In the second example (2.0.2) the system model was enriched by a second kind of message, a label set of printer names for these messages and lines connecting workstations and printers. This is reflected by the labeled graph structure specification $Ex2(AddName, PrtName)$ including the following definition of label carriers:

**Spec** Ex2(AddName,PrtName) = Ex1 +
**Sorts** *AddMsg*, *AddName*, *PrtName*
**Opns** $t : AddMsg \rightarrow WS$
$ref : AddMsg \rightarrow Printer$
$l : Line \rightarrow PrtName$
$la : AddMsg \rightarrow AddName$
**Eqns** $\forall e, e' \in Lines : (e) = s(e')$ and $l(e) = l(e') \rightarrow e = e'$

together with label alphabets $\Sigma_{AddName} = \Sigma_{PrtName} = \{laser, ink\text{-}jet, matrix, \ldots\}$ providing fix interpretations of sorts *AddName* and *PrtName* respectively. For all graphs in the transformations in example 2.0.2, except for that in the lower right corner, the equation is satisfied since either there is no assignment for $e$ and $e'$ to different elements such that $s(e) = s(e')$ or the labels of these elements are different.

The equation is ***bleachable*** since the premise $l(e) = l(e')$ ensures that the labels of the elements to which the conclusion may be evaluated in a graph structure are always equal whenever an evaluation of the premise yields true. The equation is *local* by the same arguments as for the second equation above. Similar to the conclusion above it holds that whenever different printers can be added to different workstations this can also be done in parallel (see Theorem 4.3.6). When rules are applied in some fix sequential dependency the result can always be reached using the precomputable sequential composition (see Theorem 4.3.9).

Thus, formulating transformations in a framework of (equational) graph structure specifications provides a uniform framework for explicitly statements and proofs (dealing with rule sets, a single rule, or a single rule application) concerning general system invariants as specified by appropriate conditional equations.

## 5 Conclusion

In [Löw90] algebraic graph transformations were introduced which are defined as single pushouts in a category of algebras and partial homomorphisms w.r.t. an algebraic, so-called, graph structure signature. Adding conditional equations in the flavor of [EM85] extends the single pushout approach of [Löw90] to categories of equationally defined graph structures with partial morphisms.

Fundamentally, single pushout transformations of such equationally constrained graph structures were characterized: there is a pushout in an equationally defined algebraic category with partial morphisms if and only if the pushout in the supercategory without equations does belong to that subcategory too. In this (positive) case the pushouts coincide. More precisely, this result could be obtained for arbitrary conditional equations w.r.t. some unlabeled hierarchical graph structure signature only. The requirement on graph morphisms to be label-preserving did not allow to directly rephrase the main proof used for unlabeled graph structures. This was possible for a restricted class of so-called bleachable equations only.

As it could be expected for an extending approach, corresponding theoretical statements became weaker. A notion of a local equation was defined. Local equations demand constraints only locally such that the corresponding equationally defined subcategories are closed w.r.t. the construction of coproducts. For local equations interesting results from [Löw90] carry over to the new setting.

From the practical point of view conditional equations can be considered as application conditions in the sense of [EH86] requiring that all graphs appearing in a single pushout transformation in the sense of [Löw90] are consistent. A generalization to arbitrary logical formulas instead of mere conditional equations is straightforward and leads to a notion of a *consistent graph transformation*. Main tools for results concerning graph transformations were (and will probably be) the compositionality of pushouts and the locality of certain constraints.

This notion of a consistent graph transformation contributes the common need of practical problems by aiding rule system design by an additional mechanism for both semantical control and proof support. Practically, we observe that many consistency constraints cannot be formulated as conditional equations. Consequently, arbitrary first order logic formulas are available as in the graph specification language PROGRES ([Sch91]). The development of a corresponding variety of formalisms trading off between expressive power and graph transformation results should combine flexibility and understanding in an optimal way. Even if such consistency constraints are operationally redundant they provide proof criteria by explicitly narrowing the class of possible states. In any case, they bring clarity to a specification which is particularly helpful for determining how those graph structures for which rewriting rules shall be specified are meant to look like.

Further research is encouraged by the following topics.

- Amalgamated and distributed transformations may be reconsidered in the framework of (local) conditional equations.
- Theoretical investigations may be reconsidered for the generalized notion of consistent graph structure transformations. Conditional Equations shall be generalized to arbitrary logical axioms. A corresponding notion of local constraints may be developed.
- The theoretical approach for single pushout graph transformations in categories which are incomplete w.r.t. coproducts, pushouts (arbitrary rule applications) may be unified in a purely axiomatic framework in the sense of [EHKP92].
- Different graph structure specifications may be designed for practical problems. Their advantages and implications may be discussed. Correctness notions should be provided and applied. Appropriate concepts for supporting the corresponding proofs must be developed.
- The development of a more general type of graph grammar for modelling actor systems is in preparation.

## References


[Agh88] G.A. Agha, *ACTORS: A Model of Concurrent Computation in Distributed Systems*, The MIT Press Cambridge, Massachusetts, London, England. 1988.

[BEHL87] P. Boehm, H. Ehrig, U. Hummert, and M. Löwe, *Towards distributed graph grammars*, In Ehrig et al. [ENRR87].

[BFH87] P. Böhm, H.-R. Fonio, and A. Habel, *Amalgamation of graph transformations: a synchronization mechanism*, Journal of Computer and System Science **34** (1987), 377–408.

[Cli81] W.D. Clinger, *Foundations of Actor Semantics*, Ph.D. thesis, MIT Artificial Intelligence Laboratory, AI-TR-633, May 1981.

[CR93] A. Corradini and F. Rossi, *On the power of contrext-free jungle rewriting for term rewriting systems and logic programming*, Term Graph Rewriting: Theory and Practice, John Wiley & Sons Ltd, 1993.

[EH86] H. Ehrig and A. Habel, *Graph grammars with application conditions*, The Book of L (G. Rozenberg and A. Salomaa, eds.), Springer, 1986, pp. 87–100.

[EHKP92] H. Ehrig, A. Habel, H.-J. Kreowski, and F. Parisi-Presicce, *Parallelism and concurrency in High Level Replacement Systems*, Mathematical Structures in Comp. Sci. **1** (1992), 361–404.

[Ehr79] H. Ehrig, *Introduction to the algebraic theory of graph grammars*, 1st Int. Workshop on Graph Grammars and their Application to Computer Science and Biology, Lecture Notes in Computer Science 73, Springer, 1979, pp. 1–69.

[EK80] H. Ehrig and H.-J. Kreowski, *Application of Graph Grammar Theory to Consistency, Synchronization and Scheduling in Data Base Systems*, Inform. Systems **5** (1980).

[EKL91] H. Ehrig, M. Korff, and M. Löwe, *Tutorial introduction to the algebraic approach of graph grammars based on double and single pushouts*, In Ehrig et al. [EKR91], pp. 24–37.

[EKMS+81] H. Ehrig, H.-J. Kreowski, A. Maggiolo-Schettini, B. K. Rosen, and J. Winkowski, *Transformation of structures: an algebraic approach*, Mathematical Systems Theory **14** (1981), 305–334.

[EKR91] H. Ehrig, H.-J. Kreowski, and G. Rozenberg (eds.), *4th Int. Workshop on Graph Grammars and Their Application to Computer Science, Lecture Notes in Computer Science 532*, Springer, 1991.

[EL93] H. Ehrig and M. Löwe, *Categorical Principles, Techniques and Results for High-Level Replacement Systems in Computer Science*, Applied Categorical Structures (to appear 1993).

[EM85] H. Ehrig and B. Mahr, *Fundamentals of algebraic specifications*, EACTS-Monographs in Theoretical Computer Science, vol. 6, Springer, Berlin, 1985.

[ENRR87] H. Ehrig, M. Nagl, G. Rozenberg, and A. Rosenfeld (eds.), *3rd Int. Workshop on Graph Grammars and Their Application to Computer Science, Lecture Notes in Computer Science 291*, Springer, 1987.

[EPS73] H. Ehrig, M. Pfender, and H. J. Schneider, *Graph grammars: an algebraic approach*, 14th Annual IEEE Symposium on Switching and Automata Theory, 1973, pp. 167–180.

[Hab92] A. Habel, *Hyperedge replacement: Grammars and languages*, Springer LNCS 643, Berlin, 1992.

[HP91] B. Hoffmann and D. Plump, *Implementing term rewriting by jungle evaluation*, Informatique théorique et Applications/ Theoretical Informatics and Applications (Berlin), vol. 25 (5), 1991, pp. 445–472.

[JLR90] D. Jansens, M. Lens, and G. Rozenberg, *Computation graphs for actor grammars*, Tech. report, University of Leiden, 1990.

[JR80] D. Janssens and G. Rozenberg, *On the structure of node-label controlled graph grammars*, Information Science **20** (1980), 191–216.

[JR89] D. Jansens and G. Rozenberg, *Actor grammars*, Mathematical Systems Theory **22** (1989), 75–107.

[JR91] D. Jansens and G. Rozenberg, *Structured transformations and computation graphs for actor grammars*, In Ehrig et al. [EKR91], pp. 446–460.

[Kor91] M. Korff, *Application of graph grammars to rule-based systems*, In Ehrig et al. [EKR91], pp. 505–519.

[Kor92] M. Korff, *Algebraic Transformations of Equationally Definied Graph Structures*, Tech. Report 92/32, Technical University of Berlin, 1992, ca. 120 pages.

[Kor93] M. Korff, *An Algebraic Graph Grammar Model for Actor Systems*, Tech. Report 93/12, Technical University of Berlin, 1993.

[KW87] H.-J. Kreowski and A. Wilharm, *Is parallelism already concurrency? part 2: non-sequential processes in graph grammars*, 3rd Int. Workshop on Graph Grammars and Their Application to Computer Science, Lecture Notes in Computer Science 291, Springer, 1987, pp. 361–377.

[LE91] M. Löwe and H. Ehrig, *Algebraic approach to graph transformation based on single pushout derivations*, 16th Int. Workshop on Graph Theoretic Concepts in Computer Science, Lecture Notes in Computer Science 484, Springer, 1991, pp. 338–353.

[Löw90] M. Löwe, *Extended algebraic graph transformation*, Ph.D. thesis, Technical University of Berlin, Department of Computer Science, 1990.

[Pad82] P. Padawitz, *Graph grammars and operational semantics*, Theoretical Computer Science **19** (1982), 37–58.

[Sch91] A. Schürr, *Operationales Spezifizieren mit programmierten Graphersetzungssystemen*, Deutscher Universitätsverlag GmbH, Wiesbaden, 1991.

# Parallelism in Single-Pushout Graph Rewriting[1]

*Michael Löwe* and *Jürgen Dingel*
Technische Universität Berlin, Germany

**Abstract**

**The single-pushout approach to graph transformation comes equipped with an interesting parallel composition of rules and derivations: Some parallel steps cannot be decomposed into sequences with the component rules. Thus parallelism in single-pushout rewriting is not equivalent to "independence of order", which makes up a major difference with respect to the classical algebraic approach to graph transformation based on double-pushout constructions for direct derivations.**

**But as we show in this paper, the equivalence that arises from different sequentializations contains a distinguished member which is uniquely determined up to a suitable notion of isomorphism between derivations and which realizes maximal parallelism also in single-pushout rewriting. The amount of parallelism, however, which we find in these so-called canonical sequences considerably exceeds the parallelism obtained in the double-pushout framework.**

## 1. Introduction

Both algebraic approaches to graph transformation, the classical double-pushout approach [Ehr 79] and the single-pushout approach [Löw 90/93], provide a simple and elegant notion of parallel transformation. It is just a standard transformation with a parallel rule. And parallel rules are constructed by the disjoint union of the component rules.

Due to the more general concept of redices in single-pushout rewriting, however, the two approaches considerably differ in the possibilities to sequentialize parallel steps. While the effect of parallel steps in double-pushout rewriting can always be obtained by sequences using the component rules in arbitrary order, parallel single-pushout transformations are possible which cannot be simulated by any sequence with the component rules. Thus single-pushout parallelism goes beyond the concept of "independence of order".

But as we show in this paper, major properties of double-pushout parallelism can be reestablished in the single-pushout approach. So the notions of parallel independence in both approaches satisfy comparable laws which have been presented for the classical case in [Kre 76, EKT 92]. Based on these laws, [Kre 76] showed for double-pushout graph rewriting that among all different sequentializations of a given derivation sequence there is a *canonical* one in the following sense: It is uniquely determined up to a suitable notion of isomorphism between derivations and it realizes maximal parallelism in the sense that each atomic step is performed as early as possible; i.e. earlier than in any other sequentialization.[2] These laws are given on an abstract level in [EKT 92], where, for example, it does not matter that direct derivations are given by double-pushout constructions. Instead, [EKT 92] formulate all laws as requirements for certain parallel and sequential compositions of derivation steps. However, one implicit requirement has been inherited from double-pushout rewriting on the abstract level, namely that all parallel steps must be sequentializable in any order with the component rules.

Thus the axiomatic framework of [EKT 92] is not directly applicable to the single-pushout approach. But a comprehensive analysis of single-pushout transformation on the basis of [EKT 92] becomes possible if we restrict parallel steps to those which allow all sequentializations. This can

[1]Research has been partly supported by ESPRIT Basic Research Working Group 7183 (COMPUGRAPH).

[2]Note that this notion of isomorphism of derivation sequences has only been handled implicitely in [Kre 76] and [EKT 92]. The results in [CELMR 93] showed that the right choice of this abstraction up to isomorphism is not an easy task. Hence, we introduce our isomorphism concept for derivations explicitely in sections 2 and 3.

either be done by restricting the class of redices or, less restrictively, by reducing the considered sequences to those satisfying this property for each parallel step they contain. This analysis has been carefully performed in [Din 92], where one of us showed that single-pushout rewriting satisfies the laws of [EKT 92] in these cases and, therefore, should possess unique canonical sequences up to isomorphism between derivations.

The following paper addresses the general case. We directly apply the proof techniques of [Kre 76] to single-pushout sequences and show the existence of canonical sequences. Hence the concept of canonical derivations is also relevant in rewriting approaches whose parallelism concept is not based on "independence of order".

The paper is self-contained. Section 2 introduces and recapitulates the basic notions and results of single-pushout rewriting [Löw 90/93] and provides the notion of abstract derivation which is suitable for the scope of this paper. The corresponding notions of parallel derivation, abstract parallel derivations, and parallel independence are introduced in section 3. It also provides examples showing parallel steps which are not sequentializable. Section 4 is dedicated to the proof that the equivalence of abstract sequences based on this notion of independence allows for canonical derivation sequences. The concluding remarks in section 5 compare parallelism in double- and single-pushout rewriting. By a typical example we show that the single-pushout parallelism considerably exceeds the double-pushout parallelism in general.

## 2. Single-Pushout Graph Transformation

A comprehensive overview over the theory of the algebraic approach to single-pushout graph transformation is given in [Löw 93][3]. Here, we only review the basic notions and results that are needed as a basis for the construction of canonical derivations in the following sections.

Although the general theory is applicable to all types of "graph structures", which are algebraic categories with unary operators only (see [Löw 93]), we consider only the category of labeled directed graphs as a typical member of this class here. It simplifies the presentation considerably. With the results of [Löw 93], however, it should be easy to carry all results of this paper over to arbitrary graph structures (hence to make the results applicable for example to hypergraphs).

**Definition 2.1 (Graph, Partial Graph Morphism)** A graph $G = (V, E, s, t, l)$ consists of a set of vertices V, a set of edges E, two mappings s and t which provide a source resp. a target vertex for every edge, and a mapping $l:V \cup E \rightarrow L$ which provides a label for every object of the graph in a given label alphabet L. A partial graph morphism $f:G \rightarrow H$ between graphs G and H is a pair of partial mappings $(f_V:G_V \rightarrow H_V, f_E:G_E \rightarrow H_E)$ which satisfy:

(1) If $f_E$ is defined for $e \in G_E$, $f_V$ is defined for $s^G(e)$ and $t^G(e)$ and we have (i) $f_V(s^G(e)) = s^H(f_E(e))$, (ii) $f_V(t^G(e)) = t^H(f_E(e))$, and (iii) $l^H(f_E(e)) = l^G(e)$.

(2) If $f_V$ is defined for $v \in G_V$ then $l^H(f_V(v)) = l^G(v)$.

Note that the domain of $f:G \rightarrow H$ together with the source and target mappings of G restricted to this domain provide a subgraph of G which is denoted by G(f) in the following. Thus a partial graph morphism $f:G \rightarrow H$ is just a total morphism $f_T:G(f) \rightarrow H$ from some subgraph G(f), the *domain* of f, to H.

[3]Related notions of single-pushout rewriting can be found in [Rao 84, Ken 87, Bro 91].

**Definition 2.2 (Underlying Category)** All graphs together with all partial graph morphisms form a category $\mathbf{GRA^P}$. Composition of morphisms is given by componentwise composition of partial mappings and the identities are given by pairs of (total) identities on the components.

Note that the category $\mathbf{GRA^T}$ of graphs and total graph morphisms is a subcategory of $\mathbf{GRA^P}$ having the same object class. Furthermore, we have the following results:

**Proposition 2.3 (Total and Partial Morphisms)** (1) Colimits in $\mathbf{GRA^T}$ are also colimits in $\mathbf{GRA^P}$. (2) Co-products in $\mathbf{GRA^P}$ coincide with the co-products in $\mathbf{GRA^T}$.

**Proof:** (1) follows from the fact that the inclusion functor $\subseteq:\mathbf{GRA^T} \to \mathbf{GRA^P}$ from graphs with total morphisms into the category of graphs with partial morphisms has a right-adjoint, see [Löw 93]. (2) is a consequence of (1) since the inclusion functor is surjective on objects. □

Essential for the algebraic way of single-pushout rewriting is the following fact.

**Theorem 2.4 (Pushouts in $\mathbf{GRA^P}$)** The category $\mathbf{GRA^P}$ has all pushouts. For a pair of morphisms $f:A \to B$ and $g:A \to C$, it can be constructed as the triple $(D, f^*:C \to D, g^*:B \to D)$ by the following three steps:

**1.** The common domain $A^* = A(f^* \bullet g) = A(g^* \bullet f)$ of $f^* \bullet g$ and $g^* \bullet f$ is the largest subgraph of A satisfying (i) $A^* \subseteq A(f) \cap A(g)$ and (ii) for all $x \in A^*$ and $y \in A$:$[f(x) = f(y)$ or $g(x) = g(y)]$ implies $y \in A^*$. Property (ii) means that $A^*$ is closed w.r.t. the equivalence that f and g induce on A.

**2.** $B(g^*)$, the domain of the pushout morphism $g^*$, is the largest subgraph of B contained in $B - f(A - A^*)$. Symmetrically $C(f^*)$ is the largest subgraph of C in $C - g(A - A^*)$.

**3.** If $f'$ and $g'$ denote the restrictions of f and g to $A^*$ on the domain and $B(g^*)$ resp. $C(f^*)$ on the codomains, we construct D, $f^*_T$, and $g^*_T$ as the pushout of the total morphisms $f'$ and $g'$ in $\mathbf{GRA^T}$.

**Proof:** See [Löw 93].

The construction of theorem 2.4 is depicted in diagram 2.1 and immediately provides the following corollaries for injective partial morphisms[4].

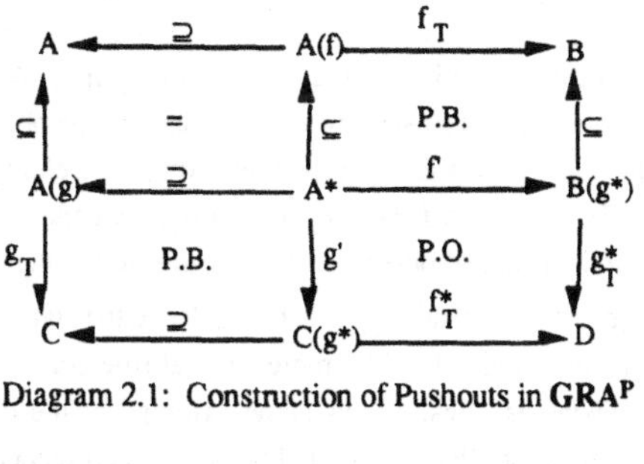


Diagram 2.1: Construction of Pushouts in $\mathbf{GRA^P}$

**Corollary 2.5 (Pushout Properties)** If $(D, f^*:C \to D, g^*:B \to D)$ is the pushout of $f:A \to B$ and $g:A \to C$ in $\mathbf{GRA^P}$, we have: (1) $f^*$ is injective if f is injective, (2) $A^* = A(f) \cap A(g)$ if both f

[4]A partial morphism is injective if [f defined for x and y and $f(x) = f(y)$] implies $x = y$.

and g are injective, and (3) if f and g are injective we obtain a total morphism $f^* \bullet g \bullet m$ for each morphism $m:X \to A$ such that $f \bullet m$ and $g \bullet m$ are total.

**Proof:** (1) By the construction in theorem 2.4, f' and g' are injective if f and g are. That $f^*$ and $g^*$ are injective follows from the corresponding property in $\mathbf{GRA^T}$, compare [Ehr 79]. (2) If f and g are injective, their induced equivalence is the diagonal on A, which immediately provides the statement 2. (3) The morphism $f \bullet m$ being total means that m is total and $m(X) \subseteq A(f)$. Symmetrically $g \bullet m$ total means $m(X) \subseteq A(g)$. Thus $m(X) \subseteq A(f) \cap A(g)$, which means $m(X) \subseteq A^* = A(f^* \bullet g)$ since f and g are injective, see 2. above. Therefore $(f^* \bullet g) \bullet m$ is total. □

The construction of pushouts in $\mathbf{GRA^P}$ is rather complex. This is mainly due to the complicated interaction of the partiality and the identification of the two morphisms f and g. This interaction reveals interesting operational effects if pushouts in $\mathbf{GRA^P}$ are taken as a model for direct derivations in graph transformation.

**Definition 2.6 (Single-Pushout Graph Transformation)** The algebraic approach to single-pushout graph transformation can be completely described in the category $\mathbf{GRA^P}$:

A *graph transformation rule* $r:L \to R$ is an **injective** morphism in $\mathbf{GRA^P}$. A *redex* $m:L \to G$ for a rule $r:L \to R$ in a mother graph G is a **total** morphism in $\mathbf{GRA^P}$, i.e. $m \in \mathbf{GRA^T}$. The *application* of a rule $r:L \to R$ at a redex $m:L \to G$ transforms the mother graph G to the daughter graph H where $(H, r^*:G \to H, m^*:R \to H)$ is the pushout of the rule r and the redex m. The application is also called *direct derivation*, $r^*$ is called *derivation morphism*, and we write G[r, m>H in this situation. *Sequential derivations* with a rule set R are sequences of rule applications, i.e. the smallest set containing: (i) G[$\mathbf{id}_G$>G for every graph G (the sequences of length 0) and (ii) G[(r, m); *X*>H for every rule application G[r, m>K and derivation sequence K[*X*>H.

**Remarks:** Note that we consider the ;-operator to be associative. Thus we omit brackets in the following and write G[(r, m); (s, n); (t, p)>H for a three step derivation or short (r, m); (s, n); (t, p) if we are not interested in the start and result graph. We use capital letters taken from the end of the alphabet as variables for derivation sequences, i.e. *V*, *W*, *X*,... The sequences of length 0 are considered to be neutral w.r.t. the ";"-operator. Hence we let *X*; **id** = *X* = **id**; *X*.

The operational idea behind these definitions in 2.6 is the following: A rule $r:L \to R$ specifies a rewriting by declaring a set of items which shall be deleted, namely the objects outside L(r), and a set of items which shall be added in an application of the rule, namely the elements outside the image of r in R, i.e. outside of r(L). The objects in L(r) resp. r(L) provide a context which can be used in the right-hand side to specify *how* the added objects shall be connected to the rest of the mother graph.

Note that due to the injectivity of the rules, $L(r) \cong r(L)$. Hence the context is not changed when the right-hand side replaces the left-hand side. The more general rule concept of [Löw 93] which also allows non-injective morphisms as rules, allows some "gluing" of the context. We consider the restricted case firstly to obtain an intuitive operational behaviour and secondly to obtain the results in section 3, which are not true for the general case (compare remarks in section 5).

Redices indicate the place in a mother graph where the effect of a rule shall take place. They are required to be total, since the pattern of a rule's left-hand side shall work as a precondition for rule application. Hence, in order for a rule to be applicable, a homomorphic image of the *whole* left-hand side must be found. Redices are quite flexible since they are not required to be injective. This issue will be important for parallel rewriting steps in section 3.

Rule application operationally performs the steps 1. - 3. of theorem 2.4. The substeps 1. and 2. erase everything from the mother graph which has a preimage under the redex in L - L(r). Note that this implies that deletion is dominant w.r.t preservation of objects. This means: if a redex identifies objects x, y ∈ L such that x ∈ L(r) and y ∈ (L - L(r)), the corresponding object in the mother graph is deleted. Besides that deletion is context-wide, i.e. if a vertex which is object of deletion has some incoming or outgoing edges, these edges are deleted as well (compare 2.4 (2))[5]. The pushout construction in the third step of proposition 2.4 does the addition of the objects in R - r(L).

Due to the complex construction of pushouts in **GRA**$^{\mathbf{P}}$, we obtain a very simple notion of rule application on the rewriting level. A single-pushout comprises all operational effects, namely deletion, addition and embedding of objects. Thus, the complexity of the underlying construction pays off on a level that handles rewriting steps as atomic entities. This becomes obvious in the straightforward theory of parallel rewriting in the single-pushout approach presented in the following section.

Note that the result of rule application is only determined up to isomorphism due to the same property of pushouts. Hence algebraic rewriting is not concerned with the transformation of concrete graphs but with the transformation of *abstract graphs* which are given by isomorphism classes of concrete graphs, i.e. $[G] = \{G' \mid G \equiv G'\}$. This equivalence on objects immediately leads to equivalence classes of derivation sequences: We say that two direct derivations G1[r, m1>H1 and G2[r, m2>H2 are equivalent if there is an isomorphism $i: G1 \to G2$ such that $i \circ m1 = m2$. As it is indicated in diagram 2.2, we obtain a unique isomorphism $i^*: H1 \to H2$ in this case which satisfies $i^* \circ r1^* = r2^* \circ i$ and $i^* \circ m1^* = m2^*$ since H1 is a pushout object and $r2^* \circ i \circ m1 = r2^* \circ m2 = m2^* \circ r$. That $i^*$ is an isomorphism directly follows from H2 being pushout as well which provides an inverse $(i^*)^{-1}$ satisfying $(i^*)^{-1} \circ r2^* = r1^* \circ i^{-1}$ and $(i^*)^{-1} \circ m2^* = m1^*$. With these preliminaries, we are able to define abstract derivation sequences precisely:

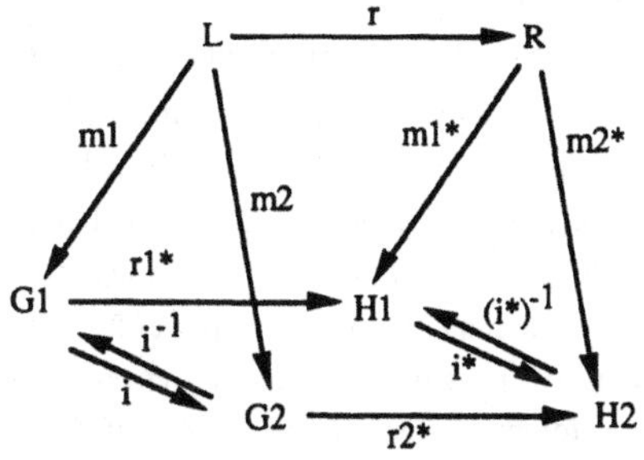


Diagram 2.2: Induced Isomorphism

**Definition 2.7 (Abstract Sequential Derivations)** Two derivations G1[*X1*>H1 and G2[*X2*>H2 are equivalent w.r.t. an isomorphism $i: G1 \to G2$, written $X1 \equiv_i X2$, if (i) *X1* and *X2* are both derivations of length 0 or (ii) *X1* = G1[(r, m1); *Y1*>H1, *X2* = G2[(r, m2); *Y2*>H2, $i \circ m1 = m2$, and $Y1 \equiv_{i^*} Y2$ where $i^*$ is the uniquely induced isomorphism of diagram 2.2. We say that *X1* and *X2* are *equivalent* if there is an isomorphism i such that $X1 \equiv_i X2$ and write $X1 \equiv^a X2$. *Abstract derivations* $[X]^a$ are equivalence classes w.r.t. $\equiv^a$, i.e. $[X]^a = \{Y \mid X \equiv^a Y\}$.

[5]In the case of non-injective morphisms as rules, deletion within a rule application is even more complex, since identifications of the rule and the redex can considerably interact.

**Remarks:** It is easy to check that $\equiv^a$ is reflexive, symmetric, and transitive using arguments based on identities, inverse isomorphisms and composition of isomorphisms. Note that equivalent derivations must be of same length *n*. In the case of equivalence G1[*X1*>H1 $\equiv_i$ G2[*X2*>H2, we obtain a family of *n*+1 isomorphisms between all graphs which represent intermediate states of the derivations. All these isomorphisms are uniquely determined by the one between the start graphs. The last one between H1 and H2 is also denoted by $i^*$ in the following.

Due to this fact, $\equiv^a$ is not a congruence w.r.t. the sequence operator ";". For a discussion of the problems which are caused by this fact compare [CEMLR 92/93]. Within the scope of this paper, we are not interested in the sequential composition of abstract derivations. Therefore $\equiv^a$ provides the right notion of equivalence for our purposes.

## 3. Independence and Parallelism in Single-Pushout Rewriting

The fundamental idea for the modelling of parallel steps in algebraic graph rewriting is the use of direct derivations with *parallel rules*. Here parallel rules r + s are just co-products of the component rules r and s. Combining redices m for a rule r and n for s into a *parallel redex* m ⊕ n, a parallel step G[(r, m) + (s, n)>H can be defined by G[r + s, m ⊕ n>H. Hence there is no additional derivation concept required for parallel steps. Precise meaning is given to these concepts by the following definition:

**Definition 3.1 (Parallel Rewriting)** The *parallel rule* r + s:L + M → R + S for rules r:L → R and s:M → S is given by the disjoint union of the components. Thus (L + M, i:L → L + M, j:M → L + M) and (R + S, e:R → R + S, f:S → R + S) are the coproducts of L and M resp. R and S. The morphism r + s is uniquely determined by the properties (r + s) ∘ i = e ∘ r and (r + s) ∘ j = f ∘ s. The rules r and s are called *components* of r + s.

The *parallel redex* m ⊕ n:L + M → G for two given redices m:L → G for r:L → R and n:M → G for s:M → S is the unique morphism satisfying (m ⊕ n) ∘ i = m and (m ⊕ n) ∘ j = n.

The *parallel application* G[(r, m) + (s, n)>H of the rule r at the redex m and the rule s at the redex n is given by the simple application of r + s at m ⊕ n, i.e. G[r + s, m ⊕ n>H.

The *parallel steps* over a rule set R is the smallest set containing: (i) (r, m) and (ii) (r, m) + *A* if (r, m) is a rule application with r ∈ R and *A* is a parallel step over R.

*Parallel derivation sequences* w.r.t. a rule set R are sequences of parallel steps over R, i.e. the smallest set containing (i) $\mathbf{id}_G$ for every graph G and (ii) *A*; *X* if *A* is a parallel step over R and *X* is a parallel sequences over R.

**Remarks:** Due to proposition 2.3 all co-product embeddings are total morphisms as well as induced parallel redices. Note that the co-product operator + as well as the redex operator ⊕ are commutative and associative; commutative in the following sense: If a rule p is the co-product object of r and s, i.e. p = r + s, then it is also the co-product object of s and r, i.e. p = s + r; and for redices m and n for r and s resp., we have m ⊕ n = n ⊕ m for the parallel rule p if we choose r + s = p = s + r. Thus from G[(r, m) + (s, n)>H, it follows that G[(s, n) + (r, m)>H with exactly the same defining diagram. Hence we omit brackets in the following and do not distinguish parallel steps if they only differ in the order of the components. Individual parallel steps are denoted by capital letters taken from the beginning of the alphabet, i.e. by *A*, *B*, *C*,... If we talk about derivation sequences in the following, we always mean parallel derivations.

The idea of abstract sequences formulated in definition 2.7 can easily be lifted to parallel derivation sequence. We only have to define equivalence of parallel steps:

**Definition 3.2 (Abstract Parallel Derivation Sequences)** Two parallel steps G1[*A1*>H1 and G2[*A2*>H2 are *equivalent* w.r.t. an isomorphism i:G1 → G2, written $A1 \equiv_i A2$, if (i) *A1* = (r,m1), *A2* = (r,m2), and i ∘ m1 = m2 or (ii) *A1* = (r,m1) + *B1*, *A2* = (r,m2) + *B2*, i ∘ m1 = m2, and $B1 \equiv_i B2$. Two parallel derivation sequences G1[*X1*>H1 and G2[*X2*>H2 are *equivalent* w.r.t. an isomorphism i:G1 → G2, written $X1 \equiv_i X2$ if (a) both *X1* and *X2* are sequences of length 0 or (b) *X1* = *A1*; *Y1*, *X2* = *A2*; *Y2*, $A1 \equiv_i A2$, and $Y1 \equiv_{i^*} Y2$ for parallel steps *A1* and *A2*, parallel sequences *X1* and *X2*, and i* the induced isomorphism[6] of the parallel rule applications *A1* resp. *A2*. *Abstract parallel derivation sequences* are equivalence classes $[X]^a$ w.r.t. this equivalence, i.e. $[X]^a = \{Y \mid X \equiv_i Y$ for some isomorphism i$\}$.

**Remark:** Note that this inductive definition allows parallel steps to be equivalent only if exactly the same rules are applied in both steps. Thus in the case that we have pairwise different rules r, s, t, u ∈ R, (r, m) + (s, n) and (t, o) + (u, p) cannot be equivalent even if r + s = t + u and m ⊕ n = o ⊕ p. This is due to (i) and (ii) in definition 3.2. This choice seems reasonable since we do not want to loose track of the applied rules within the abstraction process.

The fundamental relationship between parallel application and derivation sequences with the component rules is provided by the butterfly lemma, which is true in every category (see for example [Kre 76]):

**Lemma 3.3 (Butterfly Lemma)** Let r + s be the co-product of r and s and i, j, e, and f the corresponding co-product embeddings: Subdiagram (1) in diagram 3.1 is a pushout, if and only if there are pushouts (2)-(4) where the redices are related by m = (m ⊕ n) ∘ i and n = (m ⊕ n) ∘ j, the derivation morphisms satisfy $(r + s)^* = y \circ s^* = x \circ r^*$, and the right-hand side embeddings provide the relationship $x \circ m^* = (m \oplus n)^* \circ e$ and $y \circ n^* = (m \oplus n)^* \circ f$.

**Proof:** Due to the fact that the construction of colimits is commutative and associative.

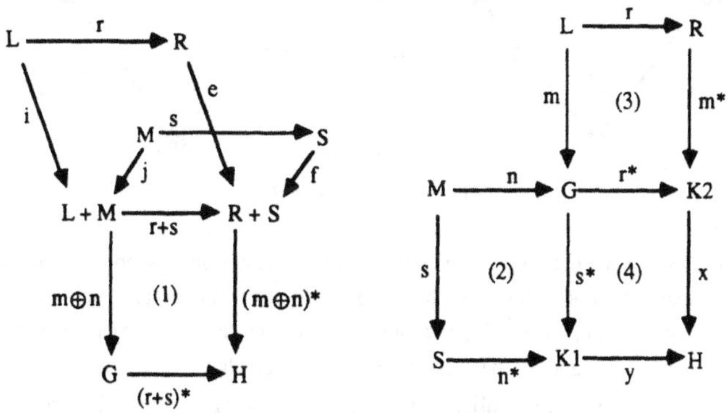


Diagram 3.1: Butterfly Lemma

[6]If G1[(r, m1) + (s, n1)>H1 and G2[(r, m2) + (s, n2)>H2 are equivalent w.r.t. an isomorphism i: G1 → G2, we have (i) i ∘ m1 = m2 and (ii) i ∘ n1 = n2. Again we obtain an induced isomorphism $i^*$: H1 → H2 since (i) and (ii) imply i ∘ (m1 ⊕ n1) = (m2 ⊕ n2). The morphism $i^*$ is again uniquely determined by the property that diagram 2.2 commutes if we substitute r by r + s, m1 by (m1 ⊕ n1), and m2 by (m2 ⊕ n2). Here we can take advantage of the fact that parallel applications are just applications of parallel rules at parallel redices.

The butterfly lemma tells us that the effect of a parallel application of r at m and s at n can be obtained by first applying both rules at their redices independently. This provides the two derivation morphisms $r^*$ and $s^*$ whose pushout (compare (4) in diagram 3.1) is the result of the parallel application.

The butterfly lemma provides the basis for the sequentialization relation which allows to rearrange the individual rule applications in a derivation sequence without changing the overall effect. The basic notion in this respect is the concept of *residual* within single-pushout rewriting. For a motivation of this concept consider again diagram 3.1.

Since the composition of pushouts provides pushouts again, the subdiagram (2)+(4) can be considered a rule application of s at $r^* \circ n$ provided that $r^* \circ n$ is a redex, i.e. is total. In this case, we call $r^* \circ n$ a *residual* of n after (r, m). If $r^* \circ n$ is a residual, the parallel step G[(r, m) + (s, n)>H could be "simulated" by the sequence G[r, m>K2 [s, $r^* \circ n$>H.

**Definition 3.4 (Residual)** Given two rule applications (r:L → R, m:L → G) and (s:M → S, n:M → G) as in diagram 3.1, we say that n has a residual after (r, m) if $r^* \circ n$ is a redex, i.e. is total. The derivation step (s, $r^* \circ n$) is also called (s, n) *at a residual after* (r, m) in the following.

The existence of a residual for (s, n) after (r, m) is sometimes also called *(parallel) independence* of (s, n) *from* (r, m). The double pushout approach, only considers a symmetric notion of independence: (r, m) and (s, n) are independent if they are independent from each other. This is mainly due to the fact that only symmetrically independent steps can be combined into a parallel step. Vice versa, the existence of a parallel step G[(r, m) + (s, n)>H in double-pushout rewriting always implies both (r, m) being independent from (s, n) and (s, n) being independent from (r, m). That this is not true in the single-pushout framework is illustrated by the following examples:

**Examples 3.5 (Residuals)** Consider the rule r in figure 3.2 as the only rule in a rule set R.

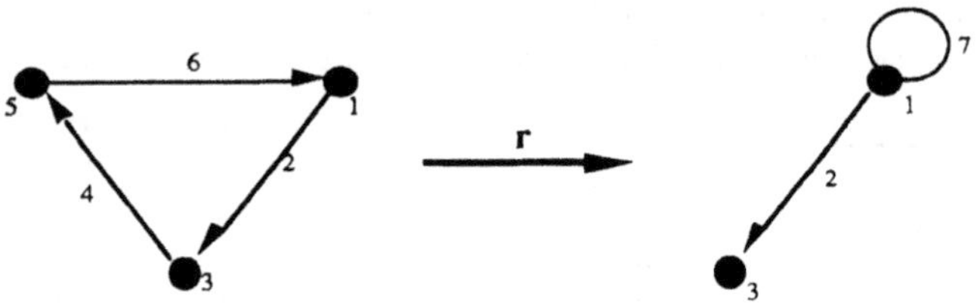


Figure 3.2: Sample Rule

Its operational effects is the deletion of one vertex and two edges and the addition of a loop. The rule morphism is indicated by corresponding object numbers. In this example L(r) = {1,2,3}, L - L(r) = {4, 5, 6}, and R - r(L) = {7}. For this example, we call L(r) the set of objects that r needs, L - L(r) the set of objects r destroys, and R - r(L) the set of objects r creates.

Figure 3.3 shows a parallel application of r + r:G → H at the redex m1 ⊕ m2 which is symmetrically independent. The individual redices m1 and m2 only overlap in needed objects. Thus $r1^* \circ m2$ and $r2^* \circ m1$ are total morphisms and therefore residuals. Figure 3.4 shows a different situation. Here the redices overlap in objects which one rule needs and the other rule destroys.

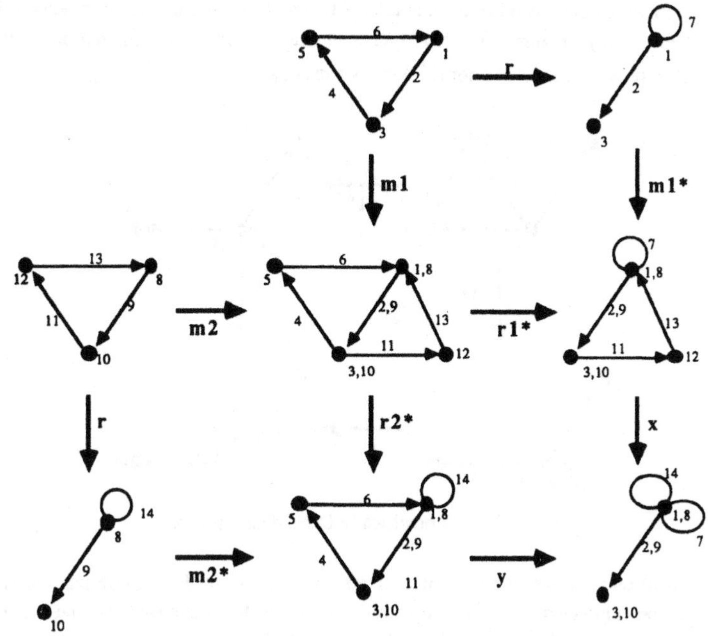


Figure 3.3: Symmetrically Independent Situation

Hence only $r1^* \circ m2$ is a residual and $r2^* \circ m1$ is not. The needed vertex (3) is destroyed by the rule application at redex m2.

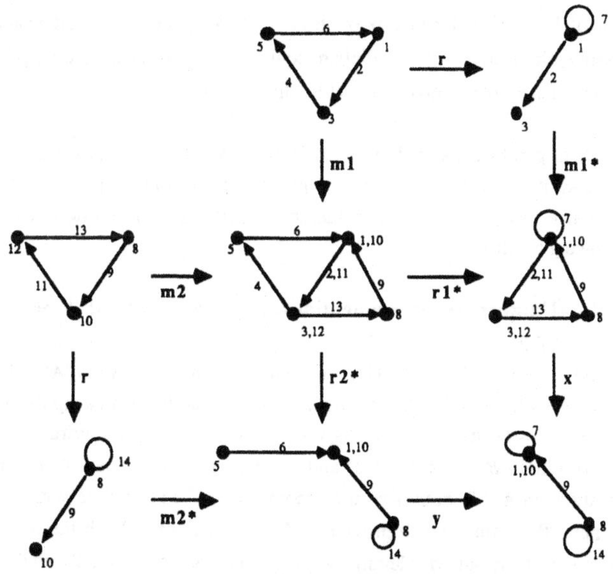


Figure 3.4: Situation of Non-Symmetrical Independence

For a situation in which there is no residual at all, the two redices have to overlap in objects that both rule applications destroy. Figure 3.5 depicts such a situation where m1 and m2 overlap in the edges (6) resp. (13). Here neither $r1^* \circ m2$ nor $r2^* \circ m1$ is a residual. □

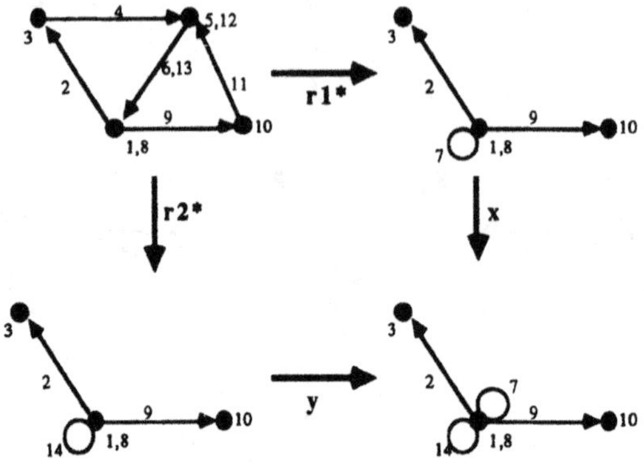


Figure 3.5: Symmetrically Dependent Situation

As we have mentioned above, the existence of residuals allows to decompose parallel steps into sequences with the components. And the rewriting at residuals does not change the overall effect of a derivation sequence. It only changes the order in which the component rules are applied. The following equivalence is the right tool if we want to abstract from these details of order in a sequence.

**Definition 3.6 (Parallelism Equivalence)** If we have derivation sequences $V = X; A + B; Y$ and $W = X; A; B'; Y$ such that $B'$ is $B$ at a residual after $A$, we call $W$ the $(X, B)$-sequentialization of $V$, written $V \rightsquigarrow_{X,B} W$ or short $V \rightsquigarrow W$ if we are not interested in the exact position of the sequentialization. The parallelism equivalence $\equiv^p$ is the smallest equivalence on the parallel derivation sequences over a given rule set R which contains $\rightsquigarrow$.

If we have $V \rightsquigarrow W$, all graphs computed in $V$ also occur in $W$. But there is exactly one intermediate graph in $W$ that does not occur in $V$. And this graph is only unique up to isomorphism due to definition 3.6. Hence $\rightsquigarrow_{X,B}$ cannot be considered a partial function on derivations. We have, however, the following result.

**Proposition 3.7 (Uniqueness of Sequentialization up to $\equiv^a$)** If $V \rightsquigarrow_{X,B} W1$ and $V \rightsquigarrow_{X,B} W2$, then $W1 \equiv^a W2$.

**Proof:** By definition 3.6, $V$ = F[$X$>G[$A$ + $B$>K[$Y$>J, $W1$ = F[$X$>G[$A$>H1[$B1$>K[$Y$>J, and $W2$ = F[$X$>G[$A$>H2[$B2$>K[$Y$>J. And $B1$ and $B2$ are both residual applications of $B$. If we let $A$ = (r, m) and $B$ = (s, n), the following diagram shows the two rewritings from G to K in $W1$ resp. $W2$. We obtain $B1 = (s, r1^* \circ n)$ and $B2 = (s, r2^* \circ n)$. Since $A; B1$ and $A; B2$ are both sequentializations of $A + B$, both derivation morphisms $s1^* \circ r1^*$ and $s2^* \circ r2^*$ coincide with $(r + s)^*$ due to the butterfly lemma which provides $s1^* \circ r1^* = s2^* \circ r2^*$. With the relationship of the right-hand embeddings of the butterfly lemma, we also obtain $s1^* \circ m1^* = s2^* \circ m2^*$. Since H1 and H2 are both pushouts of r and m, there is an isomorphism $i: H1 \rightarrow H2$ with $i \circ m1^* = m2^*$ and $i \circ r1^* =$

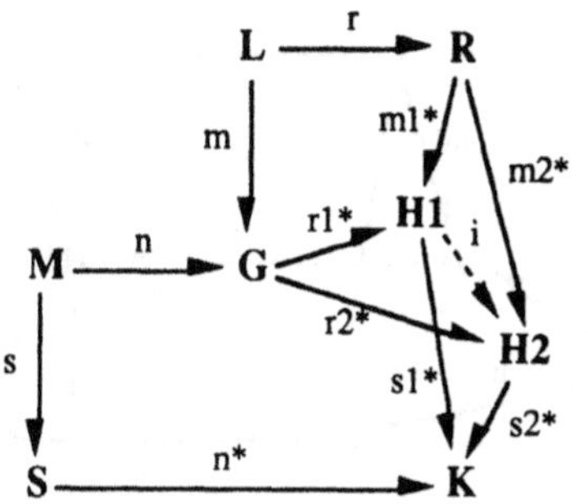


r2*. Thus i = $id_G{}^*$, compare diagram 2.2. It remains to show that $id_K$ = i*, which requires $id_K \circ s1^*$ = $s2^* \circ i$. For this, $s1^* \circ r1^* = s2^* \circ r2^* = (s2^* \circ i) \circ r1^*$ and $s1^* \circ m1^* = s2^* \circ m2^* = (s2^* \circ i) \circ m1^*$ is sufficient since H1 is pushout. Thus $W1 \equiv_{id_F} W2$. □

In the following we carefully investigate the parallelism equivalence on derivation sequences, which abstract from a pre-given order of rule application up to residuals. Especially we are interested in so-called canonical derivation sequences w.r.t. $\equiv^p$ which realize a maximal amount of parallelism. As a prerequisite for this analysis, we need the following properties of $\rightsquigarrow$.

**Proposition 3.8 (Injectivity of Sequentialization)** The $\rightsquigarrow$-relation is injective, i.e. if $V1 = X; A + B1; Y$, $V2 = X; A + B2; Y$, $W = X; A; B; Y$, and $B1$ and $B2$ apply the same (multi-) set of rules such that $V1 \rightsquigarrow_{X,B1} W$ and $V2 \rightsquigarrow_{X,B2} W$, then $B1 = B2$ and $V1 = V2$.

**Proof:** Let $A$ = (r, m) and $B$= (s, n). It follows that $B1$= (s, n1) and $B2$= (s, n2) since $B$ is residual of $B1$ and $B2$ and therefore all three use the same parallel rule s. If r* is the derivation morphism of $A$, we have $r^* \circ n1 = n = r^* \circ n2$ due to n being residual of n1 and n2. Since r* is injective (compare corollary 2.5(1)) and $r^* \circ n1$ and $r^* \circ n2$ are total, it follows that n1 = n2. Hence $B1$ = $B2$ and $V1 = V2$. □

Furthermore, there are the following compatibilities of $\rightsquigarrow$ with $\equiv^a$.

**Proposition 3.9 (Relation of $\rightsquigarrow$ with $\equiv^a$)** If $V1 \rightsquigarrow W1$ and $V1 \equiv^a V2$, then there is $W2$ such that $V2 \rightsquigarrow W2$ and $W1 \equiv^a W2$. Vice versa: if $V1 \rightsquigarrow W1$ and $W1 \equiv^a W2$, then there is $V2$ such that $V2 \rightsquigarrow W2$ and $V1 \equiv^a V2$.

**Proof:** For the first case, we obtain from $V1 \rightsquigarrow W1$ that $V1 = X1; A1 + B1; Y1$ and $W1 = X1; A1; B1'; Y1$ and from $V1 \equiv^a V2$ that there is an isomorphism i such that $X1 \equiv_i X2$, $A1 \equiv_j A2$, $B1 \equiv_j B2$, and $Y1 \equiv_k Y2$, where the isomorphism j is induced by i, $X1$, and $X2$ and k is induced by j, $A1 + B1$, and $A2 + B2$. If we let $A1$ = (r, m1), $A2$ = (r,m2), $B1$ = (s, n1), and $B2$ = (s, n2), the following diagram shows the interesting part of the whole situation: $s1^* \circ r1^*$ resp. $s2^* \circ r2^*$ are the derivation morphisms for $A1 + B1$ resp. $A2 + B2$ due to the butterfly lemma. Let h:H1 → H2 be the isomorphism induced by j. Then $r2^* \circ n2 = r2^* \circ j \circ n1 = h \circ r1^* \circ n1$. It is total since h and $r1^* \circ n1$ are total. Thus it is a residual for $B2$ after $A2$. If we call it $B2'$, we get $V2 \rightsquigarrow X2; A2; B2'; Y2 := W2$. For $W1 \equiv_i W2$, it remains to show that k is the induced isomorphism of h. This is straightforward with the butterfly lemma, compare proof of proposition 3.7.

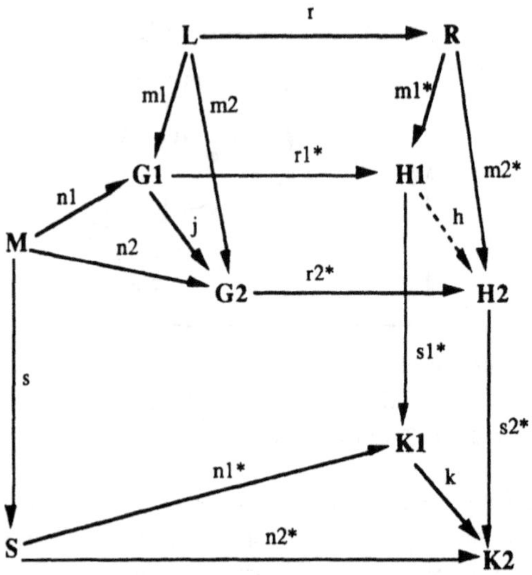


In the second case, there is the following situation: $V1 = X1; A1 + B1; Y1$, $W1 = X1; A1; B1'; Y1$, $W2 = X2; A2; B2; Y2$, and $X1 \equiv_i X2$, $A1 \equiv_j A2$, $B1' \equiv_h B2'$, and $Y1 \equiv_k Y2$, where i induces j, j induces h, and h induces k. If $A1$ = (r, m1), $A2$ = (r, m2), $B1'$ = (s, r1* ∘ n1), and $B2'$ = (s, n2), the following diagram provides an overview. $B1'$ = (s, r1* ∘ n1) is due to the fact that it is a residual of $B1$ after $A1$. Now we have n2 = h ∘ (r1* ∘ n1) = r2* ∘ (j ∘ n1) and, therefore, n2 is a residual of j ∘ n1 after $A2$. If we let (s, j ∘ n1) = $B2$, we obtain $V2 = X2; A2 + B2; Y2 \rightsquigarrow W2$. And $V1 \equiv_i V2$ because the whole diagram above commutes. □

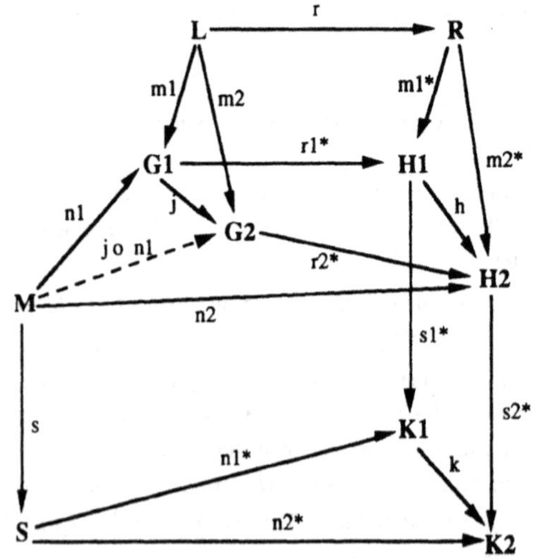

## 4. Canonical Parallel Derivation Sequences

In this section, we prove that every $\equiv^p$-equivalence class of derivation sequences contains a canonical member which is uniquely determined up to the equivalence $\equiv^a$ and which realizes maximal parallelism. Here maximal parallelism is meant in the following sense: $X \in [Y]_{\equiv^p}$ has maximal parallelism if every individual rule application in $X$ is performed earlier than in every other sequence $Z \in [Y]_{\equiv^p}$. This means that in every parallel step of $X$, *all* rule applications are performed in parallel for which the required resources have already been computed by the preceeding sequence. Thus $X$ is maximal parallel; no rule application is unnecessarily delayed in $X$.

With the remarks in section 2 and 3, it is clear that the uniqueness of these so-called *canonical parallel derivation sequences* can only be obtained up to the equivalence $\equiv^a$ since rule application is only unique up to isomorphism. And if we want to compare two equivalent derivation sequences w.r.t. the "earliness " of their rule applications, we need a concept of "same rule application up to residual". Thus we say that $B$ and $B'$ are the *same parallel steps* if $X; A + B; Y \rightsquigarrow_{X,B} X; A; B'; Y$. This concept of parallel step up to residual (-equivalence) implies that two equivalent derivation sequences contain the *same* set of rule applications.

The main idea for the construction of canonical derivation is the design of a rewrite system $>>^*$ on the derivation sequences with the following properties: Each rewrite step *shifts* at least one rule application further towards the beginning of teh sequence and the normal forms of this rewrite system are uniquely determined up to $\equiv^a$.

**Definition 4.1 (Shift)** The shift relation $>>$ (resp. $>>_{X,B}$ if we want to mention the position of the shift) on derivations is given by: $V >>_{X,B} W$ if (i) $V = X; A; B; Y$ and $W = X; A + B'; Y$ where $B$ is $B'$ at a residual after $A$ or (ii) $V = X; A; B + C; Y$ and $W = X; A + B'; C'; Y$ where $B$ is $B'$ at a residual after $A$ and $C'$ is $C$ at a residual after $B$. The *shift system* on derivation sequences is given by the reflexive and transitive closure of $>>$. We denote it by $>>^*$.

The shift relation is not a partial function on concrete derivation sequences. But if we abstract using the equivalence $\equiv^a$, we obtain:

**Proposition 4.2 (Uniqueness of Shift up to $\equiv^a$)** If $V >>_{X,B} W1$ and $V >>_{X,B} W2$, then we have $W1 \equiv^a W2$.

**Proof:** In case (i) of definition 4.2, it follows directly that $W \rightsquigarrow_{X,B} V$, and case (ii) implies the existence of $Z = X; A; B; C'; Y$ such that $V \rightsquigarrow_{(X;A),C} Z$ and $W \rightsquigarrow_{X,B} Z$. Thus uniqueness of shift is a consequence of the propositions 3.7, 3.8, and 3.9. □

**Proposition 4.3 (Relation of $>>$ with $\equiv^a$)** $V1 >> W1$ and $V1 \equiv^a V2$ implies the existence of $W2$ such that $V2 >> W2$ and $W1 \equiv^a W2$.

**Proof:** Direct consequence of propositions 3.7, 3.8, and 3.9. □

**Proposition 4.4 (Termination)** The shift system on derivation sequence is terminating.

**Proof:** For the proof, we need the following notions of *length* and *delay factor* of a derivation sequence: The length of a derivation sequence X is given by (i) length(id) = 0, (ii) length(r,m) = 1, and (iii) length(X; Y) = length(X) + length(Y). If (r, m) is a rule application in a sequence V, i.e. V =

X; A + (r, m); Y, the delay factor of (r, m) in V is given by length(X) + 1. The delay factor of a derivation sequence V is the sum of the delay factors for every individual rule application in V.

We show the proposition by proving delay(*W*) < delay(*V*) if *V* >> *W*. In the case (i) of definition 4.1, we have *V* = *X*; *A*; *B*; *Y* and *W* = *X*; *A* + *B'*; *Y*. Thus the delay factors for the rule applications in *X* and *A* coincide in *V* and *W* but the factors for alls rules in *B* and *Y* are exactly 1 less in *W* than in *V*. Since *B* is not empty, the delay of *W* is less than the delay of *V*.

In case (ii) of definition 4.1, we have *V* = *X*; *A*; *B* + *C*; *Y* and *W* = *X*; *A* + *B'*; *C'*; *Y* . Again the delays for all rules in *X*, *A*, *C*, and *Y* are the same in *V* and *W*. But the delay factors for all rules in *B* are decreased by one during the shift from *V* to *W*. □

A straightforward consequence of proposition 4.4 is that >> has normal forms, which are called canonical sequences in the following. In the rest of this section, we show that these normal forms are uniquely determined up to $\equiv^a$. Due to propositions 4.2 and 4.4, it is sufficient for this purpose to prove that all *critical pairs* of the shift relation >> are confluent.

**Lemma 4.5 (Confluence of Critical Pairs)** All critical pairs of the relation ">>" are confluent.
**Proof:** All critical pairs (*W* >> *V1*, *W* >> *V2*) are special cases of the following two patterns:
(i) *W=X;A;B+C;D+E;Y , V1=X;A+B1;C1;D+E;Y , V2=X;A;B+C+D2;E2;Y*
(ii) *W=X;A;B+C+D+E;Y, V1=X;A+B1+C1;D1+E1;Y, V2=X;A+C2+D2;B2+E2;Y*
All other critical pairs can be obtained by letting *X,Y,C*, or *E* be empty. Hence the arguments which follow for the general cases (i) and (ii) can easily be adopted to these special instances.

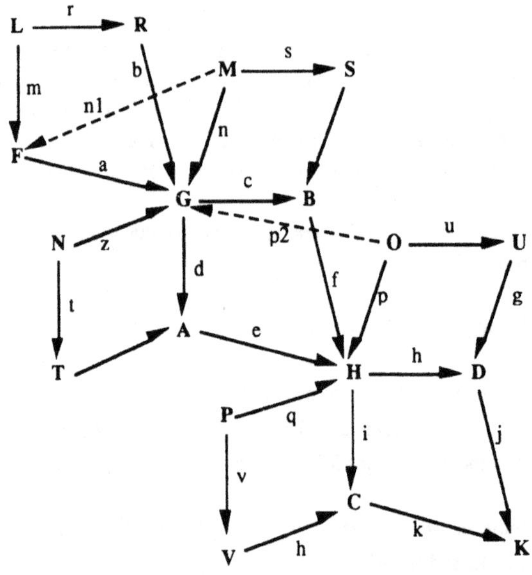


If we let *A* = (r, m), *B* = (s,n), *C* = (t, z), *D* = (u, p), and *E* = (v, q), the diagram above depicts the situation in the first case. Each quadrangle is a pushout due to the butterfly lemma and a, e ∘ d = f ∘ c, j ∘ h = k ∘ i are the derivation morphisms for *A*, *B* + *C*, and *D* + *E* resp. Since *W* >> *V1*, and *W* >> *V2* we know that (i) there is a redex n1:M → F with a ∘ n1 = n and *B1* = (s, n1), (ii) c ∘ z is a residual, used in *C1* = (t, c ∘ z), (iii) there is a redex p2:O → G with e ∘ d ∘ p2 = p = f ∘ c ∘ p2 and

*D2* = (u, p2), and (iv) $h \circ q$ is total in *E2* = (v, $h \circ q$). We want to show that *D* can be shifted in *V1*. We choose $p3 = c \circ p2$. It is a redex since $f \circ (c \circ p2)$ is total. And $f \circ p3 = f \circ c \circ p2 = p$ such that *D* is a residual of (u, p3) = *D3* after (t, $c \circ z$) which is *C1*. The morphism $h \circ q$ is total anyhow. Thus *D* can be shifted in *V1* and we get *V1* >> *X;A+B1;C1+D3;E2*. Now we show that *B* can be shifted in *V2*: We choose the redex n1 in F and obtain immediately $a \circ n1 = n$. It remains to show that *C+D2* has a residual after *B*, i.e. that $c \circ (z \oplus p2)$ is total: $c \circ (z \oplus p2) = (c \circ z) \oplus (c \circ p2)$. But $c \circ z$ is total by assumption (ii) and $c \circ p2$ is total since $f \circ c \circ p2$ is total by assumption (iii). Hence the parallel redex is total due to proposition 2.3. Thus *V2* >> *X;A+B1;C1+D3;E2* := Z. Hence *V1* >> Z, *V2* >> Z and the critical pair is confluent.

The following diagram shows the situation in the second case if we let *A* = (r, m), *B* = (s,n), *C* = (t, z), *D* = (u, p), and *E* = (v, q). In the diagram, the butterfly lemma has been applied three times such that all quadrangles in the figure are pushouts.

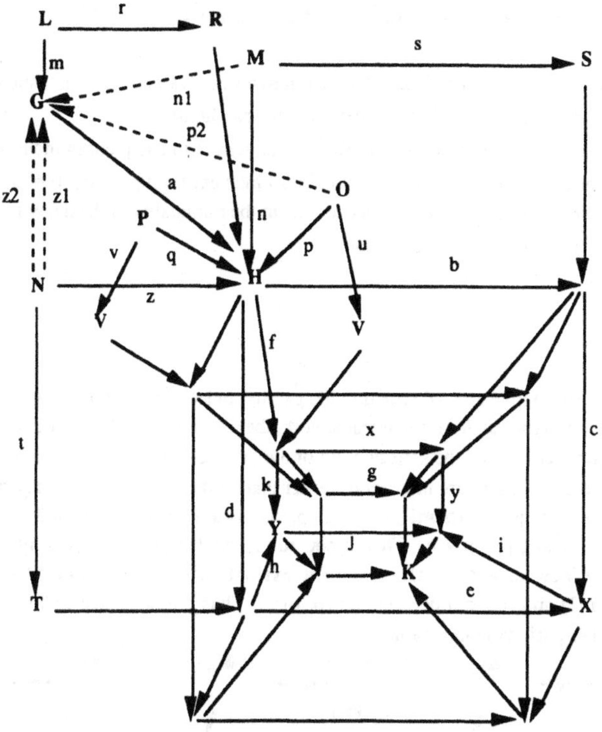


Due to the assumptions of possible shifts, we have (i) there is $n1 \oplus z1 : N + M \rightarrow G$ such that $a \circ (n1 \oplus z1) = n \oplus z$, (ii) $c \circ b \circ (p \oplus q) = e \circ d \circ (p \oplus q)$ is total, (iii) there is $z2 \oplus p2 : N + O \rightarrow G$ with $a \circ (z2 \oplus p2) = z \oplus p$, and (iv) $k \circ f \circ (n \oplus q) = h \circ d \circ (n \oplus q)$ is total.

(i) implies $a \circ n1 = n$ and $a \circ z1 = z$. (ii)implies $c \circ b \circ p$, $c \circ b \circ q$, $e \circ d \circ p$, and $e \circ d \circ q$ are total. (iii) provides that $a \circ z2 = z$ and $a \circ p2 = p$. Since a is injective and z is total, it can be concluded that $z1 = z2$. And with (iv) can be argued that $k \circ f \circ n$, $k \circ f \circ q$, $h \circ d \circ n$, and $h \circ d \circ q$ are total.

*V1* can be analysed as follows: *B1* = (s, n1), *C1* = (t, z1), *D1* = (u, $e \circ d \circ p$), and *E1* = (v, $e \circ$

d o q). And the components of $V2$ are: $B2$ = (s, k o f o n), $C2$ = (t, z2), $D2$ = (u, p2), and $E2$ = (v, k o f o q). We show that $D1$ can be shifted in $V1$: For this, $D1$ must be a residual of some $D3$ after $A+B1+C1$. This can be shown by considering the redex p2 since the derivation morphism of $A+B1+C1$ is c o b o a = e o d o a by the butterfly lemma and e o d o a o p2 = e o d o p is total by (ii). Thus (u, p2) = $D2$ has the residual $D1$ after $A+B1+C1$. It remains to show that $E1$ has a residual after $D1$, which means to show that (i o e o d) o q is total. But by assumption (ii) e o d o q is total and f o q is also total due to the implications of assumption (iv) that k o f o q is total. With corollary 2.5(3) (i o e o d) o q is total. Thus $V1$ >> $X;A+B1+C1+D2;E3;Y$, where $E3$ = (u, i o e o d o q).

Symmetrically, it can be proved that $B2$ can be shifted in $V2$ using the redex n1 of $B1$ and providing the same $E3$ as a left-over in the last step. Thus $V2$ >> $X;A+B1+C2+D2;E3;Y$. Since z1 = z2, $C1 = C2$, which provides confluence of this type of critical pair. □

**Theorem 4.6 (Confluence of Shift)** The shift relation is confluent on *abstract* derivations.
**Proof:** Consequence of propositions 4.2, 4.3, 4.4 and lemma 4.5. □

**Corollary 4.7 (Maximal Parallelism)** The parallelism equivalence $\equiv^P$ has canonical derivation sequences which are unique up to $\equiv^a$ and realize maximal parallelism.
**Proof:** Definition 4.1 implies that the equivalences on derivation sequences induced by >> and $\rightsquigarrow$ coincide. Thus the normal forms of the shift relation, which exist uniquely up to $\equiv^a$ due to proposition 4.4 and theorem 4.6, provide canonical sequences due to the minimality of the delay factor within these sequences for every individual rule application. □

## 5. Conclusions

Although parallel rule applications do not provide parallel independent applications of the component rules in general within the single-pushout framework, there are canonical derivations in this approach which can be constructed using the shift relation of the last section. The following example in figure 5.1 shows that these sequences are in general shorter than in the double-pushout rewriting. In single-pushout rewriting the sequence (p, m1); (p, m2); (p, m3) as it is shown in figure 5.1 is "shiftable" to a one step sequence using p + p + p. This is not true in double-pushout rewriting since p + p + p requires pairwise independence of all rules. The example has been constructed such that (p, m1) depends on (p, m2) and (p, m2) depends on (p, m3). Thus the sequence in figure 5.1 is already canonical in double-pushout graph grammars.

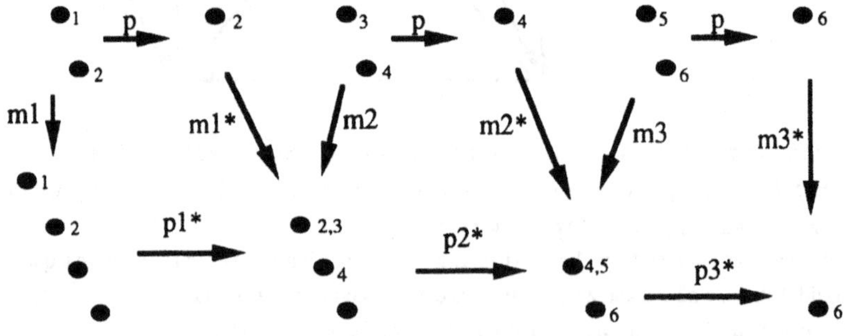


Figure 5.1: Example

Some of the interesting questions for future research that are raised by the presented paper are:

1. Is there a chance for canonical derivations if we admit non-injection rules?
2. How does the theory change if we allow the parallel composition of whole derivation sequences instead of rule applications?
3. Are there other interesting members in the equivalence classes w.r.t. $\equiv^p$ and $\equiv^a$?
4. Do canonical derivations provide a good basis for computational and true concurrent semantics for single-pushout transformations?

(1) The proof technique we have applied above heavily relies on injective rule morphisms. This is mainly due to some application of corollary 2.5 in the fourth section. Hence it does not carry over to non-injective rules. On the other hand, it is easy to provide counterexamples for the existence of canonical sequences in the presence of these rules. The criteria which could be sufficient for canonical sequence in this case, however, are far from being obvious.

(2) The butterfly lemma and the existence of derivation morphisms provide a very easy tool for the definition of parallel composition of longer sequences within single-pushout rewriting: It is just taking the pushout of the composed derivation morphisms of the individual steps. It would be interesting to investigate canonical sequences in this broader context and how they are connected to the theory presented here.

(3) Inverting the shift relation immediately provides a device to compute "lazy sequences". Their uniqueness might require an additional analysis. Interesting questions would be to compare the length and width of these sequences with the canonical ones. Other interesting members within the equivalence classes $\equiv^p$ would be those which optimize the parallelism in each step w.r.t. a constant $n$ such that *each* step optimally does not more and not less than $n$ things in parallel. This would be a nice application for parallel computing.

(4) We are currently working on the evaluation of a true concurrent semantics for double-pushout graph transformation [CEMLR 92/93]. The presented paper started to apply and generalize the ideas for single-pushout rewriting. It is an interesting question whether the canonical sequences, discussed above, contain enough information to extract the causal dependencies of abstract parallel derivations.

## References


[Bro 91] P. M. van den Broek: Algebraic graph rewriting using a single pushout,. Proc. Int. Joint Conf. on Theory and Practice of Software Development (TAPSOFT'91), Springer LNCS 493 (1991), pp.90 - 102.

[CEMLR 92] Á. Corradini, H. Ehrig, U. Montanari, M. Löwe, F. Rossi: Note on Standard Representation of Graphs and Graph Derivations. Techn. Rep. No. 92-25, TU Berlin, FB 20, 1992 (Revised version to appear in this volume).

[CEMLR 93] A. Corradini, H. Ehrig, U. Montanari, M. Löwe, F. Rossi: Abstract Graph Derivations in the Double Pushout Approach, in this volume.

[Din 92] J. Dingel: Kanonische Ableitungssequenzen im Single-Pushout-Ansatz für Graphtransformation. Techn. Rep. No. 92-36, TU Berlin, FB 20, 1992 (in English).

[Ehr 79] H. Ehrig: Introduction to the algebraic theory of graph grammars. 1st Workshop on Graph Grammars and Their Application to Computer Science, Springer LNCS 73 (1979), pp. 1-69.

[EKT 92] H. Ehrig, H.-J. Kreowski, G. Taentzer: Canonical derivations for high-level replacement systems. Techn. Report 6/92, Univ. of Bremen, FB Informatik, 1992.

[Ken 87] R. Kennaway: On "on graph rewriting". Theoret. Computer Science 52 (37-58), 1987.

[Kre 76] H.-J. Kreowski: Kanonische Ableitungssequenzen für Graphgrammatiken. Techn. Report No. 76-26, TU Berlin, FB 20, 1976.

[Löw 90] M. Löwe: Extended Algebraic Graph Transformation. PhD-thesis, TU Berlin, FB 20, 1990.

[Löw 93] M. Löwe: Algebraic approach to single-pushout graph transformation. Theoret. Computer Science 109 (181-224), 1993.

[Rao 84] J. C. Raoult: On graph rewriting. Theoret. Computer Science 32 /1-24), 1984.

# Semantics of Full Statecharts Based on Graph Rewriting*

Andrea Maggiolo-Schettini, Adriano Peron

Dipartimento di Informatica, Università di Pisa
corso Italia 40, 56125 Pisa, Italy
E_mail: {maggiolo,peron}@di.unipi.it

**Abstract**

A semantics of statecharts based on graph rewriting is presented. Statecharts are formalized as graph replacement rules. The graph of derivations gives a sequential semantics which agrees with statechart step semantics.

## 1 Introduction

The formalism of statecharts ([4]) offers interesting description facilities for reactive systems like hardware components, communication networks, computer operating systems. Statecharts have the visual appeal of formalisms like state diagrams and Petri nets, but, with respect to both of them, they offer facilities of hierarchical structuring of states and modularity, which allow high level description and stepwise development. In [5] a semantics is given in terms of steps, where a step is a set of consistent transitions from a given system configuration, which are enabled under a given external stimulus. In the present paper we propose an operational semantics of statecharts using graph rewriting. The usefulness of graph rewriting as a mathematical tool for operational semantics has been already shown (e.g. see [7]). We believe that graph rewriting can be a proper "abstract machine" for our purposes. Actually, a rich collection of results about parallelism and concurrency in graph grammar theory could be exploited in order to formally analyze properties of parallel executions of statecharts. The given semantics could also be used for implementation purposes. In details, we shall show how to translate a statechart into a set of graph productions and how to describe the step from a configuration of the statechart to another in terms of a derivation from a graph describing the start configuration to a graph describing the reached configuration. The sequential behaviour of the represented statecharts will be described by the derivation graph. In the complete version of the work (see [10]), where proofs of the propositions in this paper can be found, it is also shown how a more abstract partial ordering semantics can be obtained from that defined here by suitably mapping partial orders over transitions onto the derivation graph. This partial order semantics can be shown to coincide with that given in [11] in terms of families of configurations of Labelled Event Structures.

In section 2 we recall the definition of statecharts and their semantics, following [5]. In section 3 we introduce the basic concepts of graph rewriting we need. In section 4 we define the sequential semantics of statecharts in terms of graph rewriting.

*Research partially supported by the COMPUGRAPH Base Research Exprit Working Group n. 7183.

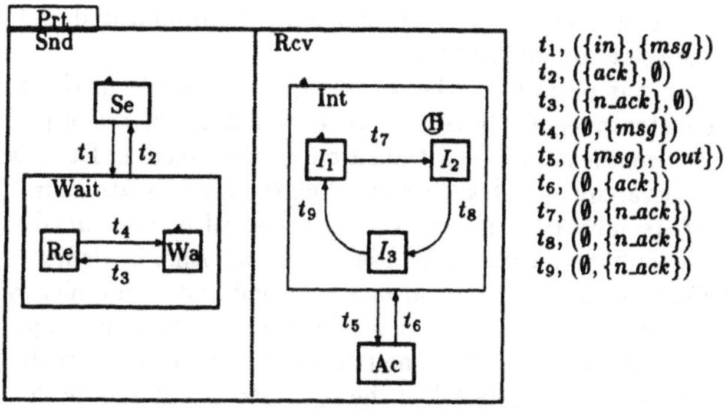


Figure 1: A specification example.

## 2 Statecharts

We introduce the main features of statecharts by showing the specification of the communication protocol of a system consisting of a sender and a receiver. It will be our running example.

The sender communicates the message (signal *msg*) to the receiver after being prompted by a sending request (signal *in*), and waits for an acknowledgement from the receiver (signal *ack*). If the acknowledgement is not communicated one unit of time after sending the message, the communication of the message is repeated (the procedure "wait and repeat" goes on until the acknowledgement is communicated). The receiver performs some internal activities and, when it is prompted by the sender, either it continues its activity or it services the communication request. In the latter case, a signal is communicated to the environment, an acknowledgement is communicated to the sender and the internal activity is resumed at the stage it was when it has been interrupted.

Statecharts is a visual formalism which enriches state-transition diagrams by a hierarchical structuring of states, explicit representation of parallelism and communication among pararallel components. The graphical convention is that states are depicted as boxes, and the box of a substate of another state is drawn inside the area of the box of that state. States are either of type *OR*, called or-states, or of type *AND*, called and-states. And-states are depicted as boxes whose area is partitioned by dashed lines. Each element of the partition is the root state of a statechart describing a "parallel component" of the and-superstate. When the system is in an or-state, it is in one and only one of its immediate substates and when it is in an and-state, it must be in all of its immediate substates simultaneously.

With reference to figure 1, state *Prt* is an and-state (all the other states are or-states) consisting of the two parallel components representing the sender (state *Snd*) and the receiver (state *Rcv*). State *Snd* is in substate *Se* when it is waiting for a sending request, it is in substates *Wait* and *Wa* when it is waiting for an acknowledgement, and it is in substates *Wait* and *Re* when it is ready to repeat the communication. State *Rcv* is in the substate *Int* when it is performing its internal activities and it is in substate *Ac* when it receiving the *msg* communication.

Transitions are represented graphically by arrows and are labelled by a pair event-action, where an event is a set of primitive and negated primitive events and the action is a set of primitive events. Primitive events are interpreted as pure signals communicated by the environment. A transition is "enabled" if the set of primitive

events of the event part of the label are currently communicated and the negated primitive event are not communicated.
When a transition $t$ is performed, the set of primitive events in the action part of the label of $t$ is instantaneously broadcast, so augmenting the set of primitive events offered by the environment. When a transition is performed and the target state is not a leaf in the hierarchy of states, then entering the target state causes "implicitly" entering some of its substates: if it is an and-state, all of its immediate substates, otherwise only one of its immediate substates is chosen, either by default or by history. The default substate of an or-state is always specified and is graphically depicted as a non labelled dangling arc leading to this substate. The substate implicitly entered by history is the most recently exited one (they are marked by an encircled H).

With reference to our example, when the state *Prt* is entered, both *Snd* and *Rcv* are entered. The default entrance for *Wait* is the state *Wa*, so when the state Wait is entered by means of the transition $t_1$ also the state *Wa* is simultaneously entered. If the state *Int* is entered by means of the transition $t_6$, then the immediate substate is chosen by history.

In the example we have exploited all the main features of statecharts, but the ability of declaring and referencing globally visible variables of specifiable abstract data types. The treatment of this feature can be found in [10].

**Definition 2.1** *A statechart $Z$ is a 9-tuple $(B, \rho, \phi, T, in, out, P, \chi, \delta, H)$ where:*

- *$B$ is the finite set of* states;
- *$\rho : B \rightarrow 2^B$ is the* hierarchy function *giving for a state the set of its immediate substates; $\rho^+$ and $\rho^*$ denote the irreflexive and reflexive transitive closure of $\rho$, resp., and $\rho$ is s.t.:*
  *–there exists a unique $b \in B$ such that $B = \rho^*(b)$ (it is called the* root *state);*
  *–$b \notin \rho^+(b)$, for all $b \in B$;*
  *–if $b, b' \in B$, $b' \notin \rho^+(b)$, $b \notin \rho^+(b')$, then $\rho^+(b) \cap \rho^+(b') \neq \emptyset$ implies $b = b'$.*
  *A state $b$ is* basic *iff $\rho(b) = \emptyset$. The* lowest common ancestor *of two states $b$ and $b'$, denoted as $LCA(b, b')$, is the state $\bar{b} \in B$ such that $\rho^*(\bar{b}) \supseteq \{b, b'\}$ and, for each $b''$ such that $\rho^*(b'') \supseteq \{b, b'\}$, $\bar{b} \in \rho^*(b'')$;*
- *$\phi : B \rightarrow \{AND, OR\}$ is the* type function;
- *$T$ is the finite set of* transitions;
- *$in, out : T \rightarrow B$ are functions giving the* target *and* source *state of a transition, respectively; it must be $\phi(LCA(t)) = OR$, for all $t \in T$ ($LCA(t)$ stands for $LCA(out(t), in(t))$); moreover $in(t) \notin \rho^*(out(t))$ and $out(t) \notin \rho^*(in(t))$;*
- *$P$ is the finite set of* primitive events; *$\neg P$ denotes the set $\{\neg e : e \in P\}$ assuming that $P \cap \neg P = \emptyset$;*
- *$\chi : T \rightarrow 2^{P \cup \neg P} \times 2^P$ is the* transition labelling function; *$Ev^+(t)$ denotes $fst(\chi(t)) \cap P$, $Ev^-(t)$ denotes $\{e : \neg e \in fst(\chi(t))\}$, and $Act(t)$ denotes $snd(\chi(t))$ (maps $fst$ and $snd$ gives the first and the second component of a pair, respectively);*
- *$\delta : \phi^{-1}(OR) \rightarrow B$, with $\delta(b) \in \rho(b)$ if $\rho(b) \neq \emptyset$ and undefined otherwise, is the* default function;
- *$H \subseteq \phi^{-1}(OR)$ is the* history set *consisting of non-basic or-states.*

In [5] statecharts have been originally endowed with a "step-semantics" which enforces a "synchrony hypothesis". The environment prompts the statechart with primitive events which are related to a discrete time domain (for instance, natural numbers $\mathcal{N}$). The history of the prompts of the environment is described as a sequence of sets of primitive events $E_1, \ldots, E_n, \ldots$, intending that the primitive events in $E_n$ are communicated at the $n$-th instant of time. At a fixed instant of time $n$ a statechart reacts to the set of communicated primitive events $E_n$ by performing a number of transitions enabled by $E_n$, so changing the set of currently entered states and possibly augmenting the set of communicated events $E_n$ by the events in the action part of the performed transitions. In this way a larger set of transitions might be enabled and a chain reaction might occur. Since events can be sensed only at the instant of time they have been communicated and transitions which are performed in the same instant of time may affect only parallel components of the statechart, the reaction to $E_n$ is finite and instantaneous. We introduce now the concepts of configuration, microstep, step and behaviour of a statechart.

**Definition 2.2** *A set of states $D \subseteq B$ is* orthogonal *iff:*
- *if $d, d' \in D$ and $LCA(d,d') \neq d$ or $LCA(d,d') \neq d'$, then $\phi(LCA(d,d')) = AND$;*
- *if $d \in D$ and $d \in \rho(d')$, for some $d' \in B$, then $d' \in D$;*
- *if $d \in D$, $\phi(d) = AND$, and $\rho(d) \cap D \neq \emptyset$, then $\rho(d) \subseteq D$.*

*An orthogonal set $D$ is* downward closed *iff $\rho(d) \cap D \neq \emptyset$, for all non basic $d \in D$.*
*A* configuration *at time $\tau$, with $\tau \in \mathcal{N}$, is a tuple $C = (D, en, hist, test)$, where:*
- *$D \subseteq B$ is an orthogonal set;*
- *$en: D \rightarrow \{0, \ldots \tau+1\}$ is the* enabling function*;*
- *$hist: B \rightarrow B \times \{0, \ldots \tau\}$, defined only for non-basic or states, is the* history function*;*
- *$test: \mathcal{N} \rightarrow 2^P$ is the* test function*;*

*The configuration $C$ is* maximal *iff $D$ is downward closed, it is* initial *iff it is maximal, $\tau = 0$, $\delta(b) \in D$ and $hist(b) = (\delta(b), 0)$, for all non-basic $b \in \phi^{-1}(OR)$, $en(b) = 1$, for all $b \in D$.*

A downward closed orthogonal set gives a maximal set of states which can be simultaneously entered consistently with the requirements that all the immediate substates of an entered and-state and only one immediate substates of an entered or-state must be entered. The enabling function gives the lowest temporal bound before which a state cannot be exited. The history function gives, for each non basic or-state, the most recently released substate and the time when it has been exited. The test function gives the set of primitive events the environment offers at each time.

**Definition 2.3** *Let $C = (D, en, hist, test)$ be a configuration at time $\tau$. A transition $t$ is* enabled *in $C$ iff $out(t) \in D$, $en(d') \leq \tau$, for all $d' \in \rho^+(LCA(t)) \cap D$, $Ev^+(t) \in test(\tau)$ and $Ev^-(t) \cap test(\tau) = \emptyset$. Two transitions $t$ and $t'$ are* event disjoint *iff $(Ev^+(t) \cup Ev^-(t)) \cap Act(t') = \emptyset$ and $(Ev^+(t') \cup Ev^-(t')) \cap Act(t) = \emptyset$; they are* structurally consistent *iff $\phi(LCA(LCA(t), LCA(t'))) = AND$.*
*A set $S$ of transitions is a* microstep *from $C$ iff each transition in $S$ is enabled in $C$ and transitions in $S$ are pairwise structurally consistent and event disjoint.*

Two transitions are structurally consistent whenever the two sets of states they affect (i.e., entered and exited states) belong to parallel components of the statechart. They are event disjoint whenever the action of a transition does not affect the enabling of the other. The requirement that transitions in a microstep are event disjoint is not in [5] and has been introduced here to guarantee that transitions in a microstep can be performed simultaneously as well as sequentially in any order.

We consider now how a configuration is changed due to the performance of a microstep. When a transition $t$ is performed, all the current substates of the source state are exited together with all the current ancestors of the source up to $LCA(t)$. States are entered which are in the downward closure of the hierarchical path from $LCA(t)$ to the target state of $t$. In the following we define the set of states which are in the path between a state $b$ and one of its substates $b'$, and we define the downward closure of a set of states with respect to default and history functions.

**Definition 2.4** *For $b, b' \in B$ with $b' \in \rho^*(b)$,* the path from $b$ to $b'$, *denoted by $B^b_{b'}$, is the set inductively defined as follows:*
- $b \in B^b_{b'}$;
- *if $x \in B^b_{b'}$, $b' \in \rho^+(x)$ and $\phi(x) = AND$, then $\rho(x) \subseteq B^b_{b'}$;*
- *if $x \in B^b_{b'}$, $x' \in \rho(x)$, $b' \in \rho^*(x')$, and $\phi(x) = OR$, then $x' \in B^b_{b'}$.*

*For a configuration $C = (D, en, hist, test)$ and $K \subseteq B$,* the downward closure of $K$ in $C$, *denoted by $\Downarrow_C K$, is inductively defined as follows*
- $K \subseteq \Downarrow_C K$;
- *if $d \in \Downarrow_C K$, $\phi(d) = AND$, then $\rho(d) \subseteq \Downarrow_C K$;*
- *if $d \in \Downarrow_C K$, $\rho(d) \cap \Downarrow_C K = \emptyset$, and $d \in H$, then $d' \in \Downarrow_C K$, with $d' = fst(hist(d))$;*
- *if $d \in \Downarrow_C K$, $\rho(d) \cap \Downarrow_C K = \emptyset$, $d$ is non-basic, $\phi(d) = OR$, and $d \notin H$, then $\delta(d) \in \Downarrow_C K$.*

We describe now how configurations are related by microsteps.

**Definition 2.5** *Let $C = (D, en, hist, test)$ be a configuration at time $\tau$ and $\Psi$ be a microstep from $C$; the configuration $(D', en', hist', test')$* reached from $C$ via $\Psi$ *is:*
- $D' = (D - \bigcup_{t\in\Psi} \rho^+(LCA(t))) \cup \bigcup_{t\in\Psi} \Downarrow_C B^{LCA(t)}_{in(t)}$;
- *$en'(d)$ = if $d \in D' - D$, then $\tau + 1$, otherwise $en(d)$, for all $d \in D'$;*
- *$hist'(d)$ = if $\rho(d) \cap (D - D') = \{b\}$, then $(b, \tau)$, otherwise $hist(d)$, for all $d \in B$;*
- *$test'(n)$ = if $n = \tau$ then $test(\tau) \cup \bigcup_{t\in\Psi} Act(t)$, otherwise $test(n)$, for all $n \in \mathcal{N}$.*

**Definition 2.6** *Let $C$ and $C'$ be two maximal configurations at time $\tau$; $S \subseteq T$ is a* step from $C$ to $C'$ *iff there exists a sequence (possibly empty) $\Psi_0, \ldots, \Psi_n$ of pairwise disjoint microsteps, a sequence $C_0, \ldots, C_{n+1}$ of maximal configurations at time $\tau$ such that $S = \bigcup_{i=0}^n \Psi_i$ is a set of pairwise structurally consistent transitions, $C_0 = C$, $C_n = C'$, $C_{i+1}$ is reached from $C_i$ via the microstep $\Psi_i$, for $1 \leq i \leq n+1$, and if $\Psi$ is a microstep from $C'$, $S \cup \Psi$ is not a set of pairwise structurally consistent transitions.* A behaviour *from an initial configuration $C$ is a sequence of steps $S_0, \ldots S_n$ such that there exists a sequence of configurations $C_0, \ldots, C_{n+1}$, with $C_0 = C$ and $S_i$ is a step from $C_i$ to $C_{i+1}$, for $0 \leq i \leq n$.*

# 3 Graph Rewriting

We adopt the *algebraic approach* ([2]) to graph rewriting. The kernel of the algebraic theory includes, besides the basic definitions of production and derivation, also some Church-Rosser and parallelism results. Given a graph to be rewritten, two production applications are "parallel independent" if their redexes are either disjoint or their intersection is preserved by both productions. In this case the two productions can be applied to the graph in any order (and in parallel as well) producing isomorphic graphs. This property can be exploited in order to give "truly concurrent semantics". The technique of graph rewriting we shall exploit in this section slightly changes the standard algebraic approach which is given in the categorical setting of graphs with color preserving morphisms. Actually, color preserving graph morphisms turn out to be too restrictive for our purposes since we have to handle variables and evaluate expressions possibly appearing in the action part of a transition. Following [6], we want to have graphs labelled over structured alphabets, namely algebras, and we consider graph morphisms which preserve the structure of the labels. In this way, one may define a signature $\Sigma$ and take advantage of algebraic techniques of specification and semantics (see [12]) in order to define the proper abstract data type (in the sense of [12], an isomorphism class of algebras). The productions are triple of graphs labelled over the free algebra of terms $T_\Sigma(X)$ for a given set of variables $X$, and the graphs to be rewritten are graphs labelled over a concrete algebra $A$ of the defined abstract data type. Therefore, having fixed a valuation of $X$ in the concrete algebra $A$, we consider a category whose objects are graphs labelled either over $T_\Sigma(X)$ or $A$. The arrows are color preserving morphisms, if both the graphs connected by the arrow are labelled over the same algebra, or morphisms which preserve the (unique) homomorphism of algebras induced by the given valuation of the set of variables, if the source of the arrow is labelled over $T_\Sigma(X)$ and the target is labelled over the concrete algebra $A$. Recently, in [3], the algebraic theory of graph grammars has been generalized to arbitrary underlaying categories, and Church-Rosser properties and Parallelism theorem have been proved to hold for a large class of categories (they are called HRL1). Also the category we shall define belongs to this class.

**Definition 3.1** *A directed graph $G$ labelled over a color alphabet $C$ is a tuple ($V_G$, $E_G$, $sc_G$, $tg_G$, $l_G$), where $V_G$ and $E_G$ are the disjoint sets of* vertices *and* edges, *respectively, $sc_G, tg_G : E_G \to V_G$ are two functions assigning source and target to edges, $l_G : V_G \cup E_G \to C$, is the function coloring vertices and edges. A* morphism *between a graph $G$ over $C$ and a graph $J$ over $C'$ is a pair of maps $f : V_G \cup E_G \to V_J \cup E_J$ and $\mu : C \to C'$ such that:*
- *$f(V_G) \subseteq V_J$ and $f(E_G) \subseteq E_J$;*
- *$sc_J(f(e)) = f(sc_G(e))$ and $tg_J(f(e)) = f(tg_G(e))$, for all $e \in E_G$;*
- *$l_J(f(v)) = \mu(l_G(v))$, for all $v \in V_G \cup E_G$.*

**Definition 3.2** *Let $\Sigma$ be a signature over a set of sorts $S$, $X$ a family of variables indexed over $S$, $A$ a $\Sigma$-algebra and $\sigma : X \to A$ a valuation. The category of graphs $\sigma$-valued on $A$, denoted by $GRAPHS_A^{T_\Sigma(X)}(\sigma)$, is defined as follows:*
- *the class of objects is given by graphs over $A$ and graphs over $T_\Sigma(X)$;*
- *the class of arrows from an object $O$ to $O'$ is the class of morphism from $O$ to $O'$ of the form $(f, \mu)$, where: if $O$ and $O'$ are graphs colored over the same alphabet, then $\mu$ is the identity on the color alphabet;*
  *if $O$ is a graph over $T_\Sigma(X)$ and $O'$ is a graph over $A$, then $\mu$ is the unique homomorphism of $\Sigma$-algebras from $T_\Sigma(X)$ to $A$ which extends the valuation $\sigma$;*
  *if $O$ is a graph over $A$ and $O'$ is a graph over $T_\Sigma(X)$, then there exists no morphism.*

**Fact 3.3** • *In the category $GRAPHS_A^{T_\Sigma(X)}(\sigma)$ a commutative square is a pushout (resp. a pullback) iff the edge and the vertex components are pushouts (resp.: pullbacks) in the category $SETS$.*
- *The category $GRAPHS_A^{T_\Sigma(X)}(\sigma)$ has coproducts, pushouts, pullbacks.*

A production $p$ is a pair of injective morphisms $b_1 : K \to L$, $b_2 : K \to R$ where $K$, $L$ and $R$ are graphs over $T_\Sigma(X)$. As usual, the production $p$ can be applied to the graph $G$ over $A$ yielding a graph $H$ over $A$ if there is a graph morphism $g : L \to G$, and

$H$ is obtained as a result of a "double pushout" construction (for the formal definition see [3]). A direct derivation from a graph $G$ to a graph $H$ via the production $p$ in the category $GRAPHS_A^{T_\Sigma(X)}(\sigma)$ will be written as $G \overset{p,\sigma}{\Longrightarrow} H$.

# 4 Statecharts semantics by graph rewriting

We shall show how a configuration of a statechart may be represented as a graph and steps by means of derivations by means of productions corresponding to transitions. We consider a fixed statechart $(B, \rho, \phi, T, in, out, P, \chi, \delta, H)$.

**Definition 4.1** *Let us take a fixed correspondence from $B$ to $\overline{B} = \{\overline{b}\colon b \in B\}$ which associates the symbol $\overline{b} \in \overline{B}$ to each $b \in B$, and assume $\overline{B} \cap B = \emptyset$. Let $S = \{state, event, time\}$ be the set of sorts; the signature $\Sigma$ over $S$ is:*
- $\Sigma_{\Lambda,state} = B \cup \overline{B} \cup \{clock\}$;
- $\Sigma_{\Lambda,event} = P \cup \{True, False\}$ *(assume $P \cap \{True, False\} = \emptyset$);*
- $\Sigma_{\Lambda,time} = \{0\}$; $\Sigma_{time,time} = \{Succ\}$; $\Sigma_{timetime,time} = \{+\}$;
- $\Sigma_{w,s} = \emptyset$ *for any other $w \in S^\star$, $s \in S$.*

*Let $Ax = \{o{+}x = x,\ x{+}y = y{+}x,\ x{+}(y{+}z) = (x{+}y){+}z,\ Succ(x{+}y) = Succ(x){+}y\}$ a set of equations. The* color alphabet *for configurations is an initial model for the specification $SP = (\Sigma, Ax)$.*

Sorts *state* and *event* contain only symbols of constants. Each symbol for states and events of the fixed statechart is then a constant. The set of symbols $\overline{B}$ is introduced in order to have a copy of the symbols of states. Sort *time* together with the set of equations Ax gives the specification of the abstract data type "natural number" (the discrete time domain) with the operation of addition. We choose as the color alphabet for configurations the concrete algebra which is the word (sub-)algebra as concerns the two sorts *state* and *event*, and the (sub-)algebra with the usual representation of naturals as concerns the sort *time*. In the following definitions of graphs, when edges are given as pairs of vertices, we shall omit to specify the source an target functions intending that the two vertices are the source and the target, respectively. The coproduct of graphs is denoted by $+$.

**Definition 4.2** *A* graph for a configuration $C = (D, en, hist, test)$ at time $\tau$, *is* $Conf(C, \tau) = Curr_C{+}Hist_C{+}Event_C{+}\ Ck(\tau)$, *where:*
- $Curr_C$ *is the graph $(V, E, sc, tg, l)$ with:*
  - $V$ *in a one-to-one correspondence $k$ with $D$; $l(v) = k(v)$, for all $v \in V$;*
  - $E = \{(v, v') : v, v' \in V, k(v') \in \rho(k(v))\}$; $l((v, v')) = en(k(v'))$, *for all* $(v, v') \in E$.
- $Hist_C$ *is the graph $(V, E, sc, tg, l)$ with:*
  - $V$ *in a one-to-one correspondence $k$ with* $\{\overline{b} : \overline{b} \in \overline{B},\ b \text{ non-basic}\} \cup \{b : b = fst(hist(b')), \text{ with } b' \in B\} \cup \{\rho(b') : b' \in \phi^{-1}(AND)\}$; $l(v) = k(v)$, *for all* $v \in V$;
  - $E = \{(v, v') : v, v' \in V,\ k(v) = \overline{b} \text{ with } \overline{b} \in \overline{B},\ k(v') \in \rho(b)\}$; $l((v, v')) = snd(hist(b))$, *with $k(v) = \overline{b}$ and $hist(b)$ defined;* $l((v, v')) = snd(hist(b''))$, *with $k(v) = \overline{b}$, $hist(b)$ undefined and $b''$ the lower or-ancestor of $b$, if any,* $l((v, v')) = 0$, *otherwise, for all* $(v, v') \in E$;
- $Event_C$ *is the graph $(V, E, sc, tg, l)$ with:*
  - $V$ *in a one-to-one correspondence $k$ with* $P \cup \{True, False\}$; $l(v) = k(v)$, *for all* $v \in V$;

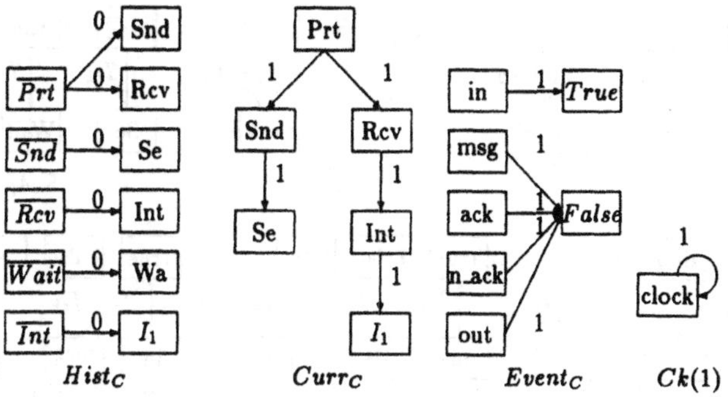


Figure 2: A configuration graph.

$-E = \{n,(v,v')) : v,v' \in V,\ n \in \mathcal{N},\ k(v) \in test(n),\ k(v') = True\} \cup \{(n,(v,v')) : v,v' \in V,\ n \in \mathcal{N},\ k(v) \in P - test(n),\ k(v') = False\}$; $sc(w) = v$, $tg(w) = v'$ *and* $l(w) = n$, *for all* $w = (n,(v,v')) \in E$.

- $Ck(\tau)$ *is the graph* $(V, E, sc, tg, l)$ *with* $V = \{v\}$, $E = \{(v,v)\}$, $l(v) = clock$, $l((v,v)) = \tau$ *for some* $v$.

*A graph G is a* configuration graph *iff there is a configuration C at time τ s.t. G is isomorphic to* $Conf(C,\tau)$; *G is* maximal *iff C is maximal, it is* initial *iff C is.*

**Example 4.3** *In figure 2 the configuration graph is shown for* $C = (D, en, hist, test)$ *at time* 1, *where* $D = \{Prt, Snd, Se, Rcv, Int, I_1\}$, $en(d) = 1$, *for all* $d \in D$, $hist(d) = (\delta(d), 0)$, *for all non basic or-states* $d \in B$, $test(1) = \{in\}$ *and* $test(n) = \emptyset$ *for* $n > 1$. *All edges labelled by times greater than 1 in* $Event_C$ *are omitted.*

In associating productions with transitions, we follow an approach which has been inspired by the principles of the Chemical Abstract Machine [1]. In the Chemical Abstract Machine [1] each change of the internal status is caused either by "reversible" or by "irreversible" reactions. Changes are as much as possible local and reversibility of reactions and nondeterminism make up for the absence of a deeper contextual information. In compliance with the principle of locality, with each transition $t$ we associate a production removing from $Curr_C$ the vertices corresponding to states in $B^{LCA(t)}_{out(t)}$ and adding to $Curr_C$ the vertices corresponding to states in $B^{LCA(t)}_{in(t)}$. Such productions are called "irreversible". Some auxiliary productions, called "deleting productions", have the function of "eroding" the graph $Curr_C$ by removing its leaves. The irreversible production can be applied when the graph is suitably eroded, namely when the set of states $B^{LCA(t)}_{out(t)}$ corresponds to a subtree of the tree represented by $Curr_C$. In other words, a vertex corresponding to a state $b$ can be removed only if all the vertices corresponding to substates of $b$ in $Curr_C$ are removed. For each deleting production, also the inverse production, called "restoring production", is provided (i.e., the production $R \leftarrow K \rightarrow L$, if $L \leftarrow K \rightarrow R$ is the deleting production). Now, the application of deleting productions does not aim at enabling the application of a particular irreversible production. Actually, suppose that an irreversible production may be applied provided that a suitable sequence of derivations via deleting productions is performed before. Suppose also that, instead, the application of a sequence of deleting productions disables the application of the irreversible production because, for instance, it removes also the source state of the corresponding transition. In this case, the reversibility of sequences of derivations via deleting productions ensures that

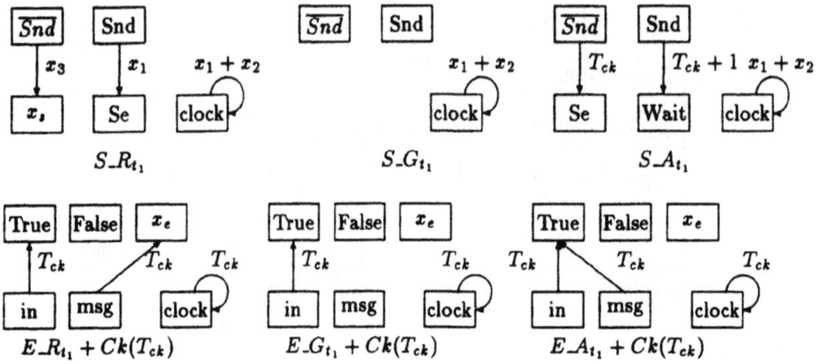


Figure 3: The production $p_{t_1}$.

the sequence of derivations can be extended offering eventually the possibility of applying the irreversible production. The task of downward closing the set of states $B_{in(t)}^{LCA(t)}$ which are added by an irreversible production $t$ is accomplished by "closure by history" and "closure by default" productions. The advantage of this kind of translation is in that the explosion of the number of transition required is avoided. We have one only production for each transition and at most a deleting production for each state, instead of having a production for each subconfiguration for the subtree of states rooted in the source of the transition.

The color alphabet for productions is the algebra of terms $T_\Sigma(X)$, where $\Sigma$ is the signature defined above and $X$ is a family of variables indexed over $S$. The family $X$ is supposed to contain a sufficient number of variables for each sort and a special variable $T_{ck}$ of sort *time*. We start with defining irreversible productions. In figure 3 the irreversible production for transition $t_1$ is shown.

**Definition 4.4** *Let $t$ be a transition and let $f : P \to X_{event}$, $f_i : B \to X_{time}$, for $i = 1,2,3$, and $g : B \to X_{state}$ injective parwise image disjoint maps. The* irreversible production *$p_t$ for $t$ (having inclusions as arrows) is*
$Ck(T_{ck}) + E\_R_t + S\_R_t \leftarrow Ck(T_{ck}) + E\_G_t + S\_G_t \rightarrow Ck(T_{ck}) + E\_A_t + S\_A_t$, *where:*

- *$E\_R_t$ is the graph $(V, E, sc, tg, l)$ with:*
  - *$V$ in a one-to-one correspondence $k$ with $Ev^+(t) \cup Ev^-(t) \cup Act(t) \cup f(Act(t) - (Ev^+(t) \cup Ev^-(t))) \cup \{True, False\}$; $l(v) = k(v)$, for all $v \in V$;*
  - *$E = \{(v,v') : v,v' \in V, k(v) \in Ev^+(t), k(v') = True\} \cup \{(v,v') : v,v' \in V, k(v) \in Ev^-(t), k(v') = False\} \cup \{(v,v') : v,v' \in V, k(v) \in Act(t) - (Ev^+(t) \cup Ev^-(t)), k(v') = f(k(v))\}$; $l(w) = T_{ck}$, for all $w \in E$;*
- *$E\_G_t$ is the subgraph of $E\_R_t$ with $V_{E\_G_t} = V_{R\_G_t}$ and $V_{E\_G_t} = E_{E\_R_t}$ - $\{(v,v') : v,v' \in V,\ k(v) \in Act(t)\}$;*
- *$E\_A(t)$ is the graph containing $E\_G_t$ with $V_{E\_A_t} = V_{E\_G_t}$ and $E_{E\_A_t} = E_{E\_G_t} \cup E'$, with $E' = \{(v,v') : v,v' \in V,\ k(v) \in Act(t),\ k(v') = True\}$; $l_{E\_A_t}(w) = T_{ck}$, for all $w \in E'$.*
- *$S\_R_t$ is the graph $(V, E, sc, tg, l)$ with:*
  - *$V$ in a one-to-one correspondence $k$ with $\{clock\} \cup B_{out(t)}^{LCA(t)} \cup g(\hat{B}) \cup \{\bar{b} : b \in B_{out(t)}^{LCA(t)}, \rho(b) \cap B_{out(t)}^{LCA(t)} \neq \emptyset\}$, with $\hat{B} = B_{out(t)}^{LCA(t)} - \{LCA(t)\}$;*
    *$l(v) = k(v)$, for all $v \in V$;*

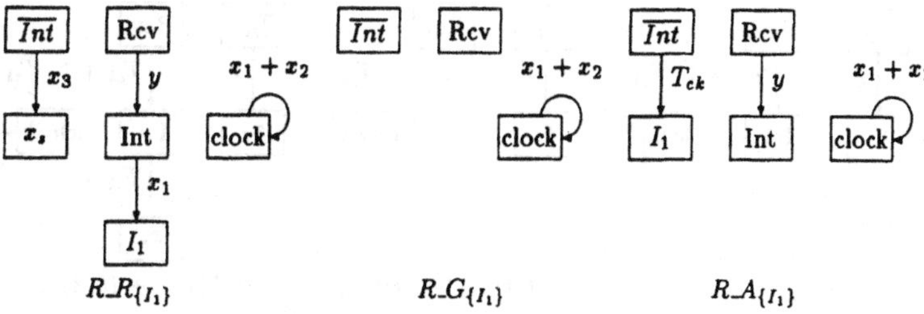


Figure 4: The deleting production $p_{\{I_1\}}$.

$-E = E_1 \cup E_2 \cup E_3$, *where:*
$E_1 = \{(v,v') : v,v' \in V, k(v) \in \rho(k(v'))\}$; $l((v,v')) = f_1(k(v'))$, *for all* $(v,v') \in E_1$;
$E_2 = \hat{B} \times \{clock\}$; $sr(w) = tg(w) = v$, *with* $k(v) = clock$, *for all* $w \in E_2$;
$l(w) = f_1(b) + f_2(b)$, *for all* $w = (b, clock) \in E_2$;
$E_3 = \{(v,v') :\ v,v' \in V,\ k(v) = \overline{b},\ k(v') = g(b)$, *for some* $b \in B^{LCA(t)}_{out(t)}\}$;
$l((v,v')) = f_3(k(v'))$, *for all* $(v,v') \in E_3$;

- $S\_G_t$ *is the subgraph of* $S\_R_t$ *with* $V_{S\_G_t} = V_{S\_R_t}$ - $(k^{-1}(\hat{B}) \cup k^{-1}(g(\hat{B})))$ *and with* $V_{S\_G_t} = E_2$;
- $S\_A_t$ *is the graph* $(V', E', sc', tg', l')$ *containing* $S\_G_t$, *with:*

$-V' = V_{S\_G_t} \cup V''$ *where* $V''$ *in a one-to-one correspondence* $k'$ *with* $(B^{LCA(t)}_{in(t)} \cup B^{LCA(t)}_{out(t)}) - \{LCA(t)\}$ *and disjoint w.r.t.* $V_{S\_G_t}$; $l'(v) = k'(v)$, *for all* $v \in V''$;
$-E' = E_{S\_G_t} \cup E_1' \cup E_2' \cup E_3'$ *where:*
$E_1' = \{(v,v') : v \in V_{S\_G_t}, v' \in V'', l_{S\_G_t}(v) = LCA(t), k'(v') \in \rho(LCA(t))\}$;
$E_2' = \{(v,v') : v,v' \in V'', k'(v), k'(v') \in B^{LCA(t)}_{in(t)}, k'(v') \in \rho(k'(v))\}$;
$l'(w) = T_{ck} + 1$, *for all* $w \in E_1' \cup E_2'$;
$E_3' = \{(v,v') : v \in V_{S\_G_t}, v' \in V'',\ l_{S\_G_t}(v) = \overline{b},\ k'(v') \in \hat{B} \cap \rho(b)\}$;
$l'(w) = T_{ck}$, *for all* $w \in E_3'$.

We define now deleting productions. A deleting production removes from $Curr_C$ only one node corresponding to a state $b$ if the immediate ancestor $b'$ of $b$ is an or-state, it removes all nodes corresponding to immediate substates of $b'$, otherwise. In figure 4 the deleting production removing a node labelled by state $I_1$ is shown.

**Definition 4.5** *For* $b, b' \in B$ *non-basic states s.t.* $b \in \rho(b)$, *let* $F = \rho(b)$ *if* $\phi(b) = AND$, *let* $F = \{b'\}$ *for some* $b' \in \rho(b)$ *otherwise. Let* $f_i$: $B \to X_{time}$, *for* $i = 1,2,3$ *and* $g : B \to X_{state}$ *be injective pairwise image disjoint maps.* The deleting production $p_F$ for $F$, *is* $(R\_R_F \leftarrow R\_G_F \rightarrow R\_A_F)$, *where*

- $R\_R_F$ *is the graph* $(V, E, sc, tg, l)$ *with:*

$-V$ *in a one-to-one correspondence* $k$ *with* $F \cup \{b, \overline{b}, b', clock\} \cup g(F)$;
$l(v) = k(v)$, *for all* $v \in V$;
$-E = E_1 \cup E_2 \cup E_3$ *where* $E_1 = \{(v,v') : v,v' \in V, k(v), k(v') \in F \cup \{b, b'\}, k(v') \in \rho(k(v))\}$; $l((v,v')) = f_1(k(v'))$, *for all* $(v,v') \in E_1$;
$E_2 = F \times \{clock\}$; $sc(w) = tg(w) = v$, *with* $k(clock) = v$, *for all* $w \in E_2$;
$l((d, clock)) = f_1(d) + f_2(d)$, *for all* $(d, clock) \in E_2$;
$E_3 = \{(v,v') : v,v' \in V, k(v) \in \overline{b}, k(v') \in g(F)\}$;
$l((v,v')) = f_3(k(v'))$, *for all* $(v,v') \in E_3$;

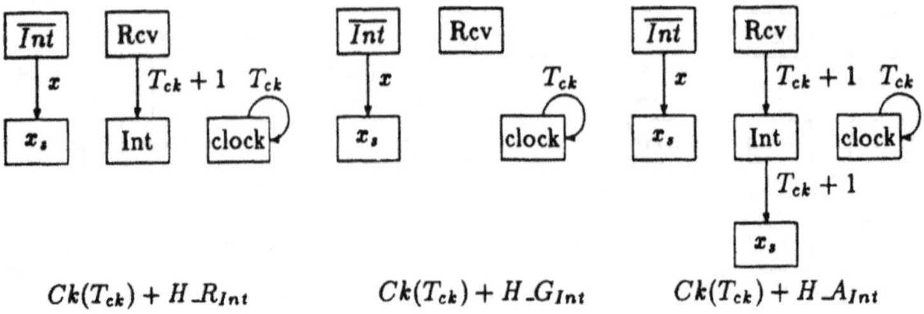


Figure 5: The closure by history production $p^h_{Int}$.

- *$R\_G_F$ subgraph of $R\_R_F$ s.t. $V_{R\_G_F} = V_{R\_R_F} - k^{-1}(F \cup \{b\} \cup g(F))$, $E_{R\_G_F} = E_2$;*
- *$R\_A_F$ is the graph $(V', E', sc', tg', l')$ containing $R\_G_F$ where:*
  *$-V' = V_{R\_G_F} \cup V''$, where $V_{R\_G_F}$ and $V''$ are disjoint, and $V''$ is in a one-to-one correspondence $k'$ with $F \cup \{b\}$; $l'(v) = k'(v)$, for all $v \in V''$;*
  *$-E' = E_{R\_G_F} \cup E'_1 \cup E'_2$ where: $E'_1 = \{(v, v') : v \in V_{R\_G_F}, v' \in V'', l_{R\_G_F}(v) = b', k'(v') = b\}$; $l'((v, v')) = f_1(k(v'))$, for all $(v, v') \in E'_1$;*
  *$E'_2 = \{(v, v') : v \in V_{R\_G_F}, v' \in V'', l_{R\_G_F}(v) = \overline{b}, k'(v') \in F\}$;*
  *$l'(w) = T_{ck}$, for all $w \in E'_2$.*

We shall define now the closure by history productions. In figure 5 the closure by history production for state *Int* is shown.

**Definition 4.6** *For $b \in H$, $b' \in B$ s.t. $b \in \rho(b')$, let $x_\tau \in X_{time}$ and $x_s \in X_{states}$. The production $p^h_b = Ck(T_{ck}) + H\_R_b \leftarrow Ck(T_{ck}) + H\_G_b \rightarrow Ck(T_{ck}) + H\_A_b$, where:*
- *$H\_R_b$ is a graph $(V, E, sc, tg, l)$ with:*
  *$-V$ in a one-to-one correspondence $k$ with $\{b', \overline{b}, b, x_s\}$; $l(v) = k(v)$, for all $v \in V$;*
  *$-E = E_1 \cup E_2$ where: $E_1 = \{(v, v') : v, v' \in V,\ k(v) = b',\ k(v') = b\}$; $l(w) = T_{ck} + 1$, for all $w \in E_1$;*
  *$E_2 = \{(v, v') : v, v' \in V,\ k(v) = \overline{b},\ k(v') = x_s\}$; $l(w) = x_\tau$, for all $w \in E_1$;*
- *$H\_G_b$ is the subgraph of $H\_R_b$ s.t. $V_{H\_G_b} = V_{H\_R_b} - k^{-1}(b)$, $E_{H\_G_b} = E_2$;*
- *$H\_A_b$ is the graph $(V', E'. sc', tg', l')$ containing $H\_R_b$ where:*
  *$-V' = V_{H\_R_b} \cup V''$ where $V''$ and $V_{H\_R_b}$ are disjoint and $V''$ is in a one-to-one correspondence $k'$ with $\{(0, x_s)\}$; $l'(v) = x_s$, for all $v \in V''$;*
  *$-E' = E_{H\_R_b} \cup E'_1$ where: $E'_1 = \{(v, v') :\ v \in V_{H\_R_b}, v' \in V'', l_{H\_R_b}(v) = b, k'(v') = (0, x_s)\}$; $l'(w) = T_{ck} + 1$, for all $w \in E'_1$.*

Finally, we define the closure by default production.

**Definition 4.7** *For a non-basic state $b \in B - H$ and $b' \in B$ s.t. $b \in \rho(b')$, let $Def = \rho(b)$ if $\phi(b) = AND$, let $Def = \{\delta(b)\}$, otherwise. The production $p^\delta_b = Ck(T_{ck}) + D\_R_b \leftarrow Ck(T_{ck}) + D\_G_b \rightarrow Ck(T_{ck}) + D\_A_b$, where*
- *$D\_R_b$ is the graph $(V, E, sc, tg, l)$ with:*
  *$-V$ in a one-to-one correspondence $k$ with $\{b', b\}$; $l(v) = k(v)$, for all $v \in V$;*
  *$-E = \{(v, v') : v, v' \in V, k(v) = b', k(v') = b\}$; $l(w) = T_{ck} + 1$, for all $w \in E$;*
- *$D\_G_b$ is the subgraph of $D\_R_b$ s.t. $V_{D\_G_b} = k^{-1}(b')$ and $E_{D\_G_b} = \emptyset$;*
- *$D\_A_b$ is the graph $(V', E'. sc', tg', l')$ containg $D\_R_b$, where:*
  *$-V' = V_{D\_R_b} \cup V''$, where $V''$ and $V_{D\_R_b}$ are disjoint and $V''$ is in a one-to-one correspondence $k'$ with $Def$; $l'(v) = k'(v)$, for all $v \in V''$;*

$-E' = E_{D\_R_b} \cup E'_1$ *with* $E'_1 = \{(v,v') : v \in V_{D\_R_b},\ v' \in V'', l_{D\_R_b}(v) = b, k'(v') \in Def\}$; $l''(w) = T_{ck}+1$, *for all* $w \in E'_1$.

Since we are also interested in a semantics expressing concurrency, the set of transitions associated with a statechart must contain also parallel irreversible productions. Given productions $p = (L \leftarrow K \rightarrow R)$ and $p' = (L' \leftarrow K' \rightarrow R')$, then the production $p+p' = (L+L' \leftarrow K+K' \rightarrow R+R')$ is the *parallel production* of $p$ and $p'$.

**Definition 4.8** *The set of productions $\mathcal{P}$ for a statechart is the union of the following sets of productions:*
- *the set of* irreversible productions $\{p_{t_1}+\ldots+p_{t_k} : t_1,\ldots,t_k$ *is a sequentialization of* $\{t_1,\ldots,t_k\}$, $t_i \in T$, $t_i \neq t_j$ *with* $i \neq j$, *for all* $1 \leq i,j \leq k$; *(we do not distinguish among different sequentializations of the same set)*;
- *the set of* reversible productions $\{p_{\{b\}}, p^{-1}_{\{b\}} : b \in \rho(b')$, *with* $b' \in \phi^{-1}(OR)\} \cup \{p_F, p_F^{-1} : F = \rho(b)$, *with* $b \in \phi^{-1}(AND)\}$.
- *the set of closure productions* $\{p^h_b : b \in H\} \cup \{p^\delta_b : b \in B - H$, *b non-basic*$\}$;
- *the clock production* $p_{ck} = (Ck(T_{ck}) \leftarrow Ck \rightarrow Ck(T_{ck}+1))$, *($Ck$ is the subgraph without edges of $Ck(T_{ck})$; a direct derivation via $p_{ck}$ is called* clock strike*)*.

*The set $\mathcal{P}$ must satisfy the requirement that, for every pair $p,p' \in \mathcal{P}$, if $p' \neq p^{-1}$, then $(Var(p) \cap Var(p')) - \{T_{ck}\} = \emptyset$, with $Var(p)$ denoting the set of variables occurring as labels of vertices or edges of the graphs of $p$.*

Now we shall establish which sequences of derivations describe the operational meaning of a statechart for a given interaction with the environment.

**Definition 4.9** *Let $G$ be a configuration graphs; the sequence of direct derivations $\mathcal{R} = G \overset{p_1,\sigma_1}{\Longrightarrow} \ldots \overset{p_n,\sigma_n}{\Longrightarrow} G'$ is a* rewriting *iff it contains no clock strike and $\sigma_i(x) = \sigma_j(x)$, for all $x \in Var(p_i) \cap Var(p_j)$ with $p_i = p_j^{-1}$ and $1 \leq i,j \leq n$.*

**Proposition 4.10** *If $G \overset{p_1,\sigma_1}{\Longrightarrow} \ldots \overset{p_n,\sigma_n}{\Longrightarrow} G'$ is a rewriting, there is a rewriting $G \overset{p_1,\sigma}{\Longrightarrow} \ldots \overset{p_n,\sigma}{\Longrightarrow} G'$, for a valuation $\sigma$ s.t. $\sigma(x) = \sigma_i(x)$, for all $x \in Var(p_i)$, with $1 \leq i \leq n$.*

The proposition above allow us to omit to specify the valuation function for direct derivations when we describe a rewriting, intending that all the derivations are with respect to the same valuation.

**Definition 4.11** *A rewriting $\mathcal{R}$ from a configuration graph $G$ with a valuation $\sigma$ is a* rewriting step *iff there is no rewriting $\mathcal{R}'$ from $G$ with a valuation $\sigma'$ s.t. $\bigcup_{t \in \mathcal{IR}} \alpha(t) \subset \bigcup_{t \in \mathcal{IR}'} \alpha(t)$, where $\mathcal{IR}$ (resp.: $\mathcal{IR}'$) is the set of irreversible productions occurring in $\mathcal{R}$ (resp.: $\mathcal{R}'$) and $\alpha$ the map giving for an irreversible production $p = p_{t_1}+\ldots+p_{t_k}$, $k \leq 1$, the set of transitions $\{t_1,\ldots,t_k\}$.*
*A maximal configuration graph is* stable *iff there is no rewriting from it involving any irreversible production. A sequence of derivations from an initial configuration graph $G$ is a* reaction *iff the sequences of derivations in between two clock strikes are rewriting steps and each clock strike is from a stable configuration graph.*

A notion of equivalence for rewritings is defined, as usual, by exploiting Church-Rosser and parallelism properties of derivations (see [2]) and by removing pairs of direct derivations via a reversible production and its inverse production.

**Definition 4.12** *Let* $\mathcal{R} = G \stackrel{p_1}{\Longrightarrow} \ldots \stackrel{p_{i-1}}{\Longrightarrow} G_{i-1} \stackrel{p_i}{\Longrightarrow} G_i \stackrel{p_{i+1}}{\Longrightarrow} G_{i+1} \stackrel{p_{i+2}}{\Longrightarrow} \ldots \stackrel{p_n}{\Longrightarrow} G'$ *be a rewriting.*

- *If* $G_{i-1} \stackrel{p_i}{\Longrightarrow} G_i$ *and* $G_i \stackrel{p_{i+1}}{\Longrightarrow} G_{i+1}$ *are sequential independent direct derivations, then* $G \stackrel{p_1}{\Longrightarrow} \ldots \stackrel{p_{i-1}}{\Longrightarrow} G_{i-1} \stackrel{p_{i+1}}{\Longrightarrow} G'_i \stackrel{p_i}{\Longrightarrow} G_{i+1} \stackrel{p_{i+2}}{\Longrightarrow} \ldots \stackrel{g_n}{\Longrightarrow} G'$ *is a* swap *of* $\mathcal{R}$ *and* $G \stackrel{p_1}{\Longrightarrow} \ldots \stackrel{p_{i-1}}{\Longrightarrow} G_{i-1} \stackrel{p_i + p_{i+1}}{\Longrightarrow} G_{i+1} \stackrel{p_{i+2}}{\Longrightarrow} \ldots \stackrel{g_n}{\Longrightarrow} G'$ *is a* parallelization *of* $\mathcal{R}$.
- *If* $p_{i+1} = p_i^{-1}$, *then* $G \stackrel{p_1}{\Longrightarrow} \ldots \stackrel{p_{i-1}}{\Longrightarrow} G_{i-1} \stackrel{p_{i+2}}{\Longrightarrow} \ldots \stackrel{g_n}{\Longrightarrow} G'$ *is a* contraction *of* $\mathcal{R}$.

*Two rewritings* $\mathcal{R}$ *and* $\mathcal{R}'$, *both from the configuration graph* $G$ *to the configuration graph* $G'$, *are* equivalent *iff there exist rewritings* $\mathcal{R}_0, \ldots, \mathcal{R}_n$, *all of them from* $G$ *to* $G'$, *such that* $\mathcal{R}_0 = \mathcal{R}$, $\mathcal{R}_n = \mathcal{R}'$, *and* $\mathcal{R}_i$ *is either a swap or a contraction or a parallelization of* $\mathcal{R}_{i-1}$, *for* $1 \leq i \leq n-1$.

A rewriting is in normal form when direct derivations via reversible productions precede all other kinds of derivations and derivations via closure productions follow all other kinds of derivations. In other words, first $Curr_C$ is eroded to allow the application of a set of productions, then these productions are applied, and only afterwards the transformations via closure productions are performed. Two direct derivations via reversible productions which are one the inverse of the other do not occur, namely only the derivations via deleting productions occur which are required to apply irreversible productions. In this way, derivations via irreversible productions are set apart from derivations via other kinds of productions. This is relevant for our idea that the semantics of a statechart is given by sequences of derivations via irreversible productions, since they are in a one-to-one correspondence with transitions.

**Definition 4.13** *A rewriting* $\mathcal{R} = G \stackrel{p_1}{\Longrightarrow} \ldots \stackrel{p_k}{\Longrightarrow} G_k \stackrel{p_{k+1}}{\Longrightarrow} \ldots \stackrel{p_n}{\Longrightarrow} G_n \stackrel{p_{n+1}}{\Longrightarrow} \ldots \stackrel{p_s}{\Longrightarrow} G_s \stackrel{p_{s+1}}{\Longrightarrow} \ldots \stackrel{p_r}{\Longrightarrow} G'$, *with* $0 \leq k \leq n \leq s \leq r$, *is in* normal form *iff:*

- $p_i$ *are restoring productions for* $1 \leq i \leq k$, $p_i$ *are deleting productions for* $k+1 \leq i \leq n$, $p_i$ *are irreversible productions for* $n+1 \leq i \leq s$, *and are closure productions for* $i \geq s+1$;
- $p_i \neq p_j^{-1}$ *for* $1 \leq i \leq k$ *and* $k+1 \leq j \leq n$.

The irreversible part of *of* $\mathcal{R}$ *is the subsequence* $G_n \stackrel{p_{n+1}}{\Longrightarrow} \ldots \stackrel{p_s}{\Longrightarrow} G_s$.

**Theorem 4.14** *For each rewriting from a configuration graph* $G$ *to a configuration graph* $G'$ *there exists a rewriting in normal form equivalent to it.*

The given semantics is correct with respect to the semantics given in [5].

**Theorem 4.15** *If* $C$ *and* $C'$ *are maximal configurations at time* $\tau$, *then* $C'$ *is reached from* $C$ *in a step* $S$ *with a sequence* $\Psi_1, \ldots, \Psi_n$ *of microsteps iff there exists a rewriting step* $\mathcal{R}$ *from a maximal configuration graph* $G$ *to a maximal configuration graph* $G'$, *in normal form and with irreversible part* $\overline{G} \stackrel{p_1}{\Longrightarrow} \ldots \stackrel{p_n}{\Longrightarrow} \overline{G}'$, *where* $G$ *and* $G'$ *are graphs isomorphic to* $Conf(C,\tau)$ *and* $Conf(C',\tau)$, *respectively, and* $\Psi_i = \alpha(p_i)$, *for* $1 \leq i \leq n$ *(*$\alpha$ *is as in 4.11).*

We introduce now the notion of derivation graph w.r.t. an initial configuration graph. If we consider a reaction whose rewriting steps are in normal form, then, due to the flow of time, it never happens that a pair of reached graphs are isomorphic. However, two configurations may represent the same situation up to time, namely they have the same set of currently entered states, the subset of currently entered states enabled to be exited is the same, the set of offered events at that time is the same as well as the history information. Configuration graphs which are "time equivalent" give rise to the same behaviour. Therefore, we shall define the "derivation graph rooted in a given initial configuration graph" as the folding of all reactions starting from the initial configuration graph. The set of vertices of the derivation graph is a subset of configuration graphs reachable from the initial one, supposed that time equivalent configuration graphs are identified. The derivation graph must abstract from details concerning derivations via productions which do not correspond to transitions. So, we shall consider only derivations in normal form and only the configuration graphs which are reached by a direct derivation via an irreversible production or from which a direct derivation via an irreversible production starts.

**Definition 4.16** *Configurations* $C = (D, en, hist, test)$ *and* $C' = (D', en', hist', test)$ *at time* $\tau$ *and* $\tau'$, *resp., are* time equivalent, *written as* $C \equiv C'$, *iff:*
• $D = D'$;
• $en(d) \leq \tau$ *iff* $en'(d) \leq \tau'$, *for all* $d \in D$;
• $(b', \tau) \in hist(b)$ *iff* $(b', \tau') \in hist'(b)$, *for all* $b \in B$;
$(b', i) \in hist(b)$ *iff* $(b', i') \in hist'(b)$, *for all* $b \in B$, *with* $i, i' \in \mathcal{N}$ $i \neq \tau$, $i' \neq \tau'$;
• $test(\tau) = test(\tau')$.
*The symbol* $\equiv$ *denotes also the time equivalence on configuration graphs induced by that on configurations.*

**Definition 4.17** *Let* $G$ *be an initial configuration graph. The* set of configuration graphs reached from $G$ *written as* $Reach(G)$ *is the set* $\{G\} \cup \{G_n$ : *there exists a reaction* $\mathcal{R} = G \stackrel{p_1}{\Longrightarrow} \ldots \stackrel{p_n}{\Longrightarrow} G_n \stackrel{p_{n+1}}{\Longrightarrow} \ldots \stackrel{p_{n+k}}{\Longrightarrow} G_{n+k}$ *such that each rewriting step is in normal form and* $p_n$ *or* $p_{n+1}$ *is an irreversible production*$\}$.
*The set* of transformations *from* $G$, *written as* $Trans(G)$, *is the set* $\{G' \stackrel{p}{\Longrightarrow} G''$: $G', G'' \in Reach(G)$, $p$ *is an irreversible production*$\} \cup \{G_n \stackrel{p_{n+1}}{\Longrightarrow} \ldots \stackrel{p_{n+k}}{\Longrightarrow} G_{n+k}$: $k \geq 1$ *and there exists a reaction* $\mathcal{R} = G \stackrel{p_1}{\Longrightarrow} \ldots \stackrel{p_n}{\Longrightarrow} G_n \stackrel{p_{n+1}}{\Longrightarrow} \ldots \stackrel{p_{n+k}}{\Longrightarrow} G_{n+k} \stackrel{p_{n+k+1}}{\Longrightarrow} \ldots \stackrel{p_{n+k+h}}{\Longrightarrow}$ $G_{n+k+h}$ *such that each rewriting step is in normal form,* $G_n$, $G_{n+k} \in Reach(G)$ *and* $p_{n+1}, \ldots, p_{n+k}$ *are not irreversible productions*$\}$.
*Two transformations* $G_1 \stackrel{p_1}{\Longrightarrow} \ldots \stackrel{p_n}{\Longrightarrow} G_n$ *and* $G'_1 \stackrel{p_1}{\Longrightarrow} \ldots \stackrel{p_n}{\Longrightarrow} G'_n$ *are* time equivalent *iff* $G_i \equiv G'_i$, *for* $1 \leq i \leq n$. *Also this equivalence is denoted by* $\equiv$.

**Definition 4.18** *Given an initial configuration graph* $G$, the derivation graph from $G$, *written as* $Der_G$, *is the edge labelled graph* $(V, E, sc, tg, l)$ *where:*
• $V = Reach(G) |_{\equiv}$ (*i.e., the quotient set of* $Reach(G)$);
• $E = Trans(G) |_{\equiv}$;
• $sc(w) = [G_1]$ (*i.e., the equivalence class of* $G_1$), $tg(w) = [G_n]$, *for all* $w \in E$ *and* $G_1 \stackrel{p_1}{\Longrightarrow} \ldots \stackrel{p_n}{\Longrightarrow} G_n \in w$.
• $l(w) = \alpha(p)$ *if there exists a direct derivation in* $w$ (*$w$ is an equivalence class*) *labelled by the irreversible production* $p$, $l(w) = clock$ *otherwise* (*$\alpha$ is as in 4.11*).

A derivation graph can be viewed as a folding of the set of behaviours of a statechart for a given initial configuration. In the following we shall formalize this correspondence.

**Definition 4.19** *A* path *in a derivation graph* $Der_G$ *is a sequence* $W = w_1, \ldots, w_n$ *of edges of* $Der_G$ *such that* $sc_{Der_G}(w_1) = [G]$, $sc_{Der_G}(w_i) = tg_{Der_G}(w_{i-1})$, *for* $2 \leq i \leq n$ *and* $l_{Der_G}(w_n) = \{clock\}$.
*An* unfolding *of* $Der_G$ *is a sequence of transition sets* $S_0, \ldots, S_h$ *such that there exists a path* $w_1, \ldots, w_n$ *and a mapping* $k : \{0, \ldots, h\} \rightarrow \{0, \ldots, n+1\}$ *satisfying*

• $k(0) = 0$, $k(i) < k(j)$, *for all* $0 \leq i < j \leq h$;
• $k(i) = j$ *implies* $l_{Der_G}(w_j) = \{clock\}$, *for all* $1 \leq i \leq h$;
• $l_{Der_G}(w_j) = \{clock\}$, *for some* $1 \leq j \leq n$, *implies there is* $1 \leq i \leq n$ *s.t.* $k(i) = j$;
• $S_i = \bigcup_{k(i)<j<k(i+1)} l_{Der_G}(w_j)$.

**Theorem 4.20** *For an initial configuration* $C$*, the sequence of transition sets* $S_0, \ldots, S_n$ *is a behaviour iff it is an unfolding of the derivation graph* $Der_G$*, with* $G$ *isomorphic to* $Conf(C, 0)$.

A remark. A derivation graph is a "transition system" whose transition relation is labelled by microsteps. Since statechart transitions in a microstep can be applied in any order as well as simultaneously, such transition system is a "distributed transition system" as defined in [8], where a logic for distributed transition systems is also provided. So from the constructed derivation graph one has a formal way to prove properties of the considered statechart.

# References


[1] Berry G., Boudol L., The Chemical Abstract Machine, Proceedings 17th Annual Symposium on Principles of Programming Languages, 1990, pp. 81-94.

[2] Ehrig, H., Introduction to the Algebraic Theory of Graph Grammars, LNCS 73, Springer, Berlin, 1979, pp. 1-69.

[3] Ehrig, H., Habel, A., Kreowski, H.-J., Parisi-Presicce, F., Parallelism and Concurrency in Hight Level Replacement Systems, Mathematical Structures in Computer Science 1 (1991), pp. 361-404..

[4] Harel, D., Statecharts: A Visual Formalism for Complex Systems, Science of Computer Programming 8 (1987), pp. 231-274.

[5] Harel, D., Pnueli, A., Schmidt, J., P., Sherman, R., On the Formal Semantics of Statecharts, Proceedings of 2nd IEEE Symposium on Logic in Computer Science, IEEE CS Press, 1987, pp. 54-64.

[6] Hess, L., Mayoh, B., Graphics and their Grammars, LNCS 291, Springer, Berlin, 1991, pp. 232-249.

[7] Janssens, D., Rozenberg, G., Actor Grammars, Mathematical Systems Theory 22 (1989), pp. 75-107.

[8] Lodaya, K., Ramanujam, R., Thiagarajan, P.S., A Logic for Distributed Transition Systems, LNCS 354, Springer, Berlin, 1989, pp. 508-522.

[9] Maggiolo-Schettini, A., Peron, A., Semantics for Statecharts Based on Graph Rewriting, in: Prinetto, P. and Camurati, P. (Eds.), Correct Hardware Design Methodologies, North-Holland, Amsterdam, pp. 91-114, 1992.

[10] Maggiolo-Schettini, A., Peron, A., Semantics for Full Statecharts Based on Graph Rewriting, Dipartimento di Matematica ed Informatica Università di Udine, Research Report UDMI/11/93/RR, 1993.

[11] Peron, A., Synchronous and Asynchronous Models for Statecharts, Dipartimento di Informatica Università di Pisa, Ph.D. Thesis, TD 21/93, 1993.

[12] Wirsing, M., Algebraic Specification, in Handbook of Theoretical Computer Science, Formal Methods and Semantics, Elsevier-MIT Press, 1990, pp. 675-788.

# Contextual Occurrence Nets and Concurrent Constraint Programming*

Ugo Montanari  Francesca Rossi

University of Pisa, Computer Science Department
Corso Italia 40, 56125 Pisa, Italy
{ugo,rossi}@di.unipi.it

**Abstract.** This paper proposes a new semantics for concurrent constraint programs. The meaning of each program is defined as a contextual net, which is just a usual net where context conditions, besides pre- and post-conditions, are allowed. Context conditions are just items which have to be present in order for an event to take place, but which are not affected by the event. They are very useful for describing situations where different events share a common resource and want to read it simultaneously. In fact, such events are concurrent in the net. The causal dependency relation of the net induces a partial order among objects in the same computation, while its mutual exclusion relation provides a way of expressing nondeterministic information. Such information can be of great help to a scheduler while trying to find an efficient execution of the program, or also to a compile-time optimizer.

## 1 Introduction

The paper proposes a new semantics for concurrent constraint (cc) programs [11]. Such programs are based on a very simple model, consisting of a collection of concurrent agents sharing a set of variables which are subject to some constraints. Each agent may perform two basic operations over the common constraint: either add a constraint, or test whether a constraint is entailed. The constraints are defined and handled by an underlying constraint system, which can be described as a set of primitive constraints and an entailment relation among subsets of them.

The semantics we propose in this paper is an extension of a partial order semantics proposed in [5, 6], where states of computations were represented by graphs, computation steps were seen as graph production applications, and each computation had an associated partial order, which was able to express the causal dependencies among the steps of such a computation. That semantics, though rather interesting and itself original for the uniform treatment of tokens and agents, as well as for the analysis of a cc program from the true-concurrency

* Research partially supported by the GRAGRA Basic Research Esprit Working Group n.7183, the ACCLAIM Basic Research Esprit Working Group n.7195 and Alenia S.p.A.

approach, was not entirely satisfactory. The main reason was that a partial order was associated to each deterministic computation, but no unique structure was associated to a (possibly nondeterministic) cc program. Therefore we could analyze the concurrency available in a cc program, but not its nondeterminism. In this paper we try to eliminate such unsatisfactory point of the approach in [5, 6].

Graph grammars are here replaced by simpler rewrite rules, which however maintain all the properties of graph grammars which are fundamental for the true-concurrency approach. In particular, one of the most important features is the possibility of expressing formally what we call "context objects", i.e., objects which are needed for a computation step to take place, but which are not affected by such step. In fact, these objects allow to model faithfully the concept of asked constraints, which is necessary if we want to model simultaneous ask operations. Therefore, our rewrite rules are context-dependent, i.e., they have a left hand side, a right hand side, and a context. A rule is applicable if both its left hand side and its context are present in the current state of the computation, and its application removes the left hand side and adds the right hand side. The evolution of each of the agents in a cc program, as well as the declarations of the program and its underlying constraint system, are all expressible by sets of such rules. In this way each computation step, i.e., the application of one of such rules, represents either the evolution of an agent, or the expansion of a declaration, or the entailment of some new token.

The set of rules obtained from a cc program are for us the starting point of a truly concurrent semantics. The idea is to construct a contextual occurrence net by starting from the initial agent and by unfolding it applying the rules in all possible ways.

A contextual net [8] is just a Petri net [9] where each event may have context conditions, besides the usual pre- and post-conditions. Any contextual net can be simulated by a suitable classical net. However, such simulation is either a less natural representation of the same situation, or it is not able to express the same amount of concurrency. In fact, in this latter case, read operations on the same resource appear to be concurrent in contextual nets, while in classical nets they are not so.

Given a contextual occurrence net, it is possible to derive three relations among its objects (either conditions or events), describing respectively the *causal dependency*, the *mutual exclusion*, and the *concurrency*. By using contextual occurrence nets to give a semantics to cc program, these relations are then interpreted as describing, respectively, the temporal precedence, the possible simultaneity, and the nondeterministic choices among steps of cc computation. In this way it is possible to see, for example, the maximal degree of both concurrency and nondeterminism available both at the program level and in the underlying constraint system.

## 2 Concurrent Constraint Programming

A cc program [11, 12, 13] is a set of agents interacting through a shared store, which is a set of constraints on some variables. The framework is parametric w.r.t. the kind of constraints that can be handled. Each step of the computation possibly adds new constraints and new variables to the store. The concurrent agents do not communicate with each other, but only with the shared store, by either checking if it entails a given constraint (ask operation) or adding a new constraint to it (tell operation). The following grammar describes the cc language we consider:
$P ::= F.A$
$F ::= p(\mathbf{x}) :: A \mid F.F$
$A ::= \mathit{success} \mid \mathit{failure} \mid \mathit{tell}(c) \rightarrow A \mid \sum_{i=1,\ldots,n} \mathit{ask}(c_i) \rightarrow A_i \mid A \parallel A \mid \exists \mathbf{x}.A \mid p(\mathbf{x})$
where $P$ is the class of programs, $F$ is the class of sequences of procedure declarations, $A$ is the class of agents, $c$ ranges over constraints, and $\mathbf{x}$ is a tuple of variables. Each procedure is defined once, thus nondeterminism is expressed via the + combinator only (which is here denoted by $\sum$). We also assume that, in $p(\mathbf{x}) :: A$, $\mathit{vars}(A) \subseteq \mathbf{x}$, where $\mathit{vars}(A)$ is the set of all variables occurring free in agent $A$. In a program $P = F.A$, $A$ is called initial agent, to be executed in the context of the set of declarations $F$.

Agent "$\sum_{i=1,\ldots,n} \mathit{ask}(c_i) \rightarrow A_i$" behaves as a set of guarded agents $A_i$ where the success of the guard is the entailment of the constraint $c_i$ by the current store. No particular order of selection of the guarded agents is assumed. Agent "tell(c) $\rightarrow$ A" adds constraint $c$ to the current store and then behaves like $A$. Agent $A_1 \parallel A_2$ behaves like $A_1$ and $A_2$ executing in parallel, agent $\exists \mathbf{x}.A$ behaves like agent $A$, except that the variables in $\mathbf{x}$ are local to $A$, and agent $p(\mathbf{x})$ is a call of procedure $p$. Given a program $P$, in the following we will refer to $Ag(P)$ as the set of all agents (and subagents) occurring in $P$.

The underlying constraint system can be described ([13]) as a *system of partial information* (derived from the *information system* introduced in [10]) of the form $< D, \vdash >$ where $D$ is a set of *tokens* (or primitive constraints) and $\vdash \subseteq \wp(D) \times D$ is the entailment relation which states which tokens are entailed by which sets of other tokens. The relation $\vdash$ has to satisfy the following axioms: $u \vdash x$ if $x \in u$ (reflexivity), and $u \vdash x$ if $v \vdash x$ and, for all $y \in v$, $u \vdash y$ (transitivity). Given $D$, $\mid D \mid$ is the set of all subsets of $D$ closed under entailment. Then, a constraint in a constraint system $< D, \vdash >$ is simply an element of $\mid D \mid$ (that is, a set of tokens).

## 3 From programs to rewrite rules

Each state of a cc computation consists of the multiset of active agents and of the already generated tokens. Each computation step will model either the evolution of a single agent, or the entailment of a new token through the $\vdash$ relation. Such a change in the state of the computation will be performed via the application of a rewrite rule. There will be as many rewrite rules as the number of agents and declarations in a program (which is finite), plus the number of pairs of the entailment relation (which can be infinite). The role of such rewrite rules coincides with that of the graph productions used in [5, 6]. Rewrite rules are simpler, while still retaining all the features of graph grammars which are necessary for our approach.

**Definition 1.** Given a program $P = F.A$ with a constraint system $< D, \vdash >$, a *state* is a multiset of elements of $Ag(P) \cup D$.■

Each item in a state involves some variables, and usually we will explicitely write these variables. For example, if $t$ involves variables $x_1, \ldots, x_n$, then we will write $t(x_1, \ldots, x_n)$. In a multiset, the multiplicity of each item is the natural number associated to it. For example, if an item $A$ appears $n$ times in the multiset, then its multiplicity is $n$ and we will write $nA$.

**Definition 2.** A *rewrite rule* has the form

$$r : L(r)(\mathbf{x}) \overset{c(r)(\mathbf{x})}{\rightsquigarrow} R(r)(\mathbf{xy})$$

where $L(r)$ is an agent, $c(r)$ is a constraint, and $R(r)$ is a state. Moreover, $\mathbf{x}$ is the tuple of variables appearing both in $L(r) \cup c(r)$ and in $R(r)$, while $\mathbf{y}$ is the tuple of variables appearing only in $R(r)$.■

The intuitive meaning of a rule is that $L(r)$ (the left hand side) is rewritten into (or replaced by) $R(r)$ (the right hand side), if $c(r)$ (the context) is present in the current state. $R(r)$ could contain some variables not appearing in $L(r)$ nor in $c(r)$ (i.e., the tuple $\mathbf{y}$). The application of $r$ would then rename such variables to constants which are different from all the others already in use. $c(r)$ is a context in the sense that is necessary for the application of the rule but it is not affected by such application. The possibility of having a context-dependent formalism is very significant if we are interested in the causal dependencies among the objects involved in a computation. In fact, rule applications with overlapping contexts but with disjoint left hand sides can be applied independently, and thus possibly simultaneously, only in a context-dependent formalism.

The cc framework is obviously context-dependent, since a constraint to be asked to the store is naturally interpreted as an object which is needed for the computation to evolve but which is not affected by such evolution. Therefore, the modelling of cc computations via a context-dependent formalism provides a more faithful description of the concurrency present in a cc program.

**Definition 3.** Consider a computation state $S_1(\mathbf{a})$ and a rule $r : L(r)(\mathbf{x}) \overset{c(r)(\mathbf{x})}{\rightsquigarrow} R(r)(\mathbf{xy})$. Suppose also that $(L(r) \cup c(r))[\mathbf{a}/\mathbf{x}] \subseteq S_1(\mathbf{a})$. Then the application of $r$ to $S_1$ is a *computation step* which yields a new computation state $S_2 = (S_1 - L(r)[\mathbf{a}/\mathbf{x}]) \cup R(r)[\mathbf{a}/\mathbf{x}][\mathbf{b}/\mathbf{y}]$, where the constants in $\mathbf{b}$ are fresh, i.e. they do not appear in $S_1$. We will write $S_1 \overset{r[\mathbf{a}/\mathbf{x}][\mathbf{b}/\mathbf{y}]}{\Longrightarrow} S_2$.■

In words, a rule $r$ can be applied to a state $S_1$ if both the left hand side of the rule and its context can be found (via a suitable substitution) in $S_1$. Then, the application of $r$ removes from $S_1$ the left hand side of $r$ and adds its right hand side.

**Definition 4.** The rules corresponding to agents, declarations, and pairs of the entailment relation are given as follows:

- $(tell(c) \rightarrow A) \leadsto c, A$
- $A_1 \parallel A_2 \leadsto A_1, A_2$
- $\exists \mathbf{x}.A \leadsto A$
- $(\sum_{i=1,\ldots,n} ask(c_i) \rightarrow A_i) \overset{c_i}{\leadsto} A_i$, for all $i = 1, \ldots, n$
- $p(\mathbf{x}) \leadsto A$, for all $p(\mathbf{x}) :: A$
- $\overset{S}{\leadsto} t$, forall $S \vdash t$ ∎

If the agent $(tell(c) \rightarrow A)$ is found in the current state, then such agent can be replaced by the agent $A$ together with the constraint $c$. In other words, $(tell(c) \rightarrow A)$ is cancelled by the current state, while both $c$ and $A$ are added. Agent $A_1 \parallel A_2$ is instead replaced by the multiset containing the two agents $A_1$ and $A_2$. Note that if $A_1 = A_2 = A$ we would still have two distinct elements, since a state is a multiset of elements. Therefore, in that case we would denote by $2A$ the new computation state. Agent $\exists \mathbf{x}.A$ is replaced by agent $A$. Agent $\sum_{i=1,\ldots,n} ask(c_i) \rightarrow A_i$ gives rise to as many rewrite rules as the number of possible nondeterministic choices. In each of such branches, say branch $i$, the whole agent is replaced by agent $A_i$ only if $c_i$ is present already in the store. Note that only this rule, which corresponds to an ask agent, needs a context. The rule for the entailment is the only one without the left hand side. The idea is that the presence of the context $S$ is enough to add the token $t$ to the current store.

Given a cc program $P = F.A$ and its underlying constraint system $< D, \vdash >$, we will call $RR(P)$ the set of rewrite rules associated to $P$, which consists of the rules corresponding to all agents in $Ag(P)$, plus the rules representing the declarations in $F$, plus those rules representing the pairs of the entailment relation.

**Example**: Consider a computation state containing agent

$$A(a_1, a_2) = ask(t_1(a_1, a_2)) \rightarrow A'(a_2)$$

and tokens $t_1(a_1, a_2)$ and $t_2(a_2, a_3)$. Consider also the rewrite rule

$$(ask(t_1(x_1, x_2)) \rightarrow A'(x_2)) \overset{t_1(x_1,x_2)}{\leadsto} A'(x_2).$$

Then, since there exists a matching between the left hand side of the rule and (a subset of) the state (via the substitution $\{a_1/x_1, a_2/x_2\}$), the rule can be applied, yielding the new state containing agent $A'(a_2)$ and tokens $t_1(a_1, a_2)$ and $t_2(a_2, a_3)$. Note that token $t_1(a_1, a_2)$ has not been cancelled by the rewrite rule, since it was in its context part. ∎

**Definition 5.** Consider a cc program $P = F.A$. A *computation* for $P$ is any sequence of computation steps

$$S_1 \stackrel{r_1[\mathbf{a_1}/\mathbf{x_1}]}{\Longrightarrow} S_2 \stackrel{r_2[\mathbf{a_2}/\mathbf{x_2}]}{\Longrightarrow} S_3 \ldots$$

such that $S_1 = \{A\}$ and $r_i \in RR(P)$, i = 1,2, .... Two computations which are the same except that different fresh constants (see Definition 3) are employed in the various steps are called *α-equivalent*.∎

## 4 Contextual nets

In classical nets, as defined for example in [9], each element of the set of conditions can be a precondition (if it belongs to the pre-set of an event) or a postcondition (if it belongs to the post-set of an event). We would now like to add the possibility, for a condition, to be considered as a *context* for an event. A context is something which is necessary for the event to be enabled, but which is not affected by the firing of that event. In other words, a context condition can be interpreted as an item which is *read without being consumed* by the event, in the same sense as preconditions can be considered being *read and consumed* and postconditions being instead simply *written*. Nets with such contexts will be called *context-dependent nets*. The formal technique we will use to introduce contexts consists of adding a new relation, beside the flow relation $F$, which we call the *context relation*. For the formal definitions about classical nets missing here we refer to [9]. A deeper and more comprehensive treatment of contextual nets can be found in [8].

**Definition 6.** A *context-dependent net* $CN$ is a quadruple $(B, E; F_1, F_2)$ where

- $B \cap E = \emptyset$, where elements of $B$ are called conditions and elements of $E$ are called events;
- $F_1 \subseteq (B \times E) \cup (E \times B)$ and it is called the flow relation;
- $F_2 \subseteq (B \times E)$ and it is called the context relation;
- $(F_1 \cup F_1^{-1}) \cap F_2 = \emptyset$.∎

**Example**: Context-dependent nets will be graphically represented in the same way as nets. I.e., conditions are circles, events are boxes, and the flow relation is represented by directed arcs from circles to boxes or viceversa. We choose to represent the context relation by undirected arcs (since the direction of such relation is unambiguous, i.e., from elements of $B$ to elements of $E$). An example of a context-dependent net can be found in Figure 1, where there are five events $e_1, \ldots, e_5$ and nine conditions $b_1, \ldots, b_9$. In particular, event $e_2$ has $b_2$ and $b_3$ as preconditions, $b_5$ as postcondition, and $b_7$ as a context. Note, however, that $b_7$ is not a context for all events. In fact, it is a precondition for $e_4$ and a context for $e_3$, for which $b_6$ is a context as well, while $b_4$ is a precondition and $b_8$ is a postcondition.∎

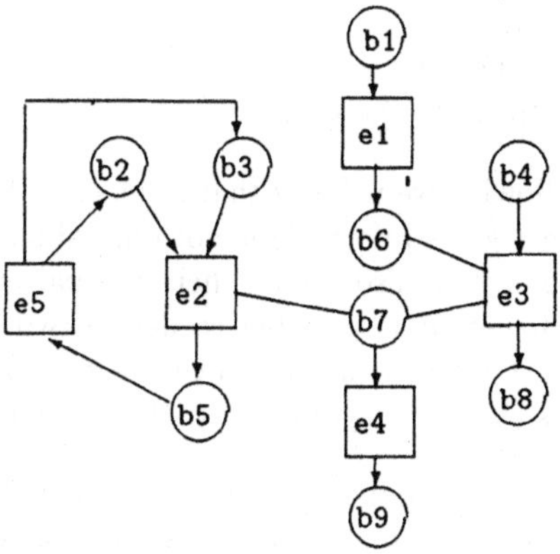


**Fig. 1.** A context-dependent net.

**Definition 7.** Given a context-dependent net $CN = (B, E; F_1, F_2)$ and an element $x \in B \cup E$,

- the *pre-set* of $x$ is the set $^\bullet x = \{y \mid F_1(y, x)\}$,
- the *post-set* of $x$ is the set $x^\bullet = \{y \mid F_1(x, y)\}$, and
- the *context* of $x$ is defined if $x \in E$ and it is the set $\widehat{x} = \{y \mid F_2(y, x)\}$.■

**Definition 8.** A case $c \subseteq B$ *enables* an event $e \in E$ iff $^\bullet e \cup \widehat{e} \subseteq c$ and $e^\bullet \subseteq {}^\bullet e \cup (B - c)$.■

**Example**: Consider again the context-dependent net in Figure 1 and the case $c = \{b_2, b_3, b_4, b_7\}$. Then $e_2$ and $e_4$ are the only enabled events. In fact, for example, $e_3$ is not enabled due to the absence of its context condition $b_6$ in the considered case.■

**Definition 9.** Given a context-dependent net $CN = (B, E; F_1, F_2)$, if $c_1, c_2 \subseteq B$ and $G \subseteq E$, then the *step* $c_1[G\rangle c_2$ is defined if:

- $\forall e \in G$, $e$ is enabled by $c_1$;
- $\forall e_1, e_2 \in G$, if $e_1 \neq e_2$, then $^\bullet e_1 \cap {}^\bullet e_2 = e_1^\bullet \cap e_2^\bullet = \emptyset$, $\widehat{e_1} \cap {}^\bullet e_2 = \widehat{e_1} \cap e_2^\bullet = \emptyset$;
- $c_2 = (c_1 - {}^\bullet G) \cup G^\bullet$.■

**Definition 10.** Given a context-dependent net $(B, E; F_1, F_2)$, a *computation* is a finite sequence of steps such that each step ends with a case which is the starting case of the subsequent step. I.e., a computation (of length $n$) is a sequence of the form $c_0[G_1\rangle c_1[G_2\rangle \ldots [G_n\rangle c_n$.■

**Definition 11.** Consider a context-dependent net $N = (B, E; F_1, F_2)$. Then we define a corresponding structure $(B \cup E, \leq_N)$, where the *dependency relation* $\leq_N$ is the minimal relation which is reflexive, transitive, and which satisfies the following conditions:

- $xF_1y$ implies $x \leq_N y$;
- $e1F_1b$ and $bF_2e2$ implies $e1 \leq_N e2$;
- $bF_2e1$ and $bF_1e_2$ implies $e1 \leq_N e2$.■

Therefore in the following we will say that $x$ depends on $y$ whenever $y \leq_N x$. A context-dependent net gives information not only about dependency (or not) of events and conditions, but also about their mutual exclusion (or conflict). In fact, given one such net, it is possible to derive two relations, besides the dependency relation just defined, all of which are sets of pairs of elements of $(B \cup E)$, which express concurrency and mutual exclusion.

**Definition 12.** Consider a context-dependent net $N = (B, E; F_1, F_2)$ and the associated dependency relation $\leq_N$. Assume also that $\leq$ is antisymmetric. Then

- The *mutual exclusion relation* $\#_N \subseteq ((B \cup E) \times (B \cup E))$ is defined as follows. First we define $x\#'y$ iff $x, y \in E$ and $\exists z \in B$ such that $zF_1x$ and $zF_1y$. Then, $\#_N$ is the minimal relation which includes $\#'$ and which is symmetric and hereditary (i.e., if $x\#_Ny$ and $x \leq z$, then $z\#_Ny$).
- The *concurrency relation* $co_N$ is just $((B \cup E) \times (B \cup E)) - (\leq_N \cup \leq_N^{-1} \cup \#_N)$.■

In words, the mutual exclusion is originated by the existence of conditions which cause more than one event, and then it is propagated downwards via the partial order. Finally, two items are concurrent if they are not dependent on each other nor mutually exclusive.

We now come to the notion of a contextual occurrence net, which is just a context-dependent net where the dependency relation is a partial order, there are no "forwards conflicts" (i.e., different conditions with a common precondition), and $\#_N$ is irreflexive. The latter requirement forbids events with mutually exclusive preconditions.

**Definition 13.** A *contextual occurrence net* is a a context-dependent net $N = (B, E; F_1, F_2)$ where

- $\leq_N$ is antisymmetric;
- $b \in B$ implies $| \; ^\bullet b \; | \leq_N 1$;
- $\#_N$ is irreflexive.■

A useful special case of a contextual occurrence net occurs when the mutual exclusion relation is empty. This means that, taken any two items in the net, they are either concurrent or dependent. Such nets are called deterministic occurrence nets.

**Definition 14.** A *deterministic contextual occurrence* net is a quadruple $N = (B, E; F_1, F_2)$ such that $N$ is a contextual occurrence net with $\#_N = \emptyset$.■

**Example**: Consider the deterministic contextual occurrence net in Figure 2 a). Then its dependency partial order can be seen in Figure 2 b).■

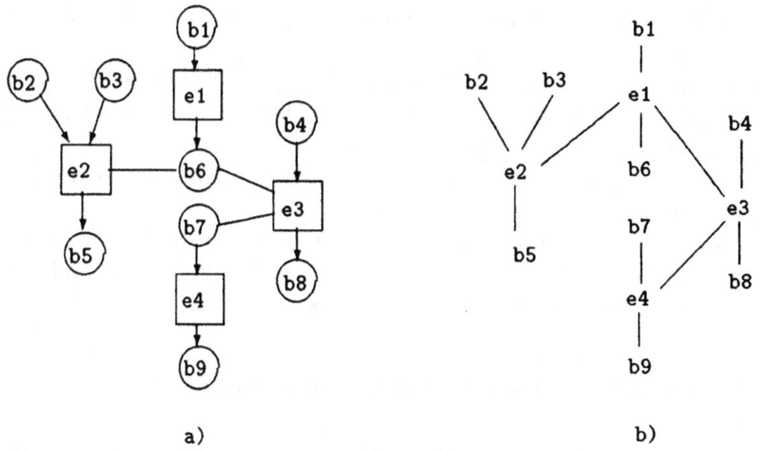


**Fig. 2.** A deterministic contextual occurrence net and the associated partial order.

**Definition 15.** Consider a contextual occurrence net $N = (B, E; F_1, F_2)$ and the associated relations $\leq$, $\#$, and *co*. Then a deterministic contextual occurrence net of $N$ is a deterministic contextual occurrence net $N' = (B', E'; F_1', F_2')$ where

- $B' \subseteq B$ and $E' \subseteq E'$;
- $F_1'$ and $F_2'$ are the restrictions to $B'$ and $E'$ of $F_1$ and $F_2$ respectively ;
- $x \in (B' \cup E')$ and $y \in (B \cup E)$ such that $y \leq x$ implies that $y \in (B' \cup E')$.■

Let us now try to relate contextual occurrence nets to cc programs. We will do that by defining a contextual process, which is just a contextual occurrence net plus a suitable mapping from the items of the net (i.e., conditions and events) to the agents of the cc program and the rules representing it.

**Definition 16.** Given a cc program P with initial agent $A$, and the associated sets of rewrite rules $RR(P)$, of agents $Ag(P)$, and of constraints $D$, consider the sets $RB = \{b\theta\}$ and $RE = \{r\theta\}$, with $b \in (Ag(P) \cup D)$, $r \in RR(P)$ and $\theta$ any substitution. Then a *contextual process* is a pair $\langle N, \pi \rangle$, where

- $N = (B, E; F_1, F_2)$ is a nondeterministic contextual occurrence net;
- $\pi : (B \cup E) \rightarrow (RB \cup RE)$ is a mapping with
  - $b \in B$ implies $\pi(b) \in RB$;
  - $e \in E$ implies $\pi(e) \in RE$;
  - consider ${}^{\circ}N = \{x \in B \mid \nexists y \in (B \cup E) \mid y \leq_N x\}$. Then $\pi({}^{\circ}N) = A$;
  - let $\pi(e) = r\theta$, with $r = L \overset{c}{\leadsto} R$. Then we have $\pi({}^{\bullet}e) = L\theta$, $\pi(\widehat{e}) = c\theta$, and $\pi(e^{\bullet}) = R\theta$. Notice that here and in the previous definition we homomorphically extended $\pi$ to a function from sets to multisets: the

multiplicity of an item in the result is simply the cardinality of its inverse image;

- for each $e \in E$, consider $\pi(e) = r\theta'_e\theta''_e$, where $\theta''_e$ replaces the variables which are in $R$ but not in $L$ nor in $c$. Then for all events $e$ and $e'$ and variables $x$ and $x'$ we have:
  - let constant $a$ occur in the initial agent $A$. Then $x\theta''_e \neq a$
  - $x\theta''_e = x'\theta''_{e'}$ implies $x = x'$ and $e = e'$

  i.e. the constants introduced by the rewrite rule instantiations associated to events must be fresh, namely they must be all different and different from the constants in the initial state.■

## 5 From rewrite rules to contextual processes

The idea is to take the set of rewrite rules $RR(P)$ associated to a given cc program $P$ and, using such rules, to incrementally construct a corresponding contextual process. Such process is able to represent all possible (deterministic) computations of the cc program $P$.

**Definition 17.** Given a cc program $P$, the pair $CP(P) = \langle (B, E; F_1, F_2), \pi\rangle$ is constructed by means of the following two inference rules:

- $A(\mathbf{a})$ initial agent of $P$
  *implies*
  $< A(\mathbf{a}), \emptyset, 1 > \in B$;
- $\{s_1, \ldots, s_n\} \subseteq B$, where $s_i$ *co* $s_j$ for all $i, j$, $i \neq j$, $i, j = 1, \ldots, n$, and $s_i =< e_i, B_i(\mathbf{a_i}), k_i >$ for all $i = 1, \ldots, n$, $r \in RR(P)$ such that $L(r) = \{B_1(\mathbf{x_1}), \ldots, B_j(\mathbf{x_j})\}$, and $c(r) = \{B_{j+1}(\mathbf{x_{j+1}}), \ldots, B_n(\mathbf{x_n})\}$, and $\exists$ a substitution $[\mathbf{a}/\mathbf{x}]$ such that $B_i(\mathbf{x_i})[\mathbf{a}/\mathbf{x}] = B_i(\mathbf{a_i})$
  *implies*
  - $e =< r[\mathbf{a}/\mathbf{x}], \{s_1, \ldots, s_n\}, 1 > \in E$,
  - $s_i F_1 e$ for all $i = 1, \ldots, j$,
  - $s_i F_2 e$ for all $i = j + 1, \ldots, n$,
  - let $h$ be the multiplicity of $B(\mathbf{x}, y_1, \ldots, y_m)$ in $R(r)$. Then for all $l = 1, \ldots, h$, $b_l = < B[\mathbf{a}/\mathbf{x}][< e, y_1 > /y_1] \ldots [< e, y_m > /y_m], e, l > \in B$, and $eF_1 b_l$.

Moreover, for any item $x =< x_1, x_2, x_3 > \in (B \cup E)$, $\pi(x) = x_1$.■

We will now try to explain informally but in great detail the above definition, since it represents the core of this paper. In this informal explanation we will never refer to the mapping $\pi$, since we know that $\pi$ always maps a triple to its first element.

The idea is to apply the rewrite rules, starting from the initial agent, in any possible way, so that different occurrences of the same rule are represented by different events of the net and generate different conditions. The technique used to achieve that consists of generating a new event representing the rule application and new conditions representing the right hand side of the rule, and by structuring each event or condition as a triple, where the first element contains the object being represented (either an agent, or a token, or a rule application), while the second element contains the whole history of the event or condition, and the third element is a number which allows us to distinguish different occurrences with the same history (we recall that a state is a multiset, while here we are generating a set of conditions). If there is only one of such occurrences, then the number will be 1, otherwise, the k-th occurrence will have number k. Note that, for the language we consider, $k$ is always either 1 or 2, since there cannot be more than two occurrences with the same history. In fact, the only rule which may generate a multiset containing different occurrences of the same element is the one associated to the agent $A \parallel A$. However, our approach can also handle languages where an agent may fork into more than two agents. Note that we don't have to handle the problem of different rules generating different occurrences of the same agent, since the identity of such occurrences is automatically made distinct by the fact that they will have different histories. Moreover, the mapping $\pi$ tells us either the rule or the object represented (with the applied substitution).

The first inference rule creates one condition which represents the initial agent $A$ of the given program $P$. Such term, $< A(\mathbf{a}), \emptyset, 1 >$, has the agent $A$ as the first element of the triple, the empty set as the second element, and the number 1 as the third element. This means that agent $A$ has no element it depends on, which is reasonable since it is the first agent of each computation. The second inference rule creates the event representing the application of a rewrite rule $r$, as well as the objects in the right hand side of $r$. To apply $r$, we have to find a set of conditions, already in $B$, which match the left hand side and the context of $r$. These are the conditions $s_i$, each of which is the triple $< e_i, B_i(\mathbf{a}), k_i >$. This means that each $s_i$ represents the object $B_i$, which can be either an agent or a token, involves constants $\mathbf{a}$, depends on term $e_i$, and it is the $k_i$-th occurrence of $B_i$. Note that if agents $B_i$ contain different sets of constants, we assume that tuple $\mathbf{a}$ contains their union. The matching condition is expressed by the substitution $[\mathbf{a}/\mathbf{x}]$, which is able to make the left hand side and the context of a rule to coincide with a subset of terms already generated. Furthermore, such conditions must be concurrent (i.e., $s_i$ *co* $s_j$). This means that they have the possibility of being all together simultaneously in a computation. With these preconditions satisfied, the inference rule creates a new event and new conditions. The event is $e =< r[\mathbf{a}/\mathbf{x}], \{s_1, \ldots, s_n\}, 1 >$: it represents the application of $r$, which depends on its left hand side and its context (this is why the second element of the triple contains the set $\{s_1, \ldots, s_n\}$). This event is then related to the conditions $s_i$ by means of the two relation $F_1$ and $F_2$. More precisely, $F_1$ is used to relate it to the conditions representing the left hand side of the rule (which are $s_1, \ldots, s_j$), while $F_2$ is used to relate it to the conditions representing the context of the rule (which are $s_{j+1}, \ldots, s_n$). The conditions which are generated represent all the objects (either tokens or agents) in the right hand side of $r$. Thus, for each of such objects, say $B$, which involves the variables in the left hand side of $r$ (i.e., $\mathbf{x}$) and possibly some other variables (i.e., $y_1, \ldots, y_m$), we create the condition $< B[\mathbf{a}/\mathbf{x}][< e, y_1 > /y_1] \ldots [< e, y_m > /y_m], e, l >$, and we make it dependent on event $e$ via relation $F_1$. This condition represents the l-th occurrence of object $B$ with variables $\mathbf{x}$ suitably substituted by the constants $\mathbf{a}$ (which is the matching needed for the application of the rule), plus the other variables, which have been renamed to contain term $e$, which is the term representing the rule application. In this way, such variables are different from any other variable ever used in the computation. The second element of the triple representing $B$ is the event $e$, because $B$ obvioulsy depends only on the rule application (and thus on the term representing such application).

**Example**: Consider the cc program $P$ consisting of the initial agent $IA = tell(c_1, c_2, c) \rightarrow A$, where
$A = (ask(c_1) \rightarrow A_1 + ask(c_2) \rightarrow success)$,
$A_1 = A_2 \parallel A_2$,
$A_2 = ask(c) \rightarrow success$,
and no declarations. The rules corresponding to such agent and all its subagents are:
$r_1 : IA \rightsquigarrow c_1, c_2, c, A$
$r_2 : A \overset{c_1}{\rightsquigarrow} A_1$
$r_3 : A \overset{c_2}{\rightsquigarrow} success$
$r_4 : A_1 \rightsquigarrow 2A_2$
$r_5 : A_2 \overset{c}{\rightsquigarrow} success$.
For simplicity sake, we assume the entailment relation to be empty (or not relevant to the constraints involved in such program). Furthermore, the variables are not taken into consideration. This program has two alternative finite computations, depending on how agent $A$ evolves. In one case, we have the parallel evolution of two occurrences of $A_2$ (even though they both ask for the same constraint $c$), and in the other case we have one computation step generating the success agent.

The process that is generated for this program is as follows (again, we will not explicitely write the mapping $\pi$, since each triple is mapped by $\pi$ to its first element). First, condition $s_1 =< IA, \emptyset, 1 >$ is generated for the initial agent $IA$. At this point, rule $r_1$ can be applied, and thus we generate event $e_1 =< r_1, \{s_1\}, 1 >$ to represent the rule application, and conditions $s_2 =< A, e_1, 1 >$, $s_3 =< c_1, e_1, 1 >$, $s_4 =< c_2, e_1, 1 >$, and $s_5 =< c, e_1, 1 >$ to represent its right hand side. Moreover, we set $s_1 F_1 e_1$, and $e_1 F_1 s_i$ for $i = 2, \ldots, 5$.

Since there are conditions representing $A$, $c_1$, and $c_2$ (i.e., $s_2$, $s_3$, and $s_4$), rules $r_2$ and $r_3$ can now be applied, and thus we have events $e_2 =< r_2, \{s_2, s_3\}, 1 >$ and $e_3 =< r_3, \{s_2, s_4\}, 1 >$, as well as conditions $s_6 =< A_1, e_2, 1 >$ and $s_7 =< success, e_3, 2 >$. Moreover, we also have $s_2 F_1 e_2$, $s_3 F_2 e_2$, $e_2 F_1 s_6$, $s_2 F_1 e_3$, $s_4 F_2 e_3$, $e_3 F_1 s_7$.

Now there is a condition representing $A_1$, thus rule $r_4$ can be applied, and we have the event $e_4 =< r_4, \{s_6\}, 1 >$, and the conditions $s_8 =< A_2, e_4, 1 >$ and $s_9 =< A_2, e_4, 2 >$. Moreover, $s_6 F_1 e_4$, $e_4 F_1 s_8$, and $e_4 F_1 s_9$.

Now both $s_8$ and $s_9$ match the left hand side of rule $r_5$, and there is a condition representing $c$ (i.e., $s_5$), therefore we can apply twice rule $r_5$, obtaining $e_5 =< r_5, \{s_5, s_8\}, 1 >$, $s_{10} =< success, e_5, 1 >$, $s_8 F_1 e_5$, $s_5 F_2 e_5$, and $e_5 F_1 s_{10}$ for the first application, and $e_6 =< r_5, \{s_5, s_9\}, 1 >$, $s_{11} =< success, e_6, 1 >$, $s_9 F_1 e_6$, $s_5 F_2 e_6$, and $e_6 F_1 s_{11}$ for the second application.

Thus we obtained the contextual process $\langle N, \pi \rangle$, where $N = (B, E; F_1, F_2)$, and $B = \{s_1, \ldots, s_{11}\}$, $E = \{e_1 \ldots, e_6\}$, and $F_1$ and $F_2$ are as defined above. Figure 3 shows such a net $N$. In $N$, it is easy to see that $e_2$ causally depends on $e_1$, since they are related by a chain of $F_1$ pairs. Similarly, $e_3$ depends on $e_1$. However, $e_3$ and $e_2$ do not depend on each other, and they are not even concurrent. In fact, thet are mutually exclusive, since they have a common precondition. This means the the rules represented by $e_2$ and $e_3$, i.e., $r_2$ and $r_3$, cannot be applied in the same computation, but only in two alternative computations. Then, we have that $e_4$ depends on $e_2$ because of a chain of $F_1$ pairs. Also, $e_5$ depends on $e_4$. Finally, $e_6$ depends on $e_4$ as well. However, $e_5$ and $e_6$ are concurrent, since they do not depend on each other and they are not mutually exclusive. Note that $e_5$ and $e_6$ have a common context condition. However, this does not generate any dependency, as desired.■

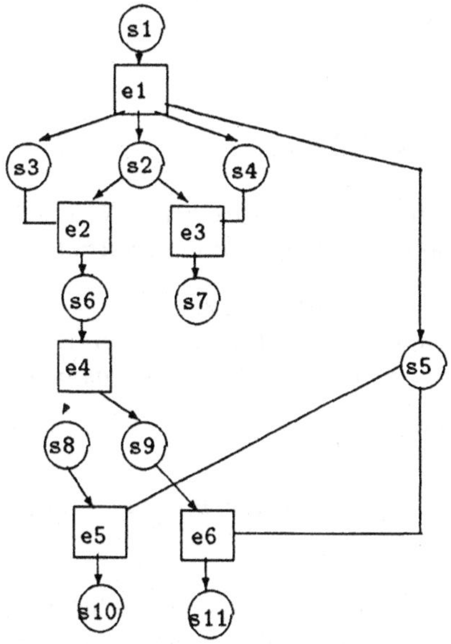


**Fig. 3.** The contextual occurrence net corresponding to a cc program.

**Theorem 18.** *Given a cc program P, consider the structure CP(P) as defined above. Then, CP(P) is a contextual process.* ■

The construction of the contextual process $CP(P)$, as described in Definition 17, is completely deterministic and independent of the order in which the rules in $RR(P)$ are selected to create new events and conditions of the net. The reason is that the set of pairs of the relations $F_1$ and $F_2$ which relate the already generated objects is not changed by the addition of new items. Such construction has been originally inspired by the one proposed in [14] to obtain a (possibly nondeterministic) occurrence net from a Petri net. Furthermore, the aim coincides, since in both cases the idea is to find a structure which represents all possible computations (of the Petri net in the case of [14], and of the given cc program in our case). However, there some differences, due to 1) the use of a set of rules instead of a net, 2) the use of context conditions, and 3) the presence of variables. A by-product of the strong relationship between the construction in [14] and ours is that, in both cases, the set of events of the process $CP(P)$, together with the dependency and mutual exclusion relations, can be seen as what is called an "event structure" in [14].

**Theorem 19 (soundness and completeness of CP(P)).** *Let P be a cc program and let* $CP(P) = \langle N, \pi \rangle$ *be the corresponding contextual process. Given a computation of P, there is:*

- *an α-equivalent computation*

$$S_1 \overset{r_1[\mathtt{a_1}/\mathtt{x_1}]}{\Longrightarrow} S_2 \overset{r_2[\mathtt{a_2}/\mathtt{x_2}]}{\Longrightarrow} S_3 \ldots$$

- *a deterministic contextual occurrence net of N, say N′, with associated partial order* $\leq'$*, and one of its linearizations (restricted to events), say* $e_1 e_2, \ldots,$

*such that* $\pi(e_i) = r_i[\mathtt{a_i}/\mathtt{x_i}]$ *for all* $i = 1, 2, \ldots$.
*Also, for any linearization* $e_1 e_2 \ldots$ *of the partial order associated to a deterministic contextual occurrence net of N, there is a computation of P*

$$S_1 \overset{r_1[\mathtt{a_1}/\mathtt{x_1}]}{\Longrightarrow} S_2 \overset{r_2[\mathtt{a_2}/\mathtt{x_2}]}{\Longrightarrow} S_3 \ldots$$

*such that, if* $e_i =< e_{i1}, e_{i2}, e_{i3} >$ *and* $\pi(e_i) = r$*, then* $r_i[\mathtt{a_i}/\mathtt{x_i}] = r$ *for all* $i = 1, \ldots$. ■

Note that there is no bijective correspondence between (α-equivalence classes of) computations of $P$ and deterministic contextual occurrence nets of $N$, since, as it is always the case in true concurrency, there may be different computations which are represented by the same deterministic partial order. There is not even a bijection between classes of computations and linearizations of partial orders associated to deterministic contextual occurrence nets. This depends on the fact that the definition of computation given by Definition 5, and usually employed in the literature, is not able to distinguish computations which use the same sequence of rules but where the items with multiplicity larger than one represent occurrences which are exchanged. This problem could be solved by adopting a different notion of computation which is based on states which are labelled sets (instead of multisets) and on the notion of "abstract derivations". This approach has been introduced in [1, 2] for a correct handling of graph derivations, and we believe that it can also be successfully used for cc computations.

## 6 Related and future work

The truly concurrent semantics we have proposed in this paper for cc programs is related to the one proposed in [5, 6]. However, there are two main differences. In [6] we associated to a program a set of partial orders, while here we introduced a contextual process. The latter is more informative, since it is able to express the choice points of the program. Moreover, the operational behaviour of cc programs is here described in terms of context-dependent rewrite rules, instead of graph productions. While being both context-dependent formalisms, the difference is reflected in the way computation steps are formally defined. In fact, here they involve a matching of the left hand side and the context, the removal of the left hand side, and the generation of the right hand side, together with its fresh variables. Instead, in the graph grammar setting a computation step is defined by means of a double pushout diagram. Therefore the notation we introduce here is simpler and more *ad hoc*.

It would be interesting to state and prove formally the relation between the two semantics. We claim that the deterministic contextual occurrence nets of a cc program $P$ we introduce here essentially coincide [2] with the occurrence nets of $P$ as defined in [5, 6]. The correspondence could possibly be made more precise if also the occurrence nets of [5, 6] (as the processes we have here) were labelled with the rewrite rules employed in the various computation steps and with the corresponding substitutions.

We believe that some results we obtained here for cc programs could be transferred to graph grammars. More precisely, the contextual process semantics could be applied to graph grammars as well: given a grammar, we could obtain a nondeterministic contextual process which represents all its derivations. Similarly to what we discussed above for cc programs, it would then be interesting to make explicit for graph grammars the relation between this semantics based on a nondeterministic contextual process, the classical theory of [3] based on the notion of parallel and sequential independence between graph production applications, and the truly concurrent semantics for deterministic computations presented in [7], which is an extension to graph grammars of the semantics in [5, 6] for cc programs. It is already clear, however, that a comprehensive theory

[2] A minor difference is that in the occurrence nets we construct in [5, 6] there are as many conditions as computation steps which "read and not consume" a context, while here only one condition is generated. However, the two different choices correspond to two different notions of process for contextual Petri nets, which are proved equivalent in [8].

of true concurrency for graph grammars will require a careful definition of graph computations, to avoid on one hand useless duplications due to the arbitrary choice of the nodes and arcs generated by productions, and on the other hand unwanted identifications due to possible endomorphisms of graphs and derivations. An ongoing development in this direction is reported in [1, 2].

A possible improvement of the semantics proposed in this paper is the explicit description, in a contextual process, of the connections among different agents and tokens due to the presence of the same variables. In other words, now a computation state is a set of conditions of the contextual process, which represent a multiset of agents and/or tokens. Instead, it would be more realistic and convenient to have, as a state, a graph whose nodes are variables and whose arcs are agents and constraints. Thus we would pass from a contextual occurrence net (plus the mapping) to a *contextual occurrence graph*, where nodes are variables, and arcs are either agents, or constraints, or links of the $F_1$ and $F_2$ relations. Again, it would be interesting to apply this extension to the class of all graph grammars.

## References


1. Corradini A., Ehrig H., Lowe M., Montanari U., Rossi F., "Standard Representation of Graphs and Graph Derivations", TU Berlin Technical Report, 1992.
2. Corradini A., Ehrig H., Lowe M., Montanari U., Rossi F., "Abstract Graph Derivations in the Double Pushout Approach", draft.
3. Ehrig H., "Introduction to the Algebraic Theory of Graph Grammars", in: *Proc. Internat. Workshop on Graph Grammars*, Springer, LNCS 73, 1978, pp. 1-81.
4. Lloyd J. W., "Foundations of Logic Programming", Springer Verlag, 1987.
5. Montanari U., Rossi F., "True Concurrency in Concurrent Constraint Programming", on *Proc. ILPS91*, MIT Press, 1991.
6. Montanari U., Rossi F., "Graph rewriting for a partial ordering semantics of concurrent constraint programming", in Theoretical Computer Science, special issue on graph grammars, Courcelle B. editor, vol.109, 1993.
7. Montanari, U., Rossi, F., "Graph Grammars as Context-Dependent Rewriting Systems: A Partial Ordering Semantics", in: J.-C. Raoult, Ed., *Proc CAAP'92*, Springer LNCS 581, pp. 232-247.
8. Montanari U., Rossi F., "Contextual nets", to appear in Acta Informatica. Also Technical Report TR-4/93, University of Pisa, CS Department, 1992.
9. Reisig W., "Petri Nets: An Introduction", Springer Verlag, EATCS Monographs on Theoretical Computer Science, 1985.
10. Scott D.S., "Domains for denotational semantics", on *Proc. ICALP*, 1982.
11. Saraswat V. A., "Concurrent Constraint Programming Languages", Ph.D. Thesis, Carnegie-Mellon University, 1989. Also 1989 ACM Dissertation Award, MIT Press, 1993.
12. Saraswat V. A., Rinard M., "Concurrent Constraint Programming", in *Proc. POPL*, ACM, 1990.
13. Saraswat V. A., Rinard M., Panangaden P., "Semantic Foundations of Concurrent Constraint Programming", in *Proc. POPL*, ACM, 1991.
14. Winskel G., "Event Structures", in *Petri nets: applications and relationships to other models of concurrency*, Springer-Verlag, LNCS 255.

# Uniform-Modelling in Graph Grammar Specifications

*M. Nagl*
*Lehrstuhl für Informatik III, RWTH Aachen*
*Ahornstr. 55, D-52056 Aachen*

**Abstract.** Building Integrated Environments in the context of Software Development /Na 94a/, Data Specification /Sc91/, and Computer Integrated Manufacturing/EWM 92, SW 92/ we ideally proceed as follows: (1) We specify the internal behavior of tools by graph grammars, (2) we edit, analyze, and (in the near future) execute such specifications thereby verifying and prototyping them, (3) we derive efficient components for specified ones by a generator machinery, and (4) we put them into a framework architecture providing for the invariant part of the environment. This paper sketches the experiences we have got by specifying tools of various environments. Doing so we have received modelling knowledge which is not only useful for one example but seems to be generally applicable. This knowledge is on three levels, namely finding reuseable spec portions, learning how to write spec portions, and structuring a spec consisting of different components. We call such a generally applicable approach 'uniform modelling'. The paper sketches three ways of uniform modelling in graph grammar specs. Whereas in the first approach we are directly dealing with a gra gra spec, in the second and third approach we are only arguing about structuring gra gra specs. At the end of the paper we give a summary and list some open problems.

*Keywords:* software development environments, data modelling environments, computer integrated manufacturing, graph grammars, specifying abstract data types, generators.

*CR classification:* D.2.6, H.2, F.4.2, D.3.4

## 1 Introduction

Since more than 10 years our research group is active in the area of integrated software development environments (SDEs, see /ACM xx/ for references), carrying out the IPSEN project (*I*ntegrated Software *P*roject *S*upport *En*vironment, /Na 94a/). In order to build integrated tools a framework architecture for an environment has been developed consisting of basic components to be used in any IPSEN-like environment and specific components to be built, configurated, and hooked into the framework in order to fulfil the specific needs of a certain environment. Before realizing these specific components the effects of tools on internal data structures are specified by graph grammars /GG xx; Na 79/. From these specifications efficient implementations of these specific components can be derived. A specification language has been defined and an environment for editing, maintaining, analyzing, and executing such graph grammar specifications is nearly completed /Sc 91, SZ 91, NS 91/. The executor of this specification environment, together with a spec and together with the above framework architecture, can be used for SDE rapid prototyping purposes. We call the IPSEN SDE an a priori environment, as all tools are new and are especially built for being integrated. Furthermore, in the context of computer aided manufacturing an a posteriori integration of existing CIM-applications is currently under work /EWM 92, SW 92/. Summing up, there is quite a lot of *experience* in building *integrated environments* in the software engineering, graph grammar spec, and CIM-area.

In general, in any of these cases *integration* on the user's side *means* that he is *handling complex configurations* of documents which, in the case of software engineering, is outlined in section 2. Different tools are working on such a complex configuration in a coordinated manner. The documents of such a configuration describe the current technical state of development of a complex system, keep track of the administration of the many components of the current state and its evolution over the time, or help to manage the processes which yield or maintain such configurations etc. At the moment we are starting with the gra gra specification of such overall integration problems in configurations and the derivation of integrator tools. We have, however, got a lot of experiences in handling the specification of integrated tools in logically separated areas of such configurations and in deriving their implementation. For such areas abstract data types are used for the internal data structures of tools. This paper gives a report on the experiences we have gained so far in specifying abstract data types.

For specification purposes we rather use graph rewriting systems than graph grammars (which are nevertheless named graph grammars, or in short gra gra, in the following paper). The aim of such specs is to formally specify, i.e. to make precise, how changes on internal documents (data structures of tools) are going to proceed. These internal documents are regarded to be graphs. As any document belongs to an abstract data type we are defining graph classes. More precisely, we use sequential and programmed graph rewriting systems on directed, attributed, node and edge labelled graphs. This paper describes our *graph grammar specification knowledge* we have got so far. It is practical knowledge about using a formal specification approach for various specification problems /En 86, Sc 86, Lew 88a, Sc 91, We 91, Ja 92/.

A gra gra spec is an operational spec, i.e. *by graph grammars* we can *program*. However, we do this an an *abstract way:* (1) We are dealing with graphs and not data structures realizing them, (2) we have complex operations at hand using graph replacement and graph tests as atomic transactions, (3) we are able to build up complex transactions by control structures /SZ 91/. (4) The declarative knowledge of a graph class is defined by the so–called schema part. Now, the specification task is to develop a graph grammar spec for any of the internal graph classes of an environment.

As many graph classes have to be specified and any of these gra gra specs is a nontrivial "program", we tried very early to learn how to do programming, i.e. to develop methodologies. A methodology does not only mean how to use the constructs of a graph grammar specification language. Moreover, it means to detect universal ways of programming which can be applied for any graph class, although every graph class looks differently as it represents a different logical (technical, administration, integration, representation) problem domain. We call such a proceeding *uniform modelling*. This paper describes three different ways of uniform modelling we have detected and applied for the specification of various graph classes.

Therefore, the overall goal of this paper is to discuss *reusability* in gra gra specs. This reusability can be detected on three levels: On product level, we shall be able to detect *reusable spec components*, which can be used for the specification of any graph class. On process level we have learned how to write the remaining specification components, i.e. *how* to do *modelling* by gra gras 'mechanically'. Finally, we shall discuss how a complex spec is structured and composed of different spec portions.

Handling complex configurations by the user of an integrated environment (e.g. SDE, CIM) also means to deal with a bunch of different languages and mutual relations of these languages. In some of these integrated problem domains these languages and their relations are not clearly defined corresponding to their syntax, semantics, and pragmatics. Therefore, our task also was to make existing languages precise or, in the case that there was no language, to precisely define such languages, before being able to realize tools. We shall learn in this paper that this business of *external modelling*, i.e. of modelling languages, is strongly connected to internal modelling. Here again, one wants to detect general principles.

This paper gives no final answers and no theoretical results. Nevertheless, we believe that it contains valuable *hints* for those who are interested in applying graph grammars for big and practical problems, as it summarizes some of our experiences of dealing with gra gra specification for many years (/Sc 75, /BBN 77/ to /Na 94a/).

The contents of this paper are as follows: In the next section we introduce the integration problem on complex configurations, taking software engineering and SDEs (as the corresponding environments) as a running example. This is done to give a motivation for the following central part of the paper. Then, in the next two sections, we outline three different uniform modelling procedures we have found so far which are called AST Graph Modelling, Metamodelling, and Layers & Parameterization, respectively. Whereas we give a detailed gra gra spec for the first approach we are only reasoning about structuring specs in the second and third approach. In the last section we summarize and give a list of open problems.

## 2 Integrated Environments and Complex Configurations

Software development, maintenance, and reuse of nontrivial software systems deal with *complex configurations*. These configurations contain product information (of the software system) and process information (of the process the result of which is the software system). SDEs have to handle these complex configurations. Furthermore, such configurations are developed, maintained, and reused by a group of software engineers rather than by single persons. This view is the starting point for any practical and intelligent support by SDEs.

A *similar situation* holds true in a CIM, a specification, or *any* other *context* where *highly structured* and *interrelated documents* are *developed* by a *complex process*. Therefore, software engineering and SDEs are only used as an example in this paper. Although we are using only a part of the overall scenario described in this section in the main sections of this paper, we first sketch the 'complete' scenario. This is necessary to understand the vision of the project and the problems which have to be solved.

Fig.1 shows a simple *software example configuration*. The requirements specification contains three documents belonging to three perspectives (functional model in SA, data model in EER, control model) in graphical representation for a survey and, furthermore, detailed descriptions (minispecs of data flow processes, control specs, attribute definitions etc.) in textual form. The architecture consists of graphical architecture diagram documents and detailed textual descriptions of components. Any software subsystem has its own representation (as diagram and text). All implementations of modules are given in textual representation (e.g. in Modula-2). Furthermore, there is a textual representation for the technical documentation. All of the above representations belong to the technical part of a software system. In addition, there are graphical/textual representations for administration items like configuration control, revision control, process control, and management and supervision of the project. Therefore, a configuration contains technical data of the product if we define the term product from the developer's or maintainer's perspective, and administration data belonging to the product, process, or project aspect. Fig.1 gives only a rough picture of the overall complexity. It is a conceptual picture and not a user interface description of existing tools.

Such a complex configuration consists of a lot of *subconfigurations* (an architecture of a subsystem together with its module implementations and corr. technical documentation as example on the technical side, the configurations' description together with its processes on the administration side etc.) down to single *documents* (the architecture diagram of a subsystem, a technical documentation of a subsystem). Documents again consist of *increments*, where an increment is derived from the syntax of the underlying language in which the document is written.

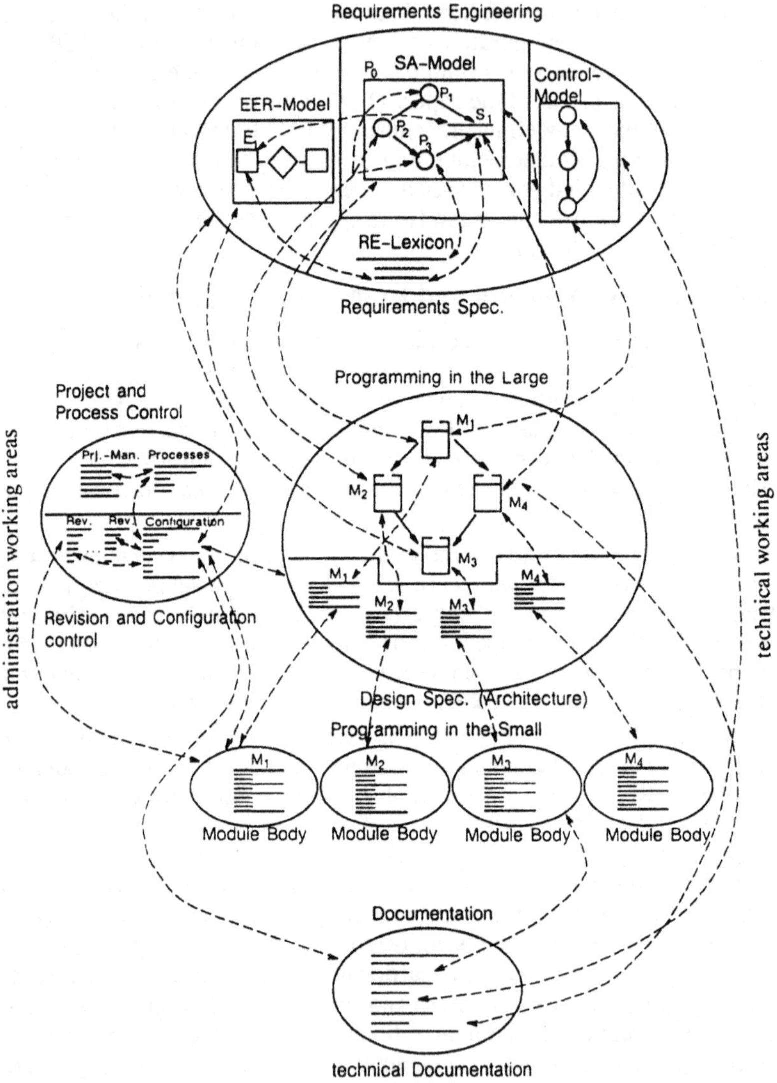


Fig.1: a complex configuration for a software system (technical product information, administrative product, process, and project information) has to be handled inside an integrated SDE

Within a document there is a large number of *fine-grained relations* between increments, which are not shown in Fig.1. Furthermore, there are a lot of fine-grained relations between increments of different documents: Objects of entity type $E_1$ are stored in the data store $S_1$, the module $M_2$ belongs to data flow processes $P_1$, $P_2$, $P_3$, a chapter of the technical documentation is dealing with the design decisions of module $M_4$, and alike. Such relations also exist on the administration side (a revision is contained in a configuration, a development process belongs to a subconfiguration, is carried out by person A, using the OOPL tool etc.). Relations between the administration side and the technical side are fine-grained (one side is an increment) and coarse-grained (the other being a document). The reason is that the administration side does only fix which documents, processes etc. occur but not how their internal structure is developed, or how creative technical processes are carried out etc.

All the items and their relations within such complex configurations have to be handled by an SDE such that tight integration within and between documents on one side and flexibility for carrying out corr. development/maintenance processes on the other side are guaranteed. One can argue that *development* and *maintenance* of complex software systems or *reuse* within the development of such systems is dramatically *facilitated* if the underlying SDE is able to handle such highly structured, fine-grained, and interrelated information /Na 94a/. Such a kind of support is not available by industrial tools today.

*Internally*, i.e. within the SDE and, therefore, to be handled by an SDE developer the situation is even more complex than shown in Fig.1. First, for any technical document the SDE user is dealing with, there is an internal *logical document* containing all the knowledge of the document (its structure, its relations, and therefore also the knowledge of the dialog the result of which is this document). For the representation of such a document there is a separate document which we call *representation document*. Translation from logical documents to representation documents is done by unparsers, the reverse direction by parsers (only for textual representations available). Furthermore, in order to handle fine-grained relations between documents internally there are *integration documents*. The administration side also contains logical documents, representation documents, and integration documents.

These different *internal documents* are *separated* because they belong to different abstractions: Logical and representation information is different because there may be more than one representation to one logical document or, vice versa, different logical documents may have one representation. Integration documents are separated from logical documents because there may be different integration mechanisms between logical documents. Internal documents for the administration side are separate from internal documents for the technical side because there may be different administrative procedures for the same technical portion of a configuration. Summing up, this separation belongs to the question what can change in the underlying models (internal logical structure, representation structure, integration mechanisms, models for the technical or administrative side). On the other side all these different internal documents *have to be integrated* by regarding all the dependencies between all of their increments.

It has been argued for some time /Na 85, En 86, Sc 86/ that *graphs* are an *appropriate* underlying *model* for describing such internal configurations because we can divide (defining separate graphs) and integrate (interrelate graphs) without leaving the graph model. The graphs occurring internally belong to graph classes (logical architecture graph class, textual representation graph class, integration graph class, logical process model graph class etc.). Although this graph classes can be very different they can be modelled uniformly. To sketch our experience in this area is the main concern of this paper. However, we concentrate in this paper on the discussion of modelling logical documents or, on the external side, on the discussion of modelling languages. Representation or integration document modelling is not regarded here. The approaches described are applicable to all kinds of graph classes.

## 3 Modelling by Abstract Syntax Graphs

The first approach of uniform modelling was called *AST graph modelling*. The essential idea is that we start with the context-free, i.e. the *tree part* of internal documents. This tree structure is often called abstract syntax tree (AST) as it is free from concrete syntax information. This context-free part is spanned by the contains- or is-part-of-relations which in most cases are derived from an external context-free syntax description, e.g. by EBNFs. This tree structure is then extended by *further* context-sensitive *relations* between tree parts, corresponding to declaring-applying-, consistent-with-relations of different internal objects (variables, types, procedures etc.) Fig.2 immediately shows the tree/nontree composition of an internal logical document. Edges down or right are tree edges, all others nontree edges. This kind of uniform internal modelling is that we have the most experience with.

The second essential idea is to extract as much declarative knowledge of a certain graph class as possible and to define it in the so-called *schema part* of a specification (not shown in the following, see /Sc91/). This schema part defines all node and edge type information allowed to occur in the graph class, similarities of node types by a class hierarchy, derived attributes of node classes, derived relations between node classes etc. The reader may imagine that this schema part is important for defining consistency constraints within a spec which, together with corresponding tools, makes it much easier to write 'correct' specs.

The third essential idea besides tree/nontree composition and schema extraction is to find *reusable spec* portions which can be used by writing any spec according to AST graph modelling. It is very obvious that these reusable spec portions belong to general tree handling and to general context-sensitive relations handling as the graphs are decomposed from these portions. According to the schema extraction idea these reusable specification portions contain a schema description. The task now for writing an AST graph spec for a certain logical document is to write the specific tree/nontree part on top of the reusable specs thereby extending the schema part appropriately.

The overall idea of AST graph modelling is to regard a certain complex structure as *one graph*. Complex increments (i.e. a module with its components) are represented by subtrees, the root node representing the complex increment (the module). Therefore, the graph structure is *flat* in contrast to hierarchical graphs /Pr 78/. The context-sensitive relations (e.g. a module to be imported has to be defined) is modelled by two tree portions (the name node in an import clause and the identifier node in the corr. exporting module) and a relation between those tree portions.

In order to see how the *specification* of a graph class (the corr. structure of the graphs and their operations) is looking like we regard an example. Fig.2 shows a portion of an *architecture graph*, i.e. an internal logical graph describing the architecture of a tiny software system. To be correct, it only exists of two modules and these two modules are still incomplete. Now we describe an operation of this graph class which corresponds to a combined editor/analyzer step, namely to insert a new name 'B' in the import list of a module A belonging to the interface description of that module and to check whether the module B is defined elsewhere. This is shown in Fig.3. We call the portion of the spec describing the operations the transaction part. Please note, that Fig.3 contains only one operation of this transaction part and does not contain the schema part. The immediate impression, however, is that such specs are rather complex.

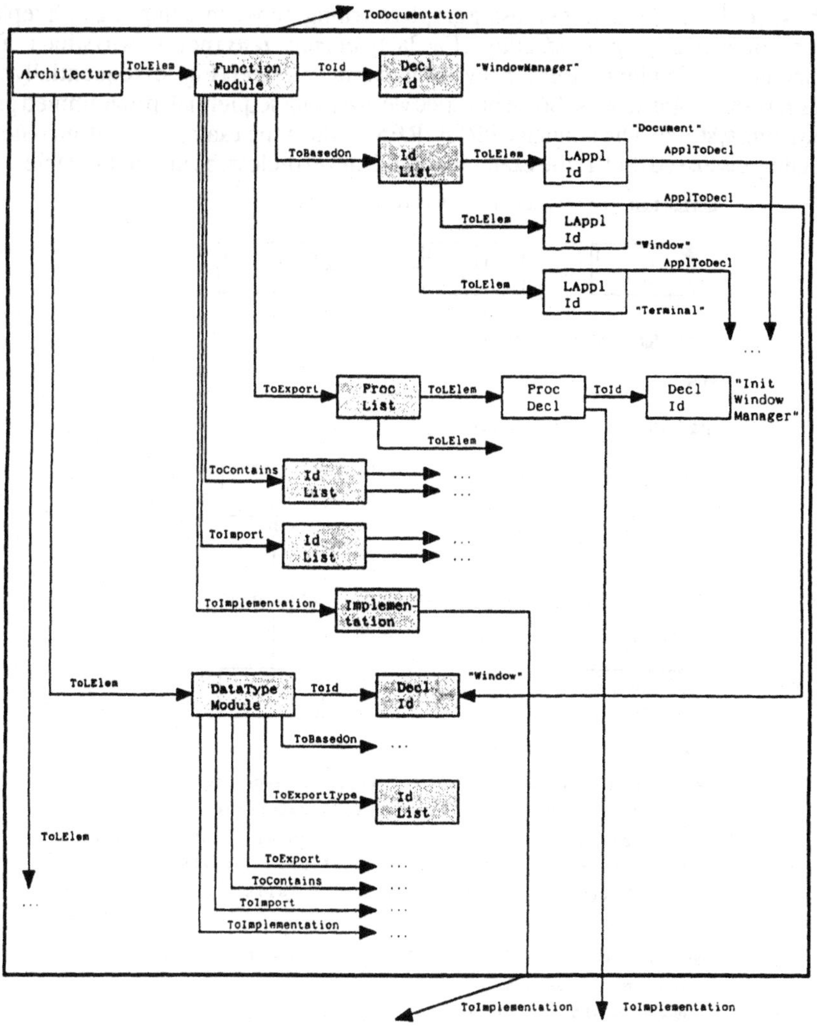


Fig.2: a portion of the architecture graph corresponding to a software system

Before discussing the spec of this transaction let us have a more careful look on the *decomposition structure* of the architecture graph of Fig.2. We see that an architecture is a list of module descriptions (represented by the `Architecture`-node and the corresponding `ToLElem`-edges). Any module is represented by a ... -`Module`-node and is the composition of items: its identifier (`ToId`-edge together with `DeclId`-node), its import clause for the interface (`ToBasedOn`-edge together with `IdList`-node) and so on. This import clause consists of different `LApplId`-nodes (applied identifier within a list). The above mentioned context-sensitive relation is expressed by an `ApplToDecl`-edge from such an applied to its defining occurrence, i.e. the corresponding module identifier node. The tree structure consists of tree increments (subtrees) representing compositions, lists, or atoms /En 86/.

The spec of the transactions is using graph rewriting steps consisting of graph replacements defined by *graph productions* (Fig.3a, b) or *graph tests* (not shown) which can be combined to complex *transactions* by using *control structures* (Fig.3c) sequentially composing other transactions. So, in our spec we are using sequential, programmed graph rewriting systems. The language PROGRES in which the example is written is not explicitely discussed /Sc91/ the example should, nevertheless, be understandable.

a) **production** InsertLApplId (ModuleId : STRING) =

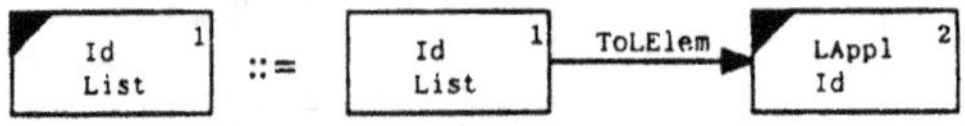


```
transfer L.Name := ModuleId
end
```

b) **production** BindToModuleId

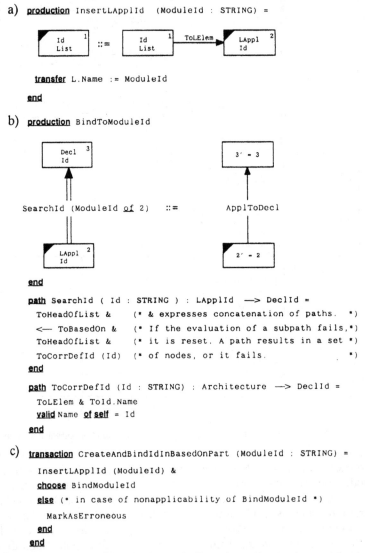

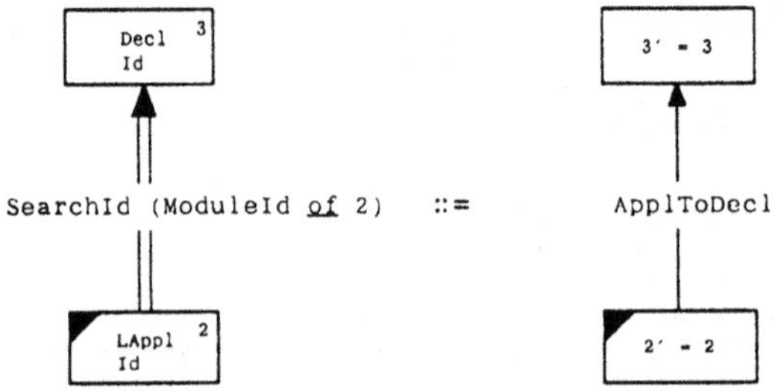


```
end
path SearchId ( Id : STRING ) : LApplId —> DeclId =
  ToHeadOfList &       (* & expresses concatenation of paths.   *)
  <— ToBasedOn &       (* If the evaluation of a subpath fails,*)
  ToHeadOfList &       (* it is reset. A path results in a set *)
  ToCorrDefId (Id)     (* of nodes, or it fails.                *)
end
path ToCorrDefId (Id : STRING) : Architecture —> DeclId =
  ToLElem & ToId.Name
  valid Name of self = Id
end
```

c) **transaction** CreateAndBindIdInBasedOnPart (ModuleId : STRING) =

```
  InsertLApplId (ModuleId) &
  choose BindModuleId
  else (* in case of nonapplicability of BindModuleId *)
    MarkAsErroneous
  end
end
```

Fig.3: a transaction describing the internal effect of an editor/analyzer tool activation

The first production inserts the name of a new module into the interface import clause (internally the BasedOn-list) by a trivial one-node replacement, the current position of the user's cursor being expressed by a marker. The name the user has put in is stored in the attribute Name of this new node 2. Identically denoted nodes are embedded identically. This is the tree part of the transaction. The context-sensitive part is given by Fig.3b using again identical embedding. The production inserts an `ApplToDecl`-edge between the two corr. nodes of the right hand side, the upper being the identifier node of the exporting module, the lower the name node in the `BasedOn` import clause. The left hand side contains an application condition, namely a path expression `SearchId`, describing a path between the two nodes of the left hand side. The production is only applied if the application condition holds true. `SearchId` is a declarative description of a path. By `ToHeadofList&<-ToBasedOn&ToHeadofList` we get the root node of the architecture graph of Fig.2 starting from the name node of the `BasedOn` clause. `ToHeadofList` here only means to traverse a `ToLElem`-edge in the reverse direction. The path `ToCorrDefId` finds the identifier node of the exporting module. It tries one path from the `Architecture`-node to the `DeclId`-node of a module after the other. If it finds a suitable node it is successful, otherwise it fails. Correspondingly, the path `SearchId` succeeds or fails. The complete transaction is given in Fig.3c. After inserting a name node a binding is tried. If it is successful an `ApplToDecl`-edge is inserted, otherwise a trivial production (not given) sets an attribute as erroneous.

*Implementation results* of the *AST graph approach* are the following (some of them, namely (1) – (3), are not restricted to AST modelling): (1) As gra gras are operational, they can be executed. An interpreter for graph grammars being a part of a gra gra spec environment (/Sc 91, NS 91, Zü 94/) has been developed. (2) As the framework architecture of this spec environment is the same as that of the IPSEN system, this interpreter and a spec together with the framework can act as a rapid prototype for an application environment. (3) Furthermore, generator tools for efficient implementations have been built, translating parts of a specification into equivalent Modula-2 code. (4) The above mentioned basic specifications for tree and nontree handling are available as efficient building blocks in Modula-2. (5) Finally, starting with an EBNF for a given language, a translator has been written, generating a framework AST graph spec from this EBNF, which is then extended by the specificator.

Fig 4 shows our short term goal for *structuring* a *graph grammar specification* for an integrated environment working on one logical document following the AST graph modelling approach. Let us start the explanation by regarding the *second layer*. The services to be specified are for editor, instrumentation, analysis, browsing, execution, and monitoring tools.

Context-free editing services change the specific tree structure according to a given EBNF. Instrumentation services (for an executable document, e.g. a module body, a conditional/unconditional breakpoint, a counter etc.) also change the tree part of the internal document according to an extension of the language syntax. Furthermore, cursor movement services, if they follow the tree structure, have to be placed here. Summing up, the left column is responsible for all tree manipulation and tree movement services to be offered by the logical document.

The middle column is responsible for nontree static services. These correspond to the context-sensitive syntax rules of the underlying language but also to extensions for instrumentation or execution or further analyses (e.g. set use, control flow analysis etc.). All these services yield the appropriate information for analyzer tools. Furthermore, here we place the services for browsing tools, as 'semantical' browsing usually means to follow traces according to context-sensitive syntax.

The right column is responsible for the services needed for execution (if the logical document is executable). Execution services need the modification of internal runtime data (stack, heap, but especially the host graph). Furthermore, execution counter information has to be set forward/backward. If the environment offers monitoring services (e.g. giving information for execution time, data consumption, or else; global or according to instrumentation) then the corresponding internal modifications have to be specified, too.

This second layer is on top of a *basic layer* for tree handling services, context-sensitive services, and dynamic semantic services, respectively. Tree handling services are like inserting a list node together with list members, a composite node together with components etc., moving up/down/right/left according to the tree structure. Context-sensitive basic services are mechanisms for identifier binding according to scope/visibility rules, type checking etc. Dynamic semantic basic services are like creating/deleting a data structure component in a runtime structure, the structure of which is described in a spec (e.g. for elaborating a variable declaration, executing a procedure call, or executing a graph production at runtime).

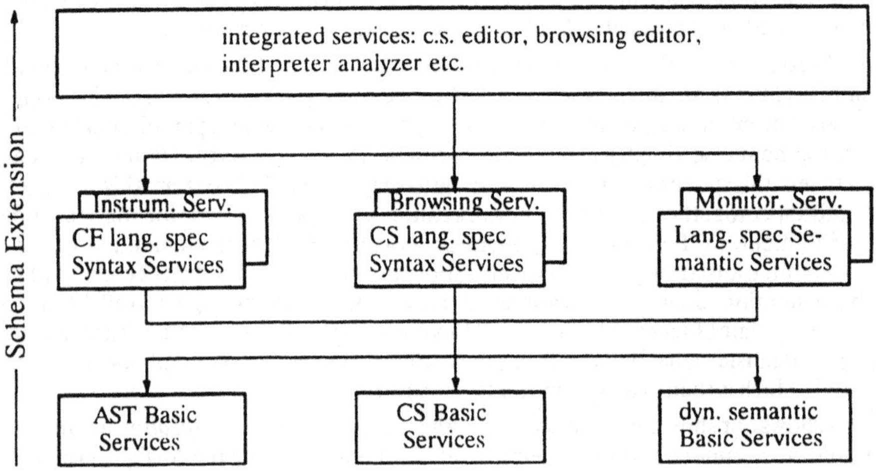


Fig.4: a methodology for writing a specification for an integrated environment on one logical document according to AST graph modelling

This basic specification layer and its correspondence to the second layer is responsible for *reuse* in *two respects* (not shown in this paper): The specification basic layer is product reuse as it has to be written only once and can be used for any specification according to the AST graph modelling approach. However, it also helps on the process side of writing/maintaining specs as the effort of writing/maintaining a spec is only on the second layer where the specifics w.r.t. a certain language have to be fixed. Furthermore, according to methodology knowledge, it is quite clear how to write this remaining part. In the best case this second layer can be generated (this is the case for the context-free language specification which is generated from a given EBNF).

The second layer offers the necessary component services for an integrated environment. On top of the second layer we find a *third layer* of *services* for *integrated tools*, as e.g. an editor which handles context-free and context-sensitive syntax rules at the same time (as the example given above, or changing a declaring identifier and all its applied occurrences etc.). A further example is an interpreter/analyzer/editor service which steps from one applied occurrence of a variable to the next for those variables which during interpretation have got no value.

According to the layers of Fig.4 there corresponds a *layering* of *graph schema information*: At the bottom all schema information is fixed which corresponds to features of tree, nontree, and dynamic semantics basic services. These schema layer is application independent. On top of that we have schema information which is language specific, namely defining node types and edge types to occur in a graph class. Similarities of node types are expressed by an inheritance structure in between.

At the end of this section let us give a remark *relating external* to *internal modelling*: Textual languages are mostly defined by EBNFs and context-sensitive constraints. This external modelling can directly be mapped on corresponding *AST graph modelling*, where EBNF portions are mapped onto tree parts and constraints on corr. nontree relations between corr. tree parts. This gives an easy 1-1 transition from external to internal level. This proceeding is not without problems, as a 'wrong' external modelling is internally copied as we shall see in the next section.

## 4 Metamodelling and Layers & Parameterization

In this section we give sketches of two further uniform modelling approaches. Because of lack of space we are only arguing about structuring a spec and do neither discuss detailed gra gra specs nor the internals of resuable spec components.

There are *different reasons* that the second and alternative uniform modelling idea, which we called *metamodelling* /Ja 92/, came up in the group: (1) When regarding logical documents with textual as well as graphical representations, we saw that it is not easy to build a graphical unparser for an internal document which was the result of a direct translation from a textual external representation. (2) When studying integrator tools for different languages (different perspectives in an integrated req. eng. language; req. eng. as source, design language as target of a transformator tool), we first had to detect similarities and differences of different languages. From these similarities we could derive transformator rules. (3) When modelling different internal documents corresponding to different languages it was quite obvious that language similarities should have a correspondence in similarities of the corresponding internal graph classes. At the time being metamodelling has its greatest advantages on the external level, i.e. for modelling languages or language correspondences.

The first of the above topics was less a question of metamodelling but a *misuse* of *AST graph modelling* as we learned later. When taking an EBNF of a textual language as the basis for internal modelling one has to be careful: One has to think about textual and graphical representations as well. It might be and it is often the case that logical items are put in a wrong place of a nesting (and correspondingly tree) hierarchy of an EBNF. Especially, relations between items, which in textual languages are represented by clauses, are expressed as inner constructs of items. This is wrong as these relations are logically on the same level as the items. A way to avoid these problems is first to think about graphical representations where items and relations are equally entitled. This then implies a modification of the EBNF one starts with. In the case that the EBNF is given (e.g. for an programming language) one takes a different internal tree structure from which it is easier to generate a textual and graphical view as well.

The *essential ideas* of *metamodelling* are the following: (1) One thinks in basic modelling concepts which are application independent and which are expressed by classes. (2) These modelling entities can be complex objects and not only nodes therefore offering hierarchical and not only flat graph structures. (3) These classes can be specialized and combined to get suitable modelling constructs resulting in a well-defined class hierarchy. (4) On top of this hierarchy certain application specific extensions are regarded. (5) The layers of the class hierarchy reflect different modelling 'dimensions'. (6) Metamodelling is not bound to a specific way of modelling. Instead, it should allow to model in different ways: Flat or hierarchical, thinking in objects as dominant items and relations as derived ones or, vice versa, in connections as dominant items and gluing points as derived ones etc.

Let us try to sketch some of the ideas of metamodelling (for a more detailed discussion see /Ja 92/). At the bottom of Fig.5 is a class describing the commonalities of all modelling entities which on the second layer is specialized into the following five classes allowing to regard the following specialized entities: A modelling entity can be a *unit* which has a denotation, or a *link* which is connected to two units in the binary link, or more units in the hyperlink case. An entity can also be a *frame* which has an inner structure. A modelling entity can, furthermore, be either an *atom*, i.e. having no *structure* or its structure being not described, or it may be a structure. (By the way it seems remarkable, that these specialized entities correspond to the definition items of hierarchical graphs /Pr 78/, namely nodes, edges, contains function, atoms or graphs).

On top of the second layer we can regard *suitable combinations* of classes in order to build up appropriate *modelling entities* for a certain group of modelling tasks. For example, it may be useful to have units (generalized nodes) which are frames (contain something) and are arbitrarily internally structured (the structure being a graph). Another combination is to have links, which are frames but the internal structure is atomic, i.e. being a label (not shown).

Alternatively, one can think of *graphtheoretic specialization* at level three, e.g. specializing links into binary or nonbinary, the binary links into directed or nondirected etc. (not shown). More interesting is the classification of graphs into single nodes, one level structures (as compositions, lists) in global structures (as trees, lattices etc.).

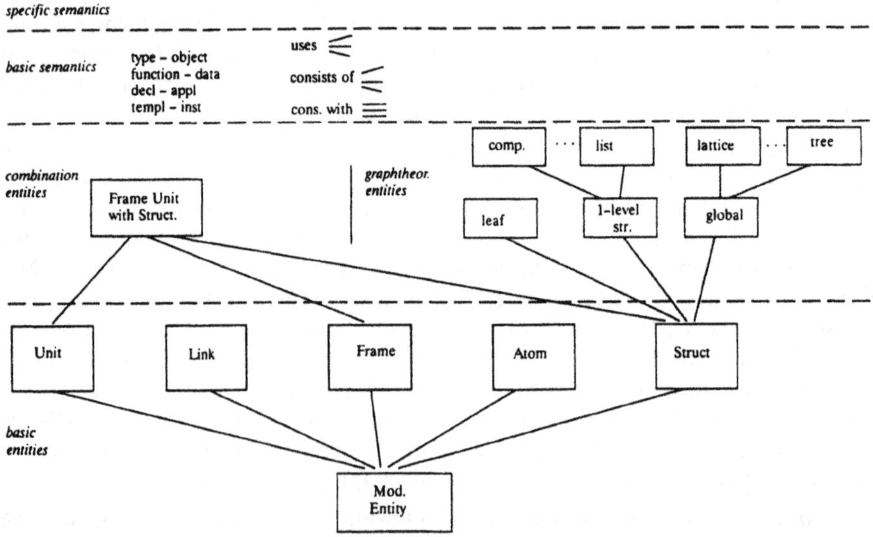


Fig.5: metamodelling yields a class hierarchy expressing different modelling dimensions

It should be clear that the *hierarchy* given so far *allows* for *different ways* of *modelling*: If we only allow atomic units and links we can think in flat node and edge–labelled graphs (as in AST modelling). On the other side, units but also links can have an inner structure thereby modelling in hierarchical graphs. Furthermore, we can think in objects as dominant modelling units and relations as derived ones, which can only be inserted if corresponding objects are at hand (as in AST modelling). Conversely, which is often used in straight-line drawings, we can think of directed link objects where the corresponding gluing points by which links are connected are derived objects.

On top of the layers regarded up to now we can think of layers expressing *basic semantics* for a certain *modelling area*. For example, if we regard architecture or programming languages and the specification of their tools we could introduce unit classes for 'types' or for 'objects', for 'functions' or for 'data', for 'declarations' or for 'applied occurrences', for 'templates' or for 'instances' etc. Appropriately, we could introduce entities for 'uses'–, 'consists of'– or 'consistent with'– relations and their specializations. On top of such layers one would define the specific classes and types for a certain architecture language or for a certain programming language, respectively.

The reader might have recognized that in the above discussion we did not speak of *external* (language) or *internal* (data structure for tools) modelling. The metamodelling class hierarchy is *applicable for both*. This is because even more than in AST graph modelling we asked for concepts for modelling as such. These are applicable for both modelling tasks.

A difference to AST graph modelling is that schema information is not a dual aspect of transaction information. In metamodelling the schema information is contained in the class hierarchy. Furthermore, the classes of the inheritance hierarchy contain the transaction part (i.e. operations on units are to create or delete them, on links are to connect or to delete a connection and, correspondingly, graph classes have appropriate operations). Especially, structures contain other more or less specialized modelling entities. The modelling task, therefore, consists of enlarging the inheritance hierarchy such that suitable modelling entities result and such that suitable graph classes are built up which consist of these entities.

The idea of the last uniform modelling approach, called *Layers & Parametrerization*, is taken from /We 91/ where it was used for consistency control in software systems. We only discuss the key idea. The value of this approach lies in the detection that when building integrated software development environments a certain environment has *a lot of parameters* (fixed models). Each parameter set yields another environment.

For the technical side of software development (c.f. Fig.1) we have (a) a certain overall lifecycle model. For a fixed lifecycle model we may (b) divide a working area (b1) into different perspectives, and/or (b2) into an overview or detailed model, and/or (b3) into information of several hierarchy levels. In any case, it has to be clear which information is contained in which perspective/representation according to a degree of detail/level of hierarchy. For these different aspects (c) detailed languages have to be developed, (d) methodologies how to use these languages have to be found, and (e) the behavior of corresponding tools (language–, methodology–dependent) has to be fixed. Furthermore (f) the integration of different languages (f1) being a part of a certain working area, or (f2) belonging to different working areas has to be determined. Correspondingly, (g) the behavior of integration tools has to be fixed.

Similar problems occur on the administration side of software engineering where we have to fix (h) a working area division as e.g. product, process, project handling, and resources control. Then again, (i) submodels as well as their languages and tools have to be fixed, (j) integration within as well as (k) between working areas has to be determined on language, methodology and on tool level. Finally, (l) integration of the technical and administration side has to be made precise.

So, we conclude that an SDE (logical separation, languages, methods, tools' behavior, representation, and their integration etc. /Na 94b/) and, therefore, also its internal spec is dependent on many parameters for models which can alternatively be set. Neither for the implementation nor for the specification we are able to realize/or specify a new SDE from the scratch. The consequence for the spec, therefore, is that we have to try to *structure a specification* such that (a) as much basic and model invariant parts have to be found as possible. They are the basic layer of a composite specification to be found at the bottom. (b) Models, which are very likely to change are inserted in the upmost levels, models which express basic assumptions (as e.g. working area division) and which are necessary to fix specific models are inserted on lower levels of the hierarchy. (c) A spec layer usually means to fix one or more models.

# 6 Summary and Future Work

We have introduced three different ways of modelling gra gra specs which are applicable not only to a certain specific problem but are general ways of structuring a spec. In the *AST graph modelling* approach we think in trees together with nontree derived relations. Hierarchical structures are represented by subgraphs connected by anchor nodes in one flat graph. Basic services for tree handling, context–sensitive mechanisms, and execution are extracted. On top of such specs language dependent specs are made precise. A schema part accompanies the transaction part. In *metamodelling*, we are more interested in general principles of modelling finding suitable modelling units and combining them to structures. The schema part as well as the transaction part is expressed in one inheritance hierarchy. The bottom of this hierarchy are general modelling units and structures. The more we come up, the more specific to a specification application we get. The hope is that, for different specs in a fixed application area, only the topmost classes have to be redefined. In the *layers & parameterization* approach we think about fixing the various models of an integrated environment at the right layer of a spec hierarchy. Thereby, we postpone fixing of models as late as possible and try to extract as much reusable specs.

The AST graph modelling approach has been applied several times to internal graph modelling of graph classes corresponding to sublanguages or languages of a working area /Sc 91/. Metamodelling has been applied to external modelling of language integration within one working area or language transformation between different working areas /Ja 92/. The layers & parameterization approach was applied to specify integrated revision and configuration control on the administration side and the integration of this administration aspect with consistency control (for a fixed scenario of languages and their integration) on the technical side /We 91/. Therefore, the above approaches have been invented and used for a certain *granularity level* of *integration*.

There are a *lot of problems* to be solved yet:

(1) The first question to be answered is whether AST graph modelling, metamodelling, and layers & paramterization approaches can be unified into one new uniform modelling approach. Alternatively, it is possible that we can show that all three approaches are inherently different. It is, for example, possible that any of those approaches is useful only for a certain granularity level of integration.

(2) We still have to find a final structure for a graph grammar specification for one document or for more than one documents within one working area (different perspectives, degrees of details, hierarchy and independence of documents).

(3) The next step is to specify integration between different sublanguages of a working area or between different languages of different working areas, i.e. the behavior of corresponding integrator tools (transformators, consistency checkers, browsers etc.). Please note, that this kind of integration appears on the technical side as well as on the administration side (c.f. Fig.1). The last step is the specification of the integration of technical and administration tools.

(4) Let us assume that all specification problems are solved. As argued above, any of these specs contains reusable spec components. For these spec components efficient realizations (e.g. in Modula-2) have to be developed: Basic services for one document, consistency handling, transformation, and overall integration of documents within or between working areas, or technical and administration integration.

(5) For the remaining specific components generators have to be developed.

The topics (2), (4), and (5) have *partially been solved* for *AST graph modelling* resulting in a specification, implementation, and generator machinery for integrated environments on one document for one working area including the representation problem. There we have impressive efficiency parameters for the process side to develop an SDE as well as for the product side /Sc 91/. However, some open questions are left. Therefore, the general problem now is to make a revision of these investigations under a new, more general, and wider perspective.

## 7 References


/ACMxx/ Proc. 1st ACM SIGSOFT/SIGPLAN Softw. Eng. Symp. on Pract. SDEs, ACM SIGPLAN Notices 19, 5 (1984), Proc. 2nd Symp., ACM SIGPLAN Notices 22, 1 (1987), Proc. 3rd Symp., ACM Softw. Eng. Notes 13, 5 (1988), Proc. 4th Symp., ACM Softw. Eng. Notes 15,6 (1990), Proc. 5th Symp., ACM Press (1992).

/BBN77/ W. Brendel/H. Bunke/M. Nagl: Syntax-directed Programming and Incremental Compilation (in German), IFB 10, 57-74, Berlin: Springer-Verlag (1977).

/ELS87/ G. Engels/C. Lewerentz/W. Schäfer: Graph Grammar Engineering: A Software Specification Method in /GG 87/, 186-201.

/ELN92/ G. Engels/C. Lewerentz/M. Nagl/W. Schäfer/A. Schürr: Experiences in Building Integrated Tools, Part 1: Tool Specification, TOSEM 1, 2, 135-167 (1992).

/En86/ G. Engels: Graphs as Central Data Structures in a Software Development Environment (in German), Diss. U. Osnabrück, VDI Fortschrittsb. 62, Düsseldorf: VDI-Verlag (1986).

/ENS84/ G. Engels/M. Nagl/W. Schäfer: On the Structure of Structure-Oriented Editors for Different Applications /ACM 84/, 190-198.

/ES89/ G. Engels/W. Schäfer: Programming Environments, Concepts and Realization (in German), Stuttgart: Teubner-Verlag (1989).

/EWM92/ W. Eversheim/M. Weck/W. Michaeli/M. Nagl/O. Spaniol: The SUKITS Project - An a posteriori Approach to Integrate CIM-Components, Proc. Ann. GI Conf., informatik aktuell, 494-504 (1992).

/GGxx/ N. Claus/H. Ehrig/G. Rozenberg (Eds.): Proc. 1st Int. Workshop on Graph Grammars and Their Appl. to Computer Science, LNCS 73 (1979), H. Ehrig/M. Nagl/G. Rozenberg (Eds.): Proc. 2nd Workshop, LNCS 153 (1983), H. Ehrig/M. Nagl/A. Rosenfeld/G. Rozenberg (Eds.): Proc. 3rd Workshop, LNCS 291 (1987), H. Ehrig/H.J. Kreowski/G. Rozenberg (Eds.): Proc. 4th Workshop, LNCS 532 (1991).

/Gö88/ H. Göttler: Graph-Grammars in Software Engineering (in German), IFB 178 (1988).

/Ja92/ Th. Janning: Integration of Languages and Tools for Requirements Engineering and Programming in the Large (in German), Diss. RWTH Aachen, Wiesbaden: Deutscher Universitätsverlag (1992).

/KSW92/ N. Kiesel/J. Schwartz/B. Westfechtel: Object and Process Management for the Integration of Heterogeneous CIM Components, Proc. Ann. GI Conf., informatik aktuell, 484–493, Springer-Verlag (1992).

/Le88a/ C. Lewerentz: Interactive Design of Large Software Systems (in German), Diss. RWTH Aachen, IFB 194, Berlin: Springer-Verlag (1988).

/Le88b/ C. Lewerentz: Extended Programming in the Large in a Software Development Environment, in /ACM 88/, 173–182.

/Na79/ M. Nagl: Graph-Grammars: Theory, Applications, and Implementation (in German), Braunschweig: Vieweg-Verlag (1979).

/Na85/ M. Nagl: Graph Technology Applied to a Software Project, in Rozenberg/Salomaa (Eds.), The Book of L, 303–322, Berlin: Springer-Verlag (1985).

/Na90/ M. Nagl: Software Engineering: Methodological Programming in the Large (in German), Berlin: Springer-Verlag (1990).

/Na94a/ M. Nagl (Ed.): Building Tightly Integrated Software Development Environments - The IPSEN Approach, to appear in LNCS.

/Na94b/ M. Nagl: Software Development Environments - Classification and Future Trends (in German), to appear in Informatik-Spektrum

/NS91/ M. Nagl/A. Schürr: A Graph Grammar Specification Environment, in /GG 91/, 599–609.

/Pr78/ T. Pratt: Definition of Programming Language Semantics Using Grammars for Hierarchical Graphs, in /GG 79/, 389–400.

/Sc75/ H.J. Schneider: Syntax-Directed Description of Incremental Compilers, LNCS 26, 192–201 (1975).

/Sc86/ W. Schäfer: An Integrated Software Development Environment: Concepts, Design, and Implementation (in German), Diss. U. Osnabrück, VDI Fortschrittsb. 57, Düsseldorf: VDI-Verlag (1986).

/Sc91/ A. Schürr: Operational Specification by Programmed Graph Rewriting Systems: Formal Definition, Applications, and Tools (in German), Diss. RWTH Aachen, Wiesbaden: Deutscher Universitäts-Verlag (1991).

/SW93/ A. Schürr/B. Westfechtel: GRAS a Graph-Oriented Data Base System for Software Engineering Applications, submitted for publication.

/SZ91/ A. Schürr/A. Zündorf: Nondeterministic Control Structures for Graph Rewriting Systems, LNCS 570, 48–62, (1991).

/We91/ B. Westfechtel: Revision and Consistency Control in an Integrated Software Environment (in German), Diss. RWTH Aachen, IFB 280, Berlin: Springer-Verlag (1991).

/Zü94/ A. Zündorf: A Graph Grammar Specification Environment - It's Use and Realization, Diss. (to appear).

# Set-Theoretic Graph Rewriting

*Jean-Claude Raoult*
*IRISA, Campus de Beaulieu*
*F-35042 RENNES CEDEX*
*raoult@irisa.univ-Rennes1.fr*

*Frédéric Voisin*
*LRI, Bâtiment 490*
*F-91405 ORSAY CEDEX*
*fv@lri.lri.fr*

**Abstract:** Considering graphs as multisets of arcs, we define their rewritings as simple set-theoretic rewritings: applying a rule consists of removing the left-hand side and adding the right-hand side. This method can simulate categorical graph rewritings, provided that no two vertices are identified. On the other hand, rewriting is possible in cases where the double push-out method does not apply, and where the single push-out method yields an unexpected result. The simplicity of our approach is illustrated by its stability under composition and by a criterion of local confluence.

## I. Introduction

Graph rewriting is now an old story (see the survey by Nagl [1979]). Given a rule $L \to R$, to rewrite a graph in which has been found an occurrence of $L$, perform the following three steps: firstly, remove $L$, possibly leaving out "dangling arcs" with deleted source or target; secondly, put in its place the right-hand side $R$ of the rule; thirdly, connect the dangling arcs to some specified vertices in the right-hand side.

The third part is the difficult one. Schneider [1970] gave two relations $\pi_+$ and $\pi_-$ between the vertices of the left and the right-hand sides. If $(x, y) \in \pi_-$ then to every incoming arc having target $x$ in $L$ should correspond an arc having same source and target $y$ in $R$. Symmetrically, if $(x, y) \in \pi_+$ then to every outgoing arc having source $x$ in $L$ should correspond an arc having same target and source $y$ in $R$. Both relations depend on the label of the arcs.

This method is very flexible, but very procedural. In 1973, Ehrig, Pfender and Schneider [1973] gave a method in which $\pi_+$ and $\pi_-$ are represented by an "interface graph" $K$ with two projections, two graph morphisms into $L$ and $R$. The construction is now categorical; it uses two push-out squares (see section III). This last method is widely known, thanks to the numerous applications published by the Berlin school, even if it is really applicable only in a restricted case, that of "fast productions": those in which $K$ is a subgraph of the left-hand side $L$. This remark has been exploited by Raoult [1984] who simplified the construction, using a single push-out square in a category of partial morphisms.

This last version has been simplified yet and improved by Kennaway [1987] and Löwe & Ehrig [1990].

In the meantime, the theory of term rewriting has been developing rapidly. Rewritings in free algebras have been known fairly early and criteria for confluence and termination have got more and more precise (cf. for instance the survey by Dershowitz and Jouannaud, in the Handbook of TCS [1990] or Rusinowitch [1989]). In algebras factored out by equations, the same problems have been tackled with some success (cf. Kirchner [1985]). For instance, techniques are known for rewriting terms containing operators that are associative and commutative (cf. Peterson & Stickel [1981]). A term built on an associative and commutative operator simply represents the multiset of its leaves, so that multisets of terms can be rewritten.

The goal of the present work is to connect graph rewritings to simple (multi)-set rewritings, following the idea that a graph is a set of arcs, or rather hyperarcs. This idea has already been investigated by Bauderon and Courcelle [1987]. They describe hypergraphs with natural numbers for vertices, and with constructions on sequences of numbers (i.e. on hyperarcs): the most important such constructions are disjoint union and vertex identification, with an explicit renumbering of vertices. A given hypergraph is represented by a term on these operators, or rather several equivalent such terms. This equivalence is described by a recursive set of axiom schemata, which turns the set of hypergraphs into an effectively presented algebra.

We develop here a slightly different point of view: vertices will be variables, of which we have a countable reserve — as in the case of first order terms. A hyperarc is now the theoretical analogue of a statement in a three address code, in which the function $f(x,y)$ is associated with a variable $z$ containing the result of the computation of $f$, so that we get an equation: $z = f(x,y)$. In this respect, our rewritings might be called equation rewritings. But we shall stick to the combinatorial point of view and denote this equation by the hyperarc $fzxy$. All this is similar to the jungles of Habel, Kreowski and Plump [1987]. In our setting, a hypergraph is simply a set of hyperarcs, while a program written in three-address code is a sequence of hyperarcs. If each variable is assigned only once, both concepts coincide.

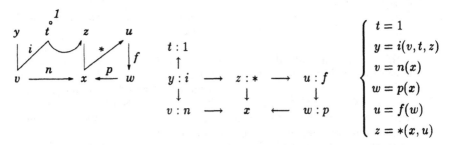

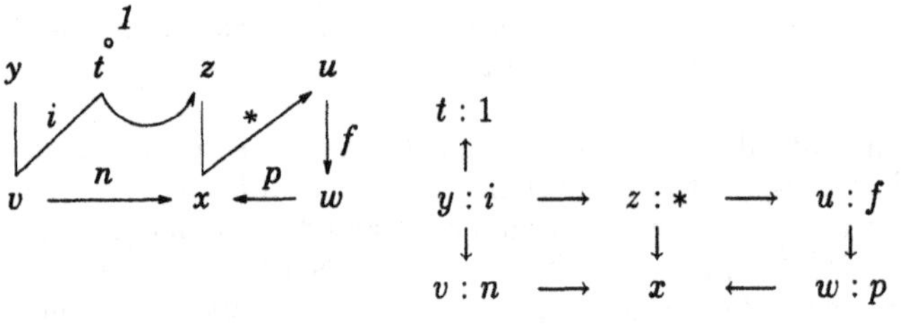


$$\begin{cases} t = 1 \\ y = i(v,t,z) \\ v = n(x) \\ w = p(x) \\ u = f(w) \\ z = *(x,u) \end{cases}$$

**Figure 1.** Three representations of the same hyper-graph. On the left is a hyper-graph (the dot denotes a unary hyperarc), in the middle a planar (oriented) graph, on the right a sequence of three adress statements. They describe the definition of the function $f(n) = n!$.

Rewriting will now consist of the first two steps only of the construction. The third and difficult task of connecting dangling arcs is implicit, taken care of automatically by the name of the vertices. Further, we shall avoid identifications of vertices: identifying two vertices is not really a rewriting. This will be closer to code optimisations, where identifications are seldom performed, for reasons of undecidability. A first consequence is that rewriting a left-hand side $L$ into a right-hand side $G$ has complexity proportional to the size of $L$ and $G$, regardless of the size of the context — which is not the case when an identification takes place. Another consequence is a property of rewritings which we feel characteristic: they can be made invertible by swapping the sides of each rule. A final consequence is the simplicity of the approach: no category theory is needed, no push-out, just set difference and sum.

In section II, we define the graphs and graph rewritings that we consider and give a few examples. In section III, we show that these rewritings are incomparable in power with the categorical graph rewritings, whether they use one or two push-out squares. The simplicity of the method is then tested in section IV in which we build a system generating the composition of two finitely generated rewritings and in section V in which we derive a test for the local confluence of the generated rewriting. Section VI is a conclusion.

## II. Definitions and notations

Graphs are usually defined by vertices and arcs. Each arc passes through two vertices: its source and its target. Hypergraphs have hyperarcs passing through non-empty sequences of vertices, i.e. elements of $X^+$ when $X$ denotes the set of vertices. If the arc $xy$ is labelled by $f$, this will be represented by the sequence $fxy$. Recall that a multiset of elements of a set $E$ is a mapping $H : E \longrightarrow \mathbb{N}$ denoted by $e \mapsto |E|_e$ for all element $e$ of $E$, or rather by a series-like notation: $H = \sum_{e \in E} |H|_e e$.

DEFINITION 1. *Given a countable set $X$ of variables, and a set $F$ of labels, a labelled hyperarc is an element of the set $FX^+$, and a (labelled) hypergraph is a multiset of labelled hyperarcs. The set $\mathbb{H}$ of all finite hypergraphs is the set $\mathbb{N}^{(FX^+)}$ where the parentheses indicate that the mappings are null almost everywhere. The set of variables occurring in the hypergraph $H$ is denoted by $X(H)$.*

An important feature of this definition is that the vertices — which are identified with variables $x, y, z$, etc. — only occur as supports for hyperarcs, and that all hyperarcs are labelled. A priori, a vertex by itself is not a hypergraph. Should this cause problems, we may assume the existence of a distinguished label $a$ meaning "is a vertex" and add to the considered hypergraph hyperarcs consisting of single vertices only: $\{ax, ay\}$ for instance would be the graph consisting of two vertices $x$ and $y$.

The sum $H + H'$, the union $H \vee H'$ (or "rounded": $H \cup H'$), the intersection $H \wedge H'$ (or "rounded": $H \cap H'$) and the difference $H - H'$ of two hypergraphs $H$ and $H'$ are defined as for any multisets. We shall avoid the "rounded" notation to remind the reader that arcs may occur several times, and that the union and intersection are actually the supremum and infimum. If we write singletons without curly brackets, hypergraphs can be written like series with integer coefficients:

$$H = c_1 t + iyvtz + nvx + pwx + fuw + *zxu$$

or simply drawn, as in figure 1.

In a *simple* hypergraph, there is no multiple hyperarc. The coefficients are 0 or 1: an hyperarc occurs or not, but if it does, it occurs only once. Simple hypergraphs are particular hypergraphs, and conversely, with each hypergraph $H$ can be associated a simple hypergraph $H'$ defined by $|H'|_{aw} = |H|_{aw} \wedge 1$: a hyperarc $aw$ occurs in $H'$ if it occurs at least once in $H$. This projection is a left inverse of the inclusion of simple hypergraphs into general hypergraphs.

Mappings $\sigma : X \to X$ extend into mappings of hyperarcs $\sigma = FX^+ \to FX^+$:

$$(fx_1 \dots x_n)\sigma = f(x_1\sigma)\dots(x_n\sigma)$$

A hypergraph morphism $H \to K$ is a mapping $\sigma : X \to X$ such that, for the extension defined above, $|H|_e \leq |K|_{\sigma(e)}$ for all hyperarc $e \in FX^+$. The domain $\mathrm{dom}(\sigma)$ of a mapping $\sigma$ is the set of variables that differ from their images. We shall only consider mappings having a finite domain. If a mapping is one-to-one it is called a vertex renaming. Two graphs deduced from one another by a vertex renaming are isomorphic, and the morphism is called an isomorphism. Beware that graph isomorphism is not preserved by adding an arbitrary context, as in the following simple example pointed to us by Detlef Plump.

*Example 1:* the graphs $axy$ and $ayx$ are clearly isomorphic. Adding the context $bxy$ yields $bxy + axy$ and $bxy + ayx$ which are not isomorphic! Only graph equality is preserved.

Rewriting relations are relations preserved by adding suitable contexts. Example 1 above shows that some care need be taken regarding the variables. For instance, if we delete a vertex $x$ labelled by $a$ and add a new vertex $y$ labelled by $b$, i.e. use the rule $ax \to by$, embedded in the context $fzy$, we should not get $ax + fzy \longrightarrow by + fzy$: the vertex $y$ which appears in the rule should remain new even in the context. This supports the following definitions.

DEFINITION 2. *A rewriting relation is a relation $G \to D \subseteq \mathbb{H} \times \mathbb{H}$ preserved by vertex renamings $\alpha$ and (disjoint) addition of a licit context:*

$$G \to D \Rightarrow G\alpha \to D\alpha$$

$$G \to D \Rightarrow G + C \to D + C \quad \textit{if } X(C) \cap X(G) \supseteq X(C) \cap X(D).$$

*A relation satisfying the more restrictive condition*

$$G \to D \Rightarrow G + C \to D + C \quad \textit{if } X(C) \cap X(G) = X(C) \cap X(D)$$

*is called a reversible rewriting.*

As its name suggests, the converse of a reversible rewriting is again a reversible rewriting: the applicability condition is symmetrical. As usual, the rewriting relation generated by a given relation (here people will say "a given system") is the least rewriting relation containing the system. We can describe it more precisely using the following propositions. The first one describes the associativity of licit context addition.

**Proposition 3.** *If $G \rightarrow D$ generates $G_1 \rightarrow D_1$ and $G_1 \rightarrow D_1$ generates $G_2 \rightarrow D_2$, then $G \rightarrow D$ generates $G_2 \rightarrow D_2$.*

*Proof:* both hypotheses amount to:

$$\begin{cases} G_1 = C_1 + G \\ D_1 = C_1 + D \end{cases} \qquad \begin{cases} G_2 = C_2 + G_1 \\ D_2 = C_2 + D_1 \end{cases}$$

with $X(G) \cap X(C_1) \supseteq X(D) \cap X(C_1)$ and $X(G_1) \cap X(C_2) \supseteq X(D_1) \cap X(C_2)$. Then by associativity of the addition, we have:

$$\begin{cases} G_2 = C_2 + C_1 + G \\ D_2 = C_2 + C_1 + D \end{cases}$$

and we only have to check the inequality on variables. By distributivity of the intersection and using

$$X(G) \cap [X(C_2) - X(C_1)] \supseteq X(D) \cap [X(C_2) - X(C_1)]$$

deduced from $X(G_1) = X(G) \cup X(C_1)$ and $X(D_1) = X(D) \cup X(C_1)$ we have

$$\begin{aligned} X(G)\cap&[X(C_1) + (X(C_2) - X(C_1))] \\ &= X(G) \cap X(C_1) + X(G) \cap [X(C_2) - X(C_1)] \\ &\supseteq X(D) \cap X(C_1) + X(D) \cap [X(C_2) - X(C_1)] \\ &= X(D) \cap [X(C_1) + (X(C_2) - X(C_1))] \end{aligned}$$

QED.

**Proposition 4.** *The rewriting generated by a relation $R$ over $\mathbb{H}$ is the relation $H \rightarrow K$ which is true if*
*(i)* $H = C + G$ *and* $K = C + D$,
*(ii)* $X(G) \cap X(C) \supseteq X(D) \cap X(C)$, *and*
*(iii)* $(G, D) \in R$ *up to vertex renaming.*

The relation *(ii)* may also be written as $[X(D) - X(G)] \cap X(C) = \emptyset$ stressing the fact that the "new" variables appearing in $D$ cannot belong to the context. Remark that given a rewriting $H \rightarrow K$ and a finite set $Y$ of variables, there exists a renaming of the variables of $X(K) - X(H)$ yielding a graph $K'$ with $H \rightarrow K'$ (because the rewriting is preserved by renamings) and $Y \cap [X(K') - X(H)] = \emptyset$: new variables may be chosen outside any finite set of variables.

Note that we need not bother about "dangling arcs". If a vertex occurs in a left-hand member but not in the right-hand member, then either it occurs also in the context graph, in which case it remains in this same context graph after rewriting, or it does not, and disappears in the result. This phenomenon does not happen with reversible rewritings.

Our method does not allow two vertices of the left-hand side to be identified in the right-hand side. Nevertheless, there is an easy variant of the definition above allowing a left-hand side to occur in a collapsed form in a context graph:

DEFINITION 5. *The collapsed rewriting generated by a relation $R$ over $\mathbb{H}$ is the relation $H \to K$ which is true if*

*(i)* $H = C + G\sigma$ *and* $K = C + D\sigma$,

*(ii)* $X(G\sigma) \cap X(C) \supseteq X(D\sigma) \cap X(C)$ *and*

*(iii)* $(G\sigma, D\sigma)$ *is deduced from* $(G, D) \in R$ *by a mapping* $\sigma : X(G) \cup X(D) \to X$ *possibly not one-to-one.*

In fact, this definition is not broader than the previous one. Indeed, define the collapsed closure of a relation $R$ over hypergraphs as the relation containing all collapsed instances $(G\sigma, D\sigma)$ of all pairs $(G, D)$ in $R$, for dom$(\sigma) \subseteq X(G)$, up to variable renaming. If $R$ is finite, its closure is also finite, and applying Definition 2 to the closure of $R$ is equivalent to applying Definition 5 to $R$ itself. Henceforth, we shall suppose the first definition, which may forbid a rewriting if the occurrence of the left-hand side is collapsed. If we want to accept it, we shall assume that we have in our system all instances of the rule schemata.

Note that hypergraphs may be considered as terms in $T(\{+\}, FX^+)$ in which the operator $+$ is associative and commutative, and where the arcs themselves are considered as terms of depth one: $fx_1 \ldots x_n = f(x_1, \ldots, x_n)$. In this setting, Definition 2 is just a particular case of term rewriting modulo AC axioms for $+$.

We close the section with a few examples.

*Example 2*: The rule $bxy + byz \to bxz + byz$ applied to $bxy + byz + bxu + buy + buz$ yields $bxz + byz + bxu + buy + buz$. This is a case where the variable $y$ does not disappear. The rule describes multiple jumps elimination in optimizing compilers, where $b$ stands for "branch" and $x$ and $y$ are the labels of the source and target instructions.

*Example 3:* The following system in which $ixuyz$ represent the conditional (if $u$ then $x$ equals $y$, else $x$ equals $z$)

$$ixuyz + \wedge uab \to ixavz + ivbyz$$
$$ixuyz + \vee uab \to ixayv + ivbyz$$
$$ixuyz + \neg ua \to ixazy$$

describes the replacement, common in compilation, of a boolean expression by a sequence of elementary tests.

*Example 4:* The following system with integer coefficients:

$$axy \to axz + azy$$
$$axy \to axy + axy = 2axy$$

generates all series-parallel graphs when started with a graph with a unique arc. This system is context-free. For hyperarc rewritings, see Habel & Kreowski [1987].

*Example 5:* The following last example generates $\mathbb{N} \times \mathbb{N}$ starting from a square $axy + axz + Eyu + Nzu$. Arcs labelled $N$, $S$, $E$, $W$ are the non-terminals on the northern, southern, eastern and western sides of each square:

$$Nxy + Ezy \to axy + azy + ays + ayv + Wxr + Nrs + Nsw + Szu + Euv + Evw$$
$$Nxy + Wyz \to axy + ayz + Wxu + Nuz$$
$$Exy + Syz \to axy + ayz + Sxu + Euz$$

These rules are visualized below.

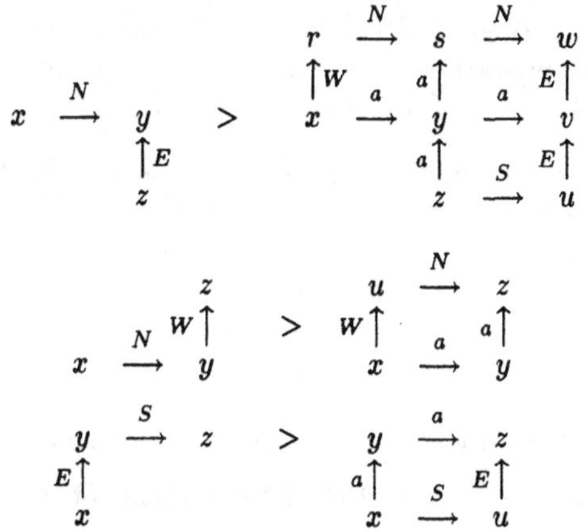


Single hyperarc replacement systems generate graphs having a bounded degree of connectivity and therefore cannot generate this grid. (cf. for instance Habel & Kreowski [1987] or Sopena [1987]).

## III. Relation with categorical rewriting methods

The so-called algebraic graph rewritings come in two flavours, according to whether they use a double push-out (cf. Ehrig & alii [1973]) or a single push-out (cf. Raoult [1984] or Löwe & Ehrig [1990]). In this section, we show that the possibilities of the set-theoretic method are incomparable with either of these.

There are several ways in which graphs may be translated into hypergraphs. One of them has been used informally in example 2, and is suitable for graphs with labelled arcs and vertices: an arc $x \to y$ having label $f$ is translated into the word $fxy$, and the fact that vertex $x$ is labelled with $a$ is rendered by $ax$. This translation stresses the fact that graphs are just particular cases of hypergraphs.

DEFINITION 6. *Let $G$ be a graph with set $X(G)$ of vertices labelled in $S$ and set $A(G)$ of arcs labelled in $F$. The canonical hypergraph $h(G)$ associated with $G$ is the set*

$$h(G) = \sum_{x \in X(G)} \{ax; x \textit{ has label } a \in S\} + \sum_{e \in A(G)} \{fxy; e = x \xrightarrow{f} y)\}$$

This corresponds to associating variables with vertices. Usually, the vertices are identified with addresses in memory and the correspondence between addresses and variables is standard. Remark: $X(h(G)) = X(G)$.

Since our method does not allow the identification of distinct vertices, it is not possible to simulate the general form of graph rewriting using two push-outs — nor *a fortiori* the general form using a single push-out. However, if a rewriting using two push-outs is one-to-one, it does correspond to a set-theoretic rewriting.

PROPOSITION 7. *Suppose that in the production $G \leftarrow K \rightarrow D$ both morphisms are one-to-one and $K$ consists only of vertices. Without loss of generality, we may assume $X(G) \cap X(D) = K$ Then a graph $L$ rewrites to $R$ according to this production only if $h(L) \rightarrow h(R)$ for the rewriting generated by the collapsed closure of the rule $h(G) \rightarrow h(D)$.*

*Proof:* In the correspondence $h$, the graph $K$ is identified with $X(h(G)) \cap X(h(D))$. The construction is shown Figure 2 below.

$$\begin{array}{ccccc} G & \overset{i}{\longleftarrow} & K & \overset{j}{\longrightarrow} & D \\ {\scriptstyle g}\downarrow & & \downarrow & & \downarrow \\ L & \longleftarrow & C & \longrightarrow & R \end{array}$$

**Figure 2.** Both squares are push-outs.

Let $g : G \rightarrow L$ be an occurrence of the left-hand side of the rule, and $C$ be the push-out complement:

$$C = [L - g(G)] + g(K)$$

Then the push-out $R$ is defined by $R = C + [D - K]$ (cf. Ehrig [1987]). These two equalities can be rewritten

$$\begin{aligned} L &= g(G) + [C - g(K)] \\ R &= D - K + g(K) + [C - g(K)] \end{aligned}$$

Define a mapping $\sigma$ by

$$\begin{cases} \sigma(x) = g(x) & \text{if } x \in h(G) \\ \sigma(x) = x & \text{otherwise} \end{cases}$$

Then the above system translates into

$$\begin{aligned} h(L) &= h(G)\sigma + h(C - g(K)) \\ h(R) &= h(D)\sigma + h(C - g(K)) \end{aligned}$$

This is a rewriting generated by the collapsed pair $(h(G)\sigma, h(D)\sigma)$, QED.

Note however that the converse is not true: there are cases where the double push-out method cannot apply because the following gluing condition is not satisfied (cf. Ehrig [1987]):

$$\text{DANG} \cup \text{IDENT} \subseteq K$$

where

$$\begin{aligned} \text{DANG} &= \{x \in X(G); (\exists a \in L - g(G)) \text{ source}(a) = g(x) \vee \text{target}(a) = g(x)\} \\ \text{IDENT} &= \{x \in G;\ (\exists y \in G)\ x \neq y \ \wedge\ g(x) = g(y)\} \end{aligned}$$

In our set-theoretic framework though, IDENT is always empty, because the pattern is isomorphic to the left-hand side; and arcs are never dangling, as shown by the following example.

*Example 6:* Let $R$ be the unique rule $axy + bxz \to cxy$. Apply it to the graph $G = axy + bxz + dzu$. We get $H = cxy + dzu$ (see the figure below).

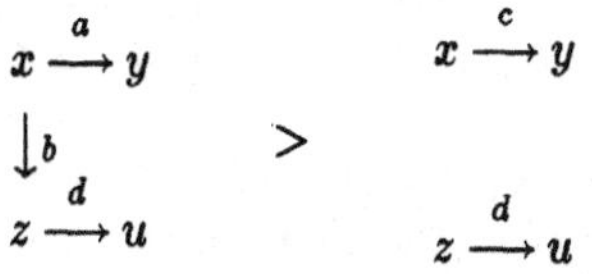

Here $z \in \mathrm{DANG} - K$. Therefore, the double push-out method cannot apply. It cannot be applied either to the graph $G' = axy + bxy + dyu$ because $y$ is at the same time dangling and identified; but the set-theoretic rewriting yields $H' = cxy + dyu$ for the collapsed rule $axy + bxy \to cxy$ (for $y = z$).

There exists a partial morphism corresponding to the set-theoretic rule, defined by $\varphi(x) = x$ and $f(y) = y$ and thus a single push-out rule $axy+bxz \xrightarrow{\varphi} cxy$. The corresponding rewriting always exists. Applying this rule to $G$ yields the graph $\{u \quad x \xrightarrow{c} y\}$ and applying it to $G'$ yields $\{x \quad u\}$. Deletion of $z$ deletes every item connected to it, while in the set-theoretic case, if an item occurs in the context, it remains in the result.

## IV. Composition of rewritings

In this section we show that given two finitely generated rewritings, the successive application of these two rewritings amounts to one step of some finitely generated rewriting, which can be deduced effectively from the two rewritings. This is not true for the parallel rewritings generated by two finite systems, nor for the rewritings — single step or parallel — generated by finite term rewriting systems.

Notice first that rewriting relations are preserved by composition.

THEOREM 8. *If* $\mathbf{R}$ *and* $\mathbf{S}$ *are two rewriting relations, then the composition* $\mathbf{R} \circ \mathbf{S}$ *is also a rewriting relation.*

*Proof:* it is clear that $\mathbf{R} \circ \mathbf{S}$ is preserved by vertex renaming. Let us check that it is preserved by addition of a licit context. Give three graphs $G$, $H$ and $K$ with $G\mathbf{R}H$ and $H\mathbf{S}K$. Let $C$ be a graph such that $[X(K) - X(G)] \cap X(C) = \emptyset$. Define a vertex renaming $\alpha$ with $\mathrm{dom}(\alpha) = X(H) - X(G) - X(K)$ such that $\alpha(x) \notin X(C)$ for all $x$ in $\mathrm{dom}(\alpha)$. Then $G\mathbf{R}(H\alpha)$ and $(H\alpha)\mathbf{S}K$ and $C$ is a licit context for both rewritings: $(G + C)\mathbf{R}(H\alpha + C)$ and $(H\alpha + C)\mathbf{S}(K + C)$. Therefore $(G + C)\mathbf{R} \circ \mathbf{S}(K + C)$, QED.

The single step of rewriting generated by a system $R$ is denoted by $\underset{R}{\longrightarrow}$.

THEOREM 9. *From two finite systems* $R$ *and* $S$ *can be deduced a finite system* $T$ *such that* $\underset{R}{\longrightarrow} \circ \underset{S}{\longrightarrow} = \underset{T}{\longrightarrow}$.

*Proof:* Without loss of generality, we assume that $R$ and $S$ are reduced to one rule, respectively $G_1 \to D_1$ and $G_2 \to D_2$.

Suppose $H \underset{R}{\longrightarrow} H' \underset{S}{\longrightarrow} H''$. This is true if and only if

$$\begin{cases} H = C_1 + G_1 \\ H' = C_1 + D_1 \end{cases} \qquad \begin{cases} H' = C_2 + G_2 \\ H'' = C_2 + D_2 \end{cases}$$

with the proviso $X(G_1) \cap X(C_1) \supseteq X(D_1) \cap X(C_1)$ and $X(G_2) \cap X(C_2) \supseteq X(D_2) \cap X(C_2)$. Set $C = C_1 \wedge C_2$, then $B_1 = C_1 - C$ and $B_2 = C_2 - C$. It is easy to check $B_1 \wedge B_2 = \emptyset$. With these notations, the two successive rewritings look as follows:

$$C + B_1 + G_1 \longrightarrow C + B_1 + D_1 = C + B_2 + G_2 \longrightarrow C + B_2 + D_2$$

so that the composed rewriting is generated by the rule $B_1 + G_1 \longrightarrow B_2 + D_2$ where $B_1$ and $B_2$ satisfy the equality on the middle graph $B_1 + D_1 = B_2 + G_2$. This graph contains $D_1$ and $G_2$, therefore also $D_1 \vee G_2$. Since $B_1 + D_1 \geq D_1 \vee G_2$ implies $B_1 \geq (D_1 \vee G_2) - D_1 = G_2 - (D_1 \wedge G_2)$, we set

$$B_1 = G_2 - (D_1 \wedge G_2) + X_1$$

where $X_1 \leq D_1 \wedge G_2$; and similarly

$$B_2 = D_1 - (D_1 \wedge G_2) + X_2$$

where $X_2 \leq D_1 \wedge G_2$. Reporting the value of $B_1$ and $B_2$ in both expressions of the middle graph and simplifying yields $X_1 = X_2$. Since $B_1 \wedge B_2 = \emptyset$ entails $X_1 \wedge X_2 = 0$, this implies $X_1 = X_2 = \emptyset$. Therefore the composed rule is

$$G_1 + G_2 - (D_1 \wedge G_2) \longrightarrow D_1 + D_2 - (D_1 \wedge G_2).$$

There remains to check the variables. We have $X(B_2 + D_2) = X(B_2) \cup X(D_2)$, hence by distributivity

$$X(C) \cap X(B_2 + D_2) = [X(C) \cap X(B_2)] \cup [X(C) \cap X(D_2)]$$

The first term is included in $X(C) \cap X(G_1)$ because $H \underset{R}{\longrightarrow} H'$ and because $X(B_2) \subseteq X(D_1)$. The second is included in $X(C) \cap X(G_2)$ because $H' \underset{S}{\longrightarrow} H''$. Likewise, setting $G_2 = D_1 \wedge G_2 + B_1$ induces a similar relation on the variables: $X(G_2) = X(D_1 \wedge G_2) \cup X(B_1)$. Now

$$X(C) \cap X(D_1 \wedge G_2) \subseteq X(C) \cap X(D_1) \subseteq X(C) \cap X(G_1)$$

because $H \underset{R}{\longrightarrow} H'$. Therefore

$$X(C) \cap X(B_2 + D_2) \subseteq X(C) \cap [X(G_1) \cup X(B_1)] \subseteq X(C) \cap X(G_1 + B_1),$$

QED.

The algorithm for deducing the generators of the composition from the generators $R$ and $S$ of the components is deduced from the proof above.

*Example 7:* using the simple rules $R = \{ax \to bx\}$ and $S = \{bx \to cx\}$, we deduce, for multigraphs $ax + by \to bx + cy$ corresponding to a disjoint union of $D_1$ and $G_2$, and $ax \to cx$ corresponding to $D_1 = G_2$.

This property of stability is false when one considers simultaneous disjoint rewritings, as shown by the following example.

*Example 8:* consider the two rules:

$$y : d \xrightarrow{f} x : a \quad > \quad y : b \xrightarrow{f} x : a$$

$$x : a \xrightarrow{f} y : b \quad > \quad x : c \xrightarrow{f} y : c$$

The first rule generates the following parallel rewriting:

$$x_1 : a \xrightarrow{f} y_1 : d \xrightarrow{f} x_2 : a \cdots x_n : a \xrightarrow{f} y_n : b \ >$$

$$x_1 : a \xrightarrow{f} y_1 : b \xrightarrow{f} x_2 : a \cdots x_n : a \xrightarrow{f} y_n : b$$

The second then generates:

$$x_1 : a \xrightarrow{f} y_1 : b \xrightarrow{f} x_2 : a \cdots x_n : a \xrightarrow{f} y_n : b \ >$$

$$x_1 : c \xrightarrow{f} y_1 : c \xrightarrow{f} x_2 : c \cdots x_n : c \xrightarrow{f} y_n : c$$

One step of this composed rule rewrites a connected graph of arbitrary size, while one step of simultaneous disjoint rewritings can only rewrite several disjoint graphs of bounded size. Note also that this example simulates a parallel term rewriting relation, and proves that these are not preserved either by composition.

## V. Local confluence

In the case of terms, Knuth & Bendix [1970] proved a criterion ensuring the local confluence of a rewriting. We present here a similar criterion in the case of graphs, which is actually easier to prove. We need beware of definitions "up to isomorphism", which may not be preserved by the addition of a context, as already noticed in section II.

*Example 9:* consider the following two rules: $ax \leftarrow ax + ay \rightarrow ay$. Are the two right-hand sides equal? No, only isomorphic. Take a simple context and apply the two rules:

$$bxy + ax \leftarrow bxy + ax + ay \rightarrow bxy + ay$$

The two results are no longer isomorphic: graph isomorphism is not preserved by embedding in an arbitrary context; graph equality only is preserved (this is first the example of section II).

However, a graph isomorphism $B \to C$ is preserved by embedding in a context whenever its restriction to variables of the context is the identity. When the graphs $B$ and $C$ are rewritten from a graph $A$, then the context of these rewritings may contain any variable of $A$; therefore these variables should not be renamed. This remark justifies the following extension of local confluence.

DEFINITION 10. *A graph rewriting relation is locally confluent if all pairs of rewritings $C \leftarrow A \rightarrow B$ meet in the following sense: there exist two graphs $D_1$ and $D_2$ such that $B \rightarrow^* D_1$ and $C \rightarrow^* D_2$ and $D_1$ is equal to $D_2$ up to a variable renaming having domain disjoint from $X(A)$.*

Therefore, this renaming is a bijection of $X(D_1) - X(A)$ onto $X(D_2) - X(A)$. For instance given the rule $ax \rightarrow ax + ay$, the graph $ax$ rewrites to $ax + ay$, but also to $ax + az$. These are not equal, only isomorphic, but the renaming involves only the new variable $y$ of the right-hand side of the rule. In this example the renaming is given by $y \mapsto z$ and $x$ is not renamed.

THEOREM 11. *For a rewriting relation to be locally confluent it is necessary and sufficient that for all pairs of generators $G_1 \rightarrow_1 D_1$ and $G_2 \rightarrow_2 D_2$ and all least graphs $G$ union of $G_1$ and $G_2$ (up to the name of variables), the pairs $H_1 \underset{1}{\longleftarrow} G \underset{2}{\longrightarrow} H_2$ meet.*

*Proof:* the necessary condition is trivial, since it is a particular case of the confluence property. To check the sufficient condition, note that if $B \underset{1}{\longleftarrow} A \underset{2}{\longrightarrow} C$ there exists some contexts $H_1$ and $H_2$ such that $A = H_1 + G_1$ and $B = H_1 + D_1$ on one hand, and $A = H_2 + G_2$ and $C = H_2 + D_2$ on the other hand. Therefore, setting $H = H_1 \wedge H_2$ we have

$$A = H + G_1 + (G_2 - G_1) = H + G_2 + (G_1 - G_2) = H + (G_1 \vee G_2)$$
$$B = H + D_1 + (G_2 - G_1)$$
$$C = H + D_2 + (G_1 - G_2)$$

The hypothesis ensures the existence of two graphs $K_1$ and $K_2$ which are isomorphic via a renaming having a domain disjoint from $X(G_1 \vee G_2) = X(G_1) \cup X(G_2)$ and such that

$$D_1 + (G_2 - G_1) \rightarrow^* K_1$$
$$D_2 + (G_1 - G_2) \rightarrow^* K_2$$

Since the composition of rewritings is a rewriting and from the remark following proposition 4, we can rename the variables of $K_1$ and $K_2$ getting $K_1'$ and $K_2'$ so that

$$X(H) \cap [X(K_1') - X(D_1 + (G_2 - G_1))] = \emptyset$$
$$X(H) \cap [X(K_2') - X(D_2 + (G_1 - G_2))] = \emptyset$$

Adding now the licit context $H$ yields

$$H + D_1 + (G_2 - G_1) \rightarrow^* H + K_1'$$
$$H + D_2 + (G_1 - G_2) \rightarrow^* H + K_2'$$

The same vertex renaming which yields $K_2'$ from $K_1'$ yields also $H + K_2$ from $H + K_1$, because the restriction of this renaming to $X(G_1) \cup X(G_2)$ is the identity, QED.

As with terms, it is useless to check the local confluence of two non overlapping left-hand sides: $D_1 + G_2 \leftarrow G_1 + G_2 \rightarrow G_1 + D_2$ obviously meet at $D_1 + D_2$.

## VI. Conclusion

The set-theoretic method of rewriting graphs has advantages and drawbacks. The main advantage is its simplicity: no categorical background, intuitive content. Left-hand sides are subtracted, right-hand sides are added. Overlapping simply amounts to non-empty intersection. The term-theoretic definition is not so simple. And in the Berlin approach, the role of intersection is held by pullbacks (this is necessary in the case of non-injective occurrences of left-hand sides).

The main characteristic of the set-theoretic method is its inability to handle identification of vertices. Whether to accept this possibility or not is not a trivial matter. In the double push-out approach, one condition for applying a (fast) rewriting rule is a restriction on the identification of two vertices. It looks more difficult to reject simple identifications like $\mathrm{I}x \rightarrow x$. But this rule deals with terms, not graphs, and in fact, the implementation of such a rule is not so straightforward: one has to look for all the "fathers" of the node labelled I and change their pointer to it. This implies visiting the whole graph, or keeping along with each node the list of all its fathers, as in Dactl (see Kennaway [1988]), where a single push-out graph rewriting is implemented in two steps: one "free" rewriting similar to our set-theoretic rewriting, and a quotient step performing some identification of vertices.

Actually, our thesis is that graph manipulations are really of two sorts: rewritings on one hand, in which each step is simple, an operation with a complexity depending only on the rule; and quotients on the other hand, which are more complex operations, since the identification of two vertices sometimes entails visiting the whole context graph. Our set-theoretic rewritings are restricted to be of the first type. They are free in some sense. Also, many such rewritings satisfy the extra condition of reversibility, which can easily be checked. This can be helpful, when dealing for instance with a data base in which one wants (sometimes) to be able to return to a previous state.

### REFERENCES

Barendregt H, van Eekelen M., Glauert J., Kennaway R., Plasmeijer M. & Sleep R.: Term Graph Rewriting, *Proc. PARLE Conference, Eindhoven, 1987, Lecture Notes in Computer Science* 259, pp 141–158, Springer Verlag(1987).

Bauderon M. & Courcelle B. : Graph Expressions and Graph Rewritings, *Mathematical Systems Theory,* vol. 20, 1987, pp. 83-127 (1987).

Dershowitz N. : Computing with rewrite Systems, in *Information and Control,* vol. 64 (2/3), pp. 122-157 (1985).

Ehrigh H., Pfender M. & Schneider H-J. : Graph Grammars: an algebraic approach, in *Proc. 14th IEEE Symp. on Switching and Automata Theory,* pp. 167-180 (1973).

Ehrig H. : Tutorial Introduction to the Algebraic Approach of Graph Grammars, in *3rd Int. Workshop on Graph Grammars and their Applications to Computer Science* , 1987, Ehrig, Nagl, Rozenberg & Rosenfeld (Eds), Lecture Notes in Computer Science 291, Springer Verlag, pp. 3-14 (1987).

Habel A & Kreowski H-J. : Some structural aspects of hypergraph languages generated by hyperedge replacement, in *Proc. STACS '87, Lecture Notes in Computer Science* 247, pp. 207–219 (1987).

Habel A, Kreowski H-J. & Plump D. : Jungle Evaluation, *"Recent Trends in Data Type Specification"*, 5th Workshop on Specification of Abstract Data Types, Gullane (Scotland), 1987, Lecture Notes in Computer Science 332, pp. 92–112, Springer Verlag(1987).

Handbook : *Handbook of Theoretical Computer Science*, Vol. B : Formal Methods and Semantics, J. van Leeuwen Ed., Elsevier 1990.

Kennaway R. : On "On Graph Rewritings", in *Theoretical Computer Science* 52, North-Holland, pp. 37–58 (1987); see also Corrigendum to On "On graph Rewritings", in *Theoretical Computer Science* 61, North-Holland, pp. 317–320 (1988).

Kennaway R. : Implementing Term Rewrite Languages in DACTL, in *Proc. of C.A.A.P.* '88, Lecture Notes in Computer Science 299, Springer Verlag, pp. 102–116 (1988).

Kirchner C. : Méthodes et outils de conception systématique d'algorithmes d'unification dans les théories équationnelles, Thèse d'Etat, Univ. Nancy (1985).

Knuth D.E. & Bendix P.B. : Simple word problems in universal algebras, *in Computational Problems in Abstract Algebra*, Leech J. ed., Pergamon Press, Braunschweig (1970).

Lafont Y. : Interaction Nets, in *Principles of Prog. Languages,* San Francisco, pp 95–108 (1990).

Löwe M. & Ehrig H. : Algebraic Approach to Graph Transformation Based on Single Pushout Derivations, in *Proc. 16th Int. Workshop on Graph Theoretic Concepts in Computer Science* (WG'90), R. H. Möhring, Lecture Notes in Computer Science 484 (1990).

Muller D. & Schupp P. [1985] The theory of ends, pushdown automata, and second order logic, Theoretical Computer Science 37, pp 51–75 (1985).

Nagl M. : A Tutorial and Bibliographical Survey on Graph Grammars, in *1st Int. Workshop on Graph Grammars and their Applications to Computer Science and Biology,* Ehrig, Claus & Rozenberg (Eds), Lecture Notes in Computer Science 79, Springer Verlag(1979).

Peterson G.E. & Stickel M.E. : Complete sets of reductions for some equational theories, in *J. ACM* 28 vol. 2, pp. 233–264 (1981).

Raoult J-C. : On Graph Rewritings, in Theoretical Computer Science 32, North-Holland, pp. 1–24 (1984).

Rusinowitch M. : *Démonstration Automatique: techniques de réécriture*, InterEditions, Paris, 1989.

Sopena E. : Combinatorial hypermap rewriting, in *Proc. RTA 87,* Bordeaux, Lecture Notes in Computer Science 256 pp. 62–73 (1987).

# On Relating Rewriting Systems and Graph Grammars to Event Structures*

Georg Schied

Institut für Informatik, Universität Stuttgart
Breitwiesenstr. 20–22, 70565 Stuttgart, Germany
schied@informatik.uni-stuttgart.de

**Abstract.** In this paper, we investigate how rewriting systems and especially graph grammars as operational models of parallel and distributed systems can be related to event structures as more abstract models. First, *distributed rewriting systems* that are based on the notion of contexts are introduced as a common framework for different kinds of rewriting systems and their parallelism properties are investigated. Then we introduce *concrete graph grammars* and show how they can be integrated into this framework for rewriting systems. A construction for the Mazurkiewicz trace language related to the derivation sequences of a distributed rewriting system is presented. Since there is a well-known relation between trace languages and event structures, this provides the link between (graph) rewriting and event structures.

## 1 Introduction

Graph grammars may be used as an operational model of parallel and distributed systems [DM87, JR89, Sch90, Sch92]: Graphs represent the states of the system and graph rewriting rules describe its dynamic behaviour. This approach has some significant advantages: Graphs are a simple and clear representation of the topological structure of a distributed system. Graph rewriting rules offer a convenient way to model dynamic changes of the topological structure of a system, e.g. the generation or deletion of processes and channels. In addition, graph rewriting provides not only a sequential but also, via the notion of parallel derivations, a concurrent execution model.

Event structures [NPW81, Win89] constitute a more abstract model of parallel and distributed systems. A system is specified by the set of events that may possibly occur and by a causality relation and a conflict relation between these events. Events that are neither causally dependent nor in conflict are concurrent, i.e. they may occur in parallel or in arbitrary order.

Causality, conflicts and concurrency are three fundamental phenomena that can also be detected in derivations of graph grammars and other rewriting systems like string or term rewriting systems.

* Supported in part by Deutsche Forschungsgemeinschaft (SFB 182, B1)

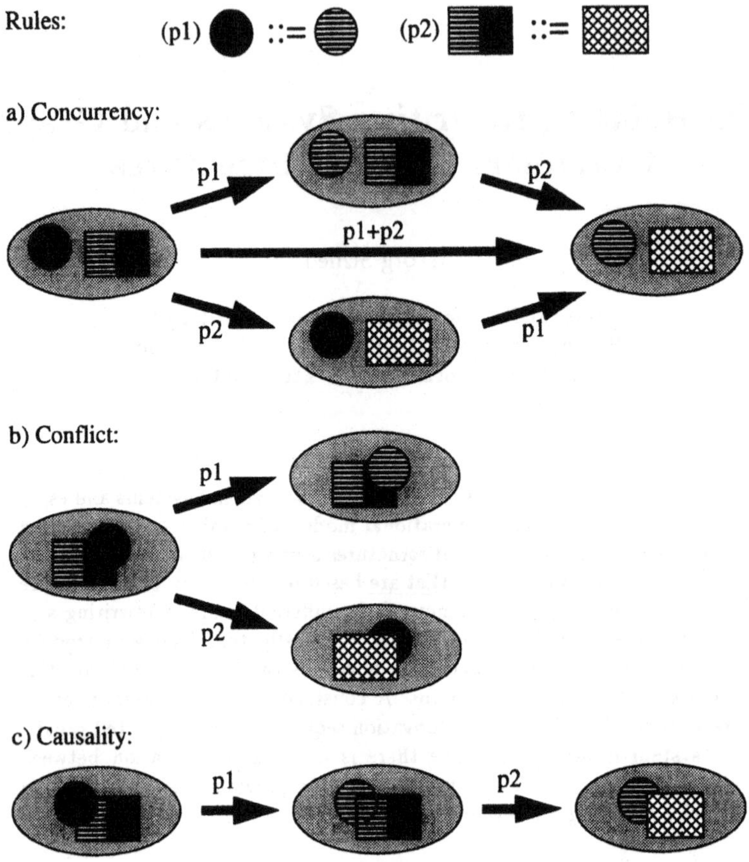


**Fig. 1.** Concurrency, conflicts, and causality in rewriting systems

- *Concurrency.* Two rules can be applied in parallel or in arbitrary order, provided they change different substructures.
- *Conflicts.* There exists a conflict between the applications of two rules if both of them try to change the same substructure. Therefore, only one of the rules can be applied.
- *Causality.* There exists a direct causal dependence between the applications of two rules if the first rule generates some substructure that is necessary to apply the second rule.

Figure 1 schematically depicts typical situations of causality, concurrency and conflicts in derivations of rewriting systems. However, these pictures are somewhat oversimplifying. This aspect will be discussed in Section 5.

If we consider the application of a rewriting rule as an event, it should be obvious that derivations of rewriting systems can be related to event structures.

This article presents a concise definition of this relation between a certain kind of graph grammars (or other rewriting systems) and event structures. In the next section, we will introduce ***distributed rewriting systems*** as a unified model for different kinds of rewriting systems, like graph grammars and term rewriting systems, and investigate properties of derivations, especially with respect to the independence of derivation steps and parallelism. In Section 3, we will introduce concrete graph grammars, an approach to graph rewriting that completely avoids any construction that is defined only "up to isomorphism", and show how they fit into our model of distributed rewriting systems. The formal construction that relates (the derivations of) a rewriting system to a Mazurkiewicz trace language and via this intermediary step to an event structure will be presented in Section 4. After that, we discuss some problems concerning asymmetric conflicts in rewriting systems.

# 2 Rewriting Systems

Considering different kinds of rewriting systems like string, graph, or term rewriting, we recognize that they obey some common principles: A rewrite rule consists of a pair $(l, r)$ of structures, called the left hand side and the right hand side. There is a derivation step from one structure $q$ to another structure $q'$ via a rule $(l, r)$, if we are able to find a context $C[\_]$, such that inserting the left hand side into the "hole" of the context ($C[l]$) yields the original structure $q$ and inserting the right hand side into the hole ($C[r]$) yields the derived structure $q'$. If we consider contexts with several holes, we immediately arrive at a notion of parallel derivation steps.

In the following sections, we will introduce ***distributed rewriting systems*** as a formalization of this idea.

## 2.1 Context systems

Contexts play a fundamental role in our model of rewriting systems. Contexts are structures with holes. Each hole can be filled with a structure by means of an insertion operator. It is possible to fill several holes simultaneously. Holes are distinguished by identifiers that are used to specify which hole is to be filled with which structure. We will distinguish between inserting a right-hand side or a left-hand side of a rewriting rule.

**Definition 1 (Assignment).** Let $Q$ be a set of structures, e.g., strings, terms or graphs, and let $S$ be a set of hole identifiers. A partial mapping $\phi : S \rightharpoonup Q \times \{\mathrm{L}, \mathrm{R}\}$ with $dom(\phi) \neq \emptyset$ is called a ***Q-assignment.*** A $Q$-assignment $\phi$ is *elementary* if $|dom(\phi)| = 1$. We denote with $\Phi_Q$ the set of all $Q$-assignments. $\phi_1 + \phi_2$ denotes that $\phi_1$ and $\phi_2$ are *disjoint* $Q$-assignments, i.e., $dom(\phi_1) \cap dom(\phi_2) = \emptyset$, and that $\phi_1 + \phi_2$ is the usual union of partial functions.

A $Q$-assignment $\phi$ attaches structures from $Q$ to some of the holes of a context and specifies whether the structure is to be inserted as a left-hand side or as a right-hand side. Henceforth, we omit the prefix $Q$ and simply speak of assignments whenever this is unambiguous.

**Definition 2 (Context system).** A $(Q, S)$-*context system* is a quadruple $CS = (\mathcal{C}, \lambda, [\ ], \xi)$ that consists of the following components:

- $\mathcal{C}$ is a set (of contexts).
- The function $\lambda : \mathcal{C} \to \mathcal{P}_{fin}(S)$ assigns a finite set of holes to every context.
- The insertion operation $[\ ]$ is a partial function $[\ ] : \mathcal{C} \times \Phi_Q \to \mathcal{C}$. We write $C[\phi]$ instead of $[\ ](C, \phi)$.
- $\xi : \mathcal{C}^{\emptyset} \to Q$ is a bijection between the set $\mathcal{C}^{\emptyset} := \{C \in \mathcal{C} \mid \lambda(C) = \emptyset\}$ of contexts without holes and the set of structures $Q$.

The following conditions must hold. We presuppose in (2) and (3) that $\phi_1$ and $\phi_2$ are disjoint assignments and in (3) that $C[\phi_1 + \phi_2]$ is defined.

(1) $C[\phi]$ defined $\Rightarrow$ $\lambda(C[\phi]) \cup dom(\phi) = \lambda(C) \ \wedge \ \lambda(C[\phi]) \cap dom(\phi) = \emptyset$
(2) $C[\phi_1 + \phi_2]$ defined $\Leftrightarrow$ $C[\phi_1]$ defined $\wedge$ $C[\phi_2]$ defined
(3) $C[\phi_1 + \phi_2] = C[\phi_1][\phi_2]$

Condition (1) requires that all holes of $C$ that are not filled by the assignment $\phi$ remain in the context $C[\phi]$ and that no new holes arise. The conditions (2) and (3) require that inserting a structure into a hole will not influence other holes. Especially, (3) states that we obtain the same result either if we simultaneously fill in $\phi_1 + \phi_2$ or if we first insert $\phi_1$ and afterwards $\phi_2$.

*Remark.* This definition of a context system is similar to that of abstract context systems introduced in [Sch92]. The difference is that we allow to distinguish between inserting a structure as a left-hand side or as a right-hand side.

**Definition 3.** A context system $CS = (\mathcal{C}, \lambda, [\ ], \xi)$ is called *definite* if for all contexts $C$ and all assignments $\phi$ the following condition holds:

(4) If $C[\phi]$ is defined and $C[\phi] = C'[\phi]$, then $C = C'$.

*Remark.* This condition seems to be rather restrictive. Nethertheless, definite context systems can be defined not only for graph rewriting, as will be shown in this paper, but, for example, also for term rewriting and C/E-nets. Simulating string rewriting by graph rewriting, we indirectly get a context system for strings, too.

## 2.2 Distributed Rewriting Systems

Let $CS$ be a $(Q, S)$-context system.

**Definition 4 (Distributed rewriting system).** A *distributed rewriting system* $(\mathcal{R}, q_0)$ (based on the context system $CS$) consists of a subset $\mathcal{R} \subseteq Q \times Q$ and an initial structure $q_0 \in Q$. We call an element $r = (q_1, q_2) \in \mathcal{R}$ a *rule*. $q_1$ is the left hand side and $q_2$ the right hand side of the rule. We call a distributed rewriting system *definite* if the underlying context system is *definite*.

**Definition 5 (Rule assignment).** Let $(\mathcal{R}, q_0)$ be a distributed rewriting system. An *$\mathcal{R}$-assignment* is a partial mapping $\psi : S \rightarrow \mathcal{R}$ (remember that $S$ is the set of hole identifiers).

If $\psi$ is an $\mathcal{R}$-assignment, then the $Q$-assignments $\psi^L$ and $\psi^R$ are defined by $\psi^L(s) := \langle q_1, \mathrm{L}\rangle$, $\psi^R(s) := \langle q_2, \mathrm{R}\rangle$, iff $\psi(s) = \langle q_1, q_2\rangle$.

**Definition 6 (Derivations).** Let $(\mathcal{R}, q_0)$ be a distributed rewriting system based on a context system $(\mathcal{C}, \lambda, [\ ], \xi)$. Then *$q' \in Q$ can be derived from $q \in Q$ via context $C$ and $\mathcal{R}$-assignment $\psi$*, written $q \overset{C,\psi}{\Longrightarrow} q'$, iff

$$q = \xi(C[\psi^L]) \quad \text{and} \quad q' = \xi(C[\psi^R]).$$

A derivation step $q \overset{C,\psi}{\Longrightarrow} q'$ is called *sequential*, if $\psi$ is an elementary assignment. Otherwise it is called a *parallel* derivation step.

This definition reflects the ideas described above: If we insert the left hand sides of the rules into the holes of the context $C$, we get the initial structure $q$; if we insert the right hand sides, we get the derived structure $q'$. For convenience we will omit the bijection $\xi$ and write $C[\phi] = q$ instead of $\xi(C[\phi]) = q$ in the following.

***Lemma 7.*** *If $q \overset{C_1,\psi}{\Longrightarrow} q_1$ and $q \overset{C_2,\psi}{\Longrightarrow} q_2$ are two derivation steps of a definite rewriting system that use the same assignment $\psi$, then $C_1 = C_2$ and $q_1 = q_2$.*

*Proof.* According to the definition of derivations we obtain $C_1[\psi^L] = q = C_2[\psi^L]$. The definiteness property immediately ensures $C_1 = C_2$. Consequently, $q_1 = C_1[\psi^R] = C_2[\psi^R] = q_2$.

### 2.3 Properties of distributed rewriting systems

The notion of contexts allows a simple definition for the concurrent and sequential independence of derivation steps.

**Definition 8 (Concurrent and sequential independence).** Two derivation steps $q \overset{C_1,\psi_1}{\Longrightarrow} q_1$ and $q \overset{C_2,\psi_2}{\Longrightarrow} q_2$ are *concurrently independent* if there exists a context $C$ with $C[\psi_2^L] = C_1$ and $C[\psi_1^L] = C_2$.

Two derivation steps $q \overset{C_1,\psi_1}{\Longrightarrow} q_1 \overset{C_2,\psi_2}{\Longrightarrow} q'$ are *sequentially independent* if there exists a context $C$ with $C[\psi_2^L] = C_1$, and $C[\psi_1^R] = C_2$.

*Remark.* In the domain of algebraic/categorical graph grammars, the term "parallel independence" is used instead of "concurrent independence" [Ehr79].

***Theorem 9 (Parallelism theorem I).*** *If $q \overset{C_1,\psi_1}{\Longrightarrow} q_1$ and $q \overset{C_2,\psi_2}{\Longrightarrow} q_2$ are concurrently independent derivation steps, then there also exist derivations*

$$q_1 \overset{C_2',\psi_2}{\Longrightarrow} q', \quad q_2 \overset{C_1',\psi_1}{\Longrightarrow} q' \quad \textit{and} \quad q \overset{C,\psi_1+\psi_2}{\Longrightarrow} q' \ .$$

*Moreover, the derivation steps* $q \overset{C_1,\psi_1}{\Longrightarrow} q_1 \overset{C_2',\psi_2}{\Longrightarrow} q'$ *and* $q \overset{C_2,\psi_2}{\Longrightarrow} q_2 \overset{C_1',\psi_1}{\Longrightarrow} q'$, *respectively, are sequentially independent. If the rewriting system is definite, then the contexts* $C$, $C_1'$ *and* $C_2'$ *(and consequently the structure* $q'$*) are uniquely determined.*

***Theorem 10 (Parallelism theorem II).*** *If* $q \overset{C_1,\psi_1}{\Longrightarrow} q_1 \overset{C_2,\psi_2}{\Longrightarrow} q'$ *are sequentially independent derivation steps, then there exist derivation steps*

$$q \overset{C_2',\psi_2}{\Longrightarrow} q_2 \overset{C_1',\psi_1}{\Longrightarrow} q' \quad \text{and} \quad q \overset{C,\psi_1+\psi_2}{\Longrightarrow} q' \ .$$

*Moreover, the derivation steps* $q \overset{C_2',\psi_2}{\Longrightarrow} q_2 \overset{C_1',\psi_1}{\Longrightarrow} q'$ *are sequentially independent and the steps* $q \overset{C_1,\psi_1}{\Longrightarrow} q_1$, $q \overset{C_2',\psi_2}{\Longrightarrow} q_2$ *are concurrently independent. If the rewriting system is definite, then the contexts* $C$, $C_1'$ *and* $C_2'$ *(and consequently the structure* $q_2$*) are uniquely determined.*

***Theorem 11 (Parallelism theorem III).*** *If* $q \overset{C,\psi_1+\psi_2}{\Longrightarrow} q'$ *is a (parallel) derivation step, then there also exist sequentially independent derivation steps*

$$q \overset{C_1,\psi_1}{\Longrightarrow} q_1 \overset{C_2',\psi_2}{\Longrightarrow} q' \quad \text{and} \quad q \overset{C_2,\psi_2}{\Longrightarrow} q_2 \overset{C_1',\psi_1}{\Longrightarrow} q' \ ,$$

*respectively. Moreover, the steps* $q \overset{C_1,\psi_1}{\Longrightarrow} q_1$ *and* $q \overset{C_2,\psi_2}{\Longrightarrow} q_2$ *are concurrently independent. If the underlying context system is definite, then* $C_1$, $C_2$, $C_1'$, *and* $C_2'$ *(and consequently* $q_1$ *and* $q_2$*) are uniquely determined.*

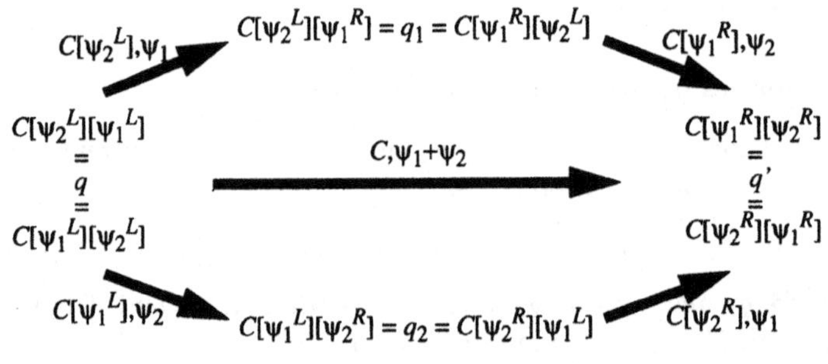


**Fig. 2.** Situation of the parallelism theorems

*Proof (for the parallelism theorems I – III).* Using the definition of concurrent and sequential independence and the property $C[\phi_1][\phi_2] = C[\phi_1 + \phi_2] = C[\phi_2][\phi_1]$, we are able to obtain the situation shown in Fig. 2 from the premises given in each of the theorems. Now it is obvious that the required derivation steps exist. The uniqueness in the case of definite rewriting systems is ensured by Lemma 7. □

## 3 Graph Grammars

We use the following kind of graphs (similar to [Rao84]) in order to avoid the separate treatment of edges and vertices. Let $L$ be a set of labels.

**Definition 12 (Graph and graph morphism).** An *(L-labelled) graph* $G = (N, s, l)$ consists of a finite set $N$ of nodes, a connection function $s : N \to N^*$, and a labelling function $l : N \to L$. If $G$ is a graph, then we denote its components with $N_G$, $s_G$, $l_G$.

A *graph morphism* $f : (N, s, l) \to (N', s', l')$ is a function $f : N \to N'$ such that $s'(f(n)) = f^*(s(n))$ and $l'(f(n)) = l(n)$ for all $n \in N$ (where $f^*$ denotes the extension of $f$ to sequences).

If $d : G \to H$ is a graph morphism, then $d^o$ denotes the corresponding surjective morphism $d^o : G \to d(G)$ with $d^o(n) = d(n)$ for all $n \in N_G$.

*Remark.* Hypergraphs and ordinary graphs are special cases of our definition of graphs. We may consider a hypergraph $(E, V, s, l)$ with $s : E \to V^*$ as a graph $(E \uplus V, s', l)$ with $s'(v) = \langle\rangle$ for all $v \in V$ and $s'(e) = s(e)$ for all $e \in E$. Similarly, ordinary graphs $(E, V, s, t, l)$ with $s, t : E \to V$ fit into our framework as graphs $(E \uplus V, s', l)$ with $s'(v) = \langle\rangle$ for all $v \in V$ and $s'(e) = \langle s(e), t(e)\rangle$ for all $e \in E$.

**Definition 13 (Graph operations).** Let $G = (N, s, l)$ and $H = (N', s', l')$ be graphs. $G \subseteq H$ ($G \subset H$) denotes that $G$ is a (proper) subgraph of $H$. We define the following operations:

$$G \cup H := \begin{cases} (N \cup N', s \cup s', l \cup l') & \text{if } \forall n \in N \cap N' : s(n) = s'(n) \ \wedge \ l(n) = l'(n) \\ \text{undefined} & \text{otherwise}, \end{cases}$$

$$G \setminus H := \begin{cases} (N \setminus N', s'', l'') & \text{if } \forall n \in N \setminus N' : s(n) \in (N \setminus N')^* \\ \text{undefined} & \text{otherwise}, \end{cases}$$

$$G \cap H := \begin{cases} (N \cap N', s''', l''') & \text{if } \forall n \in N \cap N' : s(n) = s'(n) \ \wedge \ l(n) = l'(n) \\ \text{undefined} & \text{otherwise}. \end{cases}$$

where $s''$, $l''$ and $s'''$, $l'''$ are the restrictions of $s$ and $l$ to $N \setminus N'$ and $N \cap N'$, respectively. We use also $G \setminus N$, if $N$ is a set of nodes (treating $N$ as a discrete graph). The *glueing operation* for graphs $G$, $K$, and $H$ is

$$glue(G, K, H) := \begin{cases} G \cup H & \text{if } K = G \cap H, \\ \text{undefined} & \text{otherwise}. \end{cases}$$

**Definition 14 (Interface graph).** An *interface graph* $A = (G, K)$ consists of a graph $G$ and a subgraph $K \subseteq G$. We call $G$ the body and $K$ the interface of $A$.

Interface graphs are similar to hypergraphs with sources [BC87], multi-pointed hypergraphs [HK87], or partially abstract graphs [CM91].

### 3.1 Concrete Graph Rewriting

**Definition 15 (Graph production and graph grammar).** A *graph production* $p = (L, K, R)$ consists of graphs $L$, $K$, $R$ such that $K \subset L$ and $K \subseteq R$. The nodes of $K$ are called *glueing nodes*.

A *graph grammar* $GG = (\mathcal{P}, S)$ consists of a set $\mathcal{P}$ of graph productions and an initial graph $S$ with $N_S \subseteq I\!N$ (where $I\!N$ denotes the natural numbers).

**Definition 16 (Application morphism).** A graph morphism $d : L \rightarrow G$ is called an *application morphism* for production $p = (L, K, R)$ in the graph $G$, if the following two conditions hold:

- *Identification condition.* $d(x) = d(y) \wedge x \in N_K \Rightarrow y \in N_K$.
- *Dangling condition.* $s_G(N_G \setminus d(N_L)) \subseteq (N_G \setminus d(N_L) \cup d(N_K))^*$

The identification and dangling conditions are similar to those of algebraic graph grammars [Ehr79].

**Definition 17 (Independence).** Two application morphisms $d_1 : L_1 \rightarrow G$ and $d_2 : L_2 \rightarrow G$ for productions $p_1 = (L_1, K_1, R_1)$ and $p_2 = (L_2, K_2, R_2)$ are *independent*, if $d_1(L_1) \cap d_2(L_2) \subseteq d_1(K_1) \cap d_2(K_2)$.

**Definition 18 (Derivation).** Let $G$ be a graph and $\mathcal{D} = \{(p_i, d_i) \mid i \in I\}$ be a finite set, where each $p_i = (L_i, K_i, R_i)$ is a production and $d_i : L_i \rightarrow G$ is an application morphism for $p_i$ in $G$. Let $N_{H_i} := \{d_i(n) \mid n \in N_{K_i}\} \cup \{\langle n, d_i^\circ \rangle; |; n \in N_{R_i} \setminus N_{K_i}\}$ and define $h_i : N_{R_i} \rightarrow N_{H_i}$

$$h_i(n) := \begin{cases} d_i(n) & \text{if } n \in N_{K_i} \\ \langle n, d_i^\circ \rangle & \text{if } n \in N_{R_i} \setminus N_{K_i} \end{cases}$$

$h_i$ is injective and hence uniquely determines a connection function $s_{H_i}$ and a labelling function $l_{H_i}$ such that $H_i = (N_{H_i}, s_{H_i}, l_{H_i})$ is a graph and $h_i$ extends to a graph morphism $h_i : R_i \rightarrow H_i$.

$H$ is *derived from* $G$ *via* $\mathcal{D}$, written $G \stackrel{\mathcal{D}}{\Longrightarrow} H$, if all $d_i, d_j$ are independent (for $i \neq j$) and if there exists a graph $C$ such that

$$G = glue(\bigcup_{i \in I} d_i(L_i), \bigcup_{i \in I} d_i(K_i), C) \quad \wedge \quad glue(C, \bigcup_{i \in I} d_i(K_i), \bigcup_{i \in I} H_i) = H$$

We write $G \Longrightarrow H$ if there exists a $\mathcal{D}$ such that $G \stackrel{\mathcal{D}}{\Longrightarrow} H$.

*Remark.* The graph $C$ (called *context graph* in the following) is uniquely determined as $C = G \setminus \bigcup_{i \in I} d_i(L_i \setminus K_i)$. Consequently, the derived graph $H$ is uniquely determined, too.

$d_i(L_i \setminus K_i)$ represents, for each production $p_i = (L_i, K_i, R_i)$ with application morphism $d_i$, the part of the graph $G$ that will be removed and $h_i(R_i \setminus K_i)$ is the part to be generated. $d_i(K_i) = h_i(K_i)$ is the "context sensitive" part that has to be present in order to apply the rule but is not changed.

The definition of $h_i$ guarantees that the parts to be generated for different rules are disjoint (i.e. $h_i(R_i \setminus K_i) \cap h_j(R_j \setminus K_j) = \emptyset$ if $d_i \neq d_j$). The following proposition states that the generated parts are disjoint from the remaining context graph.

***Proposition 19.*** *Let $GG = (\mathcal{P}, S)$ be a graph grammar and let $G$ and $\mathcal{D}$ be given as in the previous definition. If $S \Longrightarrow^* G$ and $C = G \setminus \bigcup_{i \in I} d_i(L_i \setminus K_i)$, then $h_j(R_j \setminus K_j) \cap C = \emptyset$ holds for all $j \in I$.*

This property essentially relies on the fact that each node $n = \langle n', d^\circ \rangle$ being generated encodes in $d^\circ$ its whole history, i.e., how it came into being.

*Remark.* If we would base the definition of derivations on a construction like $H := G \setminus d(L \setminus K) \ \uplus \ R \setminus K$ using the disjoint union $\uplus$ (e.g. defined as $A \uplus B = A \times \{0\} \cup B \times \{1\}$) the parallelism properties will not hold for concrete graphs. On the other hand, working with abstract graphs, i.e., with graphs "up to isomorphism", as usually done in the algebraic/categorical approach to graph grammars causes significant problems in relating derivations of graph grammars to event structures, because it is rather difficult to find an adequate notion of abstract derivation steps (see [CEL+93]).

The condition $K \subset L$ for graph productions $(L, K, R)$ ensures that two productions can never be applied in parallel at the same position within a graph. Especially, this excludes autoconcurrency, i.e., that the same production can be applied several times in parallel at the same position.

Our notion of graph rewriting is closely related to the algebraic double-pushout approach to graph grammars [Ehr79]. We can consider a production $p = (L, K, R)$ as a double-pushout production $(L \overset{l}{\leftarrow} K \overset{r}{\rightarrow} R)$ with the inclusions $l$ and $r$. Let $d : L \rightarrow G$ be an application morphism that satisfies the stronger identification condition $d(x) = d(y) \ \wedge \ x \neq y \ \Rightarrow \ x, y \in K$. Let $G \overset{\{(d,p)\}}{\Longrightarrow} H$ be a derivation step of the concrete graph grammar using the context graph $C$, where $G$ is derivable from the start graph $S$, i.e. $S \Longrightarrow^* G$. Then

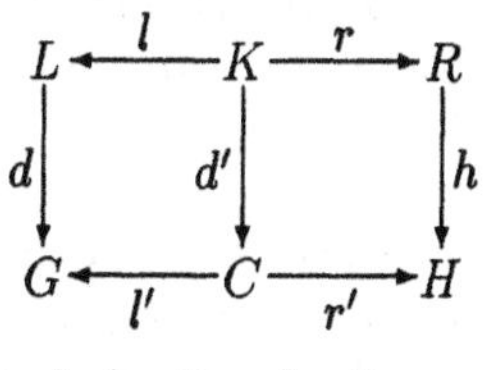


constitutes a double-pushout derivation in the category of graphs (where $l'$ : $C \rightarrow G$ and $r' : C \rightarrow H$ are inclusions and $d'$ is the restriction of $d$ to $K$).

## 3.2 Graph Grammars as Distributed Rewriting Systems

In this section we will show how concrete graph rewriting can be embedded into the framework of distributed rewriting systems.

**Definition 20 (Graph contexts).** A *graph context* (for interface graphs) $C = (G, I, D)$ consists of an interface graph $(G, I)$ and a finite set $D = \{d_i : X_i \rightarrow Y_i \mid i = 1 \ldots n\}$ of surjective graph morphisms such that for all $i \neq j$ the condition $N_{Y_i} \cap N_{Y_j} \subseteq N_G$ is satisfied. Its set of holes is $\lambda_G(C) := D$. The bijection $\xi$ between graph contexts without holes and interface graphs is defined by $\xi_G((G, I, \emptyset)) := (G, I)$. We denote with $\mathcal{C}_G$ the class of all graph contexts.

**Definition 21 (Insertion operation).** Let $C = (G, I, D)$ be a graph context. Let $\phi$ be an assignment with $\phi(d_i) = \langle (H_i, K_i), Z_i \rangle$ for $d_i \in dom(\phi)$. If $Z_i = \mathrm{L}$ let $M_i := d_i(H_i)$ and $h_i := d_i$. Otherwise, if $Z_i = \mathrm{R}$ define $N_{M_i} := \{d_i(n) \mid n \in N_{K_i}\} \cup \{\langle n, d_i^\circ \rangle \mid n \in N_{H_i} \setminus N_{K_i}\}$ and $g_i : N_{H_i} \rightarrow N_{M_i}$

$$g_i(x) := \begin{cases} d_i(x) & \text{if } x \in N_{K_i} \\ \langle x, d_i^\circ \rangle & \text{if } x \in N_{H_i} \setminus N_{K_i} \end{cases} .$$

In both cases, $g_i$ determines $s_{M_i}$ and $l_{M_i}$ such that $M_i = (N_{M_i}, s_{M_i}, l_{M_i})$ is a graph and $g_i : H_i \rightarrow M_i$ extends to a graph morphism. Let $E := dom(\phi)$. Then the insertion operation is defined by

$$C[\phi]_G := \begin{cases} (G', I, D \setminus E) \text{ if } E \subseteq D \ \wedge \ \forall d_i \in E : (dom(d_i) = H_i \\ \qquad\qquad\qquad\qquad\qquad \wedge \ d_i(K_i) = G \cap d_i(H_i)) \\ \text{undefined} \qquad \text{otherwise} \end{cases} ,$$

where $G' := glue(G, \bigcup_{d_i \in E} g_i(K_i), \bigcup_{d_i \in E} g_i(H_i))$.

*Remark.* The condition $E \subseteq D$ enforces that the assignment $\phi$ must not fill holes that are not in the context. The other conditions are a little bit tricky. In the definition of concrete graph rewriting we used information about the application position (i.e. the application morphism $d_i$) to generate new nodes. Here, using the application morphisms itselves as holes, we are able to achieve the same effect with graph contexts. The conditions $dom(d_i) = H_i$ and $d_i(K_i) = G \cap d_i(H_i)$ ensure that only the interface graph $(H_i, K_i)$ can be inserted into hole $d_i$ and that the result of the glueing operation $glue(G, d_i(K_i), d_i(H_i)$ (and consequently of $glue(G, \bigcup_{d_i \in E} g_i(K_i), \bigcup_{d_i \in E} g_i(H_i))$ ) is defined. The graph $I$ is not involved in the insertion operation. It is only needed to obtain an interface graph when all holes of a context are filled.

Based on this definition of contexts we are able to do rewriting for interface graphs. Rewriting rules are pairs $p = ((L, K), (R, K'))$ of interface graphs. We easily see from the definition of the insertion operation that $K = K'$ has to be required. Otherwise, the rule would never be applicable. So, a rewriting rule $p = ((L, K), (R, K))$ is the same as a graph production $p = (L, K, R)$ of concrete graph grammars.

**Definition 22 (Distributed graph rewriting system).** A distributed rewriting system $(\mathcal{P}, (S, I))$ for interface graphs consists of a set of graph productions and an interface graph $(S, I)$ such that $N_S \subseteq IN$.

Every concrete graph grammar is a distributed graph rewriting system since every graph $G$ can be considered as an interface graph $(G, \emptyset)$ with empty interface. It can be shown that the context-based notion of graph rewriting coincides with concrete graph rewriting as introduced in Sect.3.1.

If we directly use the set class $C_G$ of graph contexts, we cannot show that $(C_G, \lambda_G, [\ ]_G, \xi_G)$ satisfies the properties of a context system. We have to consider a specific subset of contexts for each graph rewriting system. Define $C'_G$ to be the set of graph contexts that can appear in some derivation, i.e., of all contexts $C$ such that there exists an interface graph $(G, I)$ derivable from the start graph $(S, I)$ and a derivation step $(G, I) \overset{C,\psi}{\Longrightarrow} (G', I)$ using this context.

***Proposition 23.*** *$CS_G = (C'_G, \lambda_G, [\ ]_G, \xi_G)$ constitutes a definite context system.*

***Corollary 24 (Parallelism of graph rewriting).*** *The parallelism theorems I–III (with uniqueness properties) hold for distributed rewriting systems based on the context system for graphs.*

## 4 Derivations, Trace Languages, and Event Structures

Considering the application of a rule as an event, we are able to relate rewriting systems to event structures. We will show how derivations of rewriting systems based on definite context systems are related to trace languages. As there is a well-known relation between trace languages and event structures, we obtain, via this intermediary step, a relation between distributed rewriting systems and event structures. First, we recall the definition of (prime) event structures [NPW81, Win91]:

**Definition 25 (Event structure).** An *event structure* $ES = (E, \leq, \#)$ consists of a set $E$ of *events*, a partial order $\leq \subseteq E \times E$ called the *causality relation*, and an irreflexive and symmetric relation $\# \subseteq E \times E$ called *conflict relation*, such that $\#$ is inherited via $\leq$ in the sense that $e_1 \# e_2 \wedge e_2 \leq e_3 \Rightarrow e_1 \# e_3$ Events that are neither causally ordered nor in conflict are called *concurrent*.

The *configurations* of the event structure are subsets $x \subseteq E$ which are conflict-free, i.e., $\forall e, e' \in x : \neg(e \# e')$, and downward-closed, i.e., $\forall e, e' \in x : e' \leq e \wedge e \in x \Rightarrow e' \in x$. We denote with $\mathcal{D}^o_{ES}$ the set of finite configurations of $ES$.

Let $(\mathcal{R}, q_0)$ be a definite distributed rewriting system and $\mathcal{S}$ the set of all its *initial sequential derivation sequences*, i.e., of all sequences that start with the initial state $q_0$ and that consist of sequential derivation steps only. Subsequently, we will use the variables $s$, $t$ for sequential derivation sequences and $a$, $b$ for single sequential derivation steps. Let $\leq$ denote the prefix order for sequential derivation sequences, i.e., $s \leq t$ iff $\exists s' : ss' = t$.

Derivation sequences that differ only by commuting sequentially independent derivation steps can be considered as representatives of the same concurrent computation.

**Definition 26 (Commutation equivalence).** The relation $\approx_R \subseteq \mathcal{S} \times \mathcal{S}$ is the smallest equivalence relation such that $sabt \approx_R sb'a't$, if the derivation steps $a$ and $b$ are sequentially independent and commuting these steps according to Parallelism theorem II results in the steps $b'a'$.

Define $\mathcal{T} := \mathcal{S}/_{\approx_R}$ to be the set of equivalence classes of initial sequential derivation sequences with respect to $\approx_R$. An equivalence class $[s]_{\approx_R} \in \mathcal{T}$ is called a *derivation trace*.

Let $\preceq_R \subseteq \mathcal{S} \times \mathcal{S}$ be the partial order defined by $s \preceq_R t \Leftrightarrow \exists s' \in \mathcal{S} : ss' \approx_R t$.

***Proposition 27.*** *Let $\precsim_R \subseteq \mathcal{T} \times \mathcal{T}$ be the quotient of $\preceq_R$ with respect to the equivalence relation $\approx_R$, i.e. $[s]_{\approx_R} \precsim_R [t]_{\approx_R} \Leftrightarrow s \preceq_R t$. Then $\precsim_R$ is a partial order.*

Now we introduce another relation between derivation sequences where all derivation sequences that end with the "same" derivation step (up to permutation of independent derivation steps) become equivalent.

**Definition 28 (Event equivalence).** The relation $\asymp \subseteq \mathcal{S} \times \mathcal{S}$ is the smallest equivalence relation such that

1. $sab \asymp sb'$, if the derivation steps $ab$ are sequentially independent and commuting these steps according to Parallelism theorem II results in the steps $b'a'$ (for some $a'$), and
2. $sa \asymp ta$, if $s \approx_R t$.

Let $\mathcal{E} := \{[s]_{\asymp} \mid s \in \mathcal{S}, s \text{ nonempty}\}$ be the set of equivalence classes of nonempty initial sequential derivation sequences. An equivalence class $[s]_{\asymp}$ is called a *derivation event.*

**Definition 29 (Trace language).** [Maz87, Win91] A *Mazurkiewicz trace language* $TL = (M, E, I)$ consists of a set $E$, a symmetric and irreflexive relation $I \subseteq E \times E$, called *independence relation*, and a nonempty subset $M \subseteq E^*$ such that $M$ is

- prefix closed: $se \in M \Rightarrow s \in M$ for all $s \in E^*$, $e \in E$,
- $I$-closed: $se_1e_2t \in M \wedge e_1Ie_2 \Rightarrow se_2e_1t \in M$ for all $s, t \in E^*$, $e_1, e_2 \in E$,
- coherent: $se_1 \in M \wedge se_2 \in M \wedge e_1Ie_2 \Rightarrow se_1e_2 \in M$ for all $s \in E^*$, $e_1, e_2 \in E$.

The relation $\approx_M \subset M \times M$ is the smallest equivalence relation such that $se_1e_2t \approx_M se_2e_1t$ if $e_1Ie_2$ for all $s, t \in E^*$ and $e_1, e_2 \in E$. We call an equivalence class $[s]_{\approx_M}$ a *trace* and $M_{\approx} := M/_{\approx_M}$ is the set of all traces. $\leq$ denotes the prefix relation on sequences over $E$ and the relation $\preceq_M \subseteq M \times M$ is defined by $s \preceq_M t \Leftrightarrow \exists s' \in E^* : ss' \approx_M t$. The relation $\precsim_M \subseteq M_{\approx_M} \times M_{\approx_M}$ on traces is the quotient of $\preceq_M$ with respect to the equivalence relation $\approx_M$, i.e. $[s]_{\approx_M} \precsim_M [t]_{\approx_M} \Leftrightarrow s \preceq_M t$.

Obviously, there seems to be a close relation between derivation traces of rewriting systems and Mazurkiewicz traces. We now show how to transform the derivation traces of a rewriting system into a trace language.

**Definition 30.** Define $\gamma$ to be the function $\gamma : \mathcal{S} \rightarrow \mathcal{E}^*$ with $\gamma(\epsilon) = \epsilon$ and $\gamma(sa) = \gamma(s)e$, where $e = [sa]_{\asymp} \in \mathcal{E}$. Define $\mathcal{M} := \{\gamma(s) \in \mathcal{E}^* \mid s \in \mathcal{S}\}$ to be the set of sequences of derivation events related to $\mathcal{S}$.

The relation $\mathcal{I} \subseteq \mathcal{E} \times \mathcal{E}$ is defined by $e_1 \mathcal{I} e_2$ iff for all derivation sequences $s$ and derivation steps $a$, $b$: if $sa \in e_1$ and $sb \in e_2$, then $a$ and $b$ are concurrently independent.

***Proposition 31.*** *$(\mathcal{M}, \mathcal{E}, \mathcal{I})$ is a Mazurkiewicz trace language.*

*Proof.* Obviously, $\mathcal{I}$ is symmetric and irreflexive and $\mathcal{M}$ is prefix closed, because the set $\mathcal{S}$ of initial sequential derivation traces is prefix closed. Using Parallelism theorems I and II we can show that $\mathcal{M}$ is $I$-closed and coherent.

***Proposition 32.*** *$\gamma$ is an isomorphism $\gamma : (\mathcal{S}, \leq) \rightarrow (\mathcal{M}, \leq)$ of partial orders.*

**Definition 33 (Uniform rewriting system).** A distributed rewriting system with set $\mathcal{E}$ of derivation events is *uniform*, if it satisfies the following condition: If $e_1, e_2 \in \mathcal{E}$ and $sa \in e_1$, $sb \in e_2$ such that $a$ and $b$ are concurrently independent derivation steps, then for all sequences $t$ and steps $a'$, $b'$ such that $ta' \in e_1$ and $tb' \in e_2$ the steps $a'$ and $b'$ are concurrently independent.

***Lemma 34.*** *The distributed rewriting system for graphs is uniform.*

***Lemma 35.*** *If the rewriting system is uniform, then $s \approx_R t \Leftrightarrow \gamma(s) \approx_M \gamma(t)$ for all $s, t \in \mathcal{S}$.*

***Theorem 36.*** *The partial order $(\mathcal{T}, \precsim_R)$ of derivation traces is isomorphic to the partial order $(\mathcal{M}_{\approx}, \precsim_M)$ of Mazurkiewicz traces.*

*Proof.* Lemma 35 together with Prop. 32 ensure that the quotient mapping $\gamma/_{\approx} : \mathcal{T} \rightarrow \mathcal{M}_{\approx}$, i.e. $\gamma/_{\approx}([s]_{\approx_R}) = [\gamma(s)]_{\approx_M}$ is well-defined and an isomorphism. It can easily be shown that $s \precsim_R t \Leftrightarrow \gamma(s) \precsim_M \gamma(t)$ holds. Consequently $\gamma/_{\approx}$ is an isomorphism of partial orders. □

***Theorem 37.*** *If $\mathcal{T}$ is the set of derivation traces of a definite and uniform distributed rewriting system, then there exists an event structure $ES$ such that the partial order $(D^{\circ}_{ES}, \subseteq)$ of finite configurations is isomorphic to the partial order $(\mathcal{T}, \precsim_R)$ of derivation traces.*

*Proof.* Given a Mazurkiewicz trace langage $(M, E, I)$, we can construct an event structure $ES$ such that $(M_{\approx}, \precsim_M)$ is order isomorphic to $(D^{\circ}_{ES}, \subseteq)$ (see, e.g., [Bed87, Win91]). Using Theorem 36, we get the desired result. □

# 5 Asymmetric Conflicts

Many rewriting systems, e.g., string, term, or graph rewriting systems, allow context sensitive rules. This means that a rule specifies a substructure that must be present to apply the rule, but that remains unchanged. Such context sensitive rewriting rules may lead to a situation of an asymmetric conflict [Ken93], as depicted graphically in Fig. 3 (a). The light-gray areas denote the context-sensitive parts of the left hand sides of both rules. Two string rewriting rules $r_1 = (ab, ac)$

**Fig. 3.** Asymmetric conflict and its related event structure

and $r_2 = (da, e)$ applied to the string $dab$ constitute a concrete example of an asymmetric conflict. If rule $r_1$ is applied first, then $r_2$ can be applied after that, but if rule $r_2$ is applied first, then $r_1$ is no more applicable. Our approach treats this situation as a (symmetric) conflict. The related event structure is shown in Fig.3 (b). The causality relation is depicted by thin lines, going from top to bottom, the conflict relation is drawn as a thick, hatched line (the inherited conflict line between both $r_2$-events is omitted). But this seems to be an inadequate representation of the derivation process, because the information is lost that both $r_2$-events denote the same rule application.

Neither our context-based model of distributed rewriting systems nor event structures of the kind used here are capable to deal adequately with context sensitive rewriting and asymmetric conflicts. However, if we confine graph grammars to graph productions $(L, K, R)$ where $L$ is connected (and contains at least two nodes) and $K$ is a discrete graph (i.e. $s(n) = \langle\rangle$ for all $n \in N_K$), asymmetric conflicts cannot appear due to the dangling condition for the application of productions.

# 6 Conclusion

In this paper, we introduced *distributed rewriting systems* as a unifying framework for different kinds of rewriting systems and showed how Mazurkiewicz trace languages and event structures can be related to them. This also provides a link to other non-interleaving models of concurrency, e.g. *asynchronous transition systems* [Bed87, Win91] or *concurrent transition systems* [Sta89, Sta90].

*Acknowledgement*: I am grateful to the anonymous referees and to Klaus Barthelmann, Erhard Plödereder, Wolfgang Gellerich, and Bernd Holzmüller for helpful comments on this paper.

# References


[BC87] M. Bauderon and B. Courcelle. Graph expressions and graph rewritings. *Math. Systems Theory*, 20:83–127, 1987.

[Bed87] M.A. Bednarczyk. *Categories of asynchronous systems.* PhD thesis, University of Sussex, 1987.

[CEL+93] A. Corradini, H. Ehrig, M. Löwe, U. Montanari, and F. Rossi. True concurrency in graph grammars. In *Dagstuhl Seminar-Report*, number 9301, pages 12–13, 1993.

[CM91] A. Corradini and U. Montanari. An algebra of graphs and graph rewriting. *Lecture Notes in Computer Science 530*, Springer, 1991.

[DM87] P. Degano and U. Montanari. A model for distributed systems based on graph rewriting. *Journal of the ACM*, 34(2):411–449, 1987.

[Ehr79] H. Ehrig. Introduction to the algebraic theory of graph grammars (a survey). *Lecture Notes in Computer Science 73*, pages 1–69. Springer, 1979.

[HK87] A. Habel and H.-J. Kreowski. May we introduce to you: Hyperedge replacement. *Lecture Notes in Computer Science 291*, Springer, pages 15–25, 1987.

[JR89] D. Janssens and G. Rozenberg. Actor grammars. *Mathematical Systems Theory*, 22:75–107, 1989.

[Ken93] R. Kennaway. Private communication. January, 1993.

[Maz87] A. Mazurkiewicz. Trace theory. *Lecture Notes in Computer Science 255*, pages 279–323. Springer-Verlag, 1987.

[Mes90] J. Meseguer. Rewriting as a unified model of concurrency. Technical Report SRI-CSL-90-02R, SRI International, Computer Science Laboratory, 1990.

[NPW81] M. Nielsen, G. Plotkin, and G. Winskel. Petri nets, event structures and domains. *Theoretical Computer Science*, 13:85 – 108, 1981.

[Rao84] J.C. Raoult. On graph rewritings. *Theoretical Computer Science*, 32:1–24, 1984.

[Sch90] H.J. Schneider. Describing distributed systems by categorical graph grammars. *Lecture Notes in Computer Science 411*, pages 121–135. Springer, 1990.

[Sch92] G. Schied. *Über Graphgrammatiken, eine Spezifikationsmethode für Programmiersprachen und verteilte Regelsysteme.* Arbeitsberichte des IMMD, Band 25, Nr. 2, Universität Erlangen-Nürnberg, 1992.

[Sta89] E.W. Stark. Concurrent transition systems. *Theoretical Computer Science*, 64:221–269, 1989.

[Sta90] E. W. Stark. Connections between a concrete and an abstract model of concurrent systems. *Lecture Notes in Computer Science 442*, pages 53–79. Springer, 1990.

[Win89] G. Winskel. An introduction to event structures. *Lecture Notes in Computer Science 354*, pages 364–397, Springer, 1989.

[Win91] G. Winskel. *Categories of models for concurrency.* Advanced School on the Algebraic Logical and Categorical Foundations of Concurrency, Gargnano del Garda, Italy, 1991.

# Logic Based Structure Rewriting Systems

Andy Schürr

Lehrstuhl für Informatik III
Aachen University of Technology
Ahornstr. 55, D-52074 Aachen
e-mail: andy@rwthi3.informatik.rwth-aachen.de

**Abstract:** This paper presents a new logic based framework for the formal treatment of graph rewriting systems and graph languages as special cases of rewriting systems and languages for arbitrary relational data structures. Considering its expressive power, the new formalism is intended to contain almost all forms of nonparallel algebraic as well as algorithmic graph grammar approaches as special cases. Furthermore, our formalism tries to close the gap between the operation oriented manipulation of data structures by means of graph rewrite rules and the declaration oriented description of data structures by means of logic based languages.

Nevertheless, the main motivation for the development of yet another structure rewriting approach was not to combine the advantages of graph grammar based and logic based languages but to provide us with a solid fundament for a formal definition of our own graph grammar based language PROGRES, which is a kind of visual data definition and data manipulation language. This language has been designed for specifying and rapid prototyping of complex data structures and contains many constructs which enhance its expressiveness considerably (like definition of derived properties, consistency constraints, rewrite rules with complex application conditions and embedding rules) but which were not definable within the framework of a single already existing graph grammar approach.

## 1 Introduction

Modern software systems for application areas like office automation or software engineering are usually highly interactive and deal with complex structured objects. The systematic development of these systems requires precise and readable descriptions of their desired behavior. Many specification languages have been proposed to produce formal descriptions of various aspects of software systems, such as the design of object structures, the effect of operations on objects, or the synchronization of concurrently executing tasks. Many of these languages use **special classes of graphs** as their underlying data models.

Within the research project IPSEN (an abbr. of "Integrated Project Support ENvironments") a **graph grammar based specification method** is used for modeling the internal structure of abstract data types and for producing executable specifications of their interface operations [Nag 86]. The development of such a specification consists of two closely related subtasks. The first one is to design a graph model for the corresponding object structure. The second one is to program object (graph) analyzing and modifying operations by composing sequences of subgraph tests and graph rewrite rules.

Based on our experiences with graph rewriting systems, software specification languages, and programming environments, we started the design of a **graph grammar language** and the development of an **integrated set of tools** for this language several years ago. One result of this design process is the language PROGRES - an acronym for "**PRO**grammed Graph Rewriting System" - which possesses an almost static type system as well as a formally defined syntax and semantics. Furthermore, it is the first attempt to combine graph grammars with new elements for the declaration of graph schemes and consistency constraints as well as for the definition of derived relations and attributes. As a consequence, we were forced to develop a **new framework** for the formal definition of our language which allows us to formalize

- derived graph properties and global consistency constraints,
- positive and negative pre- and postconditions for graph rewrite rules,
- deletion of nodes with unknown context,
- complex embedding transformation rules,
- and programming with parametrized graph rewrite rules.

This framework is based on an enhancement of **predicate logic with nonmonotonic reasoning** capabilities and has some similarities to underlying theories of knowledge representation languages or deductive database systems [MBJK 90, NT 89]. To summarize, the acronym PROGRES is a "trade mark" for an integrated set of tools (environment), a very high level strongly typed programming language, and its underlying formalism.

The main purpose of this paper is to introduce to the reader a new **structure rewriting formalism**, which is a generalized variant of the graph rewriting approach in [Schü 91b], whereas previously published papers had their main focus on the language PROGRES and its integrated programming environment (cf. [NS 91, Schü 91a]). Note that due to size limits a couple of definitions are presented in simplified versions and the aspect of programming with parametrized graph rewrite rules had to be omitted (cf. [Schü 91b, SZ 91] for further information on these topics).

In order to emphasize that PROGRES has mainly been developed for specifying and implementing software engineering environments, we will use the definition of a **syntax-directed editor** for the applicative language "Exp" as a running example within this paper. This language characterizes all legal input sequences for a very primitive desk calculator. The well-known "let"-operator is used to name and reuse intermediate computation results and the usual scoping rules for block structured programming languages direct the binding of applied occurrences of names to their corresponding definitions. Figure 1 defines the "Exp" syntax and figure 2 presents the text as well as the **graph representation** of a typical sentence of this language (composed of the sentence's abstract syntax tree skeleton and additional edges for binding applied identifier occurrences to their declarations). It contains one unexpanded subexpression with an undefined value represented by "(<Exp>)". As a consequence, the values of two subexpressions starting with "let x := ... " are undefined, too. And the computation of the subexpression "let y := 2 in x * y end" yields the value "16" due to the fact that "x" is bound to "8" (by the outermost "let") and "y" to "2".

| | |
|---|---|
| EXP ::= | IntLiteral \| DefExp \| ApplId \| PlusExp \| MinusExp \| MultExp \| DivExp ; |
| IntLiteral ::= | Text ; *(* Text is a sequence of digits *)* |
| DefExp ::= | "let" DeclId ":=" EXP "in" EXP "end" ; |
| DeclId ::= | Name ; *(* Name is a sequence of characters *)* |
| ApplId ::= | Name ; |
| PlusExp ::= | EXP "+" EXP ; |
| . . . | |

Fig. 1: Syntax of the language "Exp"

The rest of this paper is organized as follows:

- Section 2 repeats some basic terminology of predicate logic and contains the formal definition of (data) structures as sets of atomic formulas, with graphs being special cases of structures.
- Section 3 presents the definition of structure schemes as sets of formulas, too.
- Section 4 is dedicated to the formal definition of substructures and redex selection (a redex is that part of a structure which matches a rewrite rule's left-hand side).
- Section 5 introduces the new structure rewriting formalism itself.
- And the last section of the paper reviews the main motivation for the development of yet another structure rewriting approach, considers its relationships to similar graph/structure rewriting approaches, and lists a number of open questions.

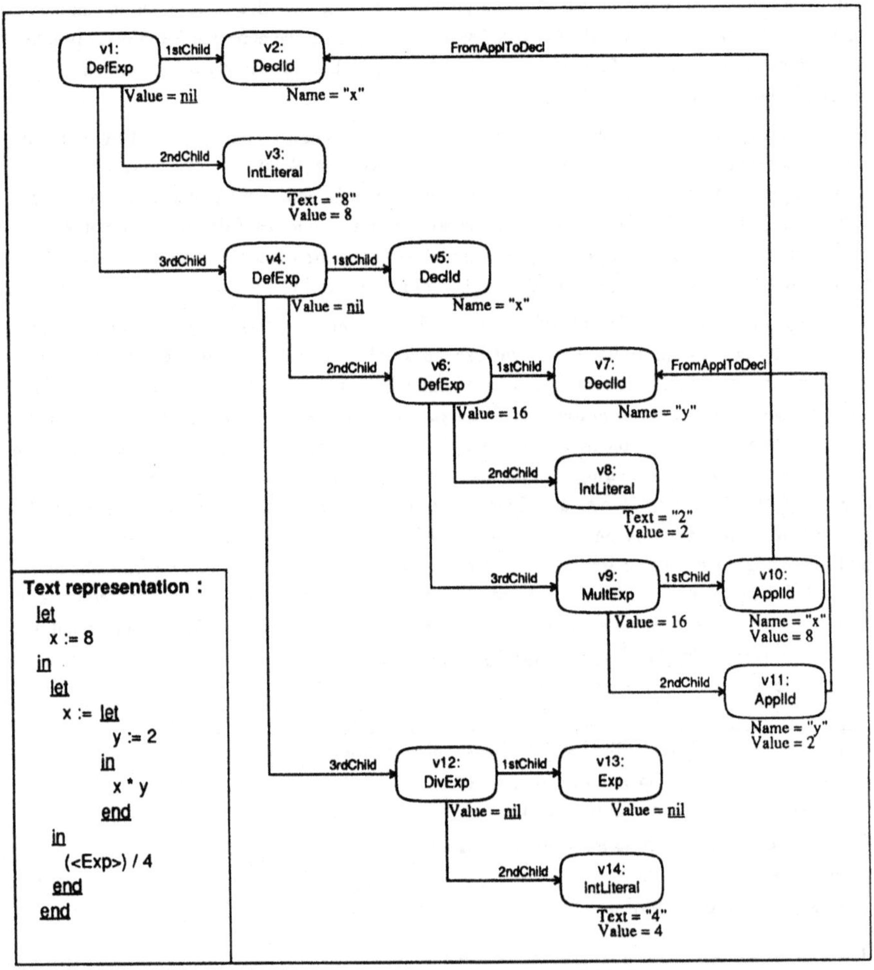


Fig. 2: Text and graph representation of an "Exp" sentence

## 2 Predicate Logic Formulas, Structures, and Graphs

This section introduces basic terminology of predicate logic as e.g. the syntax of formulas over certain alphabets of predicate symbols etc. and explains the **modelling of graphs** with directed, labeled edges and attributed, labeled nodes as special cases of structures, i.e. as sets of atomic formulas. Note that all definitions and propositions of the following sections are almost independent of the selected graph model and its encoding. Therefore, our structure rewriting approach is able to deal with about the same class of relational structures as [MW 90], may easily be instantiated with a different graph model as e.g. the hypergraphs of [Ehr 79], and should even be adaptable to the case of almost arbitrary knowledge bases as e.g. defined in [MBJK 90, NT 89].

**Definition 2.1 Signature.**

A 6-tuple $\Sigma := (\mathcal{L}_F, \mathcal{L}_P, \mathcal{L}_C, \mathcal{V}, \mathcal{W}, \mathcal{X})$ is a **signature** iff:

(1) $\mathcal{L}_F$ is an alphabet of function symbols[1].

(2) $\mathcal{L}_P$ is an alphabet of predicate symbols[2].

(3) $\mathcal{L}_C$ is an alphabet of constant symbols (for attribute values, type identifiers etc.).

(4) $\mathcal{V}$ is a special alphabet of object (node) identifiers.

(5) $\mathcal{W}$ is a special alphabet of identifiers for sets of objects (nodes).

(6) $\mathcal{X}$ is an alphabet of variables used for the construction of first order formulas. ■

**Example 2.2 Signature for an "Exp"-Graph.**

The graph signature for fig. 2 is $\Sigma := (\mathcal{L}_F, \mathcal{L}_P, \mathcal{L}_C, \mathcal{V}, \mathcal{W}, \mathcal{X})$ with:

(1) $\mathcal{L}_F := \{ +, -, \dots \}$.

(2) $\mathcal{L}_P := \{ <, \dots, rel, type \}$.

(3) $\mathcal{L}_C := \{ 0, 1, \dots,$ Integer, DefExp, … , Value, FromApplToDecl, … $\}$.

(4) $\mathcal{V} := \{ v1, v2, \dots \}$.

(5) $\mathcal{W} := \{ w1, w2, \dots \}$.

(6) $\mathcal{X} := \{ x1, x2, \dots \}$. ■

Note that the alphabet $\mathcal{L}_P$ contains two predicate symbols which are very important for the chosen encoding of graphs (there are of course many other ways of representing graphs as structures):

(1) The predicate symbol "*type*" with: "*type*(x, vl)" for "node x has label vl" or with "*type*(τ, t)" for "term τ has type t", and

(2) The predicate symbol "*rel*" with: "*rel*(x, el, y)" for "edge with label el connects source node x to target node y" or with "*rel*(x, a, τ)" for "attribute a at node x has value τ".

In the sequel we will always assume that Σ is a (graph) signature over the above mentioned alphabets.

**Definition 2.3 Σ-Term.**

**TERM(Σ)** is the set of all terms (in the usual sense) which contain function symbols of $\mathcal{L}_F$, free variables of $\mathcal{X}$, and constant symbols of $\mathcal{L}_C$, $\mathcal{V}$, and $\mathcal{W}$. ■

---

1) More precisely, a family of alphabets of symbols if we take domain and range of functions into account.
2) A family of alphabets of symbols if we take arities of predicates into account.

**Definition 2.4 Σ-Atom.**

ATOM(Σ) is the set of all atomic formulas which contain predicate symbols of $\mathcal{L}_P$ and which additionally may contain the predicate symbol "=" for expressing the equality of two Σ-terms. ■

**Definition 2.5 Σ-Formula.**

FORM(Σ) is the set of all first order predicate logic formulas with ATOM(Σ) as atomic formulas, ∧, ∨, ... as logical connectives, and ∃, ∀ as quantifiers. ■

**Definition 2.6 Derivation of Formulas.**

With Φ and Φ' being sets of Σ-formulas

$$\Phi \vdash \Phi'$$

means that all formulas of Φ' are **derivable** from the set of formulas Φ using any (consistent, complete) inference system of first order predicate logic with equality. ■

**Definition 2.7 Closed Σ-Formulas.**

The set $F \subseteq FORM(\Sigma)$ is a **closed set of Σ-formulas** (write: $F \in \mathcal{F}(\Sigma)$) ⇔

F contains only bound variables (of $\mathcal{X}$ in Σ)[3]. ■

In the following, subsets of $\mathcal{F}(\Sigma)$ will be used to represent graph schemes, scheme consistent graphs, left- and right-hand sides as well as additional pre- and postconditions of graph rewrite rules.

**Definition 2.8 Σ-Structure**

A closed set of formulas $G \in \mathcal{F}(\Sigma)$ is a **Σ-structure** (write: $G \in \mathcal{G}(\Sigma)$) ⇔

$G \subseteq ATOM(\Sigma)$ and G does not contain formulas of the form "$\tau_1 = \tau_2$". ■

**Example 2.9 An "Exp"-Graph-Structure**

The following structure is a set of formulas which has the graph G of fig. 2 as a model but which also has many other graphs as models, in which we are not interested in:

G := { *type*(v1, DefExp), *type*(v2, DeclId), *type*(v3, IntLiteral), ... ,
*rel*(v1, 1stChild, v2), *rel*(v1, 2ndChild, v3), ... , *rel*(v3, Value, 8), ... } . ■

The next section introduces a so-called "completing operator" which allows us to get rid of unwanted graph models and to reason about properties of "minimal" (graph) models on a pure syntactical level. Therefore, models of structures are not introduced formally and will not be used within the rest of this paper.

## 3 Structure Schemes and Scheme Consistent Structures

This section is mainly dedicated to the definition of **structure schemes** - the term "structure scheme" is used in the same sense as the term "database scheme" - and scheme consistent graphs. Such a scheme is a set of properties of a certain class of structures. Thus, PROGRES and its underlying theory support the separation of implicitly given integrity constraints, functional attribute dependencies etc. from graph (structure) rewrite rules. For this purpose, an extension of pure 1st-order predicate logic is necessary with respect to the capability of nonmonotonic reasoning:

---

3) Thus, we avoid any difficulties with different interpretations of free variables in different inference systems.

**Example 3.1 Nonmonotonic Reasoning.**

G := { *type*(v1, ApplId) } is a structure which has all graphs containing at least one "ApplId" node as models. Being interested in properties of "minimal" G graph models only, we would like to be able to prove

$$\forall\, x, l{:}\ type(x, l) \;\rightarrow\; (\, x = v1 \wedge l = \text{ApplId}\,),$$

i.e. that all models of G, we are interested in, contain exactly one node. One obvious solution for this problem is to add the above mentioned formula to our structure G (and further formulas for prohibiting edges in models for G etc.). But our structure rewriting formalism is not able to rewrite sets of nonatomic formulas[4] and, therefore, structures are restricted to contain atomic formulas only (see def. 2.8). Therefore, we need a mechanism which allows us to use these additional formulas without explicitly including them into our structures.

A similar problem has been extensively studied within the field of deductive database systems and has been solved by a number of quite different approaches either based on the so-called "**closed world assumption**" or using "**nonmonotonic reasoning**" (cf. e.g.[Naq 86a/b, Min 88]). The main idea of (almost) all of these approaches is to distinguish between basic facts and derived facts and to add only negations of basic facts to a rule base which are not derivable from the original set of facts. It is beyond the scope of this paper to explain nonmonotonic reasoning in more detail. Therefore, we will simply assume the existence of a "**completing operator**" $\mathcal{C}$ which adds a certain set of additional formulas to a structure such that the resulting set of formulas is consistent and "sufficiently complete" to allow us to prove the above mentioned properties by using the axioms of "pure" first-order predicate logic only. For the example mentioned above the operator $\mathcal{C}$ might have the following form:

$$\mathcal{C}(G) := G \cup \{\neg\exists\, y{:}\ rel(v, r, y) \mid v \in \mathcal{V}, r \in \mathcal{L}_C, \text{ and not exist } v'{:}\ rel(v, r, v') \in G\,\} \cup \ldots \cup^{5}$$
$$\{\forall\, x, l{:}\ type(x, l) \rightarrow (x = v_1 \vee \ldots \vee x = v_k) \mid v_1, \ldots, v_k \in \mathcal{V} \text{ are all object id in } G\,\}.$$

By means of this operator we are able to prove for instance:

$$\mathcal{C}(G) \vdash (\neg\exists\, y{:}\ rel(v1, \text{FromApplToDecl}, y)) \;\wedge\; (\forall\, x, l{:}\ type(x, l) \rightarrow x = v1\,).\ \blacksquare$$

**Definition 3.2 Σ-Structure Completing Operator.**

A function $\mathcal{C}{:}\ \mathcal{G}(\Sigma) \rightarrow \mathcal{F}(\Sigma)$ is a **Σ-structure completing operator** ⇔
For all structures $G \in \mathcal{G}(\Sigma)$ at least the following two conditions hold:

(1) $G \subseteq \mathcal{C}(G)$.

(2) $\mathcal{C}(G)$ is a consistent set of formulas. ■

**Definition 3.3 Σ-Structure Scheme.**

A tuple $S := (\Phi, \mathcal{C})$ is a **Σ-structure scheme** (write: $S \in \mathcal{S}(\Sigma)$) ⇔

(1) $\Phi \in \mathcal{F}(\Sigma)$ is a consistent set of formulas without object (set) identifiers, i.e. it contains no propositions about individual objects.

(2) $\mathcal{C}$ is a Σ-structure completing operator. ■

---

4) The problem is to find a reasonable semantics for the deletion of nonatomic formulas which is compatible with the derivation process of a first order predicate logic inference system.

5) We have omitted the definition of many additional sets of formulas which for instance force different object identifiers to represent different objects (in order to exclude final graph models which map all constants onto the same value/object; for further details see [Schü 91b]).

**Definition 3.4 Scheme Consistent Structure.**

Let S := (Φ, $\mathcal{C}$) ∈ $\mathcal{S}(\Sigma)$ a scheme. A Σ-structure G ∈ $\mathcal{G}(\Sigma)$ is **scheme consistent with respect to** S (write: G ∈ $\mathcal{G}$(S)) ⇔ $\mathcal{C}$(G) ∪ Φ is a consistent set of formulas. ■

Fig. 3 contains a cutout of the PROGRES definition of our "Exp" example's graph scheme with three different kinds of type declarations. **Node and edge type** declarations are used to introduce type labels for nodes and edges, whereas **node class** declarations are used as types of node types with common properties (for further details see [Schü 91a/b]). The class BINARY_EXP e.g. is a subclass of the classes EXP and BINARY_NODE and inherits all properties from both superclasses (nodes of a type of this class are sources of 1stChild and 2ndChild edges and have a Value attribute). And the node type MultExp defines the behavior of all nodes of this type, i.e. the directed equation for the computation of their Value attribute. This attribute is called a **derived attribute** (in contrast to an intrinsic attribute, which has a value of its own and receives new values by means of assignments within rewrite rules only), because its value is derived from values of other attributes only and will be kept consistent with these foreign attribute values (during graph transformations) by means of an incrementally working attribute evaluation process (see [KSW 93]).

**Example 3.5 A Graph Scheme for "ExpLanguage".**

A graph scheme S := (Φ, $\mathcal{C}$) for the graph G in fig. 2 has the following form:

Φ := { *type*(MultExp, BINARY_EXP) ,
∀t: *type*(t, BINARY_EXP) → *type*(t, EXP) ∧ *type*(t, BINARY_NODE) ,
∀ x, y: *rel*(x, 1stChild, y) → ∃ t: *type*(x, t) ∧ *type*(t, BINARY_NODE) ,
... ,
∀ x, y: 1stOpd(x, y) ↔ *rel*(x, 1stChild, y) ∧ ∃ t: *type*(y, t) ∧ *type*(t, EXP)
∀ x, v: *type*(x, MultExp) ∧ *rel*(x, Value, v)
→ ∃ y1, v1, y2, v2: 1stOpd(x, y1) ∧ *rel*(y1, Value, v1)
∧ 2ndOpd(x, y2) ∧ *rel*(y2, Value, v2)
∧ v = v1 * v2 } .

$\mathcal{C}$(G) := G ∪ .... ■

```
node class AST_NODE end; (* AST = Abstract Syntax Tree *)
node class BINARY_NODE is a AST_NODE end; (* AST_NODE with two children. *)
edge type 1stChild, 2ndChild: BINARY_NODE --> AST_NODE;
node class EXP is a AST_NODE;
   derived Value: integer;
end; (* The directed equations for the derived Value attribute are part of the type declarations below. *)
node class BINARY_EXP is a EXP, BINARY_NODE end;
path 1stOpd: BINARY_EXP --> EXP = -1stChild-> & instance of EXP  end;
   (* Follows a 1stChild edge and verifies that its target belongs to the subclass EXP of AST_NODE. *)
   ...
node type MultExp: BINARY_EXP;
   redef derived Value = self.1stOpd.Value * self.2ndOpd.Value;
end; (* Value of MultExp = Value of its first Operand * Value of its second operand. *)
```

Fig. 3: The "Exp" graph scheme (cutout)

## 4 Redex Selection with Additional Constraints

The main purpose of this section is to formalize the term **"redex under additional constraints"** by means of morphisms. This term is central to the definition of "structure rewriting with pre- and post-conditions" of the next section and determines which substructures of a structure, called redices, are legal "targets" for a rewriting step, i.e. match a given rule's left-hand side. Note that the embedding rules of PROGRES[6] are able to delete, copy, and redirect arbitrary large bundles of edges. We will handle these embedding rules by introducing special **object set identifiers** on a structure rewrite rule's left- and right-hand side which are allowed to match an arbitrarily large (maximal) set of object identifiers in a given redex (see def. 5.1 and example 5.2). Therefore, morphisms between structures, which select the affected substructure/redex for a rewrite rule in a structure, are neither total nor partial functions. In the general case, these **morphisms are relations** between object (set) identifiers. Furthermore, these relations are required to preserve the structure of the source graph in the target graph as usual and to take additional constraints for the embedding of the source structure (e.g a rule's left-hand side) into the target structure (e.g the structure we have to rewrite) into account.

**Definition 4.1 Redex Selecting Relation (simplified).**

With F, F' $\in \mathcal{F}(\Sigma)$ and $\mathcal{V}_F$, $\mathcal{W}_F$ and $\mathcal{V}_{F'}$, $\mathcal{W}_{F'}$ being those sets of object (set) identifiers which are used in F and F', respectively, a relation

$$u \subseteq (\mathcal{V}_F \times \mathcal{V}_{F'}) \cup (\mathcal{W}_F \times \mathcal{W}_{F'}) \cup (\mathcal{W}_F \times \mathcal{V}_{F'})$$

is a **redex selecting relation** (also called **$\Sigma$-relation**) from F to F' $\Leftrightarrow$

(1) For all $v \in \mathcal{V}_F$: $|u(v)| = 1$ (with $u(x) := \{ y \mid (x, y) \in u \}$),
i.e. every source object identifier will be mapped onto one target object identifier.

(2) For all $w \in \mathcal{W}_F$: $u(w) \subseteq \mathcal{W}_{F'}$ and $|u(w)| = 1$ or $u(w) \subseteq \mathcal{V}_{F'}$.
i.e. every source object set identifier will be mapped either onto one target object set identifier or onto a set of target object identifiers.

Let u* be the natural extension[7] of u to the domains of $\Sigma$-terms and $\Sigma$-formulas; furthermore, with $\Phi \subseteq \mathcal{F}(\Sigma)$, the following short-hand will be used in the sequel:

$$u^*(\Phi) := \{ \phi' \in \mathcal{F}(\Sigma) \mid \phi \in \Phi \text{ and } (\phi, \phi') \in u^* \} . ■$$

**Example 4.2 A Redex Selecting Relation.**

Let F, F' $\in \mathcal{F}(\Sigma)$ with $\mathcal{V}_F := \{ v1 \}$, $\mathcal{V}_{F'} := \{ v11, v12, v21, v22 \}$, $\mathcal{W}_F := \{ w1, w2 \}$, and with $\mathcal{W}_{F'} := \{\}$. Then u := { (v1, v11), (w2, v21), (w2, v22) } is a redex selecting relation from F to F':

u*({ *type*(v1, t) }) = { *type*(v11, t) } .

u*({ *rel*(v1, r, w1) }) = { } .

u*({ *type*(w2, t) }) = { *type*(v21, t), *type*(v22, t) } . ■

Note that in the general case (defined in [Schü 91b] as so-called "renaming" relations), redex selecting relations are allowed to map identifiers onto (sets of) $\Sigma$-terms. In this way, we are also able to parametrize structure rewrite rules with terms representing values for object properties (e.g. attributes of nodes) and to modify properties of already existing objects.

---

6) PROGRES embedding rules are slightly simplified variants of those belonging to the expression oriented algorithmic graph grammar approach à la [Nag 79a].

7) u* relates two terms or formulas to each other, if both are identical with the exception of u-related object (set) identifiers.

**Definition 4.3** **Σ-(Structure-)Morphism.**

Let F, F' ∈ $\mathcal{F}(\Sigma)$. A Σ-relation u from F to F' is a **Σ-morphism** from F to F' (write: u: F ⇉ F') ⇔
F' ⊢ u*(F) .

With F, F' ∈ $\mathcal{G}(\Sigma) \subseteq \mathcal{F}(\Sigma)$ being Σ-structures, u will be called **Σ-structure morphism.** ■

**Proposition 4.4** **The Category of Σ-Structures.**

Assume "∘" to be the usual composition of binary relations. Then, $\mathcal{G}(\Sigma)$ together with the family of Σ-structure morphisms defined above and "∘" is a **category** (the same holds true for the set of closed formulas $\mathcal{F}(\Sigma)$ and the family of Σ-morphisms). ■

**Proof:**

The nontrivial part of this proof is to show that the composition of two morphisms is a morphism, too. This problem is reducible to the question whether consistent renaming of node identifiers in formulas is compatible with the derivability of formulas, i.e. that consistent renaming of constants does not destroy proofs (which is always true). For further details see [Schü 91b]. ■

**Definition 4.5** **Substructure.**

G, G' ∈ $\mathcal{G}(\Sigma)$ are structures. G is a (partial) **substructure** of G' with respect to a Σ-relation u (write: $G \subseteq_u G'$) ⇔ u: G ⇉ G'.

This definition coincides with the usual meaning of substructure (or subgraph), if G and G' do not contain object set identifiers, i.e. are not left- or right-hand sides of rewrite rules. ■

**Proposition 4.6** **Soundness of Substructure Property.**

For G, G' ∈ $\mathcal{G}(\Sigma)$ being structures, the following properties are equivalent:

G is a substructure of G' with respect to a Σ-relation u ⇔ u*(G) is a subset of G'.

**Proof:** $G \subseteq_u G'$ $\Leftrightarrow_{\text{see def. 4.5}}$

u: G ⇉ G∏ $\Leftrightarrow_{\text{see def. 4.3}}$

G' ⊢ u*(G) $\Leftrightarrow_{\text{u*(G) and G' are sets of atomic formulas without "="}}$

u*(G) ⊆ G' ("⊆" denotes the inclusion for sets of formulas). ■

**Definition 4.7** **Substructure with Additional Constraints.**

S := (Φ, $\mathcal{C}$) is a Σ-structure scheme, G, G' ∈ $\mathcal{G}(S)$ are structures, and F ∈ $\mathcal{F}(\Sigma)$ is a set of constraints/formulas with references to object (set) identifiers of G only. G is a **substructure** of G' with respect to a Σ-relation u and the additional set of constraints F (write: $G \subseteq_{u,F} G'$) ⇔

(1) $G \subseteq_u G'$ , i.e. G' ⊢ u*(G) , i.e. G is a substructure of G'.

(2) u: F ∪ G ⇉ Φ ∪ $\mathcal{C}$(G') , i.e. Φ ∪ $\mathcal{C}$(G') ⊢ u*(F ∪ G) ,
i.e. we are able to prove all required additional constraints F for the embedding of G in G', using the basic facts of G' and the set of formulas generated by the completing operator $\mathcal{C}$ as well as the set of formulas Φ of the structure scheme S.

These conditions are equivalent to the existence of the diagram in fig. 4, with inclusions $i_1$ and $i_2$ being the following morphisms:

$i_1$: G ⇉ F ∪ G, because $i_1$*(G) = G ⊆ F ∪ G, i.e. F ∪ G ⊢ $i_1$*(G) .
$i_2$: G' ⇉ Φ ∪ $\mathcal{C}$(G'), because $i_2$*(G') = G' ⊆ $\mathcal{C}$(G'), i.e. Φ ∪ $\mathcal{C}$(G') ⊢ $i_2$*(G') . ■

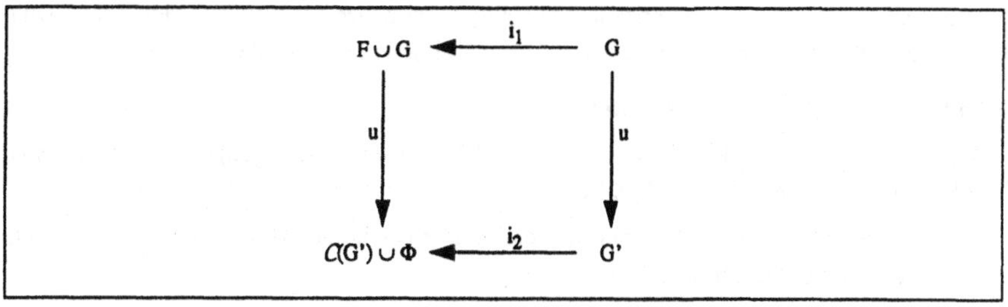


Fig. 4: Substructure (redex) selection with additional constraints

**Example 4.8 Substructure Selection.**

Let S := (Φ, 𝒞) be an extended version of the scheme belonging to fig. 3 which additionally contains the definition of a new predicate "unused":

$\Phi := \{ \forall y: \text{unused}(y) \leftrightarrow \neg \exists x: rel(x, \text{FromApplToDecl}, y), \ldots \}$ and

$\mathcal{C}(G) := G \cup \{ \neg \exists x: rel(x, r, v') \mid \text{not exists } v: rel(v, r, v') \in G \} \cup \ldots$

Furthermore, G is a S-scheme-consistent structure (for a graph with two nodes and one edge)

$G := \{ type(\text{'1}, \text{DefExp}), type(\text{'2}, \text{DeclId}), rel(\text{'1}, \text{1stChild}, \text{'2}) \}$

and F is an embedding condition for G:

$F := \{ \text{unused}(\text{'2}) \}$ .

Then G is a substructure of G' of fig. 2 with respect to F and the redex selecting relation

$u := \{ (\text{'1}, v4), (\text{'2}, v5) \}$.

**Proof:**

With

$G' := \{ type(v4, \text{DefExp}), type(v5, \text{DeclId}), rel(v4, \text{1stChild}, v5), \ldots \}$

we are able to prove that

$\mathcal{C}(G') \cup \Phi \vdash u^*(G)$ and $\neg \exists x: rel(x, \text{FromApplToDecl}, v5) \in \mathcal{C}(G')$ .

⇒

$\mathcal{C}(G') \cup \Phi \vdash \{ \text{unused}(v5) \}$ . ■

## 5 Scheme Preserving Structure Rewrite Rules

Having presented the definitions of structure schemes, scheme consistent structures, and structure morphisms, we are now prepared to introduce **structure rewrite rules** as quadruples of **sets of closed formulas** and the application of structure rewrite rules as the construction of commuting diagrams in a similar way as it is done in the algebraic graph grammar approach. But note that due to the usage of redex selecting **relations** instead of (partial) mappings, and the necessarily slightly asymmetric definition of structure rewriting (object set identifiers in a rule's left-hand side must be mapped onto a maximal set of object identifiers in a structure) the most important properties of the algebraic approach get lost:

- Our (sub-)diagrams are not pushouts in the general case.
- The application of a structure rewrite rule is not invertible in the general case.

Unfortunately, we had to pay this price for the ability to formalize complex embedding rules within the language PROGRES, which delete/copy/redirect arbitrarily large bundles of edges.

**Definition 5.1 Structure Rewrite Rule.**

A quadruple r := (AL, L, R, AR) with AL, AR ∈ $\mathcal{F}(\Sigma)$ and with L, R ∈ $\mathcal{G}(\Sigma)$ is a **structure rewrite rule** for the signature Σ (or: r ∈ $\mathcal{R}(\Sigma)$) ⇔

(1) The set of left-hand side application/embedding conditions AL contains only object (set) identifiers of the left-hand side L.

(2) The set of right-hand side application/embeddding conditions AR contains only object (set identifiers of the right-hand side R.

(3) Every set identifier of L is also a set identifier in R (set identifiers are only used for deleting otherwise dangling object references and for establishing connections between new substructures created by a rule's right-hand side and the remaining rest of the modified structure). ■

Consider the production ExpandToXIdent of fig. 5, which contains almost all important components of PROGRES productions. This production replaces an unexpanded subexpression by an applied occurrence of the identifier "x" and binds the new identifier to the nearest enclosing declaration of "x". Its execution may be divided into seven steps:

(1) **Subgraph selection**: Find any subgraph/redex within the host graph consisting of two nodes with labels DeclId and Exp, respectively.

(2) **Check preconditions**: These nodes must be connected by means of a path visibleDecl("x"). This path is a kind of derived relation with a parameter "N" and connects any Exp node to the nearest enclosing declaration with name "N"[8].

(3) **Subgraph replacement**: Erase the Exp node, create a new ApplId node and an additional ApplToDecl edge from the new node to the old DeclId node.

(4) **Embedding transformation**: Redirect any incoming edge with label 1stChild, 2ndChild, etc. from the erased node to the new node.

(5) **Attribute transfer**: Assign the value "x" to the Name attribute of the new node (which is not a derived but a so-called intrinsic attribute).

(6) **Attribute reevaluation**: Compute the new values of affected derived Value attributes at nodes v1, v4, and v12 in fig. 2 (cf. [KSW 93] for a description of our attribute evaluation algorithm).

(7) **Check postconditions**: After the application of the rewrite rule, the new node must have a defined Value attribute (this postcondition has been added for demonstration purposes only and it may be replaced by a precondition which checks whether the selected Decl node has a defined Value attribute before the application of the rule[9]).

The formal representation of this production is presented in the following example:

---

8) The definition of this path has been omitted for size limitations only and must be part of the corresponding graph scheme (cf. [Schü 91b] for the syntax of path declarations and their semantic definition by means of formulas).

9) Up to now, it's an open question which postconditions may be replaced by equivalent preconditions (see also conclusion of this paper) and postconditions have mainly been added for symmetry reasons.

```
production ExpandToXIdent
    (* Replaces any unexpanded subexpression in a graph, represented as an Exp-labeled node   *)
    (* by an applied occurrence of the identifier "x", iff a visible declaration with defined value exists. *)
=
```

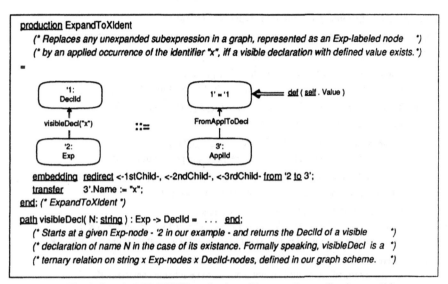

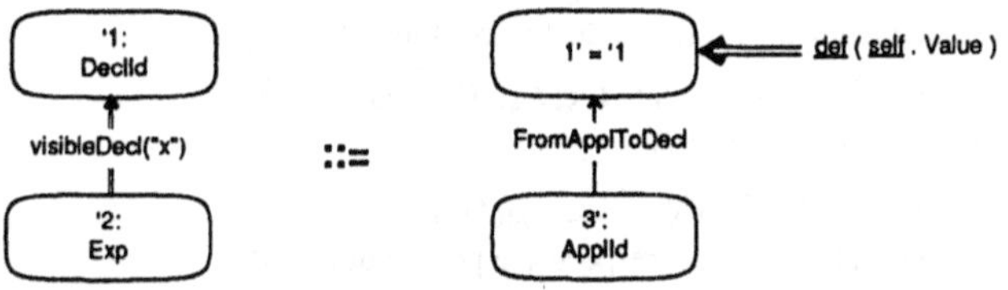


```
    embedding  redirect <-1stChild-, <-2ndChild-, <-3rdChild- from '2 to 3';
    transfer     3'.Name := "x";
end; (* ExpandToXIdent *)

path visibleDecl( N: string ) : Exp -> DeclId =  . . .  end;
    (* Starts at a given Exp-node - '2 in our example - and returns the DeclId of a visible      *)
    (* declaration of name N in the case of its existance. Formally speaking, visibleDecl is a  *)
    (* ternary relation on string x Exp-nodes x DeclId-nodes, defined in our graph scheme.     *)
```

Fig. 5: A typical PROGRES production with a complex application condition

**Example 5.2 The Graph Rewrite Rule "ExpandToXIdent".**

Let r := (AL, L, R, AR) with:

(1) AL := { visibleDecl("x", '1, '2) } .

(2) L := { *type*('1, DeclId), *type*('2, Exp),
*rel*(w1, 1stChild, '2), *rel*(w2, 2ndChild, '2), *rel*(w3, 3rdChild, '2) } .

(3) R := { *type*('1, DeclId), *type*(3', ApplId), *rel*(3', FromApplToDecl, '1),
*rel*(w1, 1stChild, 3'), *rel*(w2, 2ndChild, 3'), *rel*(w3, 3rdChild, 3'),
*rel*(3', Name, "x") } .

(4) AR := { $\exists$ v: *rel*(3', Value, v) } .

Note that w1, w2, and w3 are **node set identifiers.** They represent all three redirect rules within our production's embedding part. During the application of the rewrite rule, they will be mapped onto an arbitrarily large maximal and sometimes even empty sets of node identifiers in the host graph. Furthermore, the "encoding" of our production as a structure rewriting rule assumes that any selected Exp node is not source or target of edges with labels different from 1stChild etc. Otherwise, the application of the graph rewrite rule would attempt to produce a graph with **dangling edges** and finally fail. To overcome this problem we have to add additional formulas for each edge type r to the rule's left-hand side which cause the deletion of all potentially existing r edges at the node we have to delete. These formulas have the following form:

$$rel(\text{'}2, r, w_{rout}) \quad \text{and} \quad rel(w_{rin}, r, \text{'}2)$$

with $w_{rout}$ and $w_{rin}$ being set identifiers. In this way, we are able to overcome the problem of the algebraic double pushout approach with deletion of nodes with unknown context (note that prohibition of dangling references etc. is not part of def. 2.8 but may be enforced by appropriate formulas within a graph's scheme). ■

**Definition 5.3 Scheme Preserving Structure Rewriting.**

$S := (\Phi, \mathcal{C}) \in \mathcal{S}(\Sigma)$ and $G, G' \in \mathcal{G}(S)$. Furthermore, $r := (AL, L, R, AR) \in \mathcal{R}(\Sigma)$. The structure G' is **derivable** from G applying r (write: G ~ r ~> G') ⇔

(1) There is a morphism u: $L \xrightarrow{\sim} G$ with: $L \subseteq_{u,AL} G$,
i.e. the via u selected redex in G respects the preconditions AL.

(2) There is no morphism $\hat{u} : L \xrightarrow{\sim} G$ with: $L \subseteq_{\hat{u},AL} G$ and $u \subset \hat{u}$,
i.e. u selects a maximal subgraph in G.

(3) There is a morphism w: $R \xrightarrow{\sim} G'$ with: $R \subseteq_{w,AR} G'$,
i.e. the via w selected subgraph in G' respects the postconditions AR.

(4) The morphism w maps any new node identifier of R, which is not defined in L, onto a separate new node identifier in G', which is not defined in G.

(5) With $K := L \cap R$ the following property holds:
$v := \{ (x, y) \in u \mid x \text{ is identifier in } K \} = \{ (x, y) \in w \mid x \text{ is identifier in } K \}$,
i.e. u and w are identical with respect to the identifiers in the "gluing" structure K.

(6) There exists a graph $H \in \mathcal{G}(\Sigma)$: $G \setminus (u^*(L) \setminus v^*(K)) = H = G' \setminus (w^*(R) \setminus v^*(K))$,
i.e. the intermediate result H is not required to be scheme consistent. ■

Note that this definition of structure rewriting does not prohibit the selection of **homomorphic redices** with two identifiers o1 and o2 in L being mapped onto the same object in G, even if o1 belongs to R and o2 not. These **deleting/preserving conflicts** are resolved in favor of preserving objects (unlike to the SPO approach in [Löw 90]). Otherwise, relation v, the restriction of u to K, would no longer be a morphism from K to H. For readability reasons, these forms of redex selections may be prohibited and are prohibited in the special form of structure rewriting used within the language PROGRES.

It is a straightforward task to transform the definition of scheme preserving structure rewriting above into an **effective procedure** for the application of a rule r to a scheme consistent structure G (if consistency constraints in structure schemes as well as pre- and postconditions of r are formulas the verification of which is always possible):

(1) Assume a morphism u which selects a maximal redex in G for L w.r.t. preconditions AL.

(2) Build the "gluing" structure K as an intersection of L and R.

(3) Construct the intermediate structure $H := G \setminus (u^*(L) \setminus v^*(K))$ with v being the restriction of u onto the remaining object (set) identifiers in K.

(4) Then extend v to a relation w such that every new object identifier in R will be mapped onto a new object identifier which is neither defined in G nor in H.

(5) Thus we are able to define $G' := H \cup (w^*(R) \setminus v^*(K))$.

(6) Finally, we have to check whether G' is scheme consistent and respects postconditions AR.

It is easy to see that the construction above produces a structure G' such that

$$G \sim r \rightsquigarrow G' .$$

Note that the reevaluation of derived object properties, which are implicitly defined by sets of formulas within a scheme S, does not require explicit structure modifications (from a theoretical point of view). Furthermore, the steps (1) through (6) are equivalent to the construction of the diagram in fig. 6, the existence of which will be proved now:

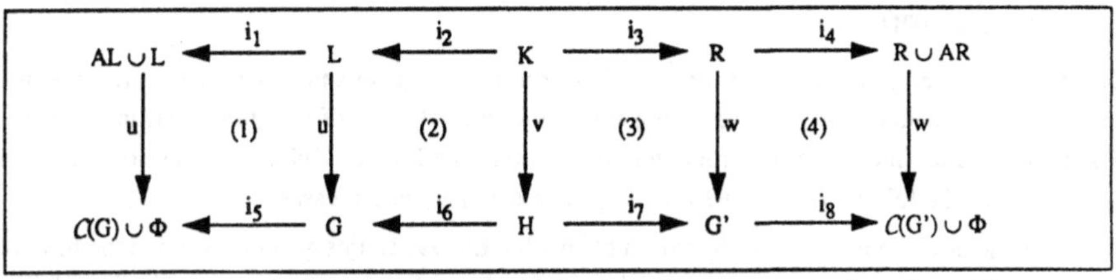


Fig. 6: Diagram for the application of a structure rewrite rule

**Proposition 5.4 Structure Rewriting is Construction of Diagrams.**

Assuming the terminology of definition 5.3 and with G ~ r ~> G' we are able to construct the diagram of fig. 6 (with $i_1, \ldots, i_8$ being inclusions) which has commuting subdiagrams (1) through (4).

**Proof:**

The existence of commuting subdiagrams (1) and (4) is guaranteed by def. 4.7. For proving the existence of commuting subdiagrams (2) and (3), we will show that step (5) above constructs a morphism v from K to H (with $i_2 \circ u = v \circ i_6$ and with $v \circ i_7 = i_3 \circ w$ being always valid, because v is a restriction of u or w onto the identifiers in K). To prove this, we start with def. 5.3, condition (1):

| | |
|---|---|
| $u: L \xrightarrow{\sim} G$ | ⇔ def. 4.3, prop. 4.6, and K ⊆ L |
| $u^*(K) \subseteq u^*(L) \subseteq G$ | ⇒ condition (5) of def. 5.3 |
| $v^*(K) \subseteq u^*(L) \subseteq G$ | ⇒ simple transformation |
| $v^*(K) \setminus (u^*(L) \setminus v^*(K)) \subseteq G \setminus (u^*(L) \setminus v^*(K))$ | ⇔ simple transformation |
| $v^*(K) \subseteq G \setminus (u^*(L) \setminus v^*(K))$ | ⇔ condition (6) of def. 5.3 |
| $v^*(K) \subseteq H$ | ⇔ def. 4.3 and prop. 4.6 |
| $v: K \xrightarrow{\sim} H$. ■ | |

**Def. 5.4.1 Direct Derivation.**

$\mathcal{R} \subseteq \mathcal{R}(\Sigma)$ is a set of structure rewrite rules and G, G' ∈ $\mathcal{G}$(S). G' is **direct derivable** from G via $\mathcal{R}$ (write: G ~ $\mathcal{R}$ ~> G') ⇔

$$\exists\, r \in \mathcal{R}: G \sim r \rightsquigarrow G'.$$

The transitive, reflexive closure of ~> is $\overset{*}{\rightsquigarrow}$ . ■

**Def. 5.4.2 Scheme Preserving Structure Rewriting System**

Σ is a signature and S ∈ $\mathcal{S}(\Sigma)$ a Σ-structure scheme. A triple gr = (S, $G_S$, $\mathcal{R}$) is a scheme preserving structure rewriting system ⇔

(1) $G_S \in \mathcal{G}(S)$, i.e. $G_S$ is a scheme consistent structure with respect to S.

(2) $\mathcal{R} \subseteq \mathcal{R}(\Sigma)$, i.e. $\mathcal{R}$ is a set of Σ-structure rewriting rules. ■

**Def. 5.4.3 Scheme Consistent Structure Language**

Let gr := (S, $G_S$, $\mathcal{R}$) be a scheme preserving structure rewriting system. The language generated by gr is defined as follows:

$$\mathcal{L}(gr) := \{ G \in \mathcal{G}(S) \mid G_S \overset{*}{\rightsquigarrow} G \}. \blacksquare$$

# 6 Conclusion

The definitions and propositions of the preceding sections constitute a **new logic based framework** for the formal treatment of graph rewriting systems as special cases of structure rewriting systems. The new calculus (including the programming constructs defined in [Schü 91b] but not explained here) contains the following structure rewriting approaches as special cases:

- the algebraic double pushout approach of [EH 86, Ehr 79] and its generalization in the form of so-called structure grammars, a special case of high-level replacement systems, in [EHKP 91],
- the programmed graph grammar approach of [Göt 88] (with the exception of certain forms of embedding rules which consist of very complex graph patterns[10]),
- the expression-oriented graph grammar approach of [Nag 79a] and the structured graph grammar approach of [KR 88] (without any restrictions),
- and finally the "programmed derivation of relational structures" approach in [MW 91] (if we omit the "pseudo parallel" interleaving operator "||" proposed there).

Furthermore, our formalism tries to close the gap between the "operation oriented" manipulation of data structures by means of rewrite rules and the "declaration oriented" description of data structures by means of logic based knowledge representation languages. In this way, both disciplines - graph grammar theory and mathematical logic theory - might be able to profit from each other:

- Structure (graph) rewrite rules might be a very comfortable and well-defined mechanism for manipulating knowledge bases (or deductive data bases; cf. [Gal 78, Ryb 87]), whereas
- many logic based techniques have been proposed for efficiently maintaining derived properties of data structures, solving constraint systems, and even for proving the correctness of certain kinds of data manipulations (see [NT 89, Hen 89, Tha 89, BMM 91]).

Being the first attempt to present such a new framework for a very general form of rewriting systems ([CMREL 91] and [Cou 91] both contain in-depth discussions of the relationships between sets of logical formulas and context-free hypergraph grammars), this paper raises a number of **open questions**:

(1) For which subset of our new approach are subdiagrams (2) and (3) of fig. 6 pushouts (beyond those subcases identified in [EHP 90])?

(2) Can we characterize a "useful" subset of rewrite rules and consistency constraints such that an effective proof procedure exists for the "are all derivable structures scheme consistent" problem (already considered in [Cou 91] for restricted forms of graph grammars and monadic second order logic formulas)?

(3) Can we develop a general procedure which transforms any set of postconditions into equivalent preconditions (the main problem are postconditions with references to implicitly derived graph properties which may be affected by the rewrite rule itself)?

The problems (2) and (3) above seem to be related to each other because we should be able to transform global consistency constraints into equivalent postconditions of individual rewrite rules (using techniques proposed in [BMM 91]). Having a procedure which translates postconditions into equivalent preconditions, problem (2) is reducible to the question whether these new preconditions are already derivable from the original sets of preconditions of rewrite rules.

---

10) Embedding based on complex graph patterns requires the usage of sets of tuples of object identifiers instead of sets of object identifiers only.

But note that the main motivation for the development of a new formalism was not to generate these problems but to provide us with a solid base for a precise description of our application oriented graph grammar language PROGRES. Therefore, we are currently spending most of our time not to find answers to the above mentioned questions but to realize an **integrated PROGRES programming environment**. Up to now, this environment has about 300.000 lines of code and offers a syntax-aided editor, an incrementally working type checker, and - not yet completed - an integrated interpreter as well as a cross-compiler to Modula-2 and C. A first (pre-)release of this environment and its underlying graph oriented database system GRAS are available as free software via anonymous ftp from ftp-server ftp.informatik.rwth-aachen.de in directories /pub/unix/PROGRES and /pub/unix/GRAS.

**Acknowledgments**: Thanks to the (two) reviewers of a preliminary form of this paper for a long list of helpful remarks and suggestions concerning the technical content of this paper and its presentation.

## References

[AA 86] Ausiello G., Atzeni P. (eds.): *Proc. Int. Conf. on Database Theory*, LNCS 243, Springer Press

[BMM 91] Bry F., Manthey R., Martens B.: *Integrity Verification in Knowledge Bases*, in: [Vor 91], 114-139

[CER 79] Claus V., Ehrig H., Rozenberg G.: *Proc. Int. Workshop on Graph-Grammars and Their Application to Computer Science and Biology*, LNCS 73, Springer Press (1979)

[Cou 91] Courcelle B.: *Graphs as Relational Structures: An Algebraic and Logical Approach*, in: [EKR 91], 238-252

[EH 86] Ehrig H., Habel A.: *Graph Grammars with Application Condition*, in: [RS 86], 87-100

[EHKP 91] Ehrig H., Habel A., Kreowski H.-J., Parisi-Presicce F.: *From Graph Grammars to High-Level Replacement Systems*, in: [EKR 91], 269-291

[Ehr 79] Ehrig H.: *Introduction to the Algebraic Theory of Graph Grammars (a Survey)*, in: [CER 79], 1-69

[EKR 91] Ehrig H., Kreowski H.-J., Rozenberg G.: *Proc. 4th Int. Workshop on Graph Grammars and Their Application to Computer Science*, LNCS 532, Springer Press (1991)

[Gal 78] Gallaire H.: *Logic and Data Bases*, Plenum Press (1978)

[Göt 88] Göttler H.: *Graphgrammatiken in der Softwaretechnik*, IFB 178, Springer Press (1988)

[Hen 89] Van Hentenryck P.: *Constraint Satisfaction in Logic Programming*, MIT Press (1989)

[KR 88] Kreowski H.-J., Rozenberg G.: *On Structured Graph Grammars: Part I and II*, Technical Report 3/88, University of Bremen, FB Mathematik/Informatik (1988)

[KSW 93] Kiesel N., Schürr A., Westfechtel B.: *GRAS, a Graph-Oriented Database System for (Software) Engineering Applications*, in: [LRJ 93], 272-286

[Löw 90] Löwe M.: *Algebraic Approach to Graph Transformation Based on Single Pushout Derivations*, TR No. 90/5, TU Berlin (1990)

[LRJ 93] Lee H.-Y., Reid Th.F., Jarzabek St. (eds.): *CASE '93 Proc. 6th Int. Workshop on Computer-Aided Software Engineering*, IEEE Computer Society Press (1993)

[MBJK 90] Mylopoulos J., Borgida A., Jarke M., Koubarkis M.: *Telos: a Language for Representing Knowledge about Information Systems*, ACM Transactions on Information Systems, Vol. 8, No. 4, acm Press (1990), 325-362

[Min 88] Minker J.: *Perspectives in Deductive Databases*, in: The Journal of Logic Programming, Elsevier Science Publ. (1988), 33-60

[MW 91] Maggiolo-Schettini A., Winkowski J.: *Programmed Derivations of Relational Structures*, in: [EKR 91], 582-598

[Nag 79a] Nagl M.: *Graph-Grammatiken*, Vieweg Press (1979)

[Nag 79b] Nagl M.: *A Tutorial and Bibliographical Survey on Graph Grammars*, in: [CER 79], 70-126

[Nag 86] Nagl M.: *Graph Technology Applied to a Software Project*, in: [RS 86], 303-322

[Naq 86a] Naqvi Sh.A.: *Some Extensions to the Closed World Assumption in Databases*, in: [AA 86], 341-348

[Naq 86b] Naqvi Sh.A.: *Negation as Failure for First-Order Queries*, in: Proc. 5th ACM SIGACT-SIGMOD Symp. on Principles of Database Systems, acm Press (1986), 114-122

[NS 91] Nagl M., Schürr A.: *A Specification Environment for Graph Grammars*, in: [EKR 91], 599-609

[NT 89] Naqvi Sh.A., Tsur Sh.: *Data and Knowledge Bases*, IEEE Computer Society Press (1989)

[RS 86] Rozenberg G., Salomaa A.: *The Book of L*, Springer Press (1986)

[Ryb 87] Rybinski H.: *On First-Order-Logic Databases*, in: ACM Transaction on Database Systems, Vol. 12, No.3, acm Press (1987), 325-349

[SB 91] Schmidt G., Berghammer R.: *Proc. Int. Workshop on Graph- Theoretic Concepts in Computer Science (WG '91)*, LNCS 570, Springer Press (1991)

[Schü 91a] Schürr A.: *PROGRES: A VHL-Language Based on Graph Grammars*, in: [EKR 91], 641-659

[Schü 91b] Schürr A.: *Operational Specifications with Programmed Graph Rewriting Systems: Formal Definitions, Applications, and Tools* (in German), Deutscher Universitätsverlag (1991)

[SZ 91] Schürr A., Zündorf A.: *Non-Deterministic Control Structures for Graph Rewriting Systems*, in: [SB 91], 48-62

[Tha 89] Thaise A. (ed.): *From Standard Logic to Logic Programming*, John Wiley & Sons Ltd. (1989)

[Vor 91] Voronkov A. (ed.): *Logic Programming*, LNCS 592, Springer Press (1991)

# Guaranteeing Safe Destructive Updates through a Type System with Uniqueness Information for Graphs

Sjaak Smetsers, Erik Barendsen, Marko van Eekelen, Rinus Plasmeijer

University of Nijmegen, Computing Science Institute
Toernooiveld 1, 6525 ED Nijmegen, The Netherlands
e-mail `sjakie@cs.kun.nl`, fax +31.80.652525.

**Abstract.** In this paper we present a type system for graph rewrite systems: *uniqueness typing*. It employs usage information to deduce whether an object is 'unique' at a certain moment, i.e. is only locally accessible. In a type of a function it can be specified that the function requires a unique argument object. The correctness of type assignment guarantees that no external access on the original object will take place in the future. The presented type system is proven to be correct. We illustrate the power of the system by defining an elegant quicksort algorithm that performs the sorting *in situ* on the data structure.

## 1 Introduction

Some operations on complex data structures (such as arrays) cannot be implemented efficiently without allowing a form of destructive updating. For convenience, we speak about those functions as 'destructively using' their arguments. In case of graph-like implementations of functional languages without any precautions, this destructive usage is dangerous: on the level of the underlying model of computation this appears when arguments are shared between two functions.

However, in some specific cases destructive updates are safe, e.g. when it is known that access on the original object is not necessary in the future. We call such an object (locally) 'unique'.

Sharing/update analysis is used to find spots where destructive updates are possible. However, some functions require that a destructive update can be done in all contexts in which the function is applied. Such updating functions are functions for file I/O, array manipulation, interfacing with existing FORTRAN or C libraries, window-based I/O and functions that require an efficient storage management (e.g. in situ sorting of a large data structure). This requirement can be explicitly specified via a type system. This paper presents a type system related to linear types: uniqueness types. The uniqueness type system is defined for graph rewrite systems. It employs usage information to deduce whether the uniqueness attribute can be assigned to a type for a subgraph. A type which has the uniqueness attribute is also called a unique type. For functions that require an object of unique type, the type system guarantees that no external access on the original object will be possible anymore. So, (depending on the use of the object in the function body) this information can be used to destructively update the unique object. A compiler can exploit uniqueness types by generating code that automatically updates unique arguments when possible. This has important consequences for the time and space behaviour of functional programs. The type system has been implemented for the lazy functional graph rewriting language Concurrent Clean. So far, it has been used for the implementation of arrays and of an efficient high-level library for screen and file I/O (see Achten *et al.* [1993]).

The structure of the paper is as follows: first graph rewrite systems are briefly introduced using standard terminology (Section 2). Then, a notion of typing is defined for graph rewrite systems in Section 3. Section 4 describes a use analysis that provides important information that is necessary to assign uniqueness attributes. How uniqueness attributes are assigned is defined in Section 5. The extension to algebraic type definitions is given in Section 6. The correctness of the type system is proven in Section 7. Section 8 illustrates how reasoning about programs with uniqueness types can be done, after which Section 9 discusses related work.

## 2 Graph rewriting

Term graph rewrite systems were introduced in Barendregt *et al.* [1987]. This section summarizes some basic notions for (term) graph rewriting as presented in Barendsen & Smetsers [1992].

### Graphs

The objects of our interest are directed graphs in which each node has a specific label. The number of outgoing edges of a node is determined by its label. In the sequel we assume that $\mathcal{N}$ is some basic set of *nodes* (infinite; one usually takes $\mathcal{N} = \mathbb{N}$), and $\Sigma$ is a (possibly infinite) set of *symbols* with *arity* in $\mathbb{N}$.

**Definition 1.** (i) A *labeled graph* (over $\langle \mathcal{N}, \Sigma \rangle$) is a triple

$$g = \langle N, symb, args \rangle$$

such that

(a) $N \subseteq \mathcal{N}$; $N$ is the set of *nodes* of $g$;
(b) $symb : N \to \Sigma$; $symb(n)$ is the *symbol* at node $n$;
(c) $args : N \to N^*$ such that $\mathrm{length}(args(n)) = \mathrm{arity}(symb(n))$.

Thus $args(n)$ specifies the outgoing edges of $n$. The $i$-th component of $args(n)$ is denoted by $args(n)_i$.

(ii) A *rooted graph* is a quadruple

$$g = \langle N, symb, args, r \rangle$$

such that $\langle N, symb, args \rangle$ is a labeled graph, and $r \in N$. The node $r$ is called the *root* of the graph $g$.

(iii) The collection of all finite rooted labeled graphs over $\langle \mathcal{N}, \Sigma \rangle$ is indicated by $\mathbb{G}$.

*Convention.* (i) $m, n, n', \ldots$ range over nodes; $g, g', h, \ldots$ range over (rooted) graphs.

(ii) If $g$ is a (rooted) graph, then its components are referred to as $N_g$, $symb_g$, $args_g$ (and $r_g$) respectively.

(iii) To simplify notation we usually write $n \in g$ instead of $n \in N_g$.

**Definition 2.** (i) A *path* in a graph $g$ is a sequence $p = (n_0, i_0, n_1, i_1, \ldots, n_\ell)$ where $n_0, n_1, \ldots, n_\ell \in g$ are nodes, and $i_0, i_1, \ldots, i_{\ell-1} \in \mathbb{N}$ are 'edge specifications' such that $n_{k+1} = args(n_k)_{i_k}$ for all $k < \ell$. In this case $p$ is said to be a *path from* $n_0$ *to* $n_\ell$ (notation $p : n_0 \leadsto n_\ell$).

(ii) Let $m, n \in g$. $m$ is *reachable from* $n$ (notation $n \leadsto m$) if $p : n \leadsto m$ for some path $p$ in $g$.

**Definition 3.** Let $g$ be a graph and $n \in g$. The *subgraph of* $g$ *at* $n$ (notation $g \mid n$) is the rooted graph $\langle N, symb, args, n \rangle$ where $N = \{m \in g \mid n \leadsto m\}$, and $symb$ and $args$ are the restrictions (to $N$) of $symb_g$ and $args_g$ respectively.

## Graph rewriting

This section introduces some notation connected with graph rewriting. For a complete operational description the reader is referred to the papers mentioned earlier.

Rewrite rules specify transformations of graphs. Each rewrite rule is represented by a special graph containing two roots. These roots determine the left-hand side (the *pattern*) and the right-hand side of the rule. Variables are represented by special 'empty nodes'. Let $R$ be some rewrite rule. A graph $g$ can be *rewritten* according to $R$ if $R$ is applicable to $g$, i.e. the pattern of $R$ *matches* $g$. A *match* $\mu$ is a mapping from the pattern of $R$ to a subgraph of $g$ that preserves the node structure. The combination of a rule and a match is called a *redex*. If a redex has been determined, the graph can be rewritten according to the structure of the right-hand side of the rule involved. This is done in three steps. Firstly, the graph is *extended* with an instance of the right-hand side of the rule. The connections from the new part with the original graph are determined by $\mu$. Then all references to the root of the redex are *redirected* to the root of the right-hand side. Finally all unreachable nodes are removed by performing *garbage collection*.

**Definition 4.** Let $\bot$ be a special symbol in $\Sigma$ with arity 0. Let $g$ be a graph.

(i) The set of *empty nodes* of $g$ (notation $g^\circ$) is the collection

$$g^\circ = \{n \in g \mid symb_g(n) = \bot\}.$$

(ii) The set of *non-empty nodes* (or *interior*) of g is denoted by $g^\bullet$. So $N_g = g^\circ \cup g^\bullet$.

(iii) $g$ is *closed* if $g^\circ = \emptyset$.

The objects on which computations are performed are closed graphs; the others are used as auxiliary objects, e.g. for defining graph rewrite rules.

**Definition 5.** (i) A *term graph rewrite rule* (or *rule* for short) is a triple $R = \langle g, l, r\rangle$ where $g$ is a (possibly open) graph, and $l, r \in g$ (called the *left root* and *right root* of $R$), such that

(a) $(g \mid l)^{\bullet} \neq \emptyset$;

(b) $(g \mid r)^{\circ} \subseteq (g \mid l)^{\circ}$.

(ii) If $symb_g(l) = \mathbf{F}$ then $R$ is said to be a *rule for* $\mathbf{F}$.

(iii) $R$ is *left-linear* if $g \mid l$ is a tree.

Here condition (1) expresses that the left-hand side of the rewrite rule should not be just a variable. Moreover condition (2) states that all variables occurring on the right-hand side of the rule should also occur on the left-hand side.

*Notation.* We will write $R \mid l$, $R \mid r$ for $g_R \mid l_R$, $g_R \mid r_R$ respectively.

**Definition 6.** Let $p, g$ be graphs. A *match* is a function $\mu : N_p \rightarrow N_g$ such that for all $n \in p^{\bullet}$

$$symb_g(\mu(n)) = symb_p(n),$$
$$args_g(\mu(n))_i = \mu(args_p(n)_i).$$

In this case we write $\mu : p \underset{\mathrm{m}}{\rightarrow} g$.

**Definition 7.** Let $g$ be a graph, and $\mathcal{R}$ a set of rewrite rules.

(i) An $\mathcal{R}$-*redex* in $g$ (or just *redex*) is a tuple $\Delta = \langle R, \mu\rangle$ where $R \in \mathcal{R}$, and $\mu : (R \mid l) \underset{\mathrm{m}}{\rightarrow} g$.

(ii) If $g'$ is the result of rewriting redex $\Delta$ in $g$ this will be denoted by $g \overset{\Delta}{\underset{\mathcal{R}}{\rightarrow}} g'$, or just $g \underset{\mathcal{R}}{\rightarrow} g'$.

(iii) Let $\Delta = \langle R, \mu\rangle$ be a redex. The *redex root* of $\Delta$ (notation $\mathrm{r}(\Delta)$) is defined by

$$\begin{aligned}\mathrm{r}(\Delta) &= \mu(r_R) \quad \text{if } r_R \in R \mid l,\\ &= r_R \qquad\quad \text{otherwise.}\end{aligned}$$

## Term graph rewrite systems

A collection of graphs and a set of rewrite rules can be combined into a (term) graph rewrite system. A special class of so-called orthogonal graph rewrite systems is the subject of further investigations.

**Definition 8.** (i) A *term graph rewrite system* (TGRS) is a tuple $\mathcal{S} = \langle \mathcal{G}, \mathcal{R}\rangle$ where $\mathcal{R}$ is a set of rewrite rules, and $\mathcal{G} \subseteq \mathbb{G}$ is a set of closed graphs which is closed under $\mathcal{R}$-reduction.

(ii) $\mathcal{S}$ is *left-linear* if each $R \in \mathcal{R}$ is left-linear.
(iii) $\mathcal{S}$ is *regular* if for each $g \in \mathcal{G}$ the $\mathcal{R}$-redexes in $g$ are pairwise disjoint.
(iv) $\mathcal{S}$ is *orthogonal* if $\mathcal{S}$ is both left-linear and regular.

It can be shown that for a large class of orthogonal TGRSs (the so-called *interference-free* systems) the Church-Rosser property holds (see Barendsen & Smetsers [1992]).

*Notation.* Let $\mathcal{S} = \langle \mathcal{G}, \mathcal{R} \rangle$ be a TGRS. $\Sigma_{\mathcal{S}}$ denotes symbols in $\Sigma$ that appear in $\mathcal{G}$ or in $\mathcal{R}$. The set of *function symbols* of $\mathcal{S}$ (notation $\Sigma_{\mathcal{F}}$) are those symbols for which there exist a rule in $\mathcal{R}$. Moreover, $\Sigma_{\mathcal{D}} = \Sigma_{\mathcal{S}} \backslash \Sigma_{\mathcal{F}}$ denotes the set of *data symbols* of $\mathcal{S}$.

# 3 Typing graphs

In this section we will define a notion of simple type assignment to graphs using a type system based on traditional systems for functional languages. The approach is similar to the one introduced in Bakel *et al.* [1992]. It is meant to illustrate the concept of 'classical' typing for graphs. In the next section a different typing system will be described.

**Definition 9.** Let $\mathbb{V}$ be a set of *type variables*, and $\mathbb{C}$ a set of *type constructors* with *arity* in $\mathbb{N}$. Write $\mathbb{C} = \mathbb{C}^0 \cup \mathbb{C}^1 \cup \ldots$ such that each $S \in \mathbb{C}^i$ has arity $i$.

(i) The set $\mathbb{T}$ of (*graph*) *types* is defined inductively as follows.

$$\begin{aligned} \alpha \in \mathbb{V} &\Rightarrow \alpha \in \mathbb{T}, \\ C \in \mathbb{C}^k, \sigma_1, \ldots, \sigma_k \in \mathbb{T} &\Rightarrow C(\sigma_1, \ldots, \sigma_k) \in \mathbb{T}, \\ \sigma, \tau \in \mathbb{T} &\Rightarrow \sigma \rightarrow \tau \in \mathbb{T}. \end{aligned}$$

(ii) The set $\mathbb{T}_{\mathrm{S}}$ of *symbol types* is defined as

$$\begin{aligned} \sigma \in \mathbb{T} &\Rightarrow \sigma \in \mathbb{T}_{\mathrm{S}}, \\ \sigma_1, \ldots, \sigma_k, \tau \in \mathbb{T} &\Rightarrow (\sigma_1, \ldots, \sigma_k) \rightarrow \tau \in \mathbb{T}_{\mathrm{S}}. \end{aligned}$$

The *arity* of a symbol type is 0 if it is introduced by the first rule. Otherwise, the arity is $k$.

*Convention.* In the sequel, $\alpha, \beta, \alpha_1, \ldots$ range over type variables; $\sigma, \tau, \tau_1, \ldots$ range over (function) types.

**Definition 10.** (i) A *substitution* is a function $* : \mathbb{V} \rightarrow \mathbb{T}$.

(ii) Let $*$ be a substitution, and $\sigma \in \mathbb{T}_{\mathrm{S}}$. The result of applying $*$ to $\sigma$ (notation $\sigma^*$) is inductively defined as follows.

$$\begin{aligned} \alpha^* &= *(\alpha), \\ (C(\sigma_1, \ldots, \sigma_k))^* &= C(\sigma_1^*, \ldots, \sigma_k^*), \\ (\sigma \rightarrow \tau)^* &= \sigma^* \rightarrow \tau^*, \\ ((\sigma_1, \ldots, \sigma_k) \rightarrow \tau)^* &= (\sigma_1^*, \ldots, \sigma_k^*) \rightarrow \tau^*. \end{aligned}$$

(iii) $\sigma$ is an *instance* of $\tau$ (notation $\sigma \subseteq \tau$) if there exists a substitution $*$ such that $\tau^* = \sigma$.

(iv) $\sigma$ and $\tau$ are *isomorphic* if $\tau^{*_1} = \sigma$ and $\sigma^{*_2} = \tau$ for some substitutions $*_1, *_2$. We will usually identify isomorphic types, i.e. types that result from each other by consistent renaming of type variables. That is, we regard types as type *schemes*.

## Applicative graph rewrite systems

In TGRS's symbols have a fixed arity. Consequently, it is impossible to use functions as arguments or to yield functions as a result. However, higher order functions can be *modeled* in TGRS's by associating to each symbol $\mathbf{S}$ with arity($\mathbf{S}$) $\geq 1$ a 0-ary constructor $\mathbf{S_0}$, and by adding a special *apply rule* (with function symbol $\mathbf{Ap}$) to the TGRS for supplying these new constructors with arguments.

For example, Combinatory Logic (CL) expressed by

$$\begin{aligned} \mathbf{S}\,xyz &\to xz(yz) \\ \mathbf{K}\,xy &\to x \\ \mathbf{I}\,x &\to x \end{aligned}$$

can be modeled in the following TGRS (using a self-explanatory linear notation).

$$\begin{aligned} \mathbf{S}(x,y,z) &\to \mathbf{Ap}(\mathbf{Ap}(x,z),\mathbf{Ap}(y,z)) \\ \mathbf{K}(x,y) &\to x \\ \mathbf{I}(x) &\to x \\ \mathbf{Ap}(\mathbf{Ap}(\mathbf{Ap}(\mathbf{S_0},x),y),z) &\to \mathbf{S}(x,y,z) \\ \mathbf{Ap}(\mathbf{Ap}(\mathbf{K_0},x),y) &\to \mathbf{K}(x,y) \\ \mathbf{Ap}(\mathbf{I_0},x) &\to \mathbf{I}(x) \end{aligned}$$

Note that each new constructor symbol introduces a new rule for $\mathbf{Ap}$.

**Definition 11.** Let $\mathcal{S} = \langle \mathcal{G}, \mathcal{R} \rangle$ be a TGRS.

(i) Let $\mathbf{S} \in \Sigma_\mathcal{S}$ with arity $\geq 1$. The above symbol $\mathbf{S_0} \in \Sigma_\mathcal{D}$ is called the *Curry variant* of $\mathbf{S}$.

(ii) The set $\Sigma_\mathcal{C} \subseteq \Sigma_\mathcal{D}$ denotes the set of Curry variants of $\Sigma_\mathcal{D}$ with arity $\geq 1$.

(iii) We say that $\mathcal{S}$ is *Curry complete* if $\mathcal{R}$ contains an $\mathbf{Ap}$-rule for each symbol $\mathbf{S}$ with arity $\geq 1$, as described above, and no other $\mathbf{Ap}$-rules.

(iv) Let $R \in \mathcal{R}$. The *principal node* of $R$ (notation p($R$) is $l_R$ if $symb(l_R) \neq$ $\mathbf{Ap}$; otherwise it is the node containing $\mathbf{S_0}$.

*Assumption.* From now on we assume that all TGRS's are Curry complete.

### Assigning types to symbols

In the rest of this section we describe how types can be assigned to graphs given a fixed type assignment to the (function and data) symbols by a so called *environment*.

Currying imposes a restriction on type environments, that is to say, the type of a Curry variant $\mathbf{S_0}$ should be related to the type assigned to $\mathbf{S}$. We also assume a standard type for the symbol **Ap** to be declared.

**Definition 12.** (i) Let $\sigma = (\sigma_1, \ldots, \sigma_k) \to \tau$ be a function type. The *curried version* of $\sigma$ (notation $\sigma^C$) is

$$\sigma^C = \sigma_1 \to (\sigma_2 \to (\cdots(\sigma_k \to \tau)\cdots)).$$

(ii) A (*type*) *environment* for $\mathcal{S}$ is a function $\mathcal{E} : \Sigma_{\mathcal{S}} \to \mathbb{T}$ such that
(a) $\mathcal{E}(\bot) = \alpha$,
(b) $\mathcal{E}(\mathbf{Ap}) = (\alpha \to \beta, \alpha) \to \beta$,
(c) $\mathcal{E}(\mathbf{S_0}) = (\mathcal{E}(\mathbf{S}))^C$.

### Algebraic data types

We consider new (basic) types to be introduced by so-called *algebraic type definitions*. In these type definitions a (possibly infinite) set of *constructor* symbols is associated with each new type $T$.

The general form of an algebraic type definition for $T$ is

$$\begin{aligned} T\,\boldsymbol{\alpha} &= C_1\,\boldsymbol{\sigma_1} \\ &= C_2\,\boldsymbol{\sigma_2} \\ &= \ldots \end{aligned}$$

Here $\boldsymbol{\alpha} \in \mathbb{V}$, and $\boldsymbol{\sigma_i} \in \mathbb{T}$ such that the variables appearing in $\boldsymbol{\sigma_i}$ are contained in $\boldsymbol{\alpha}$. Moreover, we assume that each $C_i$ is a fresh constructor symbol. E.g., the type of lists could be obtained as follows.

$$\begin{aligned} \mathrm{List}(\alpha) &= \mathbf{Cons}(\alpha, \mathrm{List}(\alpha)) \\ &= \mathrm{Nil} \end{aligned}$$

A set $\mathcal{A}$ of algebraic type definitions induces a type environment $\mathcal{E}_{\mathcal{A}}$ for all constructors introduced by $\mathcal{A}$. More specifically, Let $C_i$ be the $i^{th}$ constructor defined by some algebraic type $T$. The $\mathcal{E}_{\mathcal{A}}$ type of $C_i$ is

$$\mathcal{E}_{\mathcal{A}}(C_i) = \boldsymbol{\sigma_i} \to T\,\boldsymbol{\alpha}.$$

*Convention.* Let $\mathcal{A}$ be a set of type definitions. $\Sigma_{\mathcal{A}}$ denotes the constructor symbols that are defined via some definition in $\mathcal{A}$.

*Assumption.* In the sequel we will assume that all constructors in $\mathcal{S}$ that are not the curried variant of some other symbol, are introduced by an algebraic type definition (i.e. $\Sigma_{\mathcal{D}} \backslash \Sigma_C \subseteq \Sigma_{\mathcal{A}}$.)

**Assigning types to graphs**

**Definition 13.** Let $g = \langle N, symb, args \rangle$ be a graph.

(i) A *type assignment* to $g$ (or $g$-*typing*) is a function $\mathcal{T} : N \to \mathbb{T}$.

(ii) Let $\mathcal{T}$ be a $g$-typing, and $n \in g$. The *function type* of $n$ according to $\mathcal{T}$ (notation $\mathcal{F}_{\mathcal{T}}(n)$) is defined as

$$\mathcal{F}_{\mathcal{T}}(n) = (\mathcal{T}(n_1), \ldots, \mathcal{T}(n_l)) \to \mathcal{T}(n)$$

where $l = \text{arity}(symb(n))$, and $n_i = args(n)_i$.

(iii) Let $\mathcal{E}$ be an environment. $\mathcal{T}$ is a $g$-*typing according to* $\mathcal{E}$ if for each $n \in g$ there exists a substitution $*$ such that

$$\mathcal{F}_{\mathcal{T}}(n) = \mathcal{E}(symb(n))^*.$$

*Example 1.* Let $\mathcal{E}$ be an environment containing the following type declarations.

$$\begin{array}{ll} \mathbf{F} & : \text{List}(\beta) \to \beta, \\ \mathbf{Cons} & : (\alpha, \text{List}(\alpha)) \to \text{List}(\alpha), \\ \mathbf{Nil} & : \text{List}(\alpha), \\ \mathbf{3} & : \text{INT}. \end{array}$$

Below, a graph and its typing according to $\mathcal{E}$ are indicated.

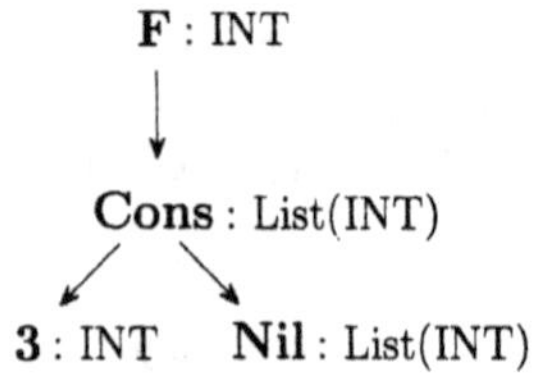


**Definition 14.** Let $\mathcal{S} = \langle \mathcal{G}, \mathcal{R} \rangle$ be a TGRS, and $\mathcal{A}$ a set of algebraic type definitions. Furthermore, let $\mathcal{E}$ be an environment for $\mathcal{S}$ .

(i) $R \in \mathcal{R}$ is *typable according to* $\mathcal{E}$ if there exist an $g_R$-typing $\mathcal{T}$ (according to $\mathcal{E}$) that meets the following requirements.

(a) $\mathcal{T}(l) = \mathcal{T}(r)$.

(b) $\mathcal{F}_{\mathcal{T}}(\text{p}(R)) = \mathcal{E}(symb(\text{p}(R)))$.

(ii) $\mathcal{R}$ is *typable* if there exists an environment $\mathcal{E}$ extending $\mathcal{E}_{\mathcal{A}}$ such that each $R \in \mathcal{R}$ is typable according to $\mathcal{E}$.

Condition (2) states that the left root node should be typed exactly with the type assigned to the root symbol by the environment. This contrasts the requirement for applicative occurrences of the function symbol.

Notice that the latter condition also provides that the abovementioned way of typing rewrite rules is essentially the same as the Mycroft type assignment system for the lambda calculus, see Mycroft [1981]. A Milner-like type assignment system (see Milner [1978]) can be obtained by stating this condition for *all* occurrences of a symbol $\mathbf{F}$ in the rule for $\mathbf{F}$.

It is possible to formulate conditions under which typing is preserved during reduction; cf. Bakel *et al.* [1992]. We will not go into this here.

# 4 Usage analysis

A first approach to a classification of 'unique' access to nodes in a graph is to count the references to each node. In practice, however, a more refined analysis is often possible. This can be achieved by taking into account the evaluation order dictated by a specific reduction strategy. This requires a nonlocal analysis of dependency of nodes in a graph. The idea is that multiple references to a node are harmless if one knows that only one of them remains at the moment of evaluation.

E.g. the standard evaluation of a conditional statement

$$\textbf{If } c \textbf{ Then } t \textbf{ Else } e$$

causes first the evaluation of the $c$ part, and subsequently evaluation of either $t$ or $e$, but not both. Hence, a single access to a node $n$ in $t$ combined with a single access to $n$ in $e$ would overall still result in a 'unique' access to $n$. It is important to note that this property only holds if execution proceeds according to the chosen strategy; it may be disturbed if one allows reduction of arbitrary redexes.

Firstly, we introduce a classification principle for function arguments. The usage analysis presented in the rest of this section will be parametric in the chosen classification.

*Assumption.* Let $\mathcal{S}$ be a TGRS.

(i) Let $\mathbf{F} \in \Sigma_{\mathcal{F}}$, say with arity $l$. Assume that $\{1, \ldots, l\}$ is divided into $k+1$ disjoint 'argument classes'

$$P, A_1, \ldots, A_k.$$

The intended meaning is that arguments occurring in $P$ are evaluated before any other argument (*primary arguments*) whereas $A_1, \ldots, A_k$ are 'alternative' groups of *secundary arguments*: during the actual evaluation, arguments belonging to different groups are never evaluated both. Furthermore, references via primary arguments to the graph are released before the graph is accessed via one of the secundary arguments.

(ii) A symbol is called *simple* if it has only primary arguments.

(iii) Each data symbol is simple.

**Remark 15.** We assume that the argument classification is consistent with each reduction rule, i.e. the way the arguments of a left-hand side are passed to functions in the corresponding right-hand side does not conflict with the respective argument classifications.

We will now describe a sharing analysis based on the above argument classification. First the argument dependency of functions is translated into dependency relations on paths and references in graphs.

**Definition 16.** Let $g \in \mathbb{G}$.

(i) The set of *references* of $g$ (notation $Ref_g$) is defined by

$$Ref_g = \{(m, i) \mid m \in g, i \leq \text{arity}(symb_g(m))\}.$$

(ii) Let $n \in g$. The set of *accesses* of $n$ (notation $acc(n)$) is

$$acc(n) = \{(m, i) \in Ref_g \mid args_g(m)_i = n\}.$$

We will now define dependency relations $\sim$ and $\lhd$.

**Definition 17.** (i) For each symbol $\mathbf{S}$ as above, and $i, j \leq l$, write $i \sim_{\mathbf{S}} j$ if $i, j$ belong to the same argument class of $\mathbf{S}$. Moreover, $i \lhd_{\mathbf{S}} j$ if $i \in P$ and $j \notin P$. The relation $\precsim_{\mathbf{S}}$ is the union of $\lhd_{\mathbf{S}}$ and $\sim_{\mathbf{S}}$.

(ii) Let $p$ be a nonempty path; say $(n, i)$ is its first reference. Then $i$ is called the *index* of $p$ (notation $\lfloor p \rfloor$).

(iii) Let $g \in \mathbb{G}$. The relation $\precsim$ is defined on nonempty paths in $g$ starting with the same node $n$, by

$$p \precsim q \ \Leftrightarrow\ \lfloor p \rfloor \precsim_{symb_g(n)} \lfloor q \rfloor.$$

**Definition 18.** Let $g \in \mathbb{G}$.

(i) Let $p = (n_0, i_0, \ldots, n_\ell)$ and $p' = (n'_0, i'_0, \ldots, n'_{\ell'})$ be paths. Then $p$ and $p'$ *join* (notation $p \wedge p'$) if $p, p'$ start in the same node and are distinct elsewhere. More precisely, $p \wedge p'$ if

$$\begin{aligned} n_0 &= n'_0, \\ n_{i+1} &\neq n'_{j+1} \quad \text{for all } i < \ell,\ j < \ell'. \end{aligned}$$

Note that in particular for paths $p$ of length 0 one has $p \wedge p$.

(ii) Let $p$ be a path in $g$, and $a \in Ref_g$. Then $p$ is *extendible* with $a$ if $p : n \rightsquigarrow m$ and $a = (m, i)$ for some $n, m, i$. The *extension* of $p$ with $a$ (defined in the obvious way) is denoted by $p * a$.

(iii) By $p \underset{a,a'}{\wedge} p'$ we will denote that $p \wedge p'$ and $p$, $p'$ are extendible with $a$, $a'$ respectively.

**Definition 19.** The relation $\precsim$ induces a relation on $Ref_g$, as follows.

$$a \precsim a' \ \Leftrightarrow\ p * a \precsim p' * a' \text{ for some acyclic } p, p' \text{ with } p \underset{a,a'}{\wedge} p'.$$

**Definition 20.** A *critical path combination* is a quadruple $p, a, p', a'$ such that $p \underset{a,a'}{\wedge} p'$, the paths $p, p'$ are acyclic, $a \neq a'$, and $\text{arg}(a) = \text{arg}(a')$.

Suppose $p \underset{a,a'}{\wedge} p'$ is a critical path combination. If $p * a \precsim p' * a'$, then the reference $a$ to $\text{arg}(a)$ might be used before $a'$ (in the $\lhd$ case) or $a, a'$ might be used in any order. The idea is now that $\text{arg}(a)$ is not allowed to be used destructively via $p * a$. This will be indicated by a suitable *labeling* of references with *use attributes* $\odot$ (for 'write use allowed') or $\otimes$ ('read access only').

A simple approach would label the reference $a$ above with $\otimes$ (see example 2). However, the usage information considered here is only important for *function* applications, in particular for parts of the graph matching the left-hand side of a rewrite rule. Since we consider systems with patterns (containing data nodes) we can be more liberal: in the above case it is sufficient that $p * a$ contains a reference labeled $\otimes$ anywhere in its 'data tail', to be made explicit below. The typing system will be such that this suffices to prevent destructive use of arg($a$) via the indicated path.

**Definition 21.** (i) The set of *use attributes* is $U = \{\odot, \otimes\}$.

(ii) Let $use : Ref_g \to U$ be a labeling. A path $p$ in $g$ is *marked* if there exist paths $p_1, p_2$ and a reference $a$ such that $p = p_1 * a * p_2$, $p_2$ is a data path, and $use(a) = \otimes$.

(iii) A labeling $use$ is a *marking* for $g$ if for each critical path combination $p \underset{a,a'}{\wedge} p'$ one has

$$p * a \underset{\sim}{\trianglelefteq} p' * a' \;\Rightarrow\; p * a \text{ is marked.}$$

There are two important examples of marking functions.

*Example 2.* Let $g$ be a graph.

(i) 'Last reference' marking is done as follows. Define $use^{\ell}$ as follows. Let $n \in g$. Then for each $a \in acc_g(n)$

$$\begin{aligned} use^{\ell}(a) &= \otimes \quad \text{if } a \underset{\sim}{\trianglelefteq} a' \text{ for some } a' \in acc_g(n) \text{ with } a' \neq a, \\ &= \odot \quad \text{otherwise.} \end{aligned}$$

Note that this definition completely specifies the function $use^{\ell}$.

(ii) Straightforward reference counting is done by considering each symbol as simple and performing last reference marking. More directly, this labeling is obtained by setting

$$\begin{aligned} use^{\mathrm{rc}}(a) &= \odot \quad \text{if arg}(a) \text{ has reference count } 1, \\ &= \otimes \quad \text{otherwise.} \end{aligned}$$

Using the standard classification of arguments of the conditional **IF**, and no specific assumptions about other symbols, the above two examples give the following markings of the displayed graph.

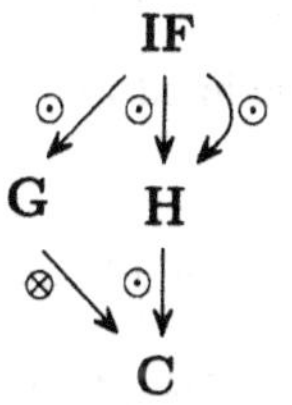


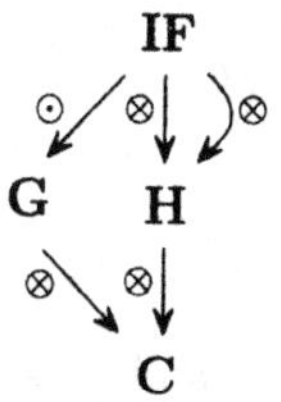


Now we can formulate which redexes are allowed to be contracted, in terms of the use function.

**Definition 22.** (i) Let $g \in \mathbb{G}$, and $m, n \in g$. Then $m$ is *local for* $n$ (*in* $g$) if

$$\forall p : r_g \rightsquigarrow m \;\; [n \in p].$$

(ii) Let $\Delta = \langle R, \mu \rangle$ be a redex in $g$. We say that $\Delta$ is *applicable* if for all $i$

$$use_g(\mu(l))_i = \odot \;\Rightarrow\; args_g(\mu(l))_i \text{ is local for } \mu(l).$$

The intention is that at least the redexes chosen by the strategy are applicable.

## 5 Uniqueness typing

### Uniqueness types

The use analysis described so far only takes the reduction strategy into account; not the particular structure of the rewrite rules. The use attributes of arguments may change during reduction, e.g. the $\odot$ attribute of a certain argument may change into a $\otimes$ after is redex has been contracted.

However, for a function $F$ that destructively uses one of its arguments it should be guaranteed that at the moment $F$ is evaluated the argument has a $\odot$ attribute. One way to ensure this is to require that this property holds at the moment the application of $F$ is built and that is remains valid during reduction.

The aim of the rest of this paper is to present a 'type system' in which the above-mentioned analysis can be performed.

The fact that a function may use one or more of its arguments destructively is expressed in its 'uniqueness type'. The syntax of these types is given in the following definition.

**Definition 23.** (i) The set $\mathbb{U}$ of *uniqueness types* is defined inductively by

$$\begin{aligned} \bullet, \times &\in \mathbb{U}, \\ u, v \in \mathbb{U} \;\Rightarrow\; u &\overset{\times}{\rightarrow} v \in \mathbb{U}, \\ u &\overset{\bullet}{\rightarrow} v \in \mathbb{U} \end{aligned}$$

(ii) The set $\mathbb{U}^{\bullet}$ of *unique types* is defined by

$$\mathbb{U}^{\bullet} = \{u \in \mathbb{U} \mid u = \bullet \text{ or } u = v \overset{\bullet}{\rightarrow} w \text{ for some } v, w \in \mathbb{U}\}.$$

Moreover, $\mathbb{U}^{\times} = \mathbb{U} \backslash \mathbb{U}^{\bullet}$.

(iii) The set $\mathbb{U}_S$ of *uniqueness symbol types* is defined as

$$\begin{aligned} u \in \mathbb{U} &\;\Rightarrow\; u \in \mathbb{U}_S, \\ u_1, \ldots, u_k, v \in \mathbb{U} &\;\Rightarrow\; (u_1, \ldots, u_k) \rightarrow v \in \mathbb{U}_S. \end{aligned}$$

The constants $\bullet$ and $\times$ represent 'unique use' and 'potentially multiple use' respectively. The arrows are annotated to distinguish unique function objects from unique objects without specified structure, and nonunique function objects from general nonunique objects. In the following example this will be illustrated.

*Example 3.* Suppose **Upd** denotes a binary function which destructively updates its first argument with its second argument. So, the intended $\mathbb{U}$-type of **Upd** is something of the form $(\bullet, \times) \rightarrow u$. It is natural to require that the uniqueness of the updated object is propagated. Thus one arrives at the following type for **Upd**.

$$\mathbf{Upd} : (\bullet, \times) \rightarrow \bullet$$

A partial application of **Upd** to some unique expression $g$ results in a function $\mathbf{Ap}(\mathbf{Upd_0}, g)$ that may not be copied. For, if copying would be allowed, then each of the applications of a copy of the function would be allowed to update the first argument $g$ destructively, as is illustrated by the expression $\mathbf{G}(\mathbf{Ap}(\mathbf{Upd_0}, g), h)$ assuming the rule

$$\mathbf{G}(f, x) \rightarrow \mathbf{Pair}(\mathbf{Ap}(f, x), \mathbf{Ap}(f, x)),$$

which is obviously unwanted.

In our type system the $\mathbb{U}$-type of the above expression $\mathbf{Ap}(\mathbf{Upd_0}, g)$ will be $\times \stackrel{\bullet}{\rightarrow} \bullet$ which will prevent it from being copied.

However, in any context in which a nonunique nonfunctional $\mathbb{U}$-type is expected it is harmless to offer a unique object. This gives rise to a subtype hierarchy specifying which types are convertible (can be *coerced*) to other types. These coercions are defined as an ordering on $\mathbb{U}$. They are not only depending on the demanded and offered types of the context but also on the way the offered object is accessed. If the use information of graphs is not taken into account, some graphs are wrongly accepted. For this reason we define a coercion relation that also depends on the use value of the reference via which the corresponding part of the graph is accessed.

**Definition 24.** The orderings $\leq^{\odot}$ and $\leq^{\otimes}$ on $\mathbb{U}$ are defined as follows.

(i) Coercions via $\odot$-references are generated by

$$\begin{aligned} \bullet &\leq^{\odot} \bullet, \\ \times &\leq^{\odot} \times, \\ \bullet &\leq^{\odot} \times, \\ u_1 \leq^{\odot} u_2, v_1 \leq^{\odot} v_2 \;\Rightarrow\; u_2 \stackrel{\bullet}{\rightarrow} v_1 &\leq^{\odot} u_1 \stackrel{\bullet}{\rightarrow} v_2, \\ u_2 \stackrel{\times}{\rightarrow} v_1 &\leq^{\odot} u_1 \stackrel{\times}{\rightarrow} v_2. \end{aligned}$$

(ii) Coercions via $\otimes$-references are the following.

$$\begin{aligned} \times &\leq^{\otimes} \times, \\ \bullet &\leq^{\otimes} \times, \\ u_1 \leq^{\odot} u_2, v_1 \leq^{\odot} v_2 \;\Rightarrow\; u_2 \stackrel{\times}{\rightarrow} v_1 &\leq^{\otimes} u_1 \stackrel{\times}{\rightarrow} v_2. \end{aligned}$$

Since we do not have type variables the notion of type instance has to be adjusted slightly. Intuitively, a type $u$ is an instance of a type $v$ if $u$ has 'more structure' than $v$. This is made precise in the following definition.

**Definition 25.** The relation $\subseteq$ on $\mathbb{U}$ is defined as:

$$\begin{aligned} \bullet \subseteq \bullet, &\quad u \overset{\bullet}{\rightarrow} v \subseteq \bullet, \\ \times \subseteq \times, &\quad u \overset{\times}{\rightarrow} v \subseteq \times, \\ u_1 \subseteq u_2, v_1 \subseteq v_2 \;\Rightarrow &\quad u_1 \overset{\bullet}{\rightarrow} v_1 \subseteq u_2 \overset{\bullet}{\rightarrow} v_2, \\ &\quad u_1 \overset{\times}{\rightarrow} v_1 \subseteq u_2 \overset{\times}{\rightarrow} v_2. \end{aligned}$$

If $u \subseteq v$ we say that $u$ is an ($\mathbb{U}$-*type*) *instance* of $v$.

## Currying

As we have seen, in some cases it can be dangerous to copy references to functions. To prevent a 'dangerous' function from being copied it is distinguished from 'safe' functions by typing it with an arrow type supplied with a $\bullet$ attribute. The observation that once a symbol has been applied to a unique argument it may not be copied anymore (see example 3) leads to the following Currying rule.

**Definition 26.** (i) Let $u \in \mathbb{U}$. The *uniqueness attribute* of $u$ (notation $[u]$ is defined as follows.

$$\begin{aligned} [u] &= \bullet, \quad \text{if } u \in \mathbb{U}^{\bullet} \\ &= \times, \quad \text{if } u \notin \mathbb{U}^{\bullet}. \end{aligned}$$

(ii) For $\mathbf{u} = (u_1, \ldots, u_k)$ and $j \leq k$ the *cumulative uniqueness attribute up to* $j$ (notation $[\mathbf{u}]_j$) is defined by

$$\begin{aligned} [\mathbf{u}]_j &= \bullet \;\; \text{if } [u_i] = \bullet \text{ for some } i \leq j, \\ &= \times \;\; \text{otherwise}. \end{aligned}$$

(iii) Let $u = (u_1, \ldots, u_k) \rightarrow v$. The set of *curried versions* of $u$ (notation $u^{\mathcal{C}}$) is

$$\begin{aligned} u^{\mathcal{C}} = \{ & u_1 \overset{\times}{\rightarrow} (u_2 \overset{[\mathbf{u}]_1}{\rightarrow} \cdots (u_k \overset{[\mathbf{u}]_{k-1}}{\rightarrow} v) \cdots), \\ & u_1 \overset{\bullet}{\rightarrow} (u_2 \overset{[\mathbf{u}]_1}{\rightarrow} \cdots (u_k \overset{[\mathbf{u}]_{k-1}}{\rightarrow} v) \cdots) \}. \end{aligned}$$

The effect of applying a (possibly curried) function to a unique argument is that the result of the application itself becomes unique. One could say that uniqueness information 'propagates upwards'.

The correspondence between a symbol (with arity $\geq 1$) and its Curry variant is given by that **Ap** rule. In contrast to the (ordinary) type system presented in section 3, **Ap** can be used with different $\mathbb{U}$ which are *not* instances of one type. To make such 'generic' functions possible we allow the type environment to contain more than one type for each symbol.

**Definition 27.** An (*applicative*) *uniqueness type environment* is a function $\mathcal{E}$ : $\Sigma \rightarrow \wp(\mathbb{U})$ such that
(1) $\mathcal{E}(\bot) = \{\bullet, \times\}$,

(2) $\mathcal{E}(\mathbf{Ap}) = \{(\times \xrightarrow{\times} \times, \times) \to \times, (\bullet \xrightarrow{\times} \times, \bullet) \to \times,$
$(\times \xrightarrow{\times} \bullet, \times) \to \bullet, (\bullet \xrightarrow{\times} \bullet, \bullet) \to \bullet,$
$(\times \xrightarrow{\bullet} \times, \times) \to \times, (\bullet \xrightarrow{\bullet} \times, \bullet) \to \times,$
$(\times \xrightarrow{\bullet} \bullet, \times) \to \bullet, (\bullet \xrightarrow{\bullet} \bullet, \bullet) \to \bullet\},$

(3) $\mathcal{E}(\mathbf{S_0}) \subseteq (\mathcal{E}(\mathbf{S}))^{\mathrm{C}}$.

Here $A^{\mathrm{C}} = \{a^{\mathrm{C}} \mid a \in A\}$.

**Assigning uniqueness types to graphs**

Assigning $\mathbb{U}$-types to graphs can be done in two ways. The first way is comparable to standard type assignment (section 3). In the second way, the use attributes of the graph as well as coercions are taken into account.

**Definition 28.** Let $g = \langle N, symb, args\rangle$ be a graph, and $\mathcal{E}$ be an environment. Furthermore, let $\mathcal{U} : N \to \mathbb{U}$.

(i) Let $n \in g$. The *function type* of $n$ (notation $\mathcal{F}_{\mathcal{U}}(n)$) is

$$\mathcal{F}_{\mathcal{U}}(n) = (\mathcal{U}(n_1), \ldots, \mathcal{U}(n_l)) \to \mathcal{U}(n),$$

where $l = \text{arity}(symb(n))$, and $n_i = args(n)_i$.

(ii) $\mathcal{U}$ is an *uniqueness typing for $g$ according to $\mathcal{E}$* if for each $n \in g$ there exists $u \in \mathcal{E}(symb(n))$ such that

$$\mathcal{F}_{\mathcal{U}}(n) \subseteq u.$$

(iii) Let *use* be the function that supplies $g$ with use attributes. $\mathcal{U}$ is an *weighted uniqueness typing for $g$ according to $\mathcal{E}$* if for each $n \in g$ there exist $u \in \mathcal{E}(symb(n))$ and $v_1, \ldots, v_k \in \mathbb{U}$ such that

$$\mathcal{U}(n_i) \leq^{u_i} v_i,$$
$$(v_1, \ldots, v_k) \to \mathcal{U}(n) \subseteq u,$$

where $n_i = args(n)_i$, and $u_i = use(n)_i$ for $i \leq k = \text{arity}(symb(n))$.

(iv) If $\mathcal{U}$ is a (weighted) uniqueness typing for $g$, then the *type of $g$* (notation $\mathcal{U}(g)$) is simply $\mathcal{U}(r_g)$.

**Definition 29.** Let $\mathcal{S} = \langle \mathcal{G}, \mathcal{R}\rangle$ be a TGRS, and $\mathcal{A}$ a set of algebraic type definitions. Furthermore, let $\mathcal{E}$ be an environment.

(i) $R \in \mathcal{R}$ is *uniqueness-typable* (according to $\mathcal{E}$) if for each $u \in \mathcal{E}(symb(l))$ there exist a function $\mathcal{U} : g_R \to \mathbb{U}$ such that

(a) $\mathcal{U}$ is a uniqueness typing for $R \mid l$,
(b) $\mathcal{U}$ is a weighted uniqueness typing for $R \mid r$,
(c) $\mathcal{U}(r) \leq^{\odot} \mathcal{U}(l)$,
(d) $\mathcal{F}_{\mathcal{U}}(\mathrm{p}(R)) = u$.

Such an $\mathcal{U}$ is called a *uniqueness typing* for $R$.

(ii) $\mathcal{R}$ is *uniqueness-typable* if there exists an environment $\mathcal{E}$ extending $\mathcal{E}_{\mathcal{A}}$, such that each $R \in \mathcal{R}$ is uniqueness-typable according to $\mathcal{E}$.

(iii) $\mathcal{S}$ is *uniqueness-typable* if there exists an uniqueness type environment $\mathcal{E}$ extending $\mathcal{E}_{\mathcal{A}}$ such that each $R \in \mathcal{R}$ as well as each $g \in \mathcal{G}$ is uniqueness-typable according to $\mathcal{E}$.

# 6 Algebraic type definitions

Since one allows pattern matching in function definitions, it is sometimes wrongly concluded that part of a pattern is unique. This appears e.g. in the following example, taking $\bullet \to \times$ for the constructor **C** and $\times \to \bullet$ for **F** with rule $\mathbf{F}(\mathbf{C}(x)) \to x$.

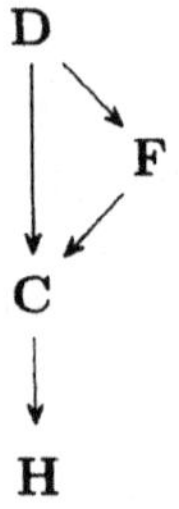


For this reason we require that (data) symbols appearing in a pattern of a rewrite rule also obey an 'upward propagation' rule, that is to say, if such a symbol expects one or more unique arguments the application itself is unique. E.g. in the above example **C** should be typed with $\bullet \to \bullet$, rejecting the given **F**-type.

Since the only symbols appearing in function patterns are constructors introduced by some algebraic type definition, the upward propagation requirement is obtained by making following assumption.

*Assumption.* Let $C \in \Sigma_{\mathcal{D}}$ with uniqueness type $(u_1, \ldots, u_k) \to v$. Then

$$u_i \in \mathbb{U}^{\bullet} \text{ for some } i \leq k \;\Rightarrow\; v \in \mathbb{U}^{\bullet}.$$

Consequently, a data object can only contain unique subparts if the object itself is unique. The fact that a symbol may have more than one environment type is also very useful for constructors. Remember, for example, the following algebraic type definition for lists.

$$\begin{aligned} \mathrm{List}(\alpha) &= \mathbf{Cons}(\alpha, \mathrm{List}(\alpha)) \\ &= \mathrm{Nil} \end{aligned}$$

A list of which the 'spine' is unique can be obtained by typing **Cons** by

$$\mathbf{Cons} : (\times, \bullet) \to \bullet.$$

A list with unique elements can be specified by assuming

$$\mathbf{Cons} : (\bullet, \bullet) \to \bullet.$$

Notice that, because of the propagation rule, the uniqueness of elements implies the uniqueness of the spine.

Allowing both types for **Cons** simultaneously in the present type system may cause type conflicts. E.g. in the rule

$$\mathbf{F}(\mathbf{Cons}(x,y)) \to x,$$

**F** can be typed with $\bullet \to \bullet$. This is wrong, as is illustrated by the following application of **F**.

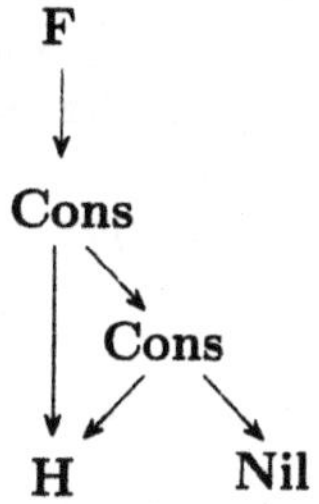


One way to solve this problem is to distinguish the different types of the constructors by introducing uniqueness *type constructors*. We only give an example.

*Example 4.* In the extended system, **Cons** can be typed as follows.

$$\mathbf{Cons} : (\bullet, \overset{\bullet}{\mathrm{List}}(\bullet)) \to \overset{\bullet}{\mathrm{List}}(\bullet),$$
$$\mathbf{Cons} : (\times, \overset{\bullet}{\mathrm{List}}(\times)) \to \overset{\bullet}{\mathrm{List}}(\times).$$

Then, a spine-unique list is typed with $\overset{\bullet}{\mathrm{List}}(\times)$ whereas the list containing also unique elements is typed with $\overset{\bullet}{\mathrm{List}}(\bullet)$.

This extension will not be elaborated here. However, to prevent incorrect type assignments we make the following assumption about type environments.

*Assumption.* If $\mathcal{E}$ is a uniqueness type environment, then the constructor types are chosen in such a way that the type conflicts mentioned above cannot occur.

## 7 Correctness

In order to show that uniqueness typing is preserved during reduction, some analysis with respect to the *use* function is needed. We focus on the relation between the uniqueness typing of a rewrite rule and the usage information of a graph before and after applying this rewrite rule. We will merely give an outline of the proof. The details will appear separately.

Fix an orthogonal TGRS $\mathcal{S} = \langle \mathcal{G}, \mathcal{R} \rangle$.

**Definition 30.** Let $\Delta = \langle R, \mu \rangle$ be a redex in $g$.

(i) Let $\mathcal{U} : R \to \mathbb{U}$. $\Delta$ is *$\mathcal{U}$-type correct* if $\mathcal{U}$ is a uniqueness typing for $R$ according to $\mathcal{E}$, and for each $n \in R \mid l$, $n \neq l$ (say $n = args(m)_i$) one has

$$\mathcal{U}(n) \in \mathbb{U}^{\bullet} \;\Rightarrow\; use_g(\mu(m))_i = \odot.$$

(ii) $\Delta$ is *type correct* if $\Delta$ is $\mathcal{U}$-type correct for some $\mathcal{U}$.

Note that the definition of 'applicable' (see 22) formulates a locality condition for the direct arguments of $\mu(l)$ only. The following result extends this property to all nodes in the matching fragment of the graph.

**Lemma 31.** *Let $\Delta$ be applicable and $\mathcal{U}$-type correct. Then for all $n \in (R \mid l) \cap (R \mid r)$ with $n \neq l$ one has*

$$\mathcal{U}(n) \in \mathbb{U}^{\bullet} \;\Rightarrow\; n \textit{ is local for } \mu(l).$$

*Proof.* For 'ordinary' reduction rules, this follows from the propagation criterion for constructors and regularity of $\mathcal{S}$. For **Ap** reduction rules, the specific form of curry types and the predefined types for **Ap** imply the result. □

**Lemma 32.** *Let $m, n \in g$ with $(m, i) \in acc_g(n)$. Suppose $n$ is on a cycle not containing $m$. Then $use_g(m)_i = \otimes$.*

*Proof.* Examine the definition of *use*. □

**Proposition 33.** *Let $\Delta = \langle R, \mu \rangle$ be applicable in $g$. Say $g \xrightarrow[R]{\Delta} h$. Suppose $\Delta$ is $\mathcal{U}$-type correct, with $\mathcal{U}(r) \in \mathbb{U}^{\bullet}$. Then*

$$acc_h(\mathrm{r}(\Delta)) \subseteq acc_g(\mu(l)).$$

*Proof Sketch.* By the following case distinction.

*Case* 1. $\mathrm{r}(\Delta) \notin \mu(R \mid l)$. Then $\mathrm{r}(\Delta)$ is fresh in $h$, so $acc_h(\mathrm{r}(\Delta)) = acc_g(\mu(l))$ after redirection.

*Case* 2. $\mathrm{r}(\Delta) = \mu(n)$, $n \in \mu(R \mid l)$. Since $\mathcal{U}(n) \in \mathbb{U}^{\bullet}$ it follows by type correctness and lemma 31 that $\mu(n)$ is local for $\mu(l)$. Hence $\mu(l) \rightsquigarrow m$ for every $(m, i) \in acc_g(\mu(n))$. Now let $(m, i) \in acc_g(\mu(n))$. We want to show that $m$ is not present in $h$. If $m \in \mu(R \mid l)$ this is easily seen. Otherwise, if $m$ would be present in $h$ (after redirection and garbage collection), then $\mu(n) \rightsquigarrow m$ ($\rightsquigarrow \mu(n)$). Hence $use_g(\mu(m'))_i = \otimes$ for any $(m', i) \in acc_R(n)$, by lemma 32, contradicting type correctness of $\Delta$. Taking the effect of redirection into account it follows that $acc_h(\mu(n)) \subseteq acc_g(\mu(l))$. □

**Proposition 34.** *Let $\Delta$ be applicable and $\mathcal{U}$-type correct in $g$; say $g \xrightarrow[R]{\Delta} h$.*

(i) *Suppose $\mathcal{U}(r) \in \mathbb{U}^{\bullet}$. Then for all $(m, i) \in acc_g(\mu(l))$ such that $m$ is present in $h$ one has*

$$use_g(m)_i = \odot \;\Rightarrow\; use_h(m)_i = \odot.$$

(ii) *Let $n \in R \mid r$ with $n \neq r$. Suppose $\mathcal{U}(n) \in \mathbb{U}^{\bullet}$. Then for all $(m,i) \in acc_R(n)$*

$$use_R(m)_i = \odot \ \Rightarrow \ use_h(\hat{m})_i = \odot,$$

*where $\hat{m}$ denotes the h-node corresponding to m.*

*Proof Sketch.* (i) Suppose $use_g(m)_i = \odot$. By proposition 33 we only have to consider $acc_g(m)$ to determine $use_h(m)_i$. If $p \underset{m,m'}{\wedge} p'$ in $h$ causing $use_h(m)_i = \otimes$, then a redirection 'above' $\mu(l)$ has taken place. This can only occur if $\mu(l)$ is on a cycle in $g$, contradicting lemma 32.

(ii) By a case distinction, distinguishing the possible positions of $n, m$. Lemma 31 is used in the case $n \in R \mid l$ and $m \notin R \mid l$. □

**Proposition 35.** *Let $\Delta$ be applicable in g; say $g \xrightarrow[\mathcal{R}]{\Delta} h$. Let $n \in g$ such that $n \notin \mu(R \mid l)$, and $n \in h$. Then for all $(m,i) \in acc_g(n)$ with m present in h one has*

$$use_g(m)_i = \odot \ \Rightarrow \ use_h(m)_i = \odot.$$

*Proof Sketch.* Suppose, towards a contradiction, $use_g(m)_i = \odot$ but $use_h(m)_i = \otimes$. Suppose this is caused by $m'$, i.e. $(m',i') \in acc_h(m)$ such that $m \sim m'$ or $m \lhd m'$, say $p \underset{m,m'}{\wedge} p'$ with $p \sim p'$ or $p \lhd p'$. Since this situation does not occur in $g$, these parts contain new nodes or new arcs. Distinguish two cases. If $\mathrm{r}(\Delta) \notin p, p'$ one arrives at a conflict with the argument classification (cf. remark 15). Assuming, on the other hand, $\mathrm{r}(\Delta) \in p$ or $\mathrm{r}(\Delta) \in p'$ leads to a contradiction with $use_g(m)_i = \odot$. □

For reduction on uniqueness-typed graphs, the above results imply a 'subject reduction' result: typing remains correct when reducing applicable redexes.

**Lemma 36.** *Let $g \in \mathcal{G}$. Suppose g is uniqueness-typable. If $\Delta$ is applicable, then $\Delta$ is type correct.*

*Proof.* Obvious. □

**Lemma 37.** (i) *Let $u, v, w \in \mathbb{U}$. Then*

$$u \leq^{\odot} v, \ v \leq^{\otimes} w \ \Rightarrow \ u \leq^{\otimes} w.$$

(ii) *Let $u, v, v' \in \mathbb{U}$. Suppose $u \leq^{\odot} v$ and $v' \subseteq v$. Then there exists $u' \in \mathbb{U}$ with $u' \subseteq u$ and $u' \leq^{\odot} v'$.*

**Theorem 38.** *Suppose $\mathcal{R}$ is uniqueness-typable according to $\mathcal{E}$. Let $\mathcal{U}$ be a uniqueness typing for g (according to $\mathcal{E}$). Furthermore, let $g \xrightarrow[\mathcal{R}]{\Delta} h$ with $\Delta$ applicable. Then there exists a uniqueness typing $\mathcal{U}'$ for h such that $\mathcal{U}'(h) = \mathcal{U}(g)$.*

*Proof.* $\mathcal{U}$ can be extended to a uniqueness typing of $h$ by defining it on the new nodes according to the type assignment to $\Delta$ (proposition 34 (ii)). The type assigned to the other nodes remains correct, as follows from propositions 34 (i, ii), 35 and lemma 37, by distinguishing the different kinds of nodes in $h$. □

# 8 Reasoning about programs with uniqueness types

Uniqueness types can be used in several contexts. When one wants to interface functional languages with imperative programs, one can assign a unique type to those arguments that are destructively updated by the imperative function. In this way file I/O and array updating can be incorporated without loosing the referential transparency. With these applications in mind it may seem that the destructive behaviour of the function has to be explicitly programmed using a non-functional programming language. However, it is of course also possible for a compiler to generate destructive updates for pure functions defined in the functional language itself. This is of great importance for improving the time-space behaviour of functional programs.

Below an example is given in a functional programming language of which it is assumed that uniqueness types are assigned on the underlying graph rewrite system (which can be derived directly from the program by removing some syntactical sugar). The language uses underlining to indicate that a type has the uniqueness attribute •. [ ] in a type denotes the List type. [ ] in a rule denotes the Nil element and [ a | b ] denotes Cons a b. ( ,. . . , ) denotes standard tupling. So, [ T ] denotes a list of type T with a unique spine.

```
qs :: [ T ] → [ T ]
qs [ ]              =  [ ]
qs [ hd | tl ]      =  (qs left) ++ [ hd | qs right ]
                       where
                       (left, right) = split tl hd

split :: [ T ] → T → ([ T ], [ T ])
split [ ] p         =  ([ ], [ ])
split [ hd | tl ] p =  ([ hd | left ], right),   if p ≥ hd
                    =  (left, [ hd | right ])
                       where
                       (left, right) = split tl p
```

Compared with the imperative quick-sort algorithm the functionally written quick-sort algorithm qs has the disadvantage that the split function has to construct new lists for its result. Now, if the function split would be defined on a spine-unique list, the construction of the new cons nodes could be done by updating the old ones. Looking at the actual difference between the old cons node given as an argument to split ([hd | tl]) and the new cons node to be constructed (either [hd | left] or [hd | right]) it can be deduced that only the tail of the cons node has to be updated. This means that the split function does not create new cons nodes at all but is actually rearranging tail pointers in such a way that the ordered list is obtained. Such *in situ* updating is essential to be able to handle large data structures efficiently.

With respect to the updating the run-time behaviour of the functional program can be similar to its imperative counterpart. However, the specified program will require a relatively large recursion stack. Both split and qs can be transformed to a tail recursive version using program transformations that also eliminate the construction of intermediate data structures. Tail recursion is usually translated into a loop on the machine code level. The applied transformation maintains the uniqueness of the types. So, for the resulting elegant functional program a compiler can generate code that is as efficient as the code for an imperatively written quick-sort algorithm. Hence, this example shows that uniqueness types solve one of the challenges set at the 1990 Dagstuhl seminar on functional languages (Johnsson [1990]).

```
qs :: [ T ] → [ T ] → [ T ]
qs [ ] tail                 =  tail
qs [ hd | tl ] tail         =  qs left [ hd | qs right tail ]
                               where
                               (left, right) = split tl hd [ ] [ ]

split :: [ T ] → T → [ T ] → [ T ] → ([ T ], [ T ])
split [ ] p left right      =  (left, right)
split [ hd | tl ] p left right =  split tl p [ hd | left ] right,  if p ≥ hd
                               =  split tl p left [ hd | right ]
```

The reasoning about the programs above implicitly made certain assumptions about the generated code. It was assumed that updating was actually done whenever this was possible. More specifically, it was assumed that updates could actually take place for all objects of the same type. Using only such very general kinds of assumptions and the uniqueness type information the storage behaviour of the functional program was deduced and improved by a program transformation. It is important that these assumptions are further formalised. Any compiler should obey the resulting formal rules such that reasoning about the time and space behaviour of a functional program is independent of a specific compiler. The programmer then can deduce whether or not it is worthwhile to use uniqueness types for those cases where the efficiency of the time-space behaviour is critical. It seems that such reasoning is relatively simple and can be applied successfully to design time and space efficient purely functional programs for many kinds of real-life applications.

## 9 Related work

The update problem is also addressed (using linear types) in Wadler [1990] and Guzmán & Hudak [1991]. Both papers use lambda calculus as basic model hence requiring a more indirect kind of analysis. With the proposed approach in this paper graphs are used directly as the objects of consideration. The presented system for uniqueness types incorporates a solution to several of the questions raised in Wadler [1990]. Uniqueness types are in a sense orthogonal to the standard type systems for functional languages. The uniqueness type system has been used successfully to support high level I/O and efficient array handling. Experience with uniqueness types has shown an important change in the use of functional languages from academic exercises to real-life programming (ranging from a window-based text editor to a relational database). The use function presented in Section 4 has been inspired by the analysis presented for *poly-lam*$_{st}$ in Guzmán & Hudak [1991] which is geared towards efficient array manipulation. They use Wadsworth's shared lambda calculus involving partly copying of lambda terms when functions are shared. In a certain sense the proposed uniqueness types are a generalisation of their single-threadedness analysis to a general graph rewriting context.

## References

Achten P.M., J.H.G. van Groningen and M.J. Plasmeijer, High Level Specification of I/O in Functional Languages, in: *Proc. of International Workshop on Functional Languages*, Glasgow, UK, Springer Verlag, 1993.

Bakel van, S, S. Smetsers and S. Brock, Partial Type Assignment in Left-Linear Term Rewriting Systems, in: *Proc. of 17th Colloqium on Trees and Algebra in Programming (CAAP'92)*, pages 300–322, Rennes, France, Springer Verlag, LNCS 581, 1992.

Barendregt H.P., M.C.J.D. van Eekelen, J.R.W. Glauert, J.R. Kennaway, M.J. Plasmeijer and M.R. Sleep, Term Graph Reduction, in: *Proc. of Parallel Architectures and Languages Europe (PARLE)*, pages 141–158, Eindhoven, The Netherlands, Springer Verlag, LNCS 259 II, 1987.

Barendsen Erik and Sjaak Smetsers, Graph Rewriting and Copying, Technical Report 92-20, University of Nijmegen, 1992.

Guzmán Juan C. and Paul. Hudak, Single-Threaded Polymorphic Lambda Calculus, in: *Proc. of Logic in Computer Science (LICS'90)*, pages 333–345, Phildelphia, IEEE Computer Society Press, 1991.

Johnsson Thomas., Discussion Summary: which analysis?, in: *Proc. of Functional Languages: Optimization For Parallelism*, pages 4–5, Dagstuhl, Germany, Dagstuhl seminar, 1990.

Milner R.A., Theory of Type Polymorphism in Programming, *Journal of Computer and System Sciences*, 17, 1978.

Mycroft A., *Abstract interpretation and optimising transformations for applicative programs*, PhD thesis, University of Edinburgh, 1981.

Wadler P., Linear types can change the world!, in: *Proc. of Working Conference on Programming Concepts and Methods*, pages 385–407, Israel, North Holland, 1990.

# Amalgamated Graph Transformations and Their Use for Specifying AGG – an Algebraic Graph Grammar System *

Gabriele Taentzer and Martin Beyer

Computer Science Department, Technical University of Berlin,
Franklinstr. 28/29, Sekr. FR 6-1, D-10587 Berlin,
e-mail: {gabi,beyer}@cs.tu-berlin.de

**Abstract.** The AGG-system is a prototype implementation of the algebraic approach to graph transformation. It consists of a flexible graph editor and a transformation component. The editor allows the graphical representation of production rules, occurrences and transformation results. The transformation component performs direct transformation steps for user–selected production rules and occurrences.

First steps towards a graph specification of an abstract version of the AGG-system are possible by using amalgamated graph transformations. AGG-states are modelled by graphs whereas AGG-operations are described by amalgamated graph transformations combining parallel and sequential rewriting of graphs.

**Keywords:** graph grammar system, algebraic graph grammars, parallel graph transformation

## 1 Introduction

Although graph grammars have been investigated for more than 20 years only few graph grammar systems exist. (For example, IPSEN [Nag87, Sch91], PAGGED [Göt87a] or GraphEd [Him91].) The algebraic graph grammar system AGG ([Bey91, LB93]) is a prototype implementation of the single pushout approach to graph rewriting. AGG is a purely graphical system, consisting of a graph editor and a transformation component. It combines the intuitiveness of graphical presentations made accessible by current workstation technology with the mathematical rigour of the algebraic approach to graph rewriting. AGG's editing and visualization features are intended to make it suitable as a tool for case studies that would soon become unwieldy if done on a pen and paper basis. This includes the ability to easily edit and navigate in graphs that are too big to be grasped at one view. E.g. several windows can be opened displaying different views of graphs.

AGG has a mouse/menu - oriented interface running on SUN workstations under X-Window. It is implemented in the object oriented language EIFFEL.

* This work has been partly supported by the ESPRIT Working Group 7183 "Computing by Graph Transformation (COMPUGRAPH II)"

Some AGG-operations may be thought of as consisting of a basic action and a context dependent number of extended actions: e.g. moving a node means changing the representation of a graph (basic action) as well as changing the contents of all windows displaying the node (extended actions). This distinction will be reflected in our graph specification of AGG-operations.

In section 3 we present very first ideas towards a graph specification of an abstract version of the AGG-system. The implementation of the AGG-system ([Bey91]) contains the data structures representing graphs, their graphical attributes and an interface to X-Window. We restrict the specification of the AGG-system to the treatment of the graph structures and the logical part of the graphical representation. AGG-states are represented as graphs where the nodes almost model AGG-objects and the edges in between show their use relation. The specification of AGG-operations is performed by graph transformations, i.e. generally the application of a number of graph productions. Those graph productions which perform the actual user interactions function as a basis for further extended actions that have also to be executed to model the whole AGG-operation. All extended actions can be done in parallel and synchronize in the basic actions. Thus, the resulting specification indicates which actions can be executed simultaneously and how they have to synchronize.

Formalizing the specification concept used to specify the AGG-operations, we present amalgamated graph transformations. This type of graph transformation combines parallel and sequential concepts of graph rewriting and is based on the amalgamation concepts introduced in [BFH87]. Parallel replacement of graphs has been investigated in several approaches to graph grammars. (See [EK76], [JRV82], [NA83], [Nag87], [BCF91], [Kre92], [DDK93], etc.) In [EK76] parallel graph grammars are presented in the algebraic framework. They can be seen as an extension of sequential grammars (see [Ehr79]) on one hand, on the other hand parallel graph grammars generalize L-systems which describe a kind of context-free parallel replacement on strings ([RS86], [RS92]). Amalgamated graph transformations are an extension of the parallel graph grammar approach in [EK76] offering additional features.

## 2 The AGG-system

The AGG-system[2] is part of an ongoing implementation effort in the Berlin graph grammar group, initiated by M. Löwe. The development of data structures for an efficient implementation of algebraic graph rewriting has been one initial motivation for the investigation of the single pushout approach (see [Löw93]).

Graphs, morphisms, production rules and occurrences are represented and manipulated in a uniform way. Therefore, AGG supports a graph model with abstraction and higher order edges. It includes the usual definition of graphs as a special case.

Higher order edges, i.e. edges between edges, are used to represent morphisms as graphs. Edges between objects of any order, e.g. between an edge and a node, are also possible. Graphs can be abstracted to nodes, bundles of edges can be abstracted to single edges. The abstraction function induces a partitioning of the graph in

[2] Note that AGG stands for algebraic graph grammars and should not be mixed up with attributed graph grammars.

***abstraction levels.*** The formal definition of this concept which can be found in e.g. [Bey92] regards graphs as a set of objects with three partial functions: the *source*, *target*, and *abstraction* function. They are totalized by including a special object $\perp$. A node is an object for which both the *source* and *target* function are undefined. There are a few rather obvious consistency conditions that are enforced by the AGG-system, e.g. the *abstraction* function has to commute with the *source* and *target* function and has to be acyclic.

The graph editor provides operations for the input and modification of graphs, different views on graphs and supports the editing and simulating of graph rewriting.

Several graphs can be edited simultaneously. Any number of windows can be opened. Each window displays (a part of) an abstraction level. An abstraction level can be shown in several windows. All windows can be used in the same way for editing or transforming graphs. Any change in the representation of a graph is immediately reflected in all appropiate windows.

Several parameters influencing the display of graphs can be adjusted in each window, like the size of node symbols, the zoom factor or the hiding of objects. It is intended to experiment with graph transformation doing the layout manually, before adding layout possibilities to the system. (See for example [Bra91] or [NP91].)

Some edit operations are especially suited for graph transformations, e.g. a gluing operation which merges two objects to a new object inheriting the incidence relations of the original objects and a copy operation drawing edges between all original nodes and edges and their copies. These operations facilitate the input of transformation rules. A simple example (cf. [LB93]) of a transformation rule is shown in figure 1.

The abstraction concept is supported conveniently by the editor. It is possible to create objects in different abstraction levels before defining the abstraction function. Another possibility is the operation "create abstraction": The user selects a number of objects belonging to one abstraction level and creates an abstraction object for them, i.e. an object is inserted into the next higher abstraction level and is defined as the abstraction of the selected objects. The created abstraction object will be a node if the selected objects include nodes.

The abstraction concept is supported by all edit operations. E.g. when objects are moved, all refinement objects as well as all incident objects are moved accordingly.

Other operations of the editor are the changing of node or edge symbols, an extensive and orthogonal grid functionality for each window, view operations, a bunch of interactive and graph based operations for the selection of objects, copy-, move- and resize-operations and more.

The transformation component implements the single-pushout approach to graph rewriting (see [Löw93]) and is illustrated by a simple example, a graph model of a queue structure. Figure 1 shows a rule which model the insertion of an element at the end (□) of the queue. The head of the queue is marked by $\triangle$.

The abstraction concept is used to present an easy to use interface to the transformation component: graphs and the left and right side of rules are abstracted to single nodes, rule morphisms and occurrences are abstracted to single edges. In this way, the higher abstraction level is used as a diagram level.

Figure 2 shows an occurrence of the insertion rule presented above on the graph level and the diagram level.

A transformation step is performed by triggering the "apply rule"-operation (via

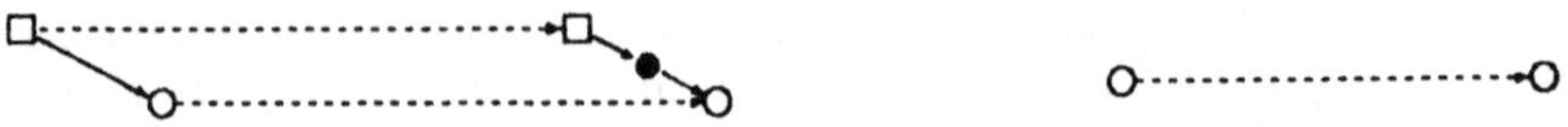

**Fig. 1.** Transformation rule: graph level and diagram level

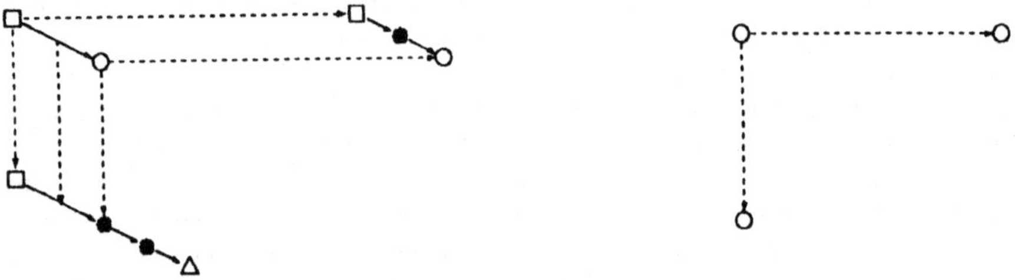

**Fig. 2.** Occurrence: graph level and diagram level

menu or keyboard) and clicking on the occurrence and the place where the derived graph is to be inserted in the window displaying the diagram level. After the execution of the transformation step, the diagram level contains the pushout diagram, the graph level includes the derived graph as well as the morphisms from the old graph and the right hand side of the rule to the derived graph (see figure 3).

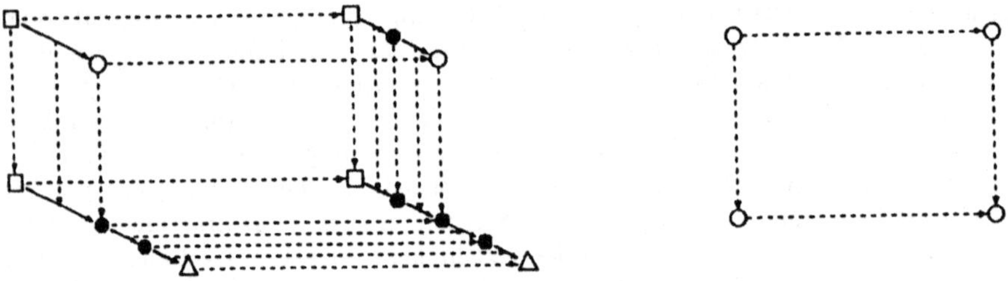

**Fig. 3.** Transformation step: graph level and diagram level

Graphs are not restricted to two abstraction levels. The abstraction concept can also be used to structure large graphs by hiding details in lower abstraction levels.

## 3 Towards a graph specification of the AGG-system

In this section we present first ideas and examples for modelling the AGG-system by graphs and graph transformations. Within this paper we restrict the model to the level of internal graph structures and logical graph representation which means that we do not care about the layout and position of graphs and windows. Principles of the modelling concept are shown by specifying the AGG-operations "insertion of a node" and "create abstraction".

The internal states of the AGG-system are modelled by graphs where important AGG-objects like windows, graphs, abstraction levels, nodes, edges, etc. are represented by nodes and the use relations between them by edges.

In figure 4 the modelling of an arbitrary state of the AGG-system is shown where the model of the initial state is bordered with dashed lines. Nodes and edges are represented as objects belonging to an abstraction level. Moreover, there is a symbol for each object and each window showing the corresponding level. Such a symbol contains the graphical representation of the object. The cursor points to the current window and abstraction level. All abstraction levels belonging to one graph form a list which is described by edges between different levels. The bottom and the top level do not have a refinement and an abstraction, resp. This is indicated by ⊥ and ⊤.

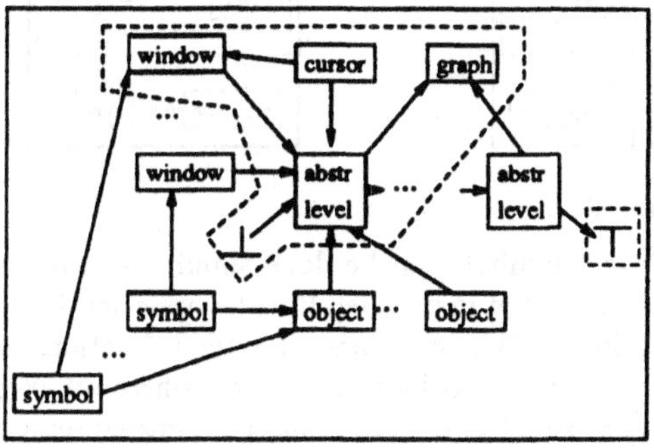


**Fig. 4.** Section of a state graph

As shown in figure 4, most of the relations between AGG-objects are 1:n-relations. For example a number of windows show one level, a number of objects belong to one level, etc.

The AGG-objects node and edge are represented as object-labelled nodes. They are distinguishable via the pointers to their source (s) and target (t)[3]. For nodes, source and target are not defined, indicated by pointers to ⊥. Modelling nodes and edges in this uniform way higher-order edges (see section 2) can be easily described (figure 5).

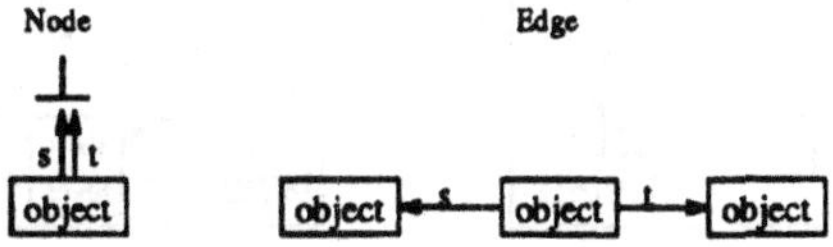


**Fig. 5.** Node and egde representation

The specification of AGG-operations is performed by graph transformations. The first example we want to discuss is the insertion of a node. What basically should be done is the addition of an object to the current abstraction level. This feature can be described by a simple graph production.

The basic action has to be extended by the insertion of a symbol for the new object for each window showing the current abstraction level. This means that there

[3] For reasons of simplicity, the representation of the abstraction function is first introduced when the abstraction concept is treated.

are $n$ extended actions if the current level is displayed in $n$ windows. In figure 6 the basic part of the graph production is bordered with dashed lines.

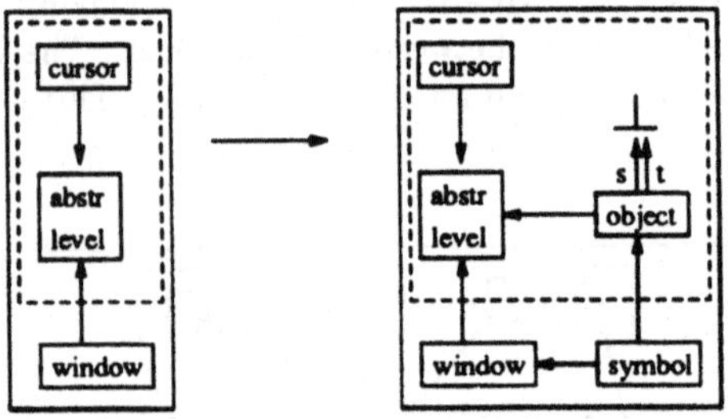


**Fig. 6.** Insertion of a node

The creation of new symbols can be done simultaneously meaning that all extended actions can be executed in parallel. But the basic action which is the insertion of a node should be done only once. Since all extended actions overlap in this basic action it can also be regarded as interface action where all extended actions have to synchronize in. For example, a node is inserted into an empty abstraction level viewed in two different windows. Another graph containing one node is not changed. The corresponding graph transformation is shown in figure 7.

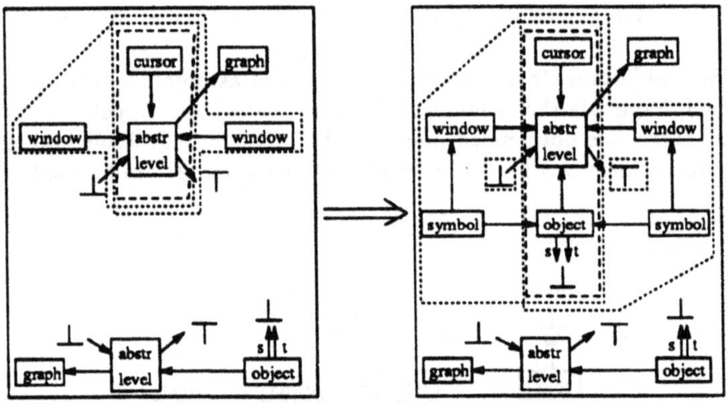


**Fig. 7.** Example of the insertion of a node

In this example, the basic action is applied once (indicated by dashed bordered lines). Since the current abstraction level is displayed in two different windows the extended action is executed twice (indicated by dotted bordered lines).

Here, performing an AGG-operation corresponds to the application of a basic action and its extended actions.

In the previous section we described the AGG-operation "create abstraction" which allows to abstract graphs to nodes and bundles of edges in the same direction to single edges. Within this paper we show only the abstraction of nodes. Specifying "create abstraction" we have to distinguish two main cases. The first one is that the current abstraction level is the top one. Then a new level has to be created which is the new top level. A new abstraction node abstracting all selected nodes has to be inserted into the new level. All the selected nodes are not yet abstracted. Such an information is described by an a-edge running from a selected node to ⊥ instead of an abstraction node. Selected objects are indicated by a selected-labelled edge pointing on the object.

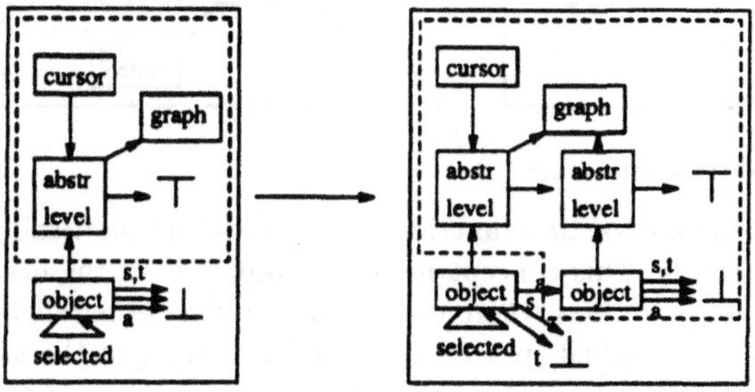


**Fig. 8.** Create abstraction - case 1

An abstraction node is created in a basic action. For each selected node an abstraction edge is inserted pointing to the abstraction node. This means that there are as many extended actions as objects are selected (figure 8).

The second case is that the abstraction level of the current level already exists. Thus, only a new abstraction node has to be inserted which is the basic action. Similar to the first case all selected nodes are joined to the abstraction node by an a-edge (figure 9).

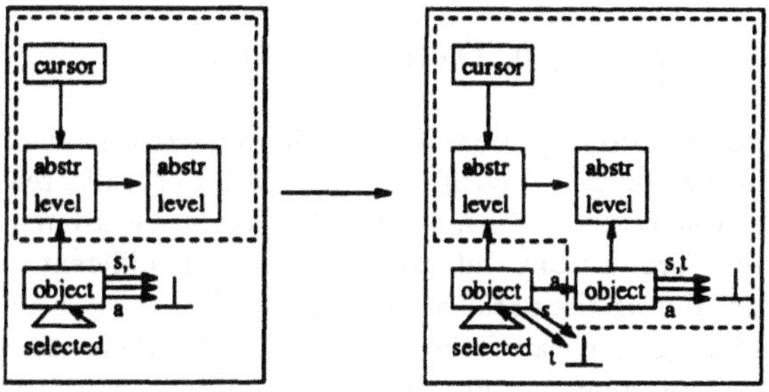


**Fig. 9.** Create abstraction - case 2.1

This production models the first kind of extended actions. The second one follows from the fact that the abstraction level of the current one can already be shown in any number of windows which is not possible in the case that the abstraction level is newly created. For each window showing the abstraction level a new symbol of the abstraction node has to be created (figure 10). Thus, for this type of AGG-operation two different kinds of extended actions are needed. But, although different kinds of extended actions are used to describe the abstraction of nodes they all can be applied in parallel due to the fact that they overlap in one basic action.

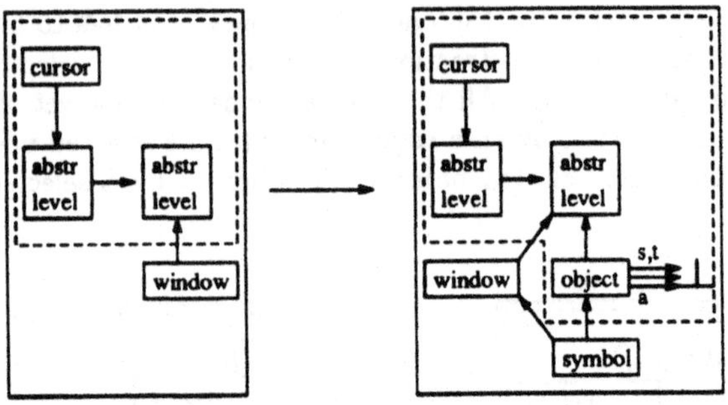


**Fig. 10.** Create abstraction - case 2.2

The input of the abstraction function is completed automatically as far as this is possible unambigously. This concept is more complex than the described ones and would lead to a set of actions where not all can be done in parallel. However, other AGG-operations as described in section 2 can be treated in the same way as this has been done with the sample AGG-operations in this section. AGG-operations like node and edge insertion and deletion, selection of objects, creation of windows and graphs and changing to another graph are already specified in [Tae92] in a similar way.

The initial state displayed in figure 4 together with sets of actions for each AGG-operation form a graph specification of the AGG-system which describes all possible state transitions of it. Moreover, the specification gives hints to what actions can be done in parallel and how they have to be synchronized.

The type of graph transformation used in this section is formalized in the next section dealing with amalgamated graph transformations.

## 4 Amalgamated Graph Transformations

Amalgamated graph transformations extend the algebraic kind of parallel graph rewriting which is described in [EK76]. That approach to parallel rewriting of graphs is a basic version where the whole mother graph has to be covered by occurrences of left hand sides of productions. Interfaces between the different occurrences have to remain constant during the parallel replacement.

For amalgamated graph transformations these restrictions are removed. There, dynamic interfaces are allowed which means that interfaces can change during a rewriting step. Moreover, the covering of the mother graph by occurrences can be partial. Only the covered part is replaced, the context is joined unchanged to the derived graph (cf. figure 7). When describing complex operations by graph transformations all graph productions involved are put together into an operational production set. These sets are similar to tables in other approaches to parallel rewriting grouping several productions together.

Note that the AGG-system implements the single-pushout approach to graph rewriting, but a first graph specification as introduced in section 3 is now formalized using amalgamated graph transformations in the double-pushout approach. Although one could see a conflict in using two different kinds of graph rewriting the main emphasis is put in the usage of graph transformations for the specification of the AGG-system, anyway. Moreover, it seems to be easy to generalize the concepts of amalgamated transformations also to partial graph morphisms ([Löw93]), at least using partial morphisms for the description of productions. This means performing amalgamated transformations in the single pushout approach to rewriting, too.

The amalgamated graph transformations build on labelled graphs and total label preserving graph morphisms which form a category GRAPH. (For this category and its properties see [Ehr79], [EHKP92], etc.) Using category GRAPH amalgamated graph transformations are one instantiation of parallel high-level replacement which is formulated in an algebraic framework based on categories in [Tae92]. Parallel high-level replacement systems generalize the transformation process known from graph grammars to other kinds of replacement systems. Such systems allow, for example, relational structures, algebraic specifications, Petri nets, etc. to be replaced. The general framework of high-level replacement systems is introduced in [EHKP92] and extended to the parallel case where different kinds of parallel transformations are introduced.

First, we review some basic concepts of the algebraic theory to graph grammars:

- Let $C_E$ and $C_N$ be fixed color alphabets for nodes and edges. A graph $G = (G_E, G_N, s_G, t_G, m_{GE}, m_{GN})$ consists of the sets of edges and nodes, $G_E$ and $G_N$, mappings $s_G, t_G : G_E \rightarrow G_N$, called source resp. target map, and the coloring maps for arcs and nodes, $m_{GE} : G_E \rightarrow C_E$ and $m_{GN} : G_N \rightarrow C_N$.
- Given two graphs $G$ and $G'$ a graph morphism $f : G \rightarrow G'$ is a pair of maps $(f_E : G_E \rightarrow G'_E, f_N : G_N \rightarrow G'_N)$ such that $f_E$ and $f_N$ are compatible with source and target mappings and label preserving.
- A production $p = (L \leftarrow K \rightarrow R)$ consists of three graphs $L$, $K$ and $R$, called left hand side, gluing graph and right hand side, resp. and two graph morphisms $l : K \rightarrow L$ and $r : K \rightarrow R$. If $l$ and $r$ are injective $p$ is called injective.
- An occurrence of the left hand side $L$ in a graph $G$ is a graph morphism. It is also called occurrence of production $p$. Production $p$ is applicable to $G$ if its occurrence satisfies the gluing condition ([Ehr79]).

**Definition 1 (Subproduction).** Given a production $p = (L \xleftarrow{l} K \xrightarrow{r} R)$ a production $s = (L_s \xleftarrow{l_s} K_s \xrightarrow{r_s} R_s)$ is called subproduction of $p$ if there are graph morphisms $l' : L_s \rightarrow L$, $k' : K_s \rightarrow K$ and $r' : R_s \rightarrow R$ called embedding of $s$ into $p$ (short $s \rightarrow p$) such that $l' \circ l_s = l \circ k'$ and $r' \circ r_s = r \circ k'$.

Subproduction $s$ is called proper w.r.t. $p$ if productions $s$ and $p$ and morphisms $l'$, $k'$ and $r'$ are injective and $l'^{-1} \circ l(K) \subseteq l_s(K_s)$ and $r'^{-1} \circ r(K) \subseteq r_s(K_s)$.

**Example 1.** In figure 6 the left and right hand side of a production are shown whereas the gluing graph is equal to the left hand side. Morphisms run between objects named alike and corresponding edges existing in between.

Modelling AGG-operations of section 3 we distinguished basic actions and extended actions. Describing extended actions by productions basic actions have to be subproductions of them.

Looking at figure 6 again, the subproduction is displayed by the part bordered with dashed lines and interpreted in the way it was done above. It can be included in the production solidly framed.

In section 3 we specified one AGG-operation by a basic action and a set of extended actions. All extended actions overlap in the basic action which should be executed only once. There, the basic action can also be regarded as an interface action between all extended actions. This schema is now generalized by allowing more than one interface between different actions and leads to operational production sets. They consist of a number of productions each equipped with a set of their subproductions. They model the possible interfaces each action can have to other actions.

**Definition 2 (Operational production set).** An operational production set $S = \{(e_k, SP_k) | k \in \mathbb{N}\}$ is a set of pairs consisting of a production $e_k$ which is also called elementary production together with a set of subproductions $SP_k$ of $e_k$.

An operational production set is called simple if each $SP_k$ in $S$ consists of just one subproduction which has only one embedding into $e_k$.

**Example 2.** We specified the second case of AGG-operation "create abstraction" in section 3 by one basic action and two extended actions. So, the corresponding simple operational production set has two pairs where each one consists of an elementary production modelling an extended action and a set of one subproduction representing the basic action.

The concept of operational production sets is more expressive than that of tables in other approaches to parallel rewriting of graphs (f.ex. in [NA83], [DDK93]) because the set of possible interfaces between productions has to be specified which cannot be done using tables.

For finding a partial covering of the mother graph $G$ we have to look for all different occurrences of the elementary productions of an operational production set $S$. This means that extended actions are done *for all* different settings in the state they can be done.

Specifying all possible interfaces in $S$ an elementary production is used only for a covering of $G$ if it provides an interface for every other production that will be applied to $G$. This concept ensures that all extended actions have to overlap in one of their interface actions pairwise.

In the case of specifying the AGG-operations in section 3 all extended actions have only one interface, the basic action. Looking for corresponding coverings all productions overlap in this single subproduction.

To get a partial covering of the mother graph all occurrences of productions found there are glued at their interfaces.

**Definition 3 (Gluing of graphs).** A *graph morphism star* $MS = (G_i \leftarrow G_{ij} \rightarrow G_j)_{1 \le i < j \le n}$ consists of graphs $G_i$, $G_j$ and $G_{ij}$ and graph morphisms $g_{ij} : G_{ij} \rightarrow G_i$ and $g_{ji} : G_{ij} \rightarrow G_j$ for all $1 \le i < j \le n$. Let $\bar{I}_G$ be the equivalence relation generated by the relation $I_G = \{(g_{ij}(x), g_{ji}(x)) \mid x \in G_{ij}, 1 \le i < j \le n\}$. The *gluing graph* $G$ of $MS$ is then defined by $G = (\biguplus_{1 \le i \le n} G_i)_{/\bar{I}_G}$, the quotient set of the disjoint union of all $G_i$. Functions $g_i : G_i \rightarrow G$ for all $1 \le i \le n$ send each element of $G_i$ to its equivalence class in $G$.

The gluing graph $G$ can be characterized to be the colimit object of $MS$ in the category GRAPH.(See [EK76], [Ehr79] and [Tae92] for more details to the gluing of graphs.)

**Example 3.** Gluing two copies of the left hand graph in figure 6 which is solidly bordered at the graph dashed bordered inside yields the graph dotted bordered within the left hand side of figure 7.

**Definition 4 (Partial dynamic covering).** Given a set of productions $P = \{p_i | p_i = (L_i \xleftarrow{l_i} K_i \xrightarrow{r_i} R_i), 1 \leq i \leq n\}$, a set of subproductions $SP = \{s_{ij} | s_{ij} = (L_{ij} \xleftarrow{l_{ij}} K_{ij} \xrightarrow{r_{ij}} R_{ij}), 1 \leq i < j \leq n\}$, graphs $G$ and $L$ and a graph morphism $g : L \to G$ a partial dynamic covering $PDCOVERING_{P,SP,L}(G)$ consists of the diagram below if the following conditions are satisfied:

1. Each production $s_{ij}$ is a subproduction of $p_i$ and $p_j$, $\forall 1 \leq i < j \leq n$.
2. Graph $L$ is the gluing graph of graph morphism star $(L_i \xleftarrow{g_{ij}} L_{ij} \xrightarrow{g_{ji}} L_j)_{1 \leq i < j \leq n}$.
3. $L_i \to G = L_i \to L \to G$, $\forall 1 \leq i \leq n$.

A spider covering is a partial dynamic covering where all $s_{ij}$, $1 \leq i < j \leq n$, are the same. A partial dynamic covering is fully synchronized if $g_i(L_i) \cap g_j(L_j) \subseteq g_i \circ g_{ij}(L_{ij}) \cup (g_i \circ l_i(K_i) \cap g_j \circ l_j(K_j)), \forall 1 \leq i < j \leq n$.

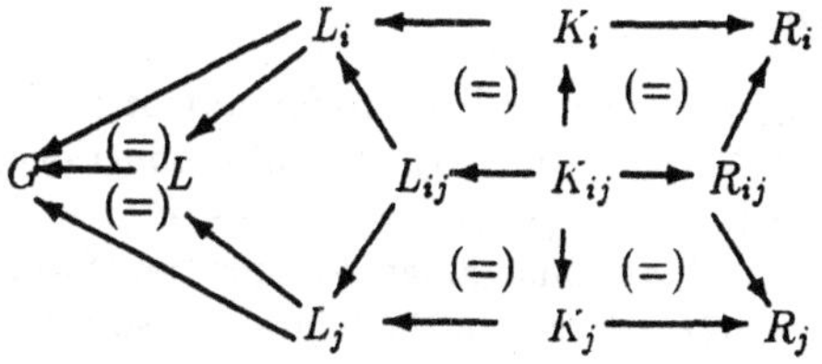


The way the construction of a partial dynamic covering is described above is now formalized in the following covering construction.

**Proposition 5 (Allquantified covering construction).** *Given an operational production set $S = \{(e_k, SP_k) | 1 \leq k \leq m\}$ and a graph $G$ there is a partial dynamic covering $PDCOVERING_{P,SP,L}(G)$ with*

1. $P = \{p_i | 1 \leq i \leq n\}$ *with* $p_i = (L_i \leftarrow K_i \to R_i) = e_k$ *for all different occurrences* $L_i \to G$ *and some* $1 \leq k \leq m$,
2. $SP = \{s_{ij} | 1 \leq i < j \leq n\}$ *with* $s_{ij} = (L_{ij} \leftarrow K_{ij} \to R_{ij}) \in SP_k \cap SP_{k'}$ *if* $p_i = e_k$ *and* $p_j = e_{k'}$, $1 \leq k, k' \leq m$,
3. $\forall 1 \leq i < j \leq n : L_{ij} \to L_i \to G = L_{ij} \to L_j \to G$
4. *$L$ is the gluing graph of* $(L_i \leftarrow L_{ij} \to L_j)_{1 \leq i < j \leq n}$.

The existence of a partial dynamic covering can be proven easily. (See [Tae92] for more details.)

In general, a partial dynamic covering need not cover the whole part of a mother graph $G$ that *can* be covered by given productions. The allquantified construction of a covering is only one way of covering the mother graph. Other covering constructions are imaginable.

The allquantified covering construction is not unique in general, but in the special case that the given operational production set is simple, as this holds in all examples of section 3, uniqueness can be proven.

**Proposition 6 (Uniqueness of special allquantified covering construction).** *Given a simple operational production set* $S = \{(e_k, SP_k) | k \in I\!N\}$ *there is a unique partial dynamic covering as constructed in proposition 5.*

*If all subproductions in* $SP_k, 1 \leq k \leq n$, *are the same the resulting covering is a spider covering.*

This result follows directly from the assumptions chosen above.

**Example 4.** All AGG-operations presented in section 3 can be described by simple operational production sets where all subproductions are the same for each operation. Therefore, all resulting coverings are spiders.

Given a partial dynamic covering a graph transformation is performed in two steps. First a new production is generated that models the synchronization of all extended actions at their interface actions. This is done by gluing the left hand sides, right hand sides and gluing graphs of the productions at their interfaces in order to get the new so-called amalgamated production. This production is then applied to the actual graph by constructing the usual double pushout as described in [Ehr79], [EHKP92], etc.

**Definition 7 (Amalgamated graph transformation).** Given a partial dynamic covering $PDCOVERING_{P,SP,L}(G)$ with sets of productions $P = \{p_i | p_i = (L_i \leftarrow K_i \rightarrow R_i), 1 \leq i \leq n\}$, $SP = \{s_{ij} | s_{ij} = (L_{ij} \leftarrow K_{ij} \rightarrow R_{ij}), 1 \leq i < j \leq n\}$ and morphism $g : L \rightarrow G$, an amalgamated graph transformation[4] $G \Longrightarrow G'$ over $PDCOVERING_{P,SP,L}(G)$ consists of the following two steps:

1. Construction of an amalgamated production $(L \xleftarrow{l} K \xrightarrow{r} R)$ by generating the gluing graphs $L$ of $(L_i \leftarrow L_{ij} \rightarrow L_j)_{1 \leq i < j \leq n}$, $K$ of $(K_i \leftarrow K_{ij} \rightarrow K_j)_{1 \leq i < j \leq n}$ and $R$ of $(R_i \leftarrow R_{ij} \rightarrow R_j)_{1 \leq i < j \leq n}$. Morphism $K \xrightarrow{l} L$ $(K \xrightarrow{r} R)$ is the unique morphism such that

$$K_i \rightarrow K \xrightarrow{l} L = K_i \rightarrow L_i \rightarrow L (K_i \rightarrow K \xrightarrow{r} R = K_i \rightarrow R_i \rightarrow R), \forall 1 \leq i < j \leq n$$

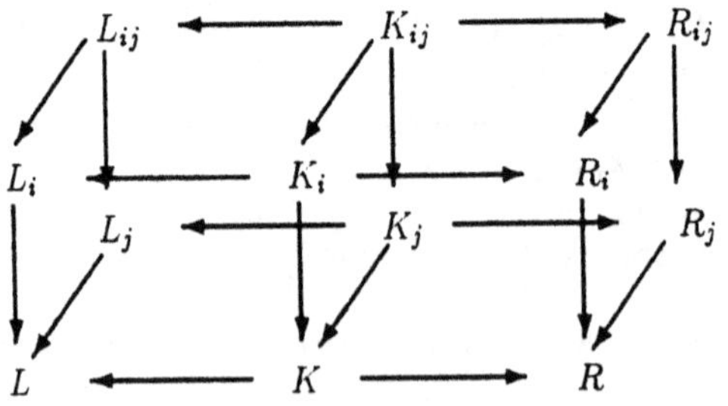


[4] Note that the kind of transformation just defined is called amalgamated *two-level derivation* in [ET92] and [Tae92].

2. Application of the amalgamated production $p = (L \xleftarrow{l} K \xrightarrow{r} R)$ to graph $G$ with occurrence $g : L \rightarrow G$ by a double pushout, i.e. constructing first the pushout-complement graph $D = G - g(L - l(K))$. The gluing graph $G'$ of $(D \leftarrow K \xrightarrow{r} R)$ is then the result of the amalgamated graph transformation.

Morphisms $K \xrightarrow{l} L$ and $K \xrightarrow{r} R$ always exist in a unique way because they are the universal morphisms out of the colimit object $K$. Obviously, $p$ is a production in the usual sense as defined above.

**Example 5.** In figure 7 all parts which lie inside the dotted lines belong to the left hand and right hand side of the new amalgamated production, resp. The gluing graph of the new production is the same as the left hand side and the morphisms in between are just inclusions.

**Proposition 8 (Injectivity of amalgamated production).** *If all subproductions $s_{ij}$ in $SP$ of a spider covering $PDCOVERING_{P,SP,L}(G)$ are proper w.r.t. their productions $p_i$ and $p_j$ in $P$ the resulting production $p$ of step 1 in definition 7 is injective.*

*Proofsketch:* The gluing of all left hand side graphs of a spider covering can be characterized as a special colimit called multiple pushout. The same holds for the right hand sides and gluing graphs. Multiple pushouts inherit injective morphisms similar to simple pushouts. Furthermore, all $s$ in $SP$ are proper w.r.t. their corresponding productions. Therefore, $K \xrightarrow{l} L$ and $K \xrightarrow{r} R$ are injective.

**Proposition 9 (Applicability of amalgamated production).** *If all productions $p_i$ in $P$ of a spider covering $PDCOVERING_{P,SP,L}(G)$ are applicable to $G$ and $PDCOVERING_{P,SP,L}(G)$ is fully synchronized the amalgamated production $p$ in definition 7 is applicable to $G$, too.*

*Proofsketch:* Consider the interesting case of two different productions $p_i$ and $p_j$ in $P$ being applicable to $G$. Their occurrences $g_i$ and $g_j$ can overlap either in the occurrence of their subproduction or their gluing graphs. First $g_i \circ g_{ij} = g_j \circ g_{ji}$ and the second case does not destroy the applicability, anyway. (For the proofs of these two propositions see [Bet92].)

The amalgamated graph transformations can be used to describe production sets which have a certain regularity. Using elementary productions for the generation of a production suitable for the given graph large sets of similar productions building a production family can be avoided. (See also [Göt87b].)

Amalgamated graph transformations combine concepts of parallel and sequential transformations by generating a new production out of elementary productions in parallel and applying this production in the usual sequential way of graph rewriting. Thus, on one hand we have the advantage of using the compact notion of operational production sets to describe complex actions, on the other hand all results which are known for sequential transformations (like Church-Rosser-Properties, Parallelism Theorem, Concurrency Theorem, Embedding Theorem, Amalgamation Theorem, etc.) hold also for the sequential part of this type of transformation.

For the specification of the AGG-system these theoretical results can give information about which AGG-operations are independent of each other and how they can be synchronized. Furthermore, the construction of operations in the transformation component from different editor operations can be supported using concepts of concurrency.

## 5 Outlook

Extensions of the AGG-system planned next include labels, finding of occurrences, completion of partial occurrences and triggering and execution of transformation sequences. Further extensions would be the implementation of the double pushout approach to graph rewriting, layouting of graphs, and more. AGG has been used in courses on the algebraic approach to graph transformation at the Technical University of Berlin. It is intended to be used also as a tool for case studies on the algebraic approach to graph transformation.

Within this paper we presented first ideas for a graph specification of internal structures of the AGG-system. It would be nice to continue this work to get a feeling how amalgamated graph transformations can be used for specifying complex operations. For a graph specification of the entire AGG-system it seems to be necessary to extend the kind of graph transformation presented to attributed transformations. The attributes would comprise all graphical information of graphs.

AGG-operations are specified by amalgamated graph transformations over partial dynamic coverings constructed in an allquantified way. For special cases uniqueness of such a construction can be shown. Moreover, other covering constructions are imaginable dependent on the problem to be specified.

Restricting the definition of subproductions to that in [BFH87] would lead to further results concerning sequentialization of amalgamated graph transformations extended to more than two productions.

Due to the two-step mechanism of the graph transformation presented it is possible to get use of the whole field of theoretical results for sequential graph transformations. Doing this efficiently, conditions which have to be checked for occurrences of productions have to be adapted to partial dynamic coverings.

**Acknowledgements:** We thank Ralph Betschko and Hartmut Ehrig for interesting discussions and the referees for valuable comments.

## References

[BCF91] D. A. Bailey, J. E. Cuny, and C. D. Fisher, *Programming with very large graphs*, In Ehrig et al. [EKR91].

[Bet92] R. Betschko, *Parallele Graphgrammatiken mit Synchronisation*, Studienarbeit, Technical University of Berlin, Dep. of Comp. Sci., 1992.

[Bey91] M. Beyer, *GAG: Ein graphischer Editor für algebraische Graphgrammatiksysteme*, Diplomarbeit, Technical University of Berlin, Dep. of Comp. Sci., 1991.

[Bey92] M. Beyer, *AGG1.0 - Tutorial*, Technical University of Berlin, Department of Computer Science, 1992.

[BFH87] P. Böhm, H.-R. Fonio, and A. Habel, *Amalgamation of graph transformations: a synchronization mechanism*, J. of Comp. and Syst. Sci. **34** (1987), 377–408.

[Bra91] F. Brandenburg, *Layout Graph Grammars: The Placement Approach*, In Ehrig et al. [EKR91], pp. 144–156.

[DDK93] G. David, F. Drewes, and H.-J. Kreowski, *Hyperedge Replacement with Rendezvous*, 1993, to appear in proc. of 12th conf. of FST and TCS'92.

[EHKP92] H. Ehrig, A. Habel, H.-J. Kreowski, and F. Parisi-Presicce, *Parallelism and concurrency in High Level Replacement Systems*, Math. Struct. in Comp. Sci. **1** (1992), 361–404.

[Ehr79] H. Ehrig, *Introduction to the algebraic theory of graph grammars*, 1st Int. Workshop on Graph Grammars and their Application to Computer Science and Biology, LNCS 73, Springer, 1979, pp. 1–69.

[EK76] H. Ehrig and H.-J. Kreowski, *Parallel Graph Grammars*, Automata,Languages, Development (A. Lindenmayer and G. Rozenberg, eds.), Amsterdam: North Holland, 1976, pp. 425–447.

[EKR91] H. Ehrig, H.-J. Kreowski, and G. Rozenberg (eds.), *4th Int. Workshop on Graph Grammars and Their Application to Computer Science, LNCS 532*, Springer, 1991.

[ET92] H. Ehrig and G. Taentzer, *From Parallel Graph Grammars to Parallel High- Lev el Replacement Systems*, Lindenmayer Systems, Springer, 1992, pp. 283–303.

[Göt87a] H. Göttler, *Graph grammars and diagram editing*, 3rd Int. Workshop on Graph Grammars and Their Application to Computer Science, LNCS 291, Springer, 1987, pp. 216–231.

[Göt87b] H. Göttler (ed.), *Graphgrammatiken in Softwareengineering*, Universität Erlangen, 1987.

[Him91] M. Himsolt, *GraphEd: An interactive tool for developing graph grammars*, In Ehrig et al. [EKR91], pp. 61–65.

[JRV82] D. Janssens, G. Rozenberg, and R. Verraedt, *On sequential and parallel node-rewriting graph grammars*, Computer Vision, Graphics and Image Processing **18** (1982), 279–304.

[Kre92] H.-J. Kreowski, *Parallel Hyperedge Replacement*, Lindenmayer Systems, Springer, 1992, pp. 271–282.

[LB93] M. Löwe and M. Beyer, *AGG — An Implementation of Algebraic Graph Rewriting*, LNCS 690, Springer, 1993, Rewriting Techniques and Applications, Fifth Int. Conf., RTA'93.

[Löw93] M. Löwe, *Algebraic approach to single-pushout graph transformation*, Theoretical Computer Science **109** (1993), 181–224.

[NA83] A. Nakamura and K. Aizawa, *On a relationship between graph L-systems and picture languages*, Theoretical Computer Science **24** (1983), 161–177.

[Nag87] M. Nagl, *A software development environment based on graph technology*, 3rd Int. Workshop on Graph Grammars and Their Application to Computer Science, LNCS 291, Springer, 1987, pp. 458–478.

[NP91] F. Newbery Paulisch, *The Design of an Extendible Graph Editor*, Ph.D. thesis, University of Karlsruhe, Department of Informatics, March 1991.

[RS86] G. Rozenberg and A. Salomaa, *The Book of L*, Springer, Berlin, 1986.

[RS92] G. Rozenberg and A. Salomaa, *Lindenmayer Systems*, Springer, 1992.

[Sch91] A. Schürr, *Operationales Spezifizieren mit programmierten Graphersetzungssystemen*, Deutscher Universitätsverlag GmbH, Wiesbaden, 1991.

[Tae92] G. Taentzer, *Parallel High-Level Replacement Systems*, Tech. Report 92/10, Technical University of Berlin, Dep. of Comp. Sci., 1992.

## List of Authors


Aizawa, K. .......... 1
Andries, M. .......... 19
Arnborg, S. .......... 37
Barendsen, E. .......... 51, 358
Barthelmann, K. .......... 71
Beyer, M. .......... 380
Corradini, A. .......... 86, 104, 119
Courcelle, B. .......... 138
Dingel, J. .......... 248
Ehrig, H. .......... 86, 104, 153
Engels, G. .......... 19
Gemis, M. .......... 170
Grabska, E. .......... 188
Janssens, D. .......... 203
Kawahara, Y. .......... 218
Korff, M. .......... 234
Kreowski, H.-J. .......... 153
Lagergren, J. .......... 138
Löwe, M. .......... 86, 104, 248
Maggiolo-Schettini, A. .......... 265
Mizoguchi, Y. .......... 218
Montanari, U. .......... 86, 104, 280
Nagl, M. .......... 296
Nakamura, A. .......... 1
Paredaens, J. .......... 170
Peelman, P. .......... 170
Peron, A. .......... 265
Plasmeijer, R. .......... 358
Raoult, J.-C. .......... 312
Rossi, F. .......... 86, 104, 280
Schied, G. .......... 71, 326
Schürr, A. .......... 341
Smetsers, S. .......... 51, 358
Taentzer, G. .......... 153, 380
van den Bussche, J. .......... 170
van Eekelen, M. .......... 358
Voisin, F. .......... 312
Wolz, D. .......... 119

# Springer-Verlag and the Environment

We at Springer-Verlag firmly believe that an international science publisher has a special obligation to the environment, and our corporate policies consistently reflect this conviction.

We also expect our business partners – paper mills, printers, packaging manufacturers, etc. – to commit themselves to using environmentally friendly materials and production processes.

The paper in this book is made from low- or no-chlorine pulp and is acid free, in conformance with international standards for paper permanency.

Printing: Weihert-Druck GmbH, Darmstadt
Binding: Buchbinderei Schäffer, Grünstadt

# Lecture Notes in Computer Science

For information about Vols. 1–704
please contact your bookseller or Springer-Verlag

Vol. 705: H. Grünbacher, R. W. Hartenstein (Eds.), Field-Programmable Gate Arrays. Proceedings, 1992. VIII, 218 pages. 1993.

Vol. 706: H. D. Rombach, V. R. Basili, R. W. Selby (Eds.), Experimental Software Engineering Issues. Proceedings, 1992. XVIII, 261 pages. 1993.

Vol. 707: O. M. Nierstrasz (Ed.), ECOOP '93 – Object-Oriented Programming. Proceedings, 1993. XI, 531 pages. 1993.

Vol. 708: C. Laugier (Ed.), Geometric Reasoning for Perception and Action. Proceedings, 1991. VIII, 281 pages. 1993.

Vol. 709: F. Dehne, J.-R. Sack, N. Santoro, S. Whitesides (Eds.), Algorithms and Data Structures. Proceedings, 1993. XII, 634 pages. 1993.

Vol. 710: Z. Ésik (Ed.), Fundamentals of Computation Theory. Proceedings, 1993. IX, 471 pages. 1993.

Vol. 711: A. M. Borzyszkowski, S. Sokołowski (Eds.), Mathematical Foundations of Computer Science 1993. Proceedings, 1993. XIII, 782 pages. 1993.

Vol. 712: P. V. Rangan (Ed.), Network and Operating System Support for Digital Audio and Video. Proceedings, 1992. X, 416 pages. 1993.

Vol. 713: G. Gottlob, A. Leitsch, D. Mundici (Eds.), Computational Logic and Proof Theory. Proceedings, 1993. XI, 348 pages. 1993.

Vol. 714: M. Bruynooghe, J. Penjam (Eds.), Programming Language Implementation and Logic Programming. Proceedings, 1993. XI, 421 pages. 1993.

Vol. 715: E. Best (Ed.), CONCUR'93. Proceedings, 1993. IX, 541 pages. 1993.

Vol. 716: A. U. Frank, I. Campari (Eds.), Spatial Information Theory. Proceedings, 1993. XI, 478 pages. 1993.

Vol. 717: I. Sommerville, M. Paul (Eds.), Software Engineering – ESEC '93. Proceedings, 1993. XII, 516 pages. 1993.

Vol. 718: J. Seberry, Y. Zheng (Eds.), Advances in Cryptology – AUSCRYPT '92. Proceedings, 1992. XIII, 543 pages. 1993.

Vol. 719: D. Chetverikov, W.G. Kropatsch (Eds.), Computer Analysis of Images and Patterns. Proceedings, 1993. XVI, 857 pages. 1993.

Vol. 720: V.Mařík, J. Lažanský, R.R. Wagner (Eds.), Database and Expert Systems Applications. Proceedings, 1993. XV, 768 pages. 1993.

Vol. 721: J. Fitch (Ed.), Design and Implementation of Symbolic Computation Systems. Proceedings, 1992. VIII, 215 pages. 1993.

Vol. 722: A. Miola (Ed.), Design and Implementation of Symbolic Computation Systems. Proceedings, 1993. XII, 384 pages. 1993.

Vol. 723: N. Aussenac, G. Boy, B. Gaines, M. Linster, J.-G. Ganascia, Y. Kodratoff (Eds.), Knowledge Acquisition for Knowledge-Based Systems. Proceedings, 1993. XIII, 446 pages. 1993. (Subseries LNAI).

Vol. 724: P. Cousot, M. Falaschi, G. Filè, A. Rauzy (Eds.), Static Analysis. Proceedings, 1993. IX, 283 pages. 1993.

Vol. 725: A. Schiper (Ed.), Distributed Algorithms. Proceedings, 1993. VIII, 325 pages. 1993.

Vol. 726: T. Lengauer (Ed.), Algorithms – ESA '93. Proceedings, 1993. IX, 419 pages. 1993

Vol. 727: M. Filgueiras, L. Damas (Eds.), Progress in Artificial Intelligence. Proceedings, 1993. X, 362 pages. 1993. (Subseries LNAI).

Vol. 728: P. Torasso (Ed.), Advances in Artificial Intelligence. Proceedings, 1993. XI, 336 pages. 1993. (Subseries LNAI).

Vol. 729: L. Donatiello, R. Nelson (Eds.), Performance Evaluation of Computer and Communication Systems. Proceedings, 1993. VIII, 675 pages. 1993.

Vol. 730: D. B. Lomet (Ed.), Foundations of Data Organization and Algorithms. Proceedings, 1993. XII, 412 pages. 1993.

Vol. 731: A. Schill (Ed.), DCE – The OSF Distributed Computing Environment. Proceedings, 1993. VIII, 285 pages. 1993.

Vol. 732: A. Bode, M. Dal Cin (Eds.), Parallel Computer Architectures. IX, 311 pages. 1993.

Vol. 733: Th. Grechenig, M. Tscheligi (Eds.), Human Computer Interaction. Proceedings, 1993. XIV, 450 pages. 1993.

Vol. 734: J. Volkert (Ed.), Parallel Computation. Proceedings, 1993. VIII, 248 pages. 1993.

Vol. 735: D. Bjørner, M. Broy, I. V. Pottosin (Eds.), Formal Methods in Programming and Their Applications. Proceedings, 1993. IX, 434 pages. 1993.

Vol. 736: R. L. Grossman, A. Nerode, A. P. Ravn, H. Rischel (Eds.), Hybrid Systems. VIII, 474 pages. 1993.

Vol. 737: J. Calmet, J. A. Campbell (Eds.), Artificial Intelligence and Symbolic Mathematical Computing. Proceedings, 1992. VIII, 305 pages. 1993.

Vol. 738: M. Weber, M. Simons, Ch. Lafontaine, The Generic Development Language Deva. XI, 246 pages. 1993.

Vol. 739: H. Imai, R. L. Rivest, T. Matsumoto (Eds.), Advances in Cryptology – ASIACRYPT '91. X, 499 pages. 1993.

Vol. 740: E. F. Brickell (Ed.), Advances in Cryptology – CRYPTO '92. Proceedings, 1992. X, 593 pages. 1993.

Vol. 741: B. Preneel, R. Govaerts, J. Vandewalle (Eds.), Computer Security and Industrial Cryptography. Proceedings, 1991. VIII, 275 pages. 1993.

Vol. 742: S. Nishio, A. Yonezawa (Eds.), Object Technologies for Advanced Software. Proceedings, 1993. X, 543 pages. 1993.

Vol. 743: S. Doshita, K. Furukawa, K. P. Jantke, T. Nishida (Eds.), Algorithmic Learning Theory. Proceedings, 1992. X, 260 pages. 1993. (Subseries LNAI)

Vol. 744: K. P. Jantke, T. Yokomori, S. Kobayashi, E. Tomita (Eds.), Algorithmic Learning Theory. Proceedings, 1993. XI, 423 pages. 1993. (Subseries LNAI)

Vol. 745: V. Roberto (Ed.), Intelligent Perceptual Systems. VIII, 378 pages. 1993. (Subseries LNAI)

Vol. 746: A. S. Tanguiane, Artificial Perception and Music Recognition. XV, 210 pages. 1993. (Subseries LNAI).

Vol. 747: M. Clarke, R. Kruse, S. Moral (Eds.), Symbolic and Quantitative Approaches to Reasoning and Uncertainty. Proceedings, 1993. X, 390 pages. 1993.

Vol. 748: R. H. Halstead Jr., T. Ito (Eds.), Parallel Symbolic Computing: Languages, Systems, and Applications. Proceedings, 1992. X, 419 pages. 1993.

Vol. 749: P. A. Fritzson (Ed.), Automated and Algorithmic Debugging. Proceedings, 1993. VIII, 369 pages. 1993.

Vol. 750: J. L. Díaz-Herrera (Ed.), Software Engineering Education. Proceedings, 1994. XII, 601 pages. 1994.

Vol. 751: B. Jähne, Spatio-Temporal Image Processing. XII, 208 pages. 1993.

Vol. 752: T. W. Finin, C. K. Nicholas, Y. Yesha (Eds.), Information and Knowledge Management. Proceedings, 1992. VII, 142 pages. 1993.

Vol. 753: L. J. Bass, J. Gornostaev, C. Unger (Eds.), Human-Computer Interaction. Proceedings, 1993. X, 388 pages. 1993.

Vol. 754: H. D. Pfeiffer, T. E. Nagle (Eds.), Conceptual Structures: Theory and Implementation. Proceedings, 1992. IX, 327 pages. 1993. (Subseries LNAI).

Vol. 755: B. Möller, H. Partsch, S. Schuman (Eds.), Formal Program Development. Proceedings. VII, 371 pages. 1993.

Vol. 756: J. Pieprzyk, B. Sadeghiyan, Design of Hashing Algorithms. XV, 194 pages. 1993.

Vol. 757: U. Banerjee, D. Gelernter, A. Nicolau, D. Padua (Eds.), Languages and Compilers for Parallel Computing. Proceedings, 1992. X, 576 pages. 1993.

Vol. 758: M. Teillaud, Towards Dynamic Randomized Algorithms in Computational Geometry. IX, 157 pages. 1993.

Vol. 759: N. R. Adam, B. K. Bhargava (Eds.), Advanced Database Systems. XV, 451 pages. 1993.

Vol. 760: S. Ceri, K. Tanaka, S. Tsur (Eds.), Deductive and Object-Oriented Databases. Proceedings, 1993. XII, 488 pages. 1993.

Vol. 761: R. K. Shyamasundar (Ed.), Foundations of Software Technology and Theoretical Computer Science. Proceedings, 1993. XIV, 456 pages. 1993.

Vol. 762: K. W. Ng, P. Raghavan, N. V. Balasubramanian, F. Y. L. Chin (Eds.), Algorithms and Computation. Proceedings, 1993. XIII, 542 pages. 1993.

Vol. 763: F. Pichler, R. Moreno Díaz (Eds.), Computer Aided Systems Theory – EUROCAST '93. Proceedings, 1993. IX, 451 pages. 1994.

Vol. 764: G. Wagner, Vivid Logic. XII, 148 pages. 1994. (Subseries LNAI).

Vol. 765: T. Helleseth (Ed.), Advances in Cryptology – EUROCRYPT '93. Proceedings, 1993. X, 467 pages. 1994.

Vol. 766: P. R. Van Loocke, The Dynamics of Concepts. XI, 340 pages. 1994. (Subseries LNAI).

Vol. 767: M. Gogolla, An Extended Entity-Relationship Model. X, 136 pages. 1994.

Vol. 768: U. Banerjee, D. Gelernter, A. Nicolau, D. Padua (Eds.), Languages and Compilers for Parallel Computing. Proceedings, 1993. XI, 655 pages. 1994.

Vol. 769: J. L. Nazareth, The Newton-Cauchy Framework. XII, 101 pages. 1994.

Vol. 770: P. Haddawy (Representing Plans Under Uncertainty. X, 129 pages. 1994. (Subseries LNAI).

Vol. 771: G. Tomas, C. W. Ueberhuber, Visualization of Scientific Parallel Programs. XI, 310 pages. 1994.

Vol. 772: B. C. Warboys (Ed.),Software Process Technology. Proceedings, 1994. IX, 275 pages. 1994.

Vol. 773: D. R. Stinson (Ed.), Advances in Cryptology – CRYPTO '93. Proceedings, 1993. X, 492 pages. 1994.

Vol. 774: M. Banâtre, P. A. Lee (Eds.), Hardware and Software Architectures for Fault Tolerance. XIII, 311 pages. 1994.

Vol. 775: P. Enjalbert, E. W. Mayr, K. W. Wagner (Eds.), STACS 94. Proceedings, 1994. XIV, 782 pages. 1994.

Vol. 776: H. J. Schneider, H. Ehrig (Eds.), Graph Transformations in Computer Science. Proceedings, 1993. VIII, 395 pages. 1994.

Vol. 777: K. von Luck, H. Marburger (Eds.), Management and Processing of Complex Data Structures. Proceedings, 1994. VII, 220 pages. 1994.

Vol. 778: M. Bonuccelli, P. Crescenzi, R. Petreschi (Eds.), Algorithms and Complexity. Proceedings, 1994. VIII, 222 pages. 1994.

Vol. 779: M. Jarke, J. Bubenko, K. Jeffery (Eds.), Advances in Database Technology — EDBT '94. Proceedings, 1994. XII, 406 pages. 1994.

Vol. 780: J. J. Joyce, C.-J. H. Seger (Eds.), Higher Order Logic Theorem Proving and Its Applications. Proceedings, 1993. X, 518 pages. 1994.

Vol. 782: J. Gutknecht (Ed.), Programming Languages and System Architectures. Proceedings, 1994. X, 344 pages. 1994.

Vol. 783: C. G. Günther (Ed.), Mobile Communications. Proceedings, 1994. XVI, 564 pages. 1994.